Smart Innovation, Systems and Technologies

Volume 229

The Smart Innovation, Systems and Technologies book series encompasses the topics of knowledge, intelligence, innovation and sustainability. The aim of the series is to make available a platform for the publication of books on all aspects of single and multi-disciplinary research on these themes in order to make the latest results available in a readily-accessible form. Volumes on interdisciplinary research combining two or more of these areas is particularly sought.

The series covers systems and paradigms that employ knowledge and intelligence in a broad sense. Its scope is systems having embedded knowledge and intelligence, which may be applied to the solution of world problems in industry, the environment and the community. It also focusses on the knowledge-transfer methodologies and innovation strategies employed to make this happen effectively. The combination of intelligent systems tools and a broad range of applications introduces a need for a synergy of disciplines from science, technology, business and the humanities. The series will include conference proceedings, edited collections, monographs, handbooks, reference books, and other relevant types of book in areas of science and technology where smart systems and technologies can offer innovative solutions.

High quality content is an essential feature for all book proposals accepted for the series. It is expected that editors of all accepted volumes will ensure that contributions are subjected to an appropriate level of reviewing process and adhere to KES quality principles.

Indexed by SCOPUS, EI Compendex, INSPEC, WTI Frankfurt eG, zbMATH, Japanese Science and Technology Agency (JST), SCImago, DBLP.

All books published in the series are submitted for consideration in Web of Science.

More information about this series at http://www.springer.com/series/8767

Jason Gu · Rajeeb Dey · Nabanita Adhikary
Editors

Communication and Control for Robotic Systems

Springer

Editors
Jason Gu
Department of Electrical and
Computer Engineering
Dalhousie University
Halifax, NS, Canada

Rajeeb Dey
Department of Electrical Engineering
National Institute of Technology Silchar
Silchar, Assam, India

Nabanita Adhikary
Department of Electrical Engineering
National Institute of Technology Silchar
Silchar, Assam, India

ISSN 2190-3018 ISSN 2190-3026 (electronic)
Smart Innovation, Systems and Technologies
ISBN 978-981-16-1776-8 ISBN 978-981-16-1777-5 (eBook)
https://doi.org/10.1007/978-981-16-1777-5

This Springer imprint is published by the registered company Springer Nature Singapore Pte Ltd.
The registered company address is: 152 Beach Road, #21-01/04 Gateway East, Singapore 189721, Singapore

Preface

This monograph is a collection of high-quality research articles disseminated as a research outcome of multiple authors in the areas of communication, computation and control of robotic systems. Through this monograph one of the objective of the project grant under the Scheme for Promotion of Academic and Research Collaboration (SPARC) funded by the Minister of Human Resource Development (MHRD) Government of India has been accomplished. This monograph has helped editors to identify and establish research collaboration within India and abroad. The Editors of the book acknowledge thanks to the MHRD for funding the collaborative research between National Institute of Technology Silchar, India and Dalhousie University, Canada. The book includes topics specific to the theme of the symposium, viz. emerging control for robotic systems, wireless communication and development of embedded systems for robotic applications. Robot control technology is widely used for space, surgery, rehabilitation, micro-machine, entertainment, underwater exploration, civil engineering, professional and domestic services, security, etc. Consequently, control will continue to play an increasing role in the areas of robotics including robot–robot and human–robot cooperation in various dynamic scenarios. Subsequently, due to changing communication infrastructure and technologies, it is highly desirable these days to propagate control commands through wireless communication channels. Thus, the inclusion of wireless communication for control of robotic systems poses serious challenges to the performance of the control system, thereby necessitating the use of cutting-edge technologies for the development and testing of control algorithms for robotic systems under such a communication environment. Moreover, real-time implementation of the control through present-day communication networks requires realizable and scalable embedded design. Contributions on basic research of control over the network as well as on relevant robotics applications are also included in this book.

This book uniquely provides a unified framework for analysis, design and deployment of the robotic applications across various engineering and non-engineering disciplines including the three primary aspects focal to the symposium, i.e. control, communication and embedded systems. The entire volume consists of thirty chapters divided into five parts dedicated respectively to Control Theories for Robotic Systems, Communication and Control for Robotic Systems, Computation

Paradigm for Robotic Systems, Power Electronics Application for Transportation and Innovations in Robotic Systems.

The first part includes six chapters on emerging control strategies for robotic applications. The articles describe modelling aspects of the robots specific to the application environment and thereby discuss the challenges involved in the modelling and the control theory. The validity of the developed theories is supported by the results obtained from the software or hardware implementation of the robotic system under consideration.

The second part consists of six chapters on emerging communication technology at the different levels of robotics and automation systems. These articles provide an easy and structured explanation on the need of integral communication technology in an attempt to achieve the challenging control performance for various robotic applications. The specific challenges in the interface of communication technology and control which cannot be solved with conventional control theory are included in the articles. This section also includes articles from networked control system with due emphasis on robotic applications.

The third part includes eight chapters on the use of intelligent- and knowledge-based computational techniques for solving highly complex, nonlinear and non-traditional robotic environment pertaining to control, communication or combination of both. The chapters give a clear theoretical integration of the computational techniques with the communication and control theories such that the readers can have a good comprehension of the theoretical and implementation basis for developing robotic applications.

The fourth part of the book includes six chapters on the recent advances on power supplies and power electronics pertaining to autonomous and robotic transportation systems with a focus on electric vehicles.

The fifth part of the book consists of four chapters describing recent innovations in various areas of robotics. These chapters mostly deal with application and development of robotic systems for agriculture, autonomous vehicles, launch vehicles, etc.

We are thankful to all the authors who helped in keeping the quality of the monograph at a high level by their contributed papers as well the reviewers for giving their time to rigorously review the papers.

The book is addressed to the research community including professors and students in robotics, control systems as well as power electronics. We hope that the book will provide useful information and inspiration to its intended readers.

Halifax, Canada — Jason Gu
Silchar, India — Rajeeb Dey
Silchar, India — Nabanita Adhikary

Contents

Editors and Contributors

About the Editors

Jason Gu is currently Professor in the Department of Electrical Engineering at Dalhousie University, Canada. Professor Jason Gu received the bachelor's degree in electrical engineering and information science from the University of Science and Technology of China, in 1992, the master's degree in biomedical engineering from Shanghai Jiao Tong University, in 1995, and the Ph.D. degree from the University of Alberta, Canada, in 2001. He is currently Full Professor of electrical and computer engineering at Dalhousie University, Canada. He is also Cross-Appointed Professor with the School of Biomedical Engineering for his multidisciplinary research work. His research interests include robotics, biomedical engineering, rehabilitation engineering, neural networks, and control. He is Fellow of the Engineering Institute of Canada. He is IEEE Canada President-elect, 2018–2019, and IEEE Canada President, 2020–2021.

Rajeeb Dey is currently Assistant Professor in the Department of Electrical Engineering at the National Institute of Technology, Silchar, India. He holds M.Tech. degree in control system engineering, from the Indian Institute of Technology, Kharagpur, India, and Ph.D. degree also in control system engineering from Jadavpur University, Kolkata, India. He is Senior Member of IEEE Control System Society, Member Institution of Engineers (India), Executive Committee Member of ACDOS (Advanced Control and Dynamic Optimization Society), and Life Member of System Society of India. His research interest includes design of robust control, optimization based on LMI techniques, time-delay systems, intelligent control, decentralized control and control applications, and biomedical control applications.

Nabanita Adhikary is currently Assistant Professor in the Department of Electrical Engineering at the National Institute of Technology, Silchar, India. She holds Ph.D. degree in control system engineering, from the Indian Institute of Technology Guwahati, India. She is Member of the IEEE Control System Society and IEEE Robotics and Automation Society. Her research interests include nonlinear control system design, robust control, time-delay control, intelligent control, robotics, and networked control systems.

Contributors

Nabanita Adhikary Department of Electrical Engineering, National Institute of Technology Silchar, Silchar, India

Sami Al Issa Department of Electronics and Electrical Engineering, IIT Guwahati, Guwahati, India;
Department of Computer Engineering and Automation, Damascus University, Damascus, Syria

Muhammad Usman Asad Department of Electrical and Computer Engineering, Dalhousie University, Halifax, Canada

U. Asfia Department of Electrical Engineering, SV National Institute of Technology Surat, Surat, India

M. Aswatha Kumar Christ (Deemed to be University), Bengaluru, Karnataka, India

Karthik Balasubramanian Design and Engineering (Electrical), Offshore Technology Development Pte., Ltd., Keppel Offshore and Marine Ltd., Singapore, Singapore

Atanu Banerjee Department of Electrical Engineering, National Institute of Technology Meghalaya, Bijni Complex, Laitumukhrah, Shillong, Meghalaya, India

V. T. Basker Vikram Sarabhai Space Centre, I.S.R.O., Thiruvananthapuram, India

Manas Kumar Bera Department of Electronics and Instrumentation Engineering, NIT Silchar, Silchar, India

Partha Bhattacharjee I. T. Group, Ministry of Science and Technology, CSIR-Central Mechanical Engineering Research Institute, Government of India, Durgapur, India

Shovan Bhaumik Department of Electrical Engineering, Indian Institute of Technology Patna, Patna, India

Vishnu Sai Bhonsle REVA University, Bangalore, Karnataka, India

Pabitra Kumar Biswas Department of Electrical and Electronics Engineering, National Institute of Technology Mizoram, Aizawl, India

Rishav Chakraborty Applied Electronics & Instrumentation Engineering, RCC Institute of Information Technology, Kolkata, West Bengal, India

Soumik Chakraborty Department of Electronics and Instrumentation Engineering, National Institute of Technology, Silchar, Cachar, Assam, India; Solar Energy Materials Research and Testing Laboratory, (Smart Lab) Department of Physics, National Institute of Technology, Silchar, Cachar, Assam, India

Chunqi Chang School of Biomedical Engineering, Health Sciences Center, Shenzhen University, Shenzhen, China

Koceila Cherfouh Department of Electrical & Computer Engineering, Dalhousie University, Halifax, Canada

Sayan Das Department of AEIE, RCCIIT, Kolkata, India

Rupak Datta Department of Mathematics, National Institute of Technology Agartala, Agartala, India

Manash Jyoti Deori Department of Electrical Engineering, National Institute of Technology Silchar, Silchar, India

Aritro Dey Department of Electrical Engineering, National Institute of Technology Durgapur, Durgapur, West Bengal, India

Naiwrita Dey Electronics and Communication Engineering, RCC Institute of Information Technology, Kolkata, West Bengal, India

Rajeeb Dey Department of Electrical Engineering, NIT, Silchar, Assam, India

Tridib Dey Applied Electronics & Instrumentation Engineering, RCC Institute of Information Technology, Kolkata, West Bengal, India

M. V. Dhekane IIST, Thiruvananthapuram, India

M. Dinesh Kumar Vikram Sarabhai Space Centre, I.S.R.O., Thiruvananthapuram, India

B. P. Divakar REVA Universtiy, Bengaluru, India

Rupak Dutta Mathematics Department, NIT Agartala, Agartala, India

Umar Farooq Department of Electrical & Computer Engineering, Dalhousie University, Halifax, Canada

Jason Gu Department of Electrical & Computer Engineering, Dalhousie University, Halifax, Canada

Santoshkumar Hampannavar REVA Universtiy, Bengaluru, India

Kudupudi Durga Harika Department of Electronics and Communication Engineering, Shri Vishnu Engineering College for Women (Autonomous), Bhimavaram, AP, India

Abul Hassan Applied Electronics & Instrumentation Engineering, RCC Institute of Information Technology, Kolkata, West Bengal, India

K. Indu Christ (Deemed to be University), Bengaluru, Karnataka, India

Sibam Jana Applied Electronics & Instrumentation Engineering, RCC Institute of Information Technology, Kolkata, West Bengal, India

Sandip Jana Academy of Scientific and Innovative Research (AcSIR), CSIR-Central Mechanical Engineering Research Institute (CSIR-CMERI) Campus, Durgapur, India;
I. T. Group, CSIR-Central Mechanical Engineering Research Institute, Govt of India, Durgapur, India

Indrani Kar Department of Electronics and Electrical Engineering, IIT Guwahati, Guwahati, India

B. Karthik Vikram Sarabhai Space Centre, I.S.R.O., Thiruvananthapuram, India

Kesavabrahmaji Karuturi ISRO Inertial Systems Unit, Thiruvananthapuram, India

Guddu Kumar Discipline of Electrical Engineering, Indian Institute of Technology Indore, Indore, India

Mano Ranjan Kumar School of Electronics Engineering, Kalinga Institute of Industrial Technology, Bhubaneswar, India

Francis Kurbah NIT Meghalaya, Shillong, India

Rajashekar P. Mandi REVA University, Bangalore, India

Poluri Sri Mannarayana Department of Electrical Engineering, National Institute of Technology Durgapur, Durgapur, West Bengal, India

Teiborlin Marngar NIT Meghalaya, Shillong, India

Shemphang Marwein NIT Meghalaya, Shillong, India

Jobin Mathew Department of Electrical Engineering, Assam Engineering College, Guwahati, India

Sunil Kumar Mishra School of Electronics Engineering, Kalinga Institute of Industrial Technology, Bhubaneswar, India

Vikash Kumar Mishra Electrical and Electronics Engineering, Government Engineering College Raipur, Raipur, India

Jiljo K. Moncy Vikram Sarabhai Space Centre, I.S.R.O., Thiruvananthapuram, India

Arpita Mukherjee CSIR-Central Mechanical Engineering Research Institute, Ministry of Science and Technology, Government of India, Durgapur, India

Ranjith G. Nair Solar Energy Materials Research and Testing Laboratory, (Smart Lab) Department of Physics, National Institute of Technology, Silchar, Cachar, Assam, India

Krishanu Nath Department of Electronics and Instrumentation Engineering, NIT Silchar, Silchar, India

Akhendra Kumar Padavala Department of Electronics and Communication Engineering, Shri Vishnu Engineering College for Women (Autonomous), Bhimavaram, AP, India

E. S. Padma Kumar Vikram Sarabhai Space Centre, I.S.R.O., Thiruvananthapuram, India

Krishna Jyothi Pallacherla Department of Electrical Engineering, National Institute of Technology Silchar, Silchar, India

Manish Patel Department of Electrical Engineering, National Institute of Technology Kurukshetra, Kurukshetra, Haryana, India

Saswati Pattnaik School of Electronics Engineering, Kalinga Institute of Industrial Technology, Bhubaneswar, India

Sudharani Potturi REVA University, Bangalore, India

Bhanu Pratap Department of Electrical Engineering, National Institute of Technology Kurukshetra, Kurukshetra, Haryana, India

R. Radhakrishnan Department of Electrical Engineering, SV National Institute of Technology Surat, Surat, India

Rahul Radhakrishnan Department of Electrical Engineering, SVNIT, Surat, India

Priyam Rai Department of Electrical Engineering, National Institute of Technology Kurukshetra, Haryana, India

B. S. Sagar REVA Universtiy, Bengaluru, India

Manika Saha Department of Electrical Engineering, Meghnad Saha Institute of Technology, Kolkata, India

K. L. N. Sai Nitish Vikram Sarabhai Space Centre, I.S.R.O., Thiruvananthapuram, India

Chiranjit Sain Department of Electrical Engineering, Siliguri Institute of Technology, Sukna, Siliguri, Darjeeling, India

Bikash Kumar Sarkar NIT Meghalaya, Shillong, India

Lalu Seban Department of Electronics and Instrumentation Engineering, National Institute of Technology, Silchar, Cachar, Assam, India

S. N. Sharma Department of Electrical Engineering, SV National Institute of Technology Surat, Surat, India

Saikat Kumar Shome CSIR-Central Mechanical Engineering Research Institute, Ministry of Science and Technology, Government of India, Durgapur, India

Abhinoy Kumar Singh Discipline of Electrical Engineering, Indian Institute of Technology Indore, Indore, India

M. Venkata Subbarao Department of Electronics and Communication Engineering, Shri Vishnu Engineering College for Women (Autonomous), Bhimavaram, AP, India

R. Swaminathan Discipline of Electrical Engineering, Indian Institute of Technology Indore, Indore, India

Sudhakar Babu Thanikanti Department of Electrical and Electronics Engineering, Chaitanya Bharathi Institute of Technology (CBIT), Hyderabad, India

Sailaja Thota REVA University, Bangalore, Karnataka, India

Surekha Thota REVA University, Bangalore, Karnataka, India

Nutan Kumar Tomar Department of Mathematics, Indian Institute of Technology Patna, Patna, India

Chintapalli Vaishnavi IIST, Thiruvananthapuram, India

Asifa Yesmin Department of Electronics and Instrumentation Engineering, NIT Silchar, Silchar, India

Control Theories for Robotic Systems

Performance Evaluation of Adaptive Dither Control Frameworks for Nonlinear Piezoactuator

Sandip Jana, Saikat Kumar Shome, Arpita Mukherjee, and Partha Bhattacharjee

Abstract Piezoelectric actuators are one of the promising industrial actuators in the precision positioning industry. However, the high quotient of nonlinearity inherently present degrades their performance. Dither control is specifically used for control of nonlinear applications through relatively simple and computationally inexpensive logic. This research focuses on performance evaluation of different adaptive dither control frameworks in the parlance of piezoelectric actuation. The results are then compared with well-established control frameworks like feed forward and cascaded feed forward and feedback topology. The observations suggest dither control as a promising controller for piezoelectric actuation under several disturbances.

1 Introduction

Piezoelectric actuators are a class of flagship actuators which has been used in micro/nano-manipulation-related industrial applications such as biological cell operator, precision engineering in PCB industry, MEMS manufacturing, space optics, etc. due to prominent advantages like precision positioning [1], micropositioning [2], nano-positioning [3], speed of response, resistance to electromagnetic interference, high torque handling capability and high force capability in [4–6]. Dithering of a nonlinear system is used in different industry applications like several areas of science and technology, including geophysics, biomedical applications, microelectronics and multidisciplinary mechatronics.

In [7], dithering is introduced in robotics nonlinear manipulator to enhance the input signal with the aid of outside noise. The occurrence of dithering is mostly observed in nonlinear dynamic systems wherein the introduction of a weak input,

S. Jana · S. K. Shome (✉) · A. Mukherjee
Academy of Scientific and Innovative Research (AcSIR), CSIR- Central Mechanical Engineering Research Institute (CSIR-CMERI) Campus, Durgapur 713209, India
e-mail: saikatkshome@cmeri.res.in

S. Jana · S. K. Shome · A. Mukherjee · P. Bhattacharjee
I. T. Group, CSIR-Central Mechanical Engineering Research Institute, Govt of India, Durgapur 713209, India

J. Gu et al. (eds.), *Communication and Control for Robotic Systems*, Smart Innovation, Systems and Technologies 229, https://doi.org/10.1007/978-981-16-1777-5_1

mostly considered as noise, leads to enhanced SNR of the process, which was first reported by [8]. Three basic criteria are responsible for this occurrence—(i) an energy activation barrier or threshold, (ii) presence of a weak periodic input and (iii) a noise source which can either be inherent in the system or added externally. Provided a system has these features, a manifestation of increased signal-to-noise ratio is observed in correspondence to the disturbance which is referred to as stochastic resonance. Accordingly, in a nutshell, dithering may be considered as purposefully varying the intensity of noise to raise the performance of the nonlinear system. Dithering has been used in several industrial applications, for example, reduction of quantization error in analog to digital conversions, geographical applications like accounting for periodicity of earth ice age and even in microelectronics semiconductor industry [9, 10]. Piezoelectric actuators have been recently explored in the domain of stochastic resonance toward improving its performance and this research put forward a comparison of different adaptive dither control approaches along with their advantages and disadvantages. In [11], a controller centered on adaptive dither in voltage mode on spectral analysis of piezoactuator is introduced and controller adaptively adjusted the dither amplitude to get optimum value and improved the system performance. The proposed controller has better sinusoidal path tracking, mixed-amplitude mixed-frequency input signal, outside disturbances like Gaussian, impulse, step response compared with the classical feedback and feed forward control.

The contribution of the paper is as follows: Different controllers been modeled in MATLAB Simulink on the nonlinear piezoelectric actuator. The studied adaptive dither controllers have then been evaluated by exposing the plant model to different real-time disturbances like error comparison, parameter sensitivity test, mixed-amplitude mixed-frequency input signal, additional disturbances with set point tracking like set point tracking in conjunction to impulse disturbances (delay 1 s, sample time 20 s), step input with impulse disturbance and zero set point tracking with Gaussian noise which are discussed subsequently. The different controllers studied in this study are as follows: C1: Feed forward traditional control (FF), C2: Cascaded feed traditional forward and feed back (FF + FB), C3: Cascaded feed traditional forward and feed back—displacement dithered, C4: Cascaded feed forward and feed back—voltage dithered, C5: Cascaded feed forward and feed back—adaptive voltage dithered.

2 RMS Error Comparison

The study has been carried using a sinusoidal wave signal as an input amplitude of 10 μm, and frequency range of signal 1 Hz which is to be followed by the piezoactuator. The RMS error of the corresponding control schemes is placed in Table 1.

It is observed that order of performance of controllers is as per the following order Adaptive Voltage Dither, FF + FB, FF shown in Fig. 1.

Table 1 RMS error of various controller

Controllers	C1	C2	C3	C4	C5
Error(nm)	6.67	0.26315	0.23277	0.23627	0.23628

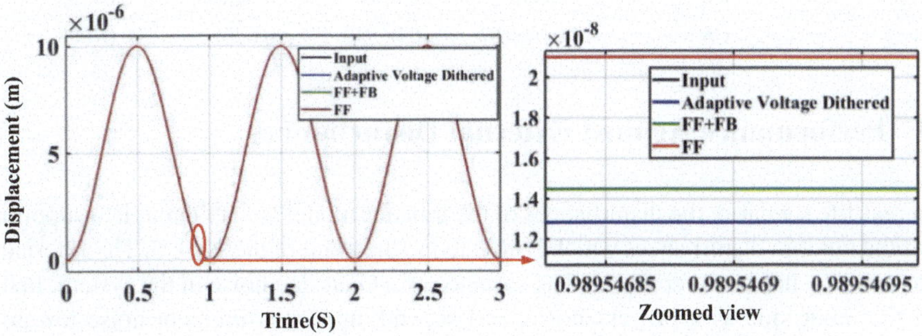

Fig. 1 Tracking output response of different controller with sinusoidal input

3 Parameter Sensitivity Test

As presented in Ref [9], the main reason for occurrence of mismatch in plant and model is accountable to the approximations while transforming a physical plant to its mathematical model. This mismatch is quite common and takes place due to several reasons like improper identification of system, or inappropriate order consideration of the system. Besides, nonlinear properties also add up to this error in approximation. In this section considers a plant model parameter mismatch of $\pm$ 2% and $\pm$ 5% in plant parameters as compared to the classical model and the RMS errors are presented in Tables 2 and 3, respectively. From Tables 2 and 3, it can be concluded that adaptive voltage dither have the better compensation of the parametric uncertainty than FF + FB. The efficiency of adaptive voltage dither over FF + FB is clearly seen as the parametric variation is increased to 5%

Table 2 RMS error for plant parameter variation of $\pm$ 2%

Variation (%)	Controller	M	D	K	T
−2	C2	0.12933	0.24121	0.17788	0.15667
2	C2	0.16073	0.13105	0.21483	0.18924
−2	C5	0.06678	0.07269	0.07067	0.07657
2	C5	0.07380	0.06841	0.07035	0.07484

Table 3 RMS error for plant parameter variation of ± 5%

Variation (%)	Controller	M	D	K	T
−5	C2	0.12855	0.20083	0.15985	0.11177
5	C2	0.09108	0.12610	0.16552	0.1882
−5	C5	0.06193	0.07403	0.07092	0.07598
5	C5	0.07869	0.06539	0.07012	0.07088

4 Performance Against External Disturbances

In real-life scenario, the disturbances of the transfer model are not enough to support for the random disturbances which is effected to the plant of the system. The external disturbance has been represented as variance of a Gaussian noise in the system, first at the input side as a process noise, and second, as a measurement noise toward the plant output. The Gaussian error values of each of the following controller are shown in Tables 4 and 5, respectively, and can be assessed through two metrics,

Table 4 RMS error for external process disturbance

Gaussian noise variance	C5(nm)	C2 (nm)
10e-22	0.25435	0.38004
10e-23	0.22815	0.40858
10e-24	0.23325	0.38373
10e-25	0.23556	0.27314
10e-26	0.23611	0.36425
10e-27	0.23623	0.26052
10e-28	0.23628	0.25366
10e-29	0.23628	0.27080
Max–min	0.0262	0.15492
Average error	0.23702	0.32434

Table 5 RMS error for external measurement disturbance

Gaussian noise (variance)	C5 (nm)	C2 (nm)
10e-22	0.24092	0.29815
10e-23	0.23688	0.27197
10e-24	0.23641	0.28242
10e-25	0.23631	0.37162
10e-26	0.23629	0.35077
10e-27	0.23629	0.38303
10e-28	0.23628	0.33925
(a) Max. value−min. value	0.00464	0.11106
Average error	0.23705	0.32817

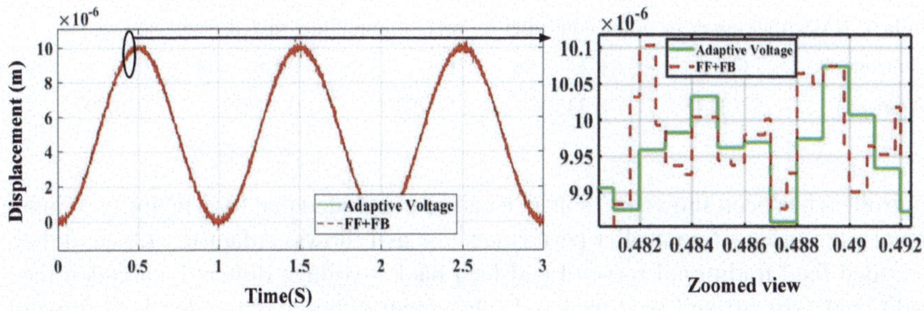

Fig. 2 Error plot with sinusoidal wave input and external measurement noise (Gaussian, variance 10e-15)

(a) variation induced due to the presence of the noise which is represented by a difference of maximum and minimum error value and (b) deviation of the average error due to noise with the best attained tracking error. From the table, it is observed that using the second metric, the adaptive voltage dither controller performance is better than FF + FB controller. A schematic of the error plot suggesting similar controller performance is shown in Fig. 2.

4.1 Gaussian Noise as Process Noise Disturbance

See Table 4.

4.2 Gaussian Noise as Measurement Noise Disturbance

See Table 5.

5 Mixed-Amplitude, Mixed-Frequency Input Test

5.1 Performance of the Controllers to Different Input Wave Signals—of Alternating Frequency and Amplitude

In this section, the piezoactuator has been subjected to different input wave signals and alteration of frequency (0.25 Hz, 0.5 Hz and 1 Hz) and alteration of amplitude (2.5 μm, 5 μm and 10 μm). For all the controllers, a common phenomenon is observed—the tracking error increases with increase in input frequency and also with an increase in input wave signal amplitude. Among the comparison of different

Table 6 RMS tracking error of the controller for mixed-amplitude, mixed-frequency signal

Controllers	C1	C2	C3	C4	C5
Error(nm)	6.3145	0.33506	0.30774	0.30705	0.30703

controllers has been shown in Table 6, it can be concluded that for a majority of input signals, the order of controller performance is as follows: Adaptive voltage dither, cascaded feed traditional forward and feed back—voltage dithered, cascaded feed traditional forward and feed back—displacement dithered, cascaded feed forward and feed back (FF + FB), and feed forward traditional control (FF).

5.2 *Performance Evaluation of the Controllers to a Mixed-Amplitude, Mixed-Frequency Input Wave Signal*

For a single mixed-amplitude, mixed-frequency signal of as shown in Fig. 3, the RMS tracking error has been tabulated in Table 6, in increasing order of controller effectiveness.

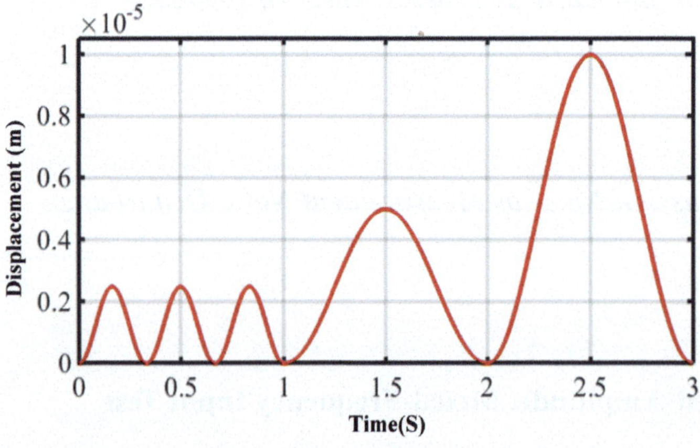

Fig. 3 Mixed-amplitude, mixed-frequency input wave signal

6 Closed-Loop Set Point Tracking with Impulse Disturbances (Delay 1 s, Sample Time 20 s)

The response of the controllers to set point tracking against an impulse disturbance of width 20 s at t = 20 s is shown in Fig. 4. Adaptive voltage dither system is noticed to have the least settling time of 20.44 s. Traditional FF + FB control settles at

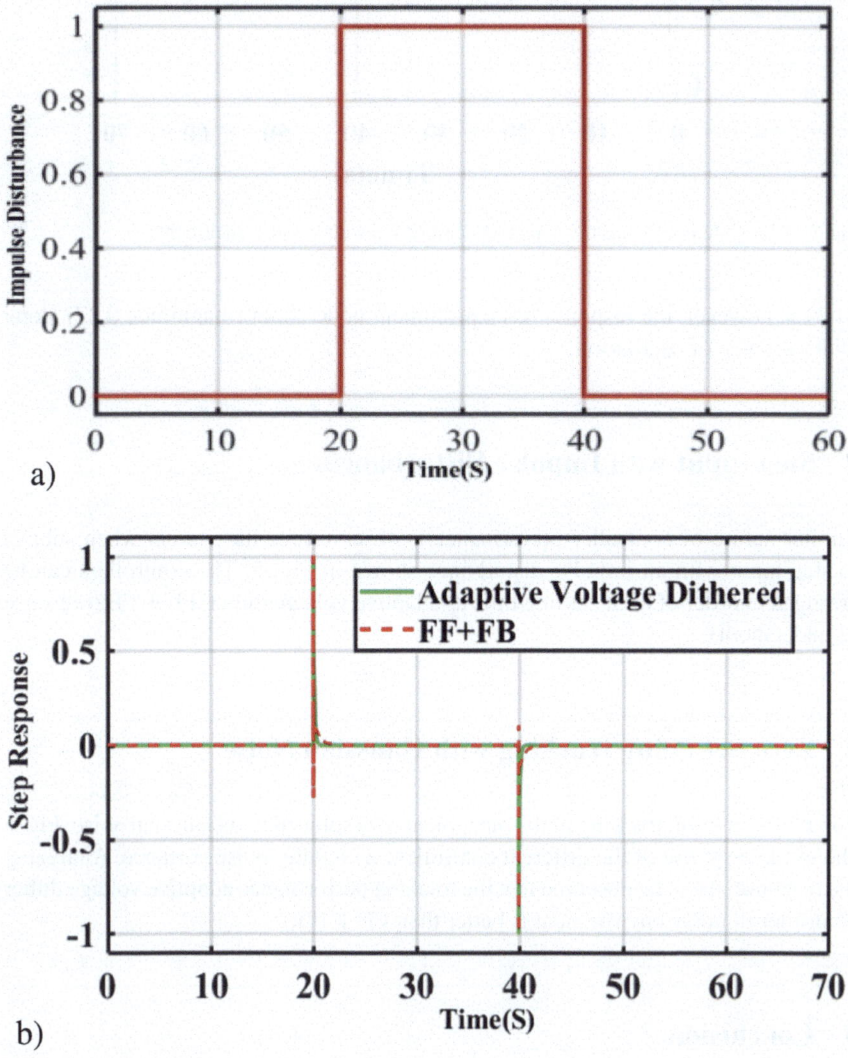

Fig. 4 **a** Impulse disturbance of 20 to 40 s, **b** set point traking performance of the controllers for impulse disturbances

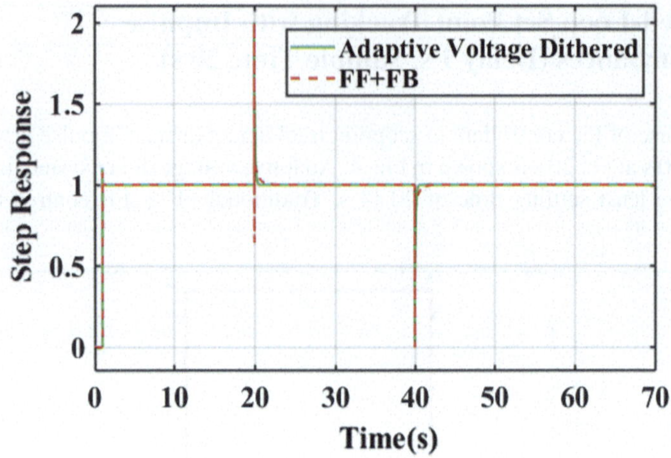

Fig. 5 PZA system response for step input conjunction with impulse disturbance

21.31 s; however, the response has a prominent undershoot (magnitude 0.27) along with presence of oscillation.

7 Step Input with Impulse Disturbances

Similar nature of controller performance is observed for the system when subject to step input with an impulse disturbance shown in Fig. 5. The controllers can be arranged in order of least settling time as adaptive voltage dither, FF + FB (presence of undershoot).

8 Zero Set Point Tracking with Gaussian Noise

For a zero set point tracking of the piezoplant, it is subjected to Gaussian noise. Fig. 6 shows the response of the different controllers with unity noise variance. Analyzing the response, it can be observed that the tracking performance adaptive voltage dither controller in noisy environment is better than FF + FB.

9 Conclusion

A comparison of performance of the different dither-based controllers has been carried out under similar conditions. Performance evaluation suggests that a

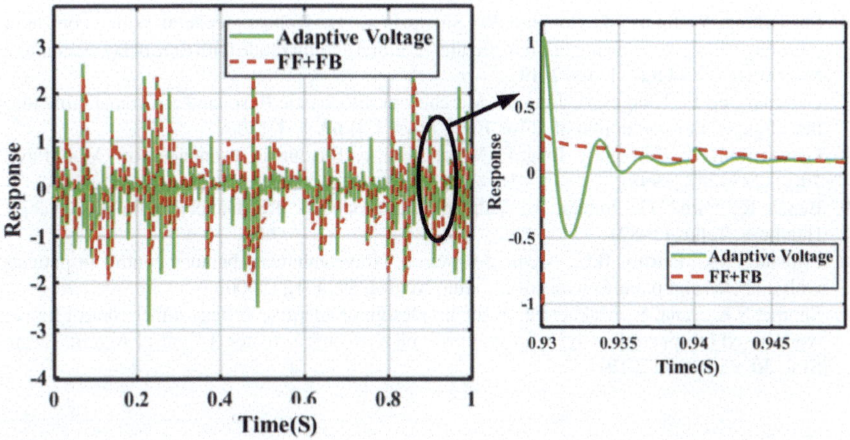

Fig. 6 PZA system response for zero set point tracking with Gaussian response of variance one

controller performs better under certain circumstances like parameter sensitivity test, performance against external disturbances, set point tracking with disturbances outperforming the others and can be chosen depending upon the application. The adaptive voltage dither shows best performance over parametric sensitivity and external disturbances with the other controller, and it is also noticed to have the least settling time with respect to set point tracking with impulse disturbance.

Acknowledgements The authors gratefully acknowledge the funding received from Indo French Center for the Promotion of Advanced Research (IFCPAR/CEFIPRA) for carrying out the project work.

References

1. Nguyen, M.L., Chen, X.: "MPC Inspired Dynamical output feedback and adaptive feedforward control applied to piezo-actuated positioning systems." IEEE Trans. Ind. Electron. **67**(5), 3921–3931, doi: https://doi.org/10.1109/TIE.2019.2916356(2020)
2. Nafea, M., Mohamed, Z., Ali, M.S.M., Mehranzamir, K., Rehman, T.: "Hybrid PSO-Tuned PID and hysteresis-observer based control for piezoelectric micropositioning stages." IEEE International Conference on Smart Instrumentation, Measurement and Application (ICSIMA), Kuala Lumpur, Malaysia, 2019, pp. 1–6 (2019)
3. Xie, S., Ren, J.: "Linearization of recurrent-neural-network- based models for predictive control of nano-positioning systems using data-driven koopman operators." IEEE Access **8**, 147077–147088, doi: https://doi.org/10.1109/ACCESS.2020.3013935 (2020).
4. Lambert, P., Valentini, A., Lagrange, B.: Delcambre P.D Design and performances of a one-degree-of-freedom guided nano-actuator. Robot. Comput. Integr. Manuf. **19**, 89–98 (2003)
5. Hii, K.F., Vallance, R.R., Mengu, M.P.: Design, operation, and motion characteristics of a precise piezoelectric linear motor. Precis. Eng. **34**, 231–241 (2010)

6. Shengdong, Y., Jinyu, M., Hongtao, W., Shengzheng, K.: Robust precision motion control of piezoelectric actuators using fast nonsingular terminal sliding mode with time delay estimation. Measure. Control **52**, 11–19 (2019)
7. Gammaitoni, L., Hänggi, P., Jung, P., Marchesoni, Stochastic Resonance: A remarkable idea that changed our perception of noise. Euro. Phys. J. B **69**, 1–3, (2009)
8. Gammaitoni, L., Hänggi, P., Jung, P., Marchesoni, F.: Stochastic resonance. Rev. Mod. Phys. **70**(1), 223–287 (1998)
9. Benzi, R., Parisi, G., Sutera, A., Vulpiani, A.: Stochastic Resonancein Climatic Change. Hamburg, Tellus (1982)
10. Priplata, A.A., Patritti, B.L., Niemi, J.B., et al.: Noise-enhanced balance control in patients with diabetes and patients with stroke. Ann. Neurol. **59**, 4–12 (2006)
11. Shome, S.K., Jana, S., Mukherjee, A., et al.: Design of adaptive voltage dither control framework based on spectral analysis for nonlinear piezoelectric actuator. J Control Autom. Electr. Syst. **30**, 954–969 (2019)

Comparative Performance Study of Different Controllers for Nonlinear Piezoelectric Stack Actuator

Sandip Jana, Saikat Kumar Shome, Arpita Mukherjee, and Partha Bhattacharjee

Abstract Robust and adaptive nonlinear controllers play an indispensable role in industrial applications targeting nanopositioning. Internal model control has been widely used in compensation of hysteresis, creep and vibration for piezoelectric actuators along with least-square estimator regimes. However, as real-time scenarios and presence of uncertainties, disturbances play a crucial factor in determining the efficacy of the chosen controller, a comparative analysis is very effective in this situation. This research focuses on an extensive comparative analysis of two controllers, namely, modified-internal model control (M-IMC) and recursive least-square estimator (RLSE) under several test cases. Results put forward which controller is effective under what scenario that enhances the overall precision positioning efficiency of the system.

1 Introduction

Piezoelectric actuators are a class of flagship actuators which has been used in micro/nano-manipulation-related industrial applications such as biological cell operator, precision engineering in PCB industry, MEMS manufacturing, space optics, etc. due to prominent advantages like precision positioning, speed of response, resistance to electromagnetic interference, high torque handling capability and high force capability in [1–3]. In [4] paper, adaptive feed forward control and recursive least-square estimator (RLSE) with adaptive feed forward (FF) controller are presented that performs better for parameter uncertainty of the system. The PSO and BFO-based redesigned internal model control (M-IMC) with PID controller is proposed in [5] for nonlinear piezoactuator which allows good set point tracking and disturbance

S. Jana · S. K. Shome (✉) · A. Mukherjee
Academy of Scientific and Innovative Research (AcSIR), CSIR- Central Mechanical Engineering Research Institute (CSIR-CMERI) Campus, Durgapur 713209, India
e-mail: saikatkshome@cmeri.res.in

S. Jana · S. K. Shome · A. Mukherjee · P. Bhattacharjee
Ministry of Science and Technology, CSIR-Central Mechanical Engineering Research Institute, Government of India, Durgapur 713209, India

J. Gu et al. (eds.), *Communication and Control for Robotic Systems*, Smart Innovation, Systems and Technologies 229, https://doi.org/10.1007/978-981-16-1777-5_2

rejection properties. Internal model control is one of the simplest control strategies being applied for several applications. PID-based IMC control strategy which is the most practically used controller is one of the reasons of its widespread acceptance and has hardware implementation reliability. A tuning method of IMC to control overshoot in a speed controlling system—a crucial subsystem in the hydro-electric power plant is elaborated in [6]. In this research, PSO-based M-IMC controller has improved path tracking and disturbance rejection performance compare with the classical IMC controller. Piezoactuators are widely used for precision and positioning control at micro-nanometer level in several industrial applications and have different advantages like higher accuracy, quick response, and enhanced torque. An improved PID-based IMC controller is proposed for nonlinear hysteretic piezoelectric actuator in [7] and result proved that controller performance improved over the traditional control. The piezoactuator is modeled as a second-order system by experimentally identified parameter and M-IMC controller is implemented in time delay environment approximated using first-order Padé expansion. A Smith predictor-based M-IMC is also proposed in [8] where M-IMC controller decreased overshoot and settling time of the response compared to traditional IMC and PID designs.

Contributions and problems addressed in the paper are as follows: Different controllers have been performed in MATLAB Simulink on the nonlinear piezoelectric actuator mentioned in Sect. 1.2. Presented controller has been explored by exposing the plant model to different disturbances like error comparison, parameter sensitivity test, mixed-amplitude, mixed-frequency input signal, external disturbances with set point tracking like zero path tracking with impulse disturbances (delay 1 s, sample time 20 s), zero path tracking with step input conjunction with impulse disturbances and zero set point tracking with Gaussian noise. The controller performance result suggests that controller's performance depends upon the various application of the system.

2 RMS Error Comparison

The study has been carried using a sinusoidal wave signal as an input amplitude of 10 μm, and frequency range of signal 1 Hz which is to be followed by the piezoactuator. The RMS error of the corresponding control schemes is placed in Table 1. C1: Feed forward traditional control (FF), C2: Cascaded feed traditional forward and feed back (FF + FB), C3: Recursive least-square estimator (RLSE), C4: Modified-internal model control (M-IMC).

Table 1 Tracking RMS error of different control paradigms

Controllers	C1	C2	C3	C4
Error(nm)	6.67	0.26315	0.19405	0.16577

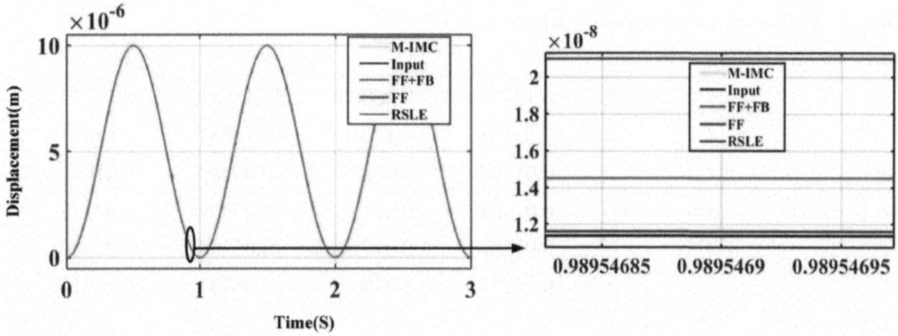

Fig. 1 Tracking performance plot of different adaptive controller with sinusoidal input response

It is observed that order of performance of controllers is as per the following order M-IMC, RLSE, FF + FB, FF as shown in Fig. 1.

3 Parameter Sensitivity Test

As illustrated in Ref [6], the main reason for occurrence of mismatch in plant and model is accountable to the approximations while transforming a physical plant to its mathematical model. This mismatch is quite common and takes place due to several reasons like improper identification of system, or inappropriate order consideration of the system. Besides, nonlinear properties also add up to this error in approximation. The following section considers a plant model parameter mismatch of ± 2% and ± 5% in plant parameters as compared to the traditional model and the RMS errors are presented in Tables 2 and 3, respectively. It has also been observed earlier that M-IMC offers the best RMS tracking error (0.16577 nm) among the observed controllers. Accordingly, to determine the deviation from this value, Tables 2 and 3 are presented which contains the difference of the RMS errors for plant model mismatch with the M-IMC controller value. From Tables 2 and 3, it can be concluded that M-IMC and RLSE have the better compensation of the parametric uncertainty, which is followed

Table 2 RMS error for plant parameter variation of ± 2%

Variation (%)	Controller	M	D	K	T
−2	C2	0.12933	0.24121	0.17788	0.15667
2	C2	0.16073	0.13105	0.21483	0.18924
−2	C4	0.0209	0.0553	0.00947	0.00453
2	C4	0.00368	0.01266	0.001	0.0047
−2	C3	0.02309	0.02365	0.02660	0.02917
2	C3	0.03375	0.03312	0.02993	0.02748

Table 3 RMS error for plant parameter variation of ± 5%

Variation (%)	Controller	M	D	K	T
−5	C2	0.12855	0.20083	0.15985	0.11177
5	C2	0.09108	0.12610	0.16552	0.1882
−5	C4	0.01868	0.00259	0.00480	0.01096
5	C4	0.01881	0.01178	0.00407	0.02133
−5	C3	0.01588	0.01716	0.02398	0.03050
5	C3	0.04259	0.04091	0.03231	0.02612

by FF + FB and FF. The efficiency of M-IMC over RLSE is clearly seen as the parametric variation is increased to 5%

4 Performance Against External Disturbances

Transfer model of disturbances is not enough to withstand for the disturbances which is effect the plant of the system in real life environment. The external disturbance has been represented as variance of a Gaussian noise in the system, first at the input side as a process noise, and second, as a measurement noise toward the plant output. The error values of Gaussian noise in each of the controller are shown in Tables 4 and 5, respectively, and can be assessed through two metrics, a) variation induced due to the presence of the noise which is represented by a difference of maximum and minimum error value and b) deviation of the average error due to noise with the best attained tracking error of M-IMC controller is 0.16577 nm. From the table, it is observed that using the second metric, the performance of the controllers toward eliminating noise is as follows RLSE, M-IMC, FF + FB and FF. A schematic of the error plot suggesting similar controller performance is shown in Fig. 2.

Table 4 RMS error for external process disturbance

Gaussian noise variance	C4 (nm)	C3 (nm)	C2 (nm)
10e-22	0.19762	0.19406	0.3800
10e-23	0.19316	0.19406	0.4085
10e-24	0.20276	0.19406	0.3837
10e-25	0.20666	0.19406	0.2731
10e-26	0.21949	0.19406	0.3642
10e-27	0.21455	0.19406	0.2605
10e-28	0.20431	0.19406	0.2536
10e-29	0.19686	0.19406	0.2708
Max–min	0.02633	-	0.1549
average	0.2044	0.19406	0.3243

Table 5 RMS error for external measurement disturbance

Gaussian noise (variance)	C4 (nm)	C3 (nm)	C2(nm)
10e-22	0.21649	0.19419	0.2981
10e-23	0.18903	0.19404	0.2719
10e-24	0.20680	0.19405	0.2824
10e-25	0.19391	0.19406	0.3716
10e-26	0.20290	0.19400	0.3507
10e-27	0.19158	0.19400	0.3830
10e-28	0.19154	0.19405	0.3392
(a) Max. value – Min. value	0.02746	0.000190	0.1110
Average	0.19889	0.19405	0.3281

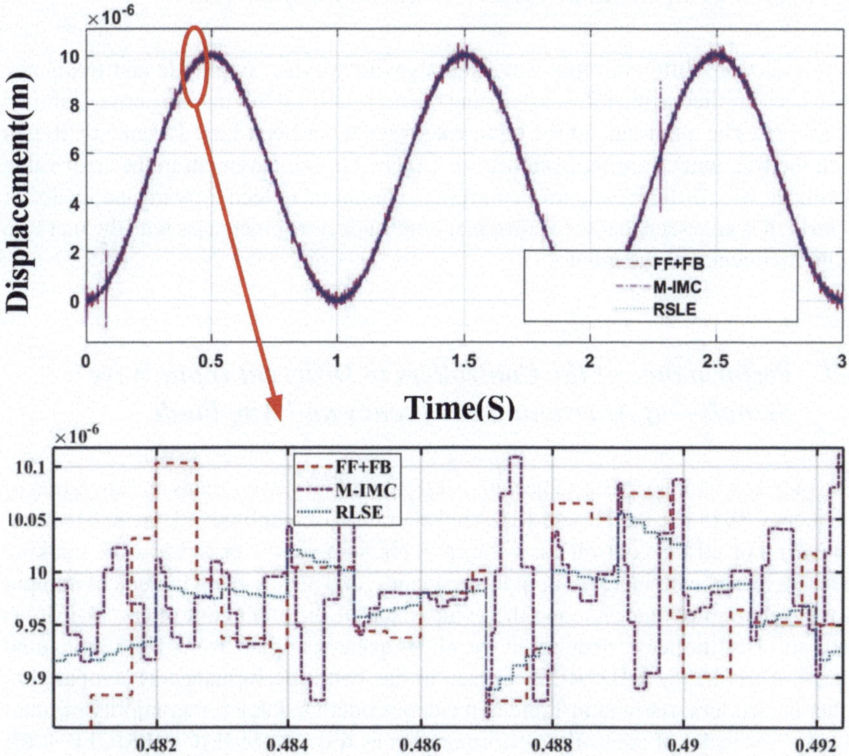

Fig. 2 Analysis of error plot with sinusoidal input and external measurement noise environment (Gaussian, Variance 10e-15)

4.1 *Gaussian Noise as Process Noise Disturbance*

See Table 4.

4.2 *Gaussian Noise as Measurement Noise Disturbance*

See Table 5.

5 Mixed-Amplitude, Mixed-Frequency Input Test

In this section, different input wave signals with varying amplitude and frequency have been applied to the PZA system and the correlative RMS tracking error is shown in Table 6. The amplitude of the input wave signal has been varied from 5 to 20 μm with the frequency varying from 0.25 to 1.0 Hz. An improvement in the error value is noticed with different adaptive controllers compared to feed forward and feedback control; it is also seen that for a particular amplitude, error increases with the increase in the frequency of the input.

5.1 *Performance of the Controllers to Different Input Wave Signals—of Alternating Frequency and Amplitude*

The piezoactuator has been subjected to different input wave signals and alteration of frequency (0.25 Hz, 0.5 Hz and 1 Hz) and alteration of amplitude (2.5 μm, 5 μm and 10 μm). For all the controllers, a common phenomenon is observed—the tracking error increases with increase in input frequency, and also with an increment in input wave signal amplitude. Among the comparison of different controllers, M-IMC is seen to offer the least error value for all frequencies, apart from 1 Hz. For input signal of frequency 1 Hz, RLSE is seen to has better performance as compared to other controllers. As seen in Table 6, it can be concluded that for a majority of input signals, the order of controller performance is as follows: M-IMC, RLSE, FF + FB and FF.

Table 6 RMS error of the controller a multi-amplitude, multi-frequency signal

C1(nm)	C2(nm)	C4(nm)	C3(nm)
6.3145	0.33506	0.16118	0.14479

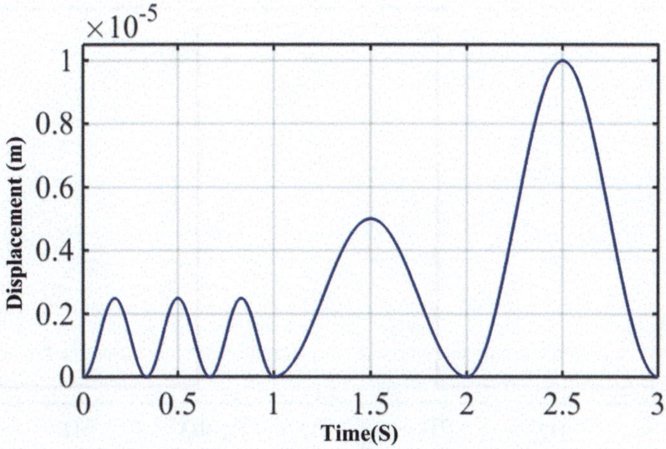

Fig. 3 Mixed-amplitude, mixed-frequency input sinusoidal wave signal

5.2 *Performance Evaluation of the Controllers to a Mixed-Amplitudes, Mixed-Frequency Input Wave Signal*

For a single mixed-amplitude, mixed-frequency signal of as shown in Fig. 3, the RMS tracking error has been tabulated in Table 6, in increasing order of controller effectiveness.

6 Zero Path Tracking Conjunction with Impulse Disturbances (Delay 1 s, Sample Time 20 s)

The response of the controllers to set point tracking against an impulse disturbance of width 20 s at $t = 20$ s is shown in Fig. 4. FF + FB control settles at 21.31 s; however, the response has a prominent undershoot (magnitude 0.27) along with presence of oscillation. The overshoot value for RLSE starts from a magnitude of 0.8 (less than one) and the response settles at 26.9 s, without any oscillation. M-IMC is noticed to have the broad settling time of 32.5 s.

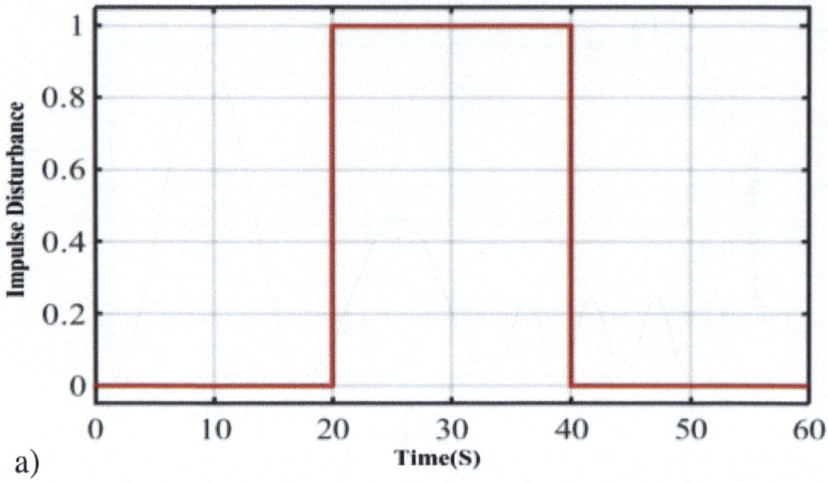

a)

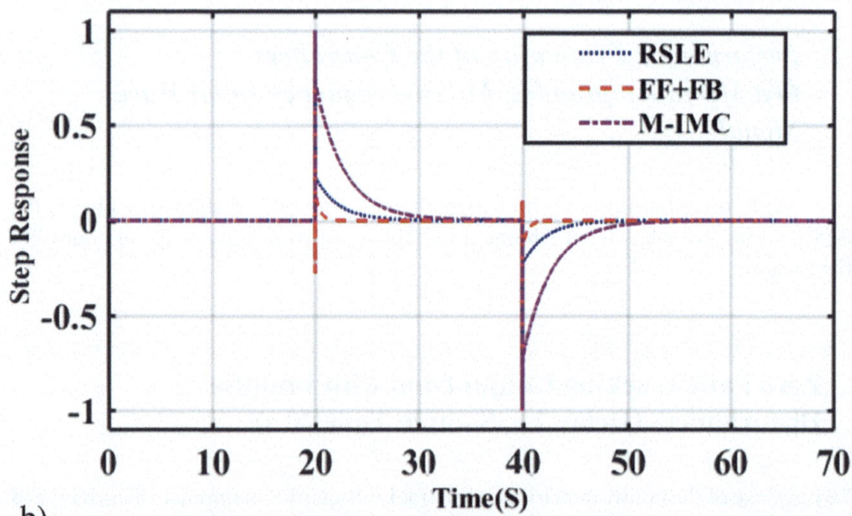

b)

Fig. 4 **a** Impulse disturbance of 20 to 40 s, **b** zero path traking performance of the RSLE,M-IMC and FF + FB controllers for impulse disturbances

7 Zero Path Tracking with Step Input Conjunction with Impulse Disturbances

Similar nature of controller performance is observed for the system when subject to step input with an impulse disturbance shown in Fig. 5. The controllers can be arranged in order of least settling time as FF + FB (presence of undershoot), RLSE and M-IMC.

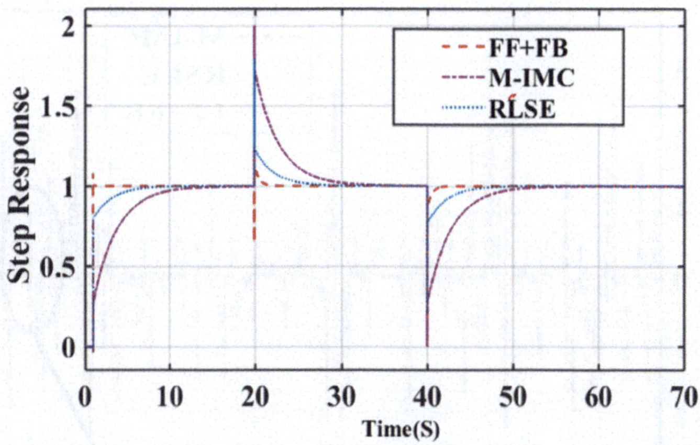

Fig. 5 System response of step input with impulse disturbance for piezoelectric stack actuator

8 Zero Set Point Tracking with Gaussian Noise

For a zero set point tracking of the piezoplant, it is subjected to Gaussian noise. Figure 6 shows the response of the different controllers with unity noise variance. Analyzing the response, it can be observed that the controller tracking performance with Gaussian noise can be arranged in the following order RLSE, M-IMC, FF + FB.

9 Conclusion

This study is focused on comparative analysis of two predominant classes of piezoelectric stack actuator and the results have been considered for different test cases.

Result indicates that a specific type of controller outperforms the other under certain circumstances and needs to be chosen depending upon the application. Modified IMC is observed to lend considerable improvement in performance for sinusoidal motion tracking and parametric uncertainty while RLSE shows a very good performance for disturbance rejection of external Gaussian noise, both as process and as measurement noise, and FF + FB controller has the least settling time with respect to set point tracking with impulse disturbance.

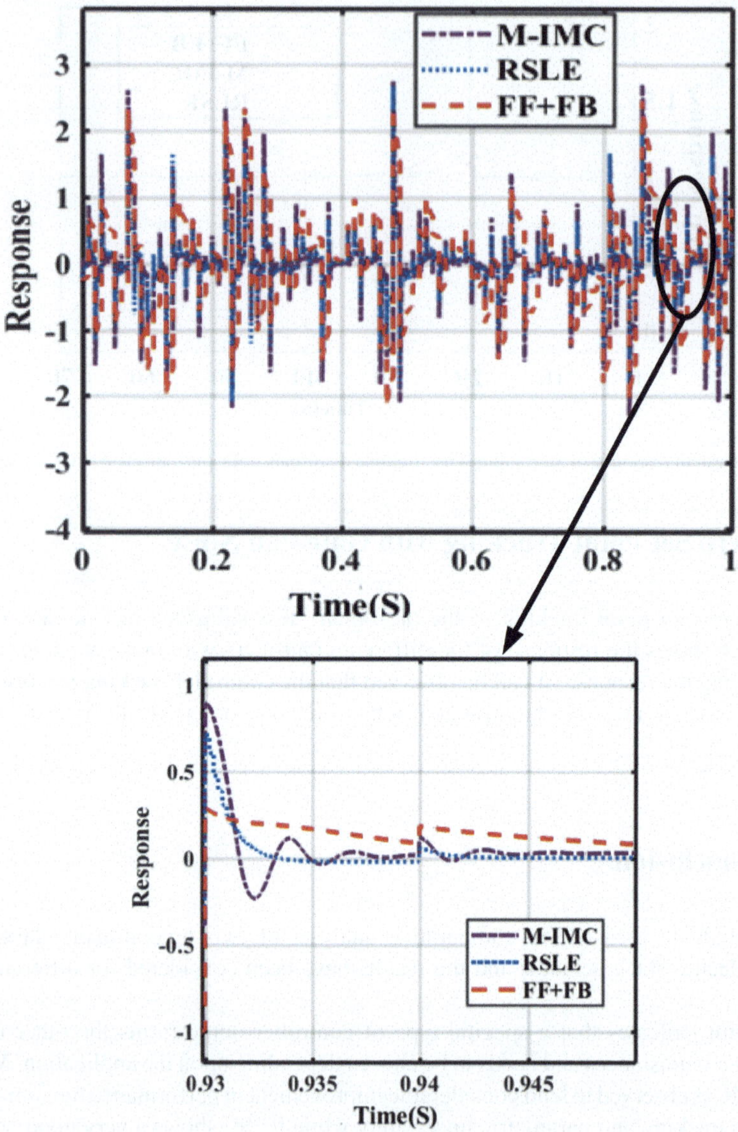

Fig. 6 PZA system response for zero set point tracking with Gaussian response of variance one

References

1. Lambert, P.,. Valentini, A., Lagrange, B.: Delcambre P.D design and performances of a one-degree-of-freedom guided nano-actuator. Robot. Comput. Integr. Manuf. **19**, 89–98 (2003)
2. Hii, K.F., Vallance, R.R., Mengu, M.P.: Design, operation, and motion characteristics of a precise piezoelectric linear motor. Precis. Eng. **34**, 231–241 (2010)

3. Shengdong, Y., Jinyu, M., Hongtao, W., Shengzheng, K.: Robust precision motion control of piezoelectric actuators using fast nonsingular terminal sliding mode with time delay estimation. Measure. Control **52**, 11–19, (2019)
4. Shome, S.K., Mukherjee, A., Karmakar, P., Datta, U.: Adaptive Feedforward controller of piezoelectric actuator for micronano positioning" in Sadhana, Indian Academy of Science, Springer, 2018. Issue—Academy proceedings in Engineering Science, Vol. 43(10) (2018)
5. Shome, S.K., Jana, S., Mukherjee, A., Bhattarcharjee, P., Datta, U.: Bio inspired modified internal model control approach for improved disturbance rejection of piezo micro manipulator. Stud. Inf. Control **27**, 295–306 (2018). https://doi.org/10.24846/v27i3y201805
6. Naik, K.A., Srikanth, P., Negi, P.: IMC tuned PID governor controller for hydro power plant with water hammer effect[J]". Procedia Technol. **4**(4) (2012)
7. Shome, S.K., Jana, S., Mukherjee, A., Bhattacharjee, P., Datta, U.: Improved internal model control based closed loopcontroller design for second order piezoelectric system with dead time. In: Proceedings of 2018 8th IEEE India International Conference on Power Electronics; Jaipur, India. pp. 1–6 (2010)
8. Shome, S., Jana, S., Mukherjee, A., Bhattarcharjee, P.: Model based control for second order piezo actuator system with hysteresis in time delay environment. Turk. J. Elec. Eng. Comp. Sci. https://doi.org/10.3906/elk-1907-28 (2019)

PID Controller Design for Reference Tracking of Single Link Manipulator

Sayan Das and Naiwrita Dey

Abstract This paper is aimed to design a PID controller for reference signal tracking. Design problem is formulated to control the angular position of a single-link manipulator at any particular desired set value. Implementation of the overall feedback control system and the proposed controller has been carried out in Python language. Additive uncertainty is considered with the transfer function model. Simulation results illustrate the usefulness of the proposed controller for angular velocity tracking in presence of uncertainty. Average mean square error and settling time have been calculated. Overall stability of the feedback system has been tested by obtaining the eigenvalues of the closed-loop system.

1 Introduction

Robotics is getting more and more popular in industrial field as it reduces manpower required for doing heavy jobs. Main types of robots used in industrial applications are "multiple-link manipulators" with different degrees of freedom. These kinds of robots can move according to their DOF and complete many tasks. Single-link manipulator model is made out of one fixed base, on which the whole system resides, a revolute joint and an arm, which can move freely on the x–y plane. DC motor or servomotors are mainly used to rotate the arms of manipulators. Here, a simple DC motor with high torque constant is used to move the arm of the manipulator. First, the dynamic model of that motor is designed [1]. Then, control logic is applied on the motor to move the arm to a specific angular position [2–4]. For tuning control parameters, simple tuning method is used [5]. Nowadays, in many industrial applications, rigid link manipulators cannot satisfy the needs. So, "Flexible link manipulators" have become an option. In comparison with rigid body manipulators, flexible manipulators have many advantages including higher efficiency, more flexibility, high load capacity,

S. Das
Department of AEIE, RCCIIT, Kolkata, India

N. Dey (✉)
Department of ECE, RCCIIT, Kolkata, India

J. Gu et al. (eds.), *Communication and Control for Robotic Systems*, Smart Innovation, Systems and Technologies 229, https://doi.org/10.1007/978-981-16-1777-5_3

low energy consumption and larger application field. Flexible link manipulators have wide applications in aerospace engineering, precision instrument production, construction and many other fields [6]. With every practical system, there remains some uncertainty with the parameter values of that system. So, the control model needs to be more accurate to induce its controllability on those uncertain models. There are various control methods like "Robust control" or "Fractional control" methods to deal with parameter uncertainty or external disturbances [7] which are available for controlling these types of manipulators [8–10]. This paper also presents results of controlling an uncertain model of rigid body single-link manipulator. The stability of the system has been taken in account at the time of testing uncertain models.

This paper presents simulations of dynamic behavior of a rigid body single-link manipulator controlled by simple PID logic. PID control is a very popular control method applicable for controlling linear systems. By tuning its parameters, desired controlled output can be generated.

2 Mathematical Modeling of Single-Link Manipulator

The physical system consists of a fixed base, a DC motor and an arm. The arm is movable in x–y plane. Here, the arm acts as the "mechanical load" of the motor. The mathematical modeling of a single-link manipulator can be constructed by modeling the motor part, which is giving driving torque to the manipulator and then the arm part.

2.1 *Transfer Function Model of the System*

DC motor converts electrical energy into mechanical energy. It has a permanent magnet that produces magnetic field around the armature circuit. According to "Fleming's Left-hand rule" that magnetic field along with the electric field of the armature inductance creates rotary motion of the rotor. When the motor spins with its full speed, a "Back EMF" is generated that decreases the applied armature voltage, which decreases the current flowing in the armature circuit. At points M_1 and M_2, mechanical load is connected. Figures 1 and 2 represent the armature circuit of DC motor and diagram of single link manipulator respectively.

Parameters:

$$m = \text{Mass of the arm}$$

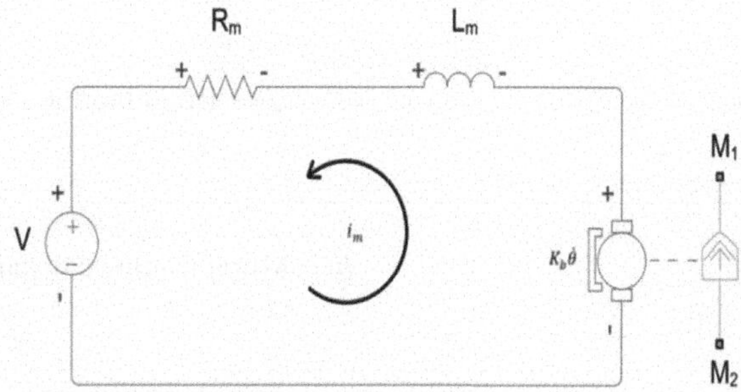

Fig. 1 DC motor armature circuit

$$g = \text{Gravitational acceleration}$$

$$l = \text{Length of the arm}$$

$$V = \text{Applied armature voltage,}$$

$$i_m = \text{Armature current,}$$

$$R_m = \text{Armature resistance,}$$

$$L_m = \text{Armature inductance,}$$

$$K_b = \text{Back emf constant,}$$

$$\theta = \text{Angular position of rotor,}$$

$$\dot{\theta} = \text{Angular velocity of rotor,}$$

$$J = \text{Rotor inertia},$$

$$b = \text{Motor friction constant},$$

$$K_T = \text{Motor torque constant},$$

$$T_m = \text{Torque produced by motor}$$

Applying KVL to the armature circuit,

$$V = L_m \frac{di_m}{dt} + K_b \dot{\theta} + R_m i_m \tag{1}$$

Taking Laplace transform of Eq. (1),

$$V(s) = L_m s.I_m(s) + K_b s.\theta(s) + R_m.I_m(s)$$
$$=> V(s) = I_m(s)[R_m + L_m s] + K_b s.\theta(s)$$

So,

$$I_m(s) = \frac{V(s) - K_b s.\theta(s)}{(R_m + L_m s)} \tag{2}$$

Torque produced by rotor,

$$T_m = K_T.i_m$$

Taking Laplace transform of it,

$$T_m(s) = K_T.I_m(s)$$

Kinematics:

For the arm in Fig. 2,

$$x = \frac{l}{2}\cos\theta$$

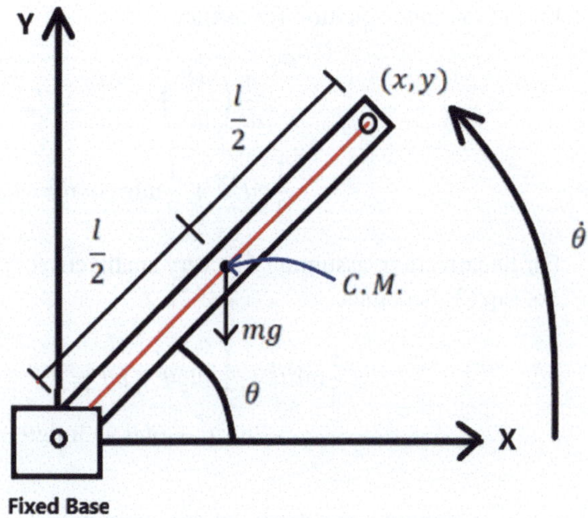

Fig. 2 Single-link manipulator

$$y = \frac{l}{2}\sin\theta$$

Velocity along x and y axis is, respectively,

$$v_x = \dot{x} = -\frac{l}{2}\dot{\theta}\sin\theta$$

$$v_y = \dot{y} = \frac{l}{2}\dot{\theta}\cos\theta$$

Overall velocity of the arm is,

$$v = \sqrt{v_x^2 + v_y^2} = \frac{l}{2}\dot{\theta}$$

Now, the kinetic energy of the system,

$$E_k = \frac{1}{2}\left(mv^2 + I\dot{\theta}^2\right) = \frac{1}{6}ml^2\dot{\theta}^2$$

Potential energy of the system,

$$E_p = \frac{1}{2}\text{mgl}\sin\theta$$

So, Lagrangian of the system is,

$$\mathcal{L} = E_k - E_p = \frac{1}{6}\text{ml}^2\dot{\theta}^2 - \frac{1}{2}\text{mgl}\sin\theta$$

From Lagrange equation for torque,

$$\frac{d}{dt}\left(\frac{\partial \mathcal{L}}{\partial \dot{\theta}}\right) - \frac{\partial \mathcal{L}}{\partial \theta} = T_m$$

$$\frac{1}{3}ml^2\ddot{\theta} + \frac{1}{2}\text{mlg}\cos\theta = T_m \tag{3}$$

For linearization, assuming θ is very small, $\cos\theta \approx 1$.
So, Eq. (3) becomes,

$$\frac{1}{3}\text{ml}^2\ddot{\theta} + \frac{1}{2}\text{mlg}\theta + b\dot{\theta} = T_m$$
$$=> 2\text{ml}^2\ddot{\theta} + 6b\dot{\theta} + 3\text{mlg}\theta = 6T_m$$

where $b\dot{\theta}$ = damping part from the driving motor.
Dynamic equation of the arm can be expressed as,

$$J\ddot{\theta} + 6b\dot{\theta} + p\theta = 6T_m \tag{4}$$

where, $J = 2\text{ml}^2$, $p = 3\text{mlg}$.
Taking Laplace transform of Eq. (4),

$$\left\{Js^2\theta(s) - \theta(0) - \dot{\theta}(0)\right\} + 6bs.\theta(s) + p\theta(s)$$
$$= 6T_m(s) = 6K_T.I_m(s)$$
$$=> \theta(s)\left[Js^2 + 6bs + p\right] = 6K_T.I_m(s)$$
$$=> I_m(s) = \frac{\theta(s).\left[Js^2 + 6bs + p\right]}{6K_T} \tag{5}$$

Comparing (2) and (5), transfer function $Y(s)$ can be obtained,

$$Y(s) = \frac{\theta(s)}{V(s)} = \frac{6K_T}{\left[(R_m + L_m s)\left(Js^2 + 6bs + p\right) + s.K_b.K_T\right]} \tag{6}$$

State-space form of the above system has been obtained as,

$$\frac{\text{d}}{\text{dt}}\begin{pmatrix}\theta \\ \dot{\theta} \\ i_m\end{pmatrix} = \begin{pmatrix}0 & 1 & 0 \\ -\frac{p}{J} & \frac{-6b}{J} & \frac{6K_T}{J} \\ 0 & \frac{-K_b}{J} & \frac{-R_m}{L_m}\end{pmatrix}.\begin{pmatrix}\theta \\ \dot{\theta} \\ i_m\end{pmatrix} + \begin{pmatrix}0 \\ 0 \\ \frac{1}{L_m}\end{pmatrix}.V \tag{7}$$

$$y = \begin{pmatrix}0\,1\,0 \\ 0\,0\,0 \\ 0\,0\,0\end{pmatrix}\begin{pmatrix}\theta \\ \dot{\theta} \\ i_m\end{pmatrix}$$

2.2 *Uncertainty Modeling of the System*

The system represented by the transfer function $Y(s)$ cannot be considered as the practical system. Practical systems can have different uncertainties which can deviate the nature of the practical system from the ideal one. Such uncertainties come mainly from external disturbances or real parameter variations. Here, the components of the system have some variation or uncertainty in their parameter values. So, parameters like m, R_m and L_m contributing to the transfer function $Y(s)$ have some uncertainty in their value. The practical parameters of the system can be written as:

$m_p = m + \Delta_{\hat{m}}$, $R_{mp} = R_m + \Delta_{R_m}$, $L_{mp} = L_m + \Delta_{L_m}$, $b_p = b + \Delta_b$ where $\Delta_r =$ uncertainty of the parameter "r" and $r_p =$ practical value of the parameter "r." So the explicit form of state-space equation of the uncertain single-link manipulator model becomes:

Finally, the transfer function of the system with uncertainty is given by,

$$Y_U(s) = \frac{6K_T}{\left[\left(R_{mp} + L_{mp}s\right)\left(2m_p l^2 s^2 + 6bs + 3m_p gl\right) + s.K_b.K_T\right]} \tag{8}$$

3 Controller Design

3.1 *Block Diagram of Single-Link Manipulator Angular Position Closed-Loop System*

The closed loop angular position control system for single link manipulator is shown in Fig. 3.

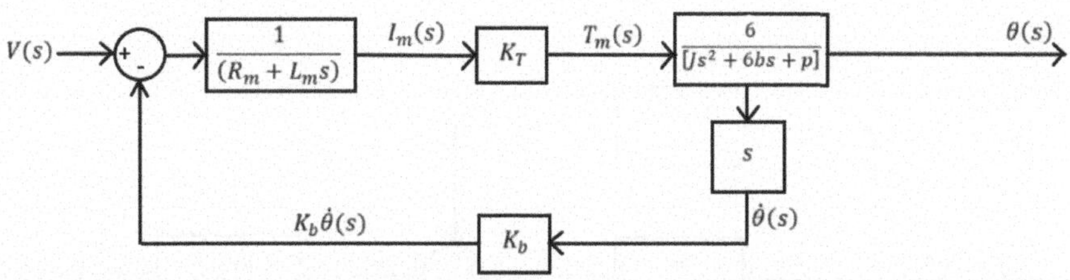

Fig. 3 Single-link manipulator angular position closed-loop block diagram

3.2 Implementing PID Control to the Nominal Manipulator System

The term PID stands for "Proportional Integral Derivative." PID controllers are flexible, simple and easy to implement. By using proper optimization, PID controller can control and regulate time-domain response of wide range of processes. The time-domain equation of PID controller is as follows:

$$u(t) = u_{\text{bias}} + K_p e(t) + K_i \int e(t)dt + K_d \frac{de(t)}{dt} \tag{9}$$

where $u(t)$ = controller output, u_{bias} = initial value of controlled output (let $u_{\text{bias}} = 0$ for this case), $e(t)$ = error term, i.e., difference between set point (SP) and process variable (PV), K_p = proportional gain, $K_i = \frac{K_p}{\tau_i}$ (τ_i is integral time constant), and $K_d = K_p \tau_d$ (τ_d is derivative time constant). Figure 4 shows the closed loop angular position control system with PID controller.

Laplace transforming (8),

$$U(s) = E(s)\left[K_p + \frac{K_i}{s} + K_d s\right]$$

Control transfer function,

$$C(s) = \frac{U(s)}{E(s)} = K_p + \frac{K_i}{s} + K_d s$$

So, the overall open-loop transfer function of the controlled model becomes (Fig. 4),

$$G(s) = Y(s).C(s)$$

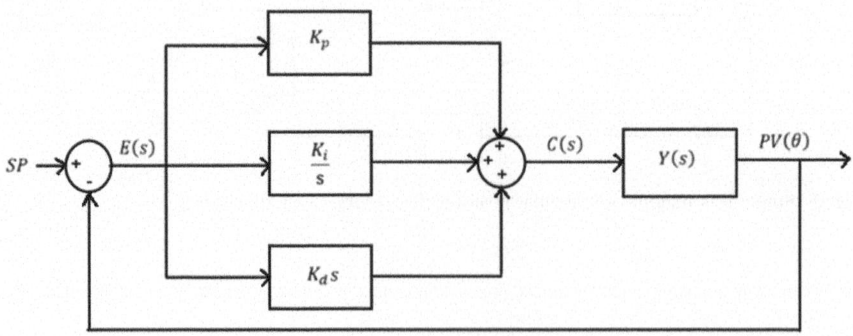

Fig. 4 Block diagram of PID control in single-link manipulator for controlling angular position

Table 1 Tuning PID parameters

Parameter change	Rise time (s)	Maximum overshoot (%)	Settling time (s)	Steady-state error
Increasing K_p	Decrease	Increase	Small change	Decrease
Increasing K_i	Decrease	Increase	Increase	Eliminate
Increasing K_d	Small change	Decrease	Decrease	Small change

Tuning PID parameters: To get desired response, the control parameters like K_p, K_i and K_d should be tuned properly. By checking the output response plot, it can be pointed out values of which parameters should be increased or decreased. Table 1 shows the same.

4 Implementation of Proposed Controller in Python

Python, as an object-oriented programming language, is simple and versatile. All the simulations presented in this paper have been done in Python language. For simulating system response in different control environments, Python's dedicated control system library is used [11]. For some mathematical purposes, different available modules are used. All the programming has been done using "Anaconda Navigator" and "Jupyter Notebook."

Anaconda is an open-source software which directly comes with different Python and R language IDEs or platforms preinstalled. Anaconda is best used for machine learning and data analysis tasks. The software also comes with common modules and libraries installed but Anaconda's Python package does not include control system library by default, so that it is need to be installed first using "Anaconda Prompt." By writing "pip install control" command in the prompt, the library can be installed easily as shown in Fig. 5.

Then, after opening Anaconda Navigator, Jupyter notebook is to be selected, and it will automatically open Jupyter notebook in the browser at predefined path. There is also another Python IDE "Spyder" is available. Anaconda makes Jupyter notebook and Spyder databases interconnected, so anyone can easily switch in between them.

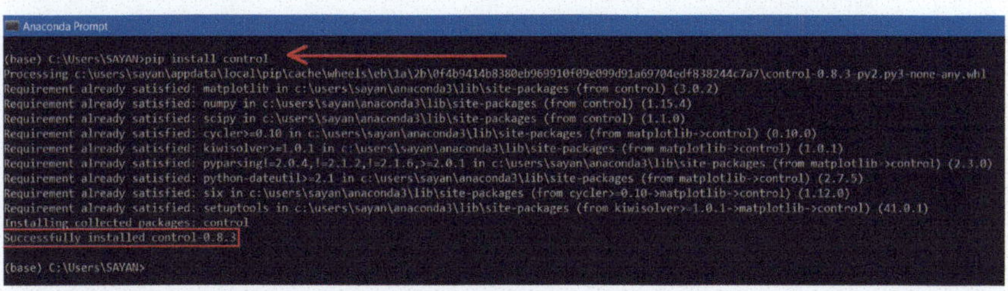

Fig. 5 Installing control system library in Anaconda

The following figure shows the Anaconda Navigator and Jupyter notebook interface. Figures 6 and 7 show the navigator and editor interface respectively.

Jupyter notebook is preferable for testing programs as it allows users to run different sections of program at any time. So, it is easy to find errors in the code.

Python control systems library is a Python package that emulates maximum functions from MATLAB's control system toolbox and implements techniques for analyzing any control system object.

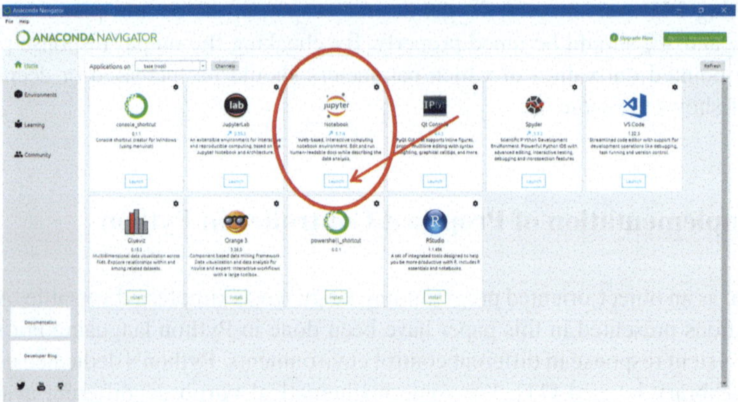

Fig. 6 Launching Jupyter notebook from Anaconda Navigator

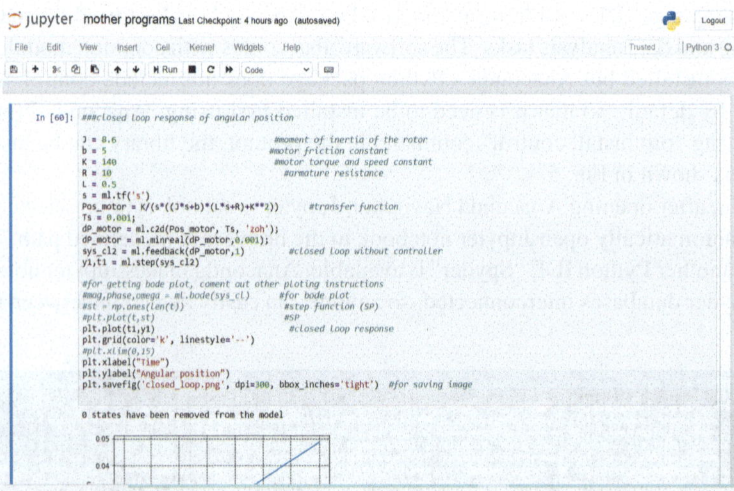

Fig. 7 Jupyter notebook interface

4.1 Flowchart of the Simulation Program

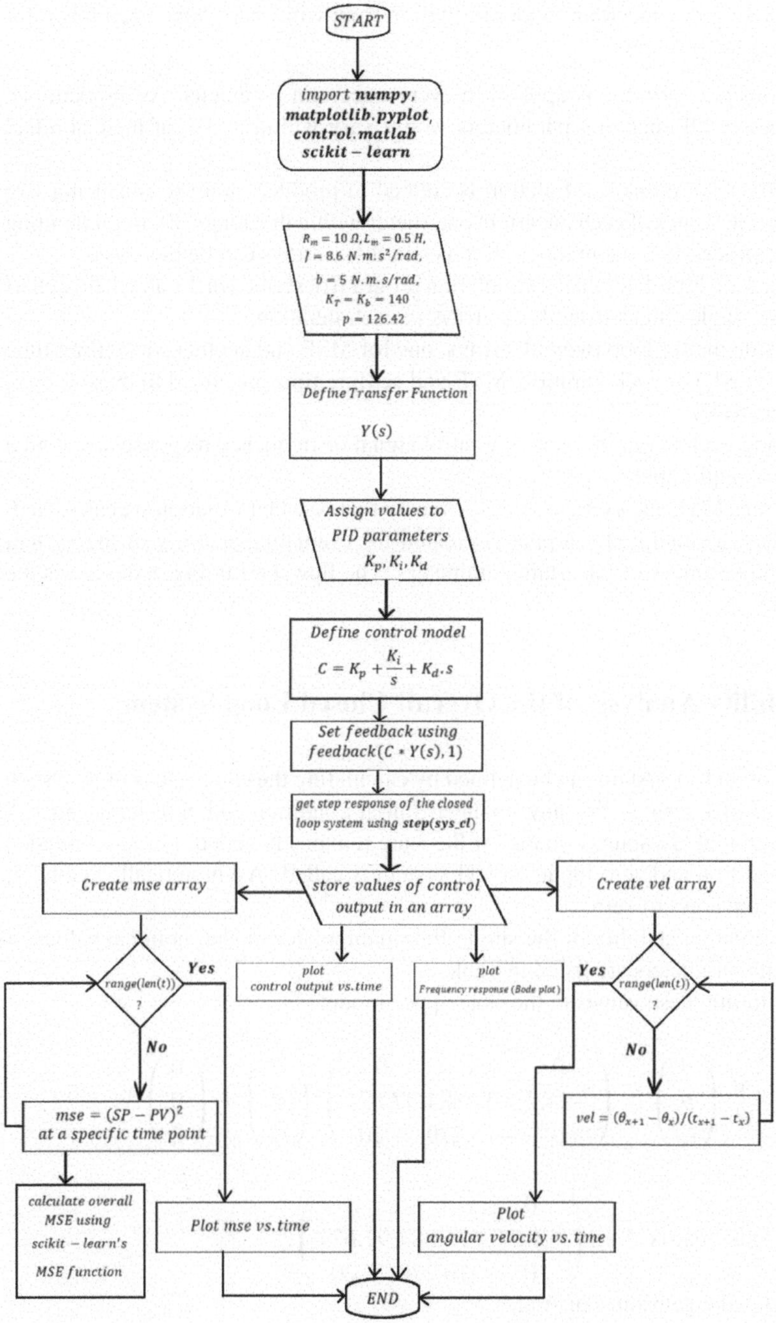

4.2 Programming for Simulating Uncertain Model

For simulating the uncertain model in Python, following steps have been taken for developing the program:

1. A constant variation is applied to every uncertain parameter. As an example, values of all uncertain parameters would vary within $\pm$ 5% of their nominal value.
2. As stated in point 1, a function is created to produce an array consisting five different values of each uncertain parameter within that range. By implementing multiprocessing, the number of values in those arrays can be increased.
3. A for loop is built to run the simulation program described in the above flowchart using single values from those arrays in each iteration.
4. Outside the for loop two null arrays, one for MSE and another for settling time is created. For each iteration, MSE and settling time are added to those arrays, respectively.
5. During each iteration, plots of control signal or frequency response are plotted as per requirement.
6. Average MSE and average settling time for that constant variation are calculated.
7. At last, a small code snippet is created for evaluating stability of the system for those values of uncertain parameters. The flowchart in Fig. 8 describes the same.

5 Stability Analysis of the Overall Closed Loop System

Stability of an LTI system can be defined by calculating the eigenvalues of the "state matrix" of that system. For any bounded initial condition and zero input, an LTI system is called "Lyapunov stable" if the state remains bounded. For any bounded initial condition and zero input, an LTI system is called "Asymptotically stable" if the state converges to zero.

For evaluating stability of the single-link manipulator model, nominal values of the system parameters are listed in Table 2.

Substituting these values in the state-space model (7),

$$\frac{d}{dt}\begin{pmatrix}\theta\\ \dot{\theta}\\ i_m\end{pmatrix} = \begin{pmatrix}0 & 1 & 0\\ -14.7 & -3.4884 & 97.674\\ 0 & -16.279 & -20\end{pmatrix}.\begin{pmatrix}\theta\\ \dot{\theta}\\ i_m\end{pmatrix} + \begin{pmatrix}0\\ 0\\ 2\end{pmatrix}.V$$

Here, state matrix $A = \begin{pmatrix}0 & 1 & 0\\ -14.7 & -3.4884 & 97.674\\ 0 & -16.279 & -20\end{pmatrix}$.

Calculated eigenvalues of A are:

$\lambda_1 = -11.6562 + 39.1731i$, $\lambda_2 = -11.6562 - 39.1731i$ and

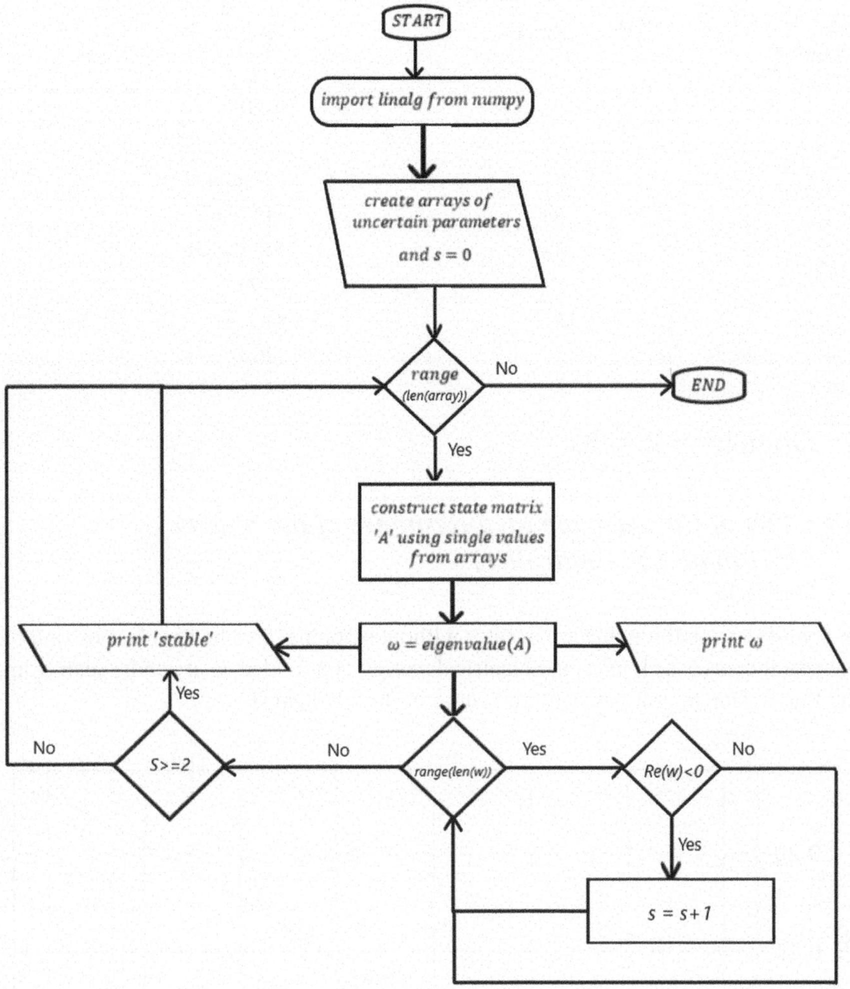

Fig. 8 Flowchart of program for evaluating stability of the system

$$\lambda_3 = -0.176006$$

So,

1. Two eigenvalues λ_1 and λ_2 are complex conjugates, and λ_3 is a negative real number.
2. $Re(\lambda_1) = Re(\lambda_2) < 0$ and $Re(\lambda_3) < 0$.

The negative real part of the eigenvalues implies the asymptotic stability of the overall closed-loop system.

Table 2 Nominal values of parameters

Parameter	Value
R_m	10 Ω
L_m	0.5 H
m	1 kg
b	5 N.m.s/rad
K_T	140 N.m/A
K_b	140 V.s/rad
l	2.07 m
J	8.6
p	126.42

6 Simulation Results

6.1 *Plot of Closed-Loop Step Response of the System (Without Any Controller)*

In open-loop condition, the arm rotates without stopping, and the angular position of the arm increases as long as the electrical energy is provided, as it can be seen from the Fig. 9. Due to arm's weight, the curve is slightly bent (Fig. 9).

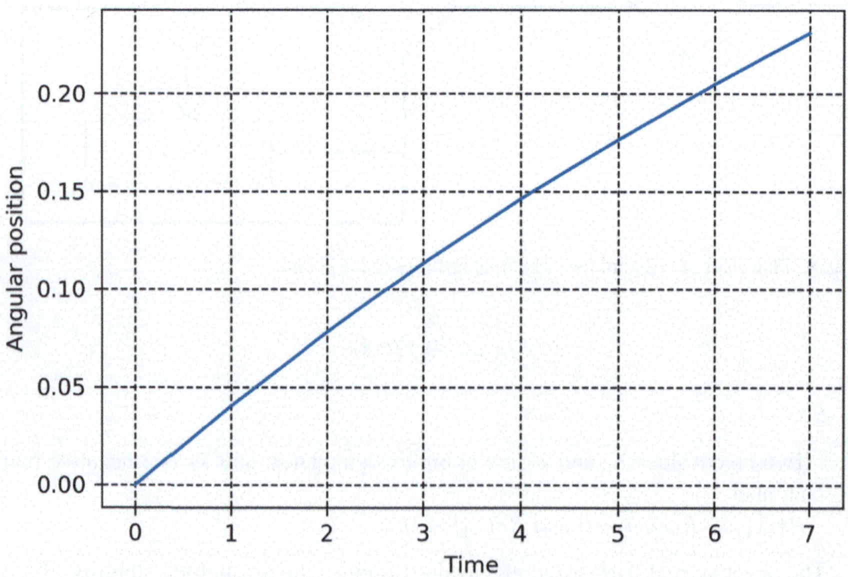

Fig. 9 Open-loop step response of the system

Table 3 Values of PID parameters, MSE and settling time for nominal system

PID parameters	MSE	Settling time (s)
$K_p = 100$ $K_i = 50$ $K_d = 1$	0.01536	1.9768

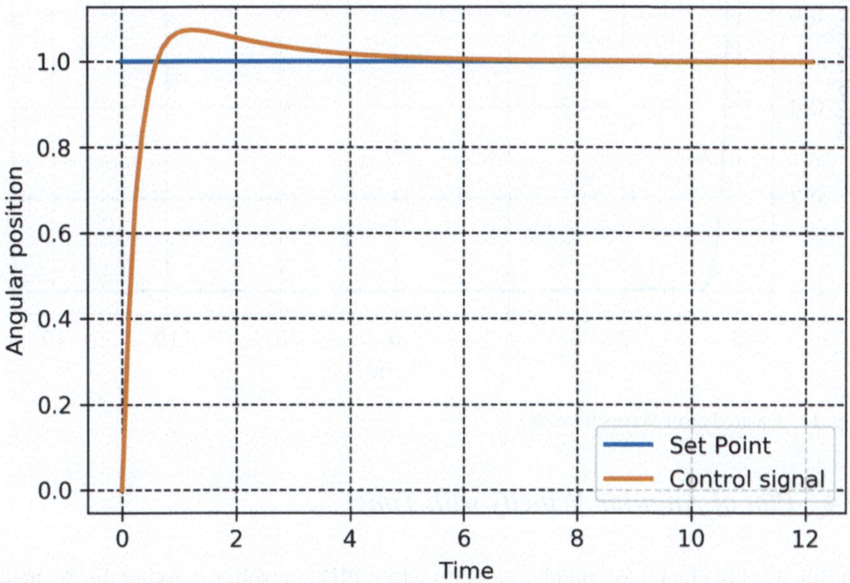

Fig. 10 PID controlled step response of the system

6.2 *Plot of PID Controlled Step Response of the Nominal System*

In case of choosing PID gains, "Manual tuning" has been taken into account. Such a case study with the values of PID parameters is used for simulating; MSE and settling time are presented in Table 3.

Figure 10 shows how perfectly the implemented PID controller fixes the angular position of the arm according to the set point.

6.3 *Plot of Squared Error with Time of the Controlled Response*

Figure 11 shows how the squared error of the PID controlled system (i.e., SP-PV) is made zero with time.

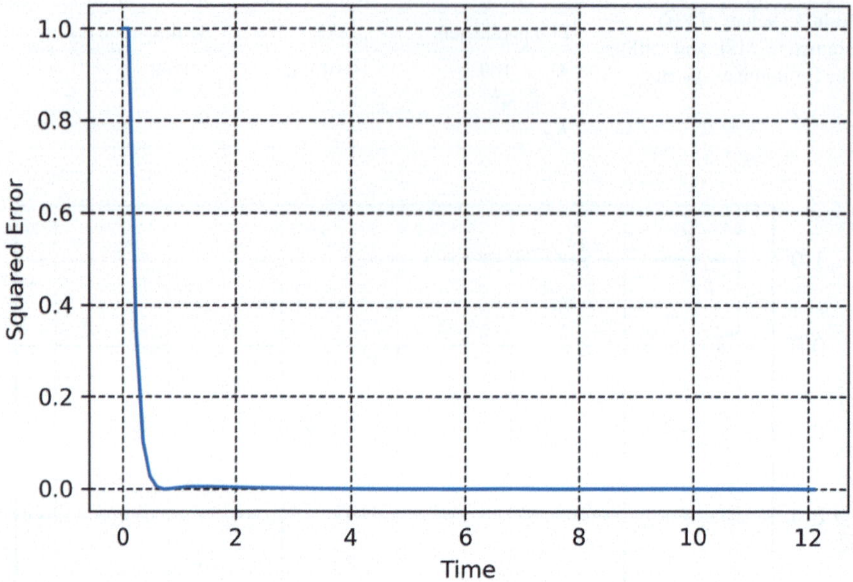

Fig. 11 Squared error versus time plot

6.4 *Plot of Angular Velocity with Time*

In Fig. 12, the change of angular velocity while PID controller is fixing the desired angular position is shown. At around 1.7 to 1.8 s, the angular velocity goes slightly negative and then settles to 0 as angular position is fixed.

6.5 *Frequency Response of the Uncertain Model*

For simulation, two different values for Δ are taken:

Different case studies for these values of uncertainties are shown in Tables 4 and 5 for considering different values of Δ. Figure 13 depicts the frequency response of the uncertain system.

For $\Delta = \pm 5\%$

For $\Delta = \pm 7\%$

From the obtained results, it can be said that parameter uncertainty within ± 7% of their value is permeable for the model, the model remains stable, and the angular position reference command is tracked within allowable range of mean square error.

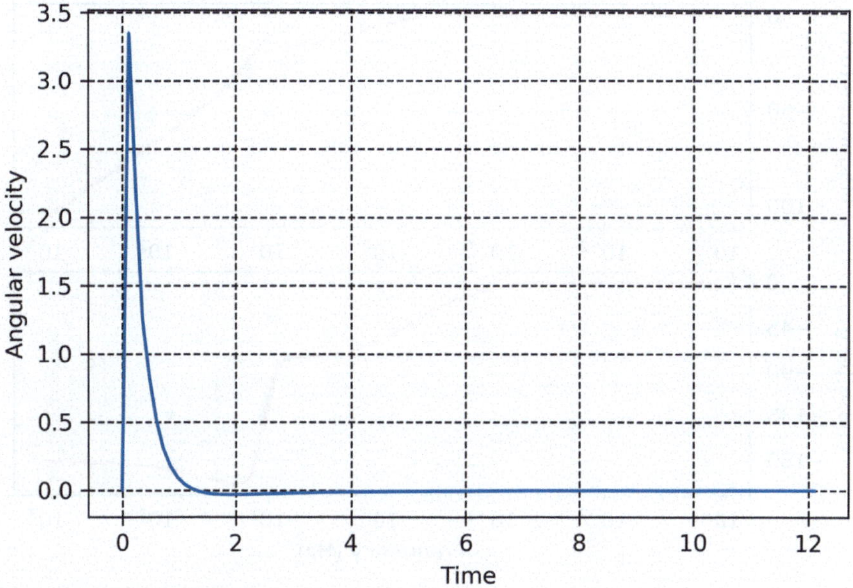

Fig. 12 Angular velocity versus time plot

Table 4 Values of PID parameters, MSE and Settling time for uncertainty, $\Delta = \pm 5\%$

PID parameters	MSE	Average settling time (sec)	Stability
$K_p = 25$ $K_i = 50$ $K_d = 50$	0.0234	2.3781	Stable
$K_p = 50$ $K_i = 80$ $K_d = 2$	0.02707	1.8247	Stable
$K_p = 100$ $K_i = 50$ $K_d = 1$	0.01161	1.5589	Stable

Table 5 Values of PID parameters, MSE and Settling time for uncertainty, $\Delta = \pm 7\%$

PID parameters	Average MSE	Average settling time (s)	Stability
$K_p = 25$ $K_i = 50$ $K_d = 5$	0.02330	1.09128	Stable
$K_p = 50$ $K_i = 50$ $K_d = 2$	0.02705	1.8247	Stable

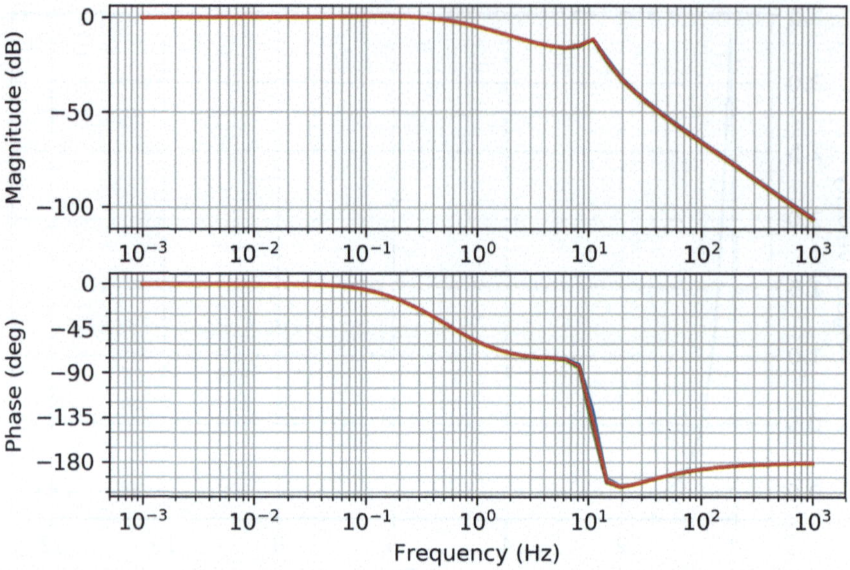

Fig. 13 Bode plot of the uncertain model ($\Delta = \pm\, 5\%$, $K_p = 100$, $K_i = 50$ and $K_d = 1$)

7 Conclusion

In this paper, PID controller is designed and implemented to control the angular position of a single-link manipulator system which is a very popular robotic application. The manipulator system has been modeled considering additive uncertainty which makes the system closer to practical system. Simulation experimentation is performed for different values of uncertain parameters. Mean square error is calculated as performance metric along with settling time. As the closed-loop system stability is a major concern, the calculated eigenvalues containing negative real part assures the same for this present work. Application of control system toolbox in Jupyter notebook is another important feature of this work. This work can be further easily extended to embed the simulated controller on a Linux-based system for standalone application.

References

1. Deb PB, Saha O, Sajan S (2017) Dynamic model analysis of a dc motor in MATLAB. Int J Sci Eng Res 8(3)
2. Yime E, Villa JL, Paez J (2014) Design of a brushed DC motors PID controller for development of low-cost robotic applications. III international congress of engineering mechatronics and automation (CIIMA). https://doi.org/10.1109/ciima.2014.6983471
3. Ulm, Steve F (2020) DC motor position control. Computer and electrical engineering technology & information systems and technology senior design projects

4. Elsrogy WM, Fkirin MA, Hassan MAM (2013) Speed control of DC motor using PID controller based on artificial intelligence techniques. International conference on control, decision and information technologies (CoDIT). doi: https://doi.org/10.1109/codit.2013.6689543
5. Abdulameer, A., Sulaiman, M., Aras, M., Saleem, D.: Tuning methods of PID controller for DC motor speed control. Indonesian J Elec Eng Comput Sci **3**, 343 (2016)
6. Ata AA, Haraz EH, Rizk AEA, Hanna SN (2012) Kinematic analysis of a single link flexible manipulator. IEEE Int Conf Ind Technol. https://doi.org/10.1109/icit.2012.6210045
7. Fareh R (2019) Control of a single flexible link manipulator using fractional active disturbance rejection control. 6th international conference on control, decision and information technologies (CoDIT), pp 900–905. https://doi.org/10.1109/codit.2019.8820708
8. Zhang H, Gao X, Xu G (2019) Research on Improved PD control of flexible manipulator. Chinese control and decision conference (CCDC). https://doi.org/10.1109/ccdc.2019.8832942
9. Zhang C, Yang T, Sun N, Zhang J (2019) A simple control method of single-link flexible manipulators. 2019 3rd international symposium on autonomous systems (ISAS). https://doi.org/10.1109/isass.2019.8757711
10. Yan Z, F., Lai X, S., Meng Q, T., Wu M, Ft.: A Novel Robust Control Method for Motion Control of Uncertain Single-Link Flexible-Joint Manipuſator. IEEE Transactions on Systems, Man, and Cybernetics: Systems, pp. 1–8 (2019)
11. Python Control Systems Library Documentation page, https://python-control.readthedocs.io/en/latest/control.html, last accessed 2020/08/06

Performance Comparison Between Higher-Order Sliding Mode and Fixed Boundary Layer Sliding Mode Controller for a 10-DoF Bipedal Robot

Koceila Cherfouh, Jason Gu, Umar Farooq, Muhammad Usman Asad, Rajeeb Dey, Nabanita Adhikary, and Chunqi Chang

Abstract In this paper, a performance comparison between the boundary layer sliding mode control (BLSMC) and the higher-order sliding mode controllers (HOSMC) is conducted. Sliding mode control (SMC) is known for its robustness to inaccuracies in the system model and external disturbances. However, SMC tends to have an undesirable chattering phenomenon causing low control accuracy and high-power consumption. The BLSMC and HOSMC are two approaches studied by researchers to solve the chattering problem. The performance comparison and results are obtained through simulations of two systems, a 4 degrees of freedom (DoF) robotic manipulator and a 10-DoF bipedal robot.

1 Introduction

Sliding mode control is a widely studied nonlinear control technique. It is known to have good robustness to external disturbances. However, this controller often has a chattering problem, which makes it non-ideal for real-world applications. The chattering problem causes high-power losses and actuator wear. Many solutions are known in the literature like the fixed and varying boundary layer SMC and the high-order SMC like the super-twisting SMC or higher-order sliding mode control.

K. Cherfouh (✉) · J. Gu · U. Farooq · M. U. Asad
Department of Electrical & Computer Engineering, Dalhousie University, Halifax, Canada
e-mail: kc389925@dal.ca

J. Gu
e-mail: jason.gu@dal.ca

R. Dey · N. Adhikary
Department of Electrical Engineering, NIT, Silchar, India

C. Chang
School of Biomedical Engineering, Health Sciences Center, Shenzhen University, Shenzhen 518060, China
e-mail: cqchang@szu.edu.cn

J. Gu et al. (eds.), *Communication and Control for Robotic Systems*, Smart Innovation, Systems and Technologies 229, https://doi.org/10.1007/978-981-16-1777-5_4

Previously, SMC was applied on a variety of nonlinear systems to remedy model uncertainties and disturbances [1–4]. In some cases, however, the chattering problem is not ideal for real-world applications. BLSMC and HOSMC, a solution to the chattering problem, have been applied and tested on a variety of physical and electrical systems ranging from the simple pendulum system to the highly nonlinear buck converter or high DoF manipulators [5–12]. The BLSMC and HOSMC proved to be effective methods in suppressing the chattering problem of the traditional SMC.

Sliding mode control has been extensively researched and applied to robotic manipulators. It is the SMC's robustness feature that makes it a desirable option when applying it to highly uncertain nonlinear systems like robotic manipulators. To remedy the chattering problem of the SMC, researchers applied various alterations of SMC techniques to manipulator systems. Fixed and time-varying boundary layer SMC was applied to manipulators to suppress the chattering effect [13–15]. The time-varying boundary layer SMC achieved better trajectory tracking than the fixed BLSMC. SOSMC was also implemented for rigid and constrained manipulators to attenuate the chattering [16, 14]. Researchers also used HOSMC to high uncertainty systems like a space manipulator [17]. Results show that the HOSMC has excellent adaptability to large uncertainties and disturbances. Purwar [18] demonstrates that HOSMC has greater robustness than fuzzy SMC.

Furthermore, researchers have also applied the SMC to high degrees of freedom biped robots for both the single-support and double-support phase [19]. Researchers have suggested methods to smooth the control input and attenuate the chattering to solve the chattering problem. BLSMC was applied to the biped, and the system converged to the boundary layer in finite time [20]. The BLSMC, however, requires to be designed to suppress the chattering effect while minimizing the steady-state error. Oza et al. [21] proposed a second-order twisting SMC combined with a boundary layer to smoothen the control input and remove any potential chattering. Robust and stable trajectory tracking of the biped walking gait was achieved.

This paper presents a performance comparison of these controllers in terms of state response and control input requirements. Additionally, we will apply the SMC, boundary layer SMC (BLSMC), second-order SMC (SOSMC), and third-order SMC (TOSMC) to two different systems: a 4-DoF robotic manipulator and a 10-DoF bipedal robot. We will use these two systems to compare: 1- the performance of position tracking using a step input on the robotic manipulator, 2- the robustness performance of all presented controllers to adapt to randomly generated uncertainties and disturbances and, 3- the trajectory tracking using the 3D spring-loaded inverted pendulum gait on the bipedal robot [22].

In Sect. 2, we talk about the derivation of the dynamics models of our two systems. Afterward, we present the mathematical representation of the four studied sliding controllers. In Sect. 3, we will compare the first-order SMC, BLSMC, SOSMC, and third-order SMC (TOSMC) on the position tracking and the trajectory tracking problem, and we will also compare their robustness to random model uncertainties and disturbances. A similar rise time and no overshoot requirement were met by all the controllers to make an even comparison in performance. The performance was measured by the integral of absolute error (IAE), integral of absolute error multiplied by time (ITAE), and integral of the error squared (ISE) index, as well as the input response and the potential to handle plant disturbances.

2 Simulation

2.1 Dynamics Model

To test the BLSMC and the SOSMC, we derive the dynamics equations of two nonlinear systems, a 4-DoF robotic manipulator and a 10-DoF manipulator. This was done using the Euler–Lagrange method:

$$\mathcal{L}(\theta, \dot{\theta}) = \text{KE}(\theta, \dot{\theta}) - \text{PE}(\theta) \tag{1}$$

where KE is the total kinetic energy of the system and PE, the total potential energy of the system. The equations of motion can then be derived as follows

$$\tau_i = \frac{d}{dt}\frac{\partial \mathcal{L}}{\partial \dot{\theta}_i} - \frac{\partial \mathcal{L}}{\partial \theta_i} \tag{2}$$

The resulting equations of motion can then be formulated as:

where M is the mass matrix, C the Coriolis matrix, G the gravitational matrix, τ the input, and δ added noise to simulate real-world uncertainties and disturbances.

$$M(\theta)\ddot{\theta} + C(\dot{\theta}, \theta) + G(\theta) = \tau + \delta \tag{3}$$

This computation was done using the Euler–Lagrange MATLAB toolbox [23]. The equations of motion of the system are however computed and presented in the form

$$\ddot{\theta} = f(\dot{\theta}, \theta) + g(\theta)\tau + \delta \tag{4}$$

2.2 Controller Design

First-Order Sliding mode controller. To compare the performance of the controllers, we first begin by implementing a sliding mode controller for the two plants. This will be used as a reference benchmark for the BLSMC and the SOSMC. The surface was designed as

$$s = \left(\frac{d}{dt} + \lambda\right)\theta_e \tag{5}$$

where s is the sliding surface vector, λ a positive constant vector, and θ_e the error vector defined as

$$\theta_e = \theta_{\text{measured}} - \theta_{\text{desired}} \tag{6}$$

where θ_{desired} represents the vector of desired position for each joint, and θ_{measured} the vector of measured position.

In order to ensure stability, we verify the Lyapunov stability criterion

$$V = \frac{1}{2}s^T s \tag{7}$$

$$\dot{V} = s\dot{s} \leq 0 \tag{8}$$

$$\dot{V} = s\left(\ddot{\theta} + \lambda\dot{\theta}\right) \tag{9}$$

$$\dot{V} = s\left(f\left(\dot{\theta}, \theta\right) + g(\theta)\tau + \delta + \lambda\dot{\theta}\right) \tag{10}$$

the control input vector to satisfy the condition in Eq. (8) is thus as follows

$$\tau = g^{-1}(\theta)\left(-f\left(\dot{\theta}, \theta\right) - \lambda\dot{\theta} - k \cdot \text{sign}(s)\right) \tag{11}$$

where

$$\text{sign}(s) = \begin{cases} -1, & s < 0 \\ 0, & s = 0 \\ 1, & s > 0 \end{cases} \tag{12}$$

Substituting Eq. (10) into Eq. (11), we obtain

$$\dot{V} = -k|s| + s\delta \leq 0 \tag{13}$$

As long as (8) is satisfied, the sliding mode will be maintained; the system is asymptotically stable and will converge to its equilibrium point. The control input gain must satisfy the condition $k > \max|(\delta) \cdot \text{sign}(s)|$, to maintain stability $s\dot{s} \leq 0$.

Fixed Boundary Layer Sliding Mode Controller. The second controller which we will discuss is the BLSMC. This controller is like the first-order SMC and satisfies all Eqs. (5–9); however, in this algorithm, we introduce a boundary layer to eliminate the chattering phenomenon. The high-frequency chattering can be eliminated by replacing the $sign(s)$ function with a saturation function. The control law is now defined as

$$\tau = g^{-1}(\theta)\left(-f(\dot{\theta}, \theta) - \lambda\dot{\theta} - k \cdot \mathrm{sat}(s)\right) \tag{14}$$

where

$$\mathrm{sat}(s) = \begin{cases} -1, & s < -1 \\ \frac{s}{\alpha}, & -1 \leq s < 1 \\ 1, & s \geq 1 \end{cases} \tag{15}$$

where α is a positive constant.

Super-Twisting Second-Order Sliding Mode Controller. The super-twisting second-order sliding mode controller is widely used to suppress the chattering problem of the first-order SMC. The control law for this algorithm is straightforward. The sliding surface will be designed as in Eq. (5).

$$s = \left(\frac{d}{dt} + \lambda\right)\theta_e$$

The continuous control law can then be defined as

$$\tau = \lambda\sqrt{|s|}\,\mathrm{sign}(s) + \omega \tag{16}$$

where ω can be defined as

$$\omega = \beta \int \mathrm{sign}(s)\mathrm{d}t \tag{17}$$

and where β is a positive constant vector. The resulting control law can then be defined as an addition of the equivalent control and sliding continuous control

$$\tau = g^{-1}(\theta)\left(-f(\dot{\theta}, \theta) - \lambda\dot{\theta} - \lambda\sqrt{|s|}\,\mathrm{sign}(s) - \beta \int \mathrm{sign}(s)\mathrm{d}t\right) \tag{18}$$

Higher-Order Sliding Mode Controller. The higher-order sliding mode controller is an alternative option which we discuss to eliminate the chattering problem of the first-order SMC. First, we must identify the relative degree of the system N. A sliding system is defined as Nth order when

$$\frac{\mathrm{d}}{\mathrm{d}u}s^{(i)} = 0, \quad \forall i = 1, 2, 3, N - 1 \tag{19}$$

Following Eq. (19), we find that both the manipulator and bipedal systems can be defined as third-order systems provided that we define the sliding surface and its derivatives as

$$s = \theta \tag{20}$$

$$\dot{s} = \dot{\theta} \tag{21}$$

$$\ddot{s} = \ddot{\theta} = f(\dot{\theta}, \theta) + g(\theta)\tau \tag{22}$$

$$\dddot{s} = \frac{d}{dt}\left(f(\dot{\theta}, \theta) + g(\theta)\tau\right) = f'(\dot{\theta}, \theta, \tau) + g'(\theta)\tau' \tag{23}$$

By computing the third derivative of the sliding surface, we obtain a virtual control input τ'. To maintain asymptotic stability of the system, we defined the virtual control input as

$$\tau'(t) = g'(\theta)^{-1}\left(-f'(\dot{\theta}, \theta, u) - k_1 s - k_2\dot{s} - k_3\ddot{s} - k_4 sign(\ddot{s})\right) \tag{24}$$

The integral of the virtual control input is then used as the control input of the system

$$\tau(t) = \int \tau'(t)\mathrm{d}t \tag{25}$$

3 Results and Discussion

The results section will provide the comparison results between the presented controllers in Sect. 2.2. We first compare the position tracking on the 4-DoF manipulator using the IAE, ITAE, and ISE index. We then discuss the robustness of each of the controllers by varying the magnitude of the plant disturbance. Lastly, we will compare the trajectory tracking problem on the 10-DoF bipedal robot. Control input response and computation time will also be discussion points in this comparison.

3.1 Position Tracking on 4-DoF Robotic Manipulator

In this section, the position tracking comparison between each of the mentioned controllers in Sect. 2.2 is presented. To conduct a fair performance comparison of position tracking on the robotic manipulator, the presented controllers were tuned to have a similar state response rise time between 0.8 ~ 0.9 s and no overshoot when the desired point is at the highest point (180 deg). The algorithms were tested on four different set points 10, 60, 135, and 180 deg for 2 s (200 samples at dt = 0.01 s). The plant noise was defined as

$$\delta = a \cdot \mathrm{rand}() \tag{26}$$

Table 1 Position tracking performance comparison of the presented sliding mode controllers when applied to a 4-DoF robotic manipulator with added plant noise $a = 0.5$

Controller	Set Point (deg)	IAE	ITAE	ISE
SMC	10	9.0019	7.3554	46.7350
	60	23.2619	9.3392	726.3252
	135	54.8543	20.7097	4174.7
	180	74.8583	26.9990	7922.7
BLSMC	10	3.4127	1.1361	17.9859
	60	22.3875	7.2035	777.9500
	135	57.4505	19.1008	4849.2
	180	81.6818	27.8441	9466.8
SOSMC	10	2.7426	0.5894	17.2276
	60	22.4849	5.8372	916.3204
	135	61.9900	18.9838	5787.7
	180	89.7340	29.5730	11,202
TOSMC	10	5.1584	2.2494	30.7253
	60	27.4873	9.5050	1049.5
	135	60.7691	20.0856	5285.8
	180	80.7782	26.5151	9380.6

where a is a positive constant, and rand() is a MATLAB which returns a uniformly distributed random number in the range of [0,1]. Given that there are four joints, the IAE, the ITAE and the ISE presented in Table 1 and represent the maximum value between all four joints of the system.

As shown in Table 1, in low set points, the BLSMC and SOSMC have the best performance in IAE and ITAE. Typically, the BLSMC tends to have a greater IAE and ITAE. This may be due to the short 2 s testing period, the disturbance, and the chattering. In higher set points, however, the SMC has good IAE and ITAE compared to the BLSMC. A disadvantage of the BLSMC is the high steady-state error due to the saturation function. The chattering observed in the SMC is eliminated in the other control techniques, as shown in the control inputs over time plots in Figs. 1, 2, 3, and 4.

Overall, the SOSMC has the best performance in terms of IAE and ITAE. The disadvantage of this technique shows in its high ISE and high set point state response. The ISE of the SOSMC indicates that the controller takes longer to activate, resulting in a smoother step input state response. On higher set points, however, the SOSMC tends to have a poor performance.

In high set points, the first-order SMC and TOSMC have the best IAE, ITAE, and ISE performance. The TOSMC has the best ITAE resulting in an accurate and low steady-state error. However, just as the SOSMC, the TOSMC has high ISE in low-to-medium set points, implying a slower activation time and a smooth state response.

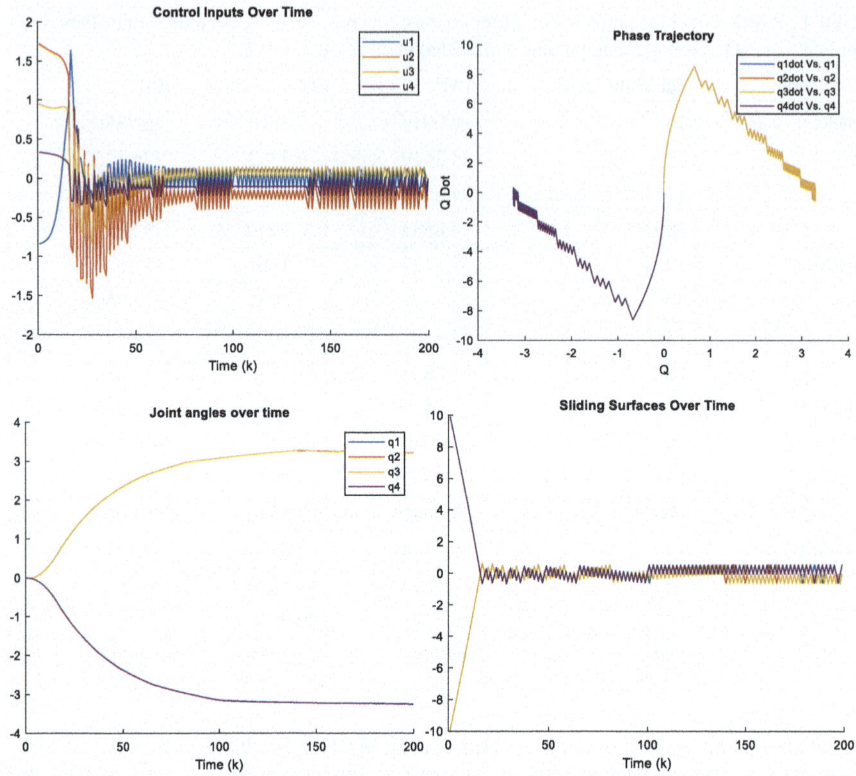

Fig. 1 SMC performance comparison. Here we display the state response, the control input, the phase trajectory, and the sliding surface over time

This smooth state response can be observed in the phase trajectory plots of Figs. 3 and 4.

The presented algorithms all present a smooth control law as opposed to the first-order SMC. Compared to the SOSMC and the TOSMC, the BLSMC has an aggressive control input over time response. In real-world applications, this may not be desirable due to hardware and mechanical specifications. On the other hand, both the SOSMC and TOSMC yield a smooth control input which puts less stress on mechanical actuators.

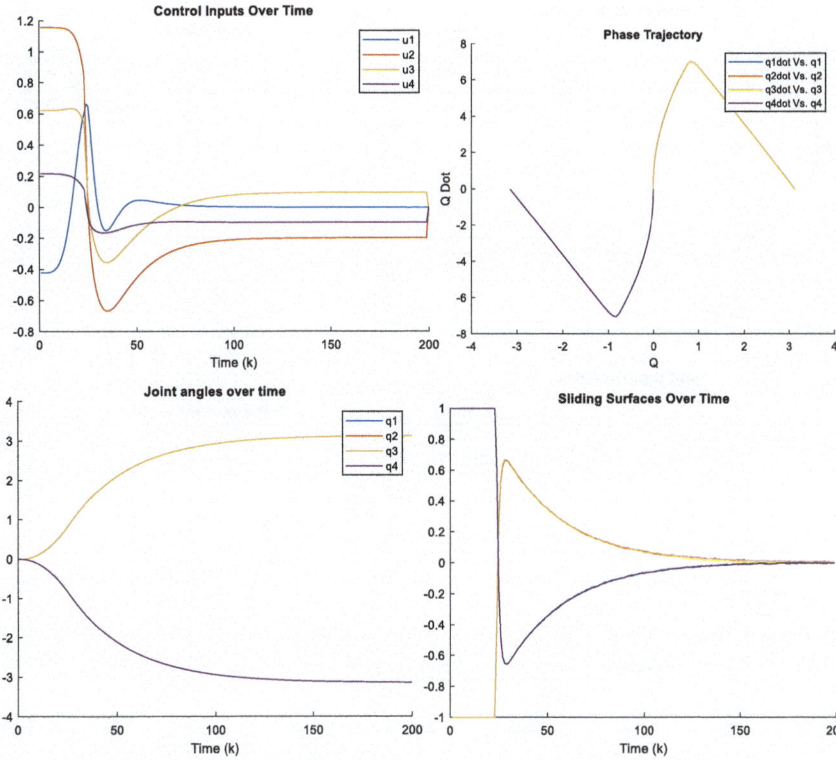

Fig. 2 BLSMC performance comparison. Here we display the state response, the control input, the phase trajectory, and the sliding surface over time

3.2 *Robustness Performance Comparison*

In this section, we will compare each of the controllers' robustness when various disturbance magnitudes are added to the 4-DoF robotic manipulator plant. The robustness of each of the controllers is measured by 1- achieving steady-state, 2- the IAE and ITAE index over 10 s dt = 0.01 s over a range of set points.

As we see in Fig. 5, the IAE and ITAE at $a = 0.5$ show that all controllers perform well and as described in Sect. 3.1. The low random disturbance input does not appear to affect the performance of the controllers. Figure 6 shows the IAE and ITAE at $a = 50$. In this case, long settling time and noticeable oscillations were observed with the SOSMC. However, with some time, all controllers were able to stabilize to the settling point. Lastly, Fig. 7 shows the IAE and ITAE at $a = 100$. In this case, the BLSMC could not converge to the set point. The SOSMC converged to the desired point accurately but had violent oscillations and had a much longer settling time. The SMC converges quickly to the desired point but has a high-frequency oscillation at the settling point and is less accurate than the SOMC at maintaining the set point.

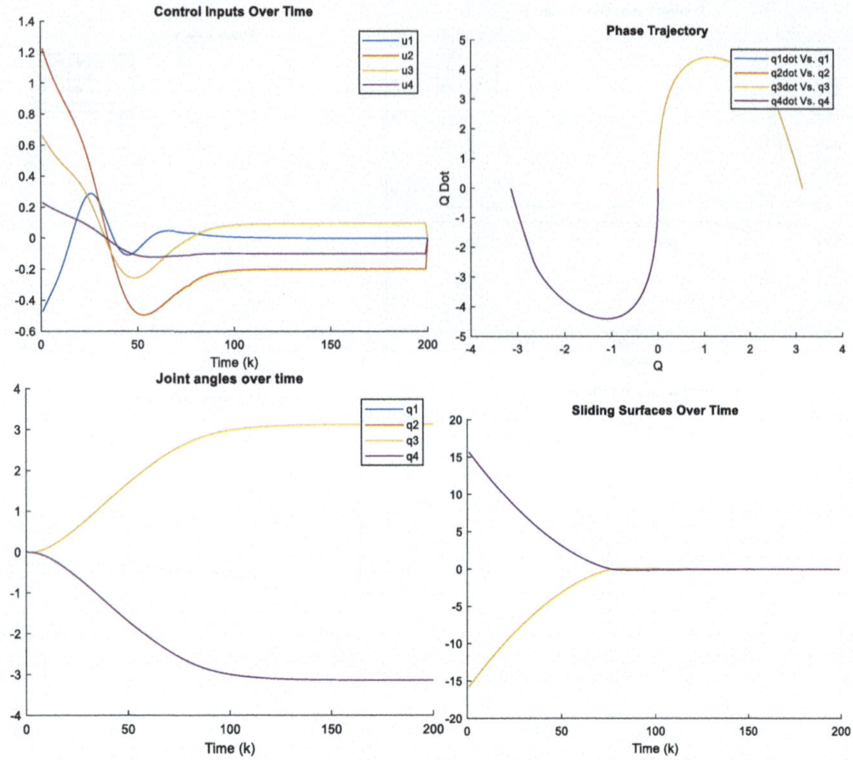

Fig. 3 SOSMC performance comparison. Here we display the state response, the control input, the phase trajectory, and the sliding surface over time

Lastly, the TOSMC performs by far the best in this robustness test. Throughout these tests, the TOSMC maintained excellent performance in both IAE, ITAE, and ISE. It converged quickly and accurately to the specified settling point. Note that the SOSMC and TOSMC's control inputs are not as attenuated when compensating for a high disturbance. We observe higher chattering with higher disturbances (Fig. 8).

3.3 Trajectory Tracking on 10-DOF Bipedal Robot

This section presents the results obtained by applying the SMC, BLSMC, and SOSMC on a 10-DoF bipedal robot. The controller design procedure is as described in Sect. 2.2. All controllers were tuned to have a similar rise time (0.8 ~ 0.9 s) and no overshoot. The presented results show each of the controllers' trajectory tracking performance during the motion of one step forward by the bipedal robot. The 3D spring-loaded inverted pendulum gait was used. The state trajectories are a function of the system's states. For simplicity, the knee trajectory was kept at 0. Therefore,

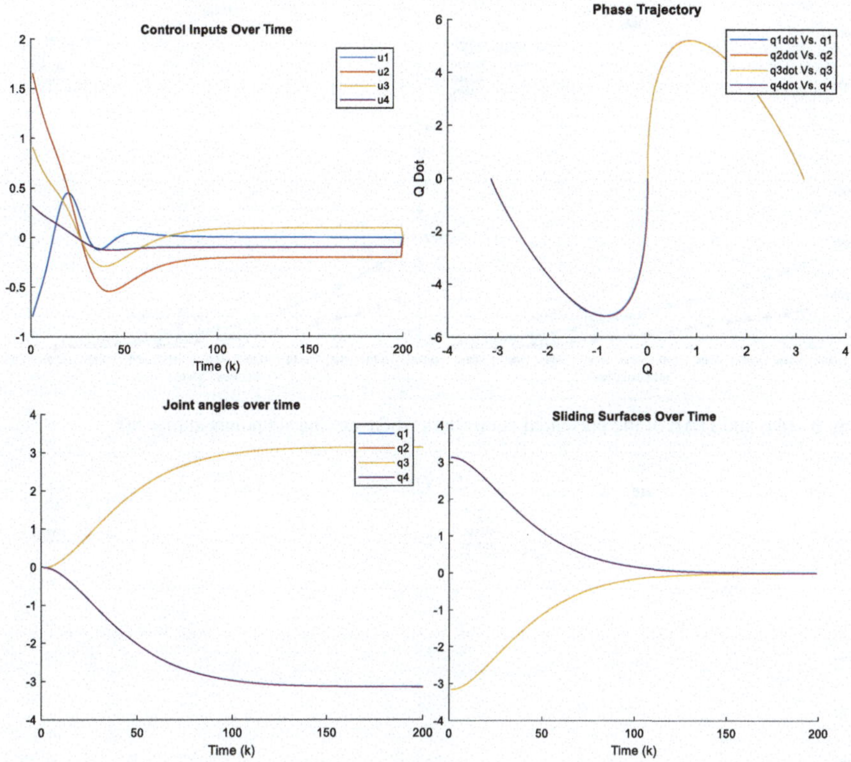

Fig. 4 TOSMC performance comparison. Here we display the state response, the control input, the phase trajectory, and the sliding surface over time

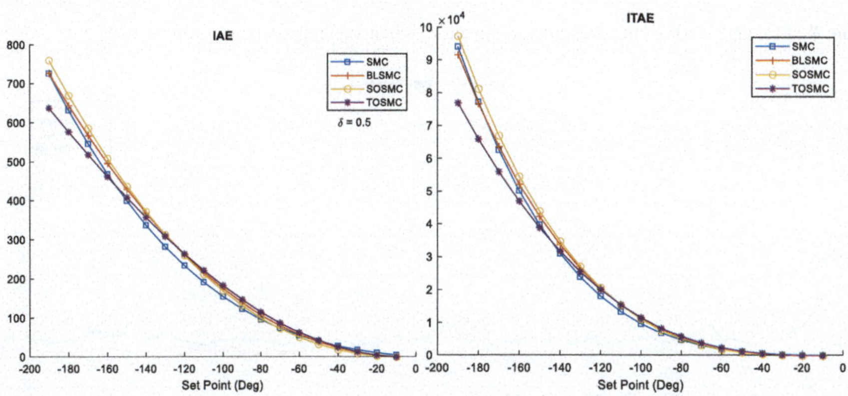

Fig. 5 IAE and ITAE of the presented controllers when varying set points at $a = 0.5$

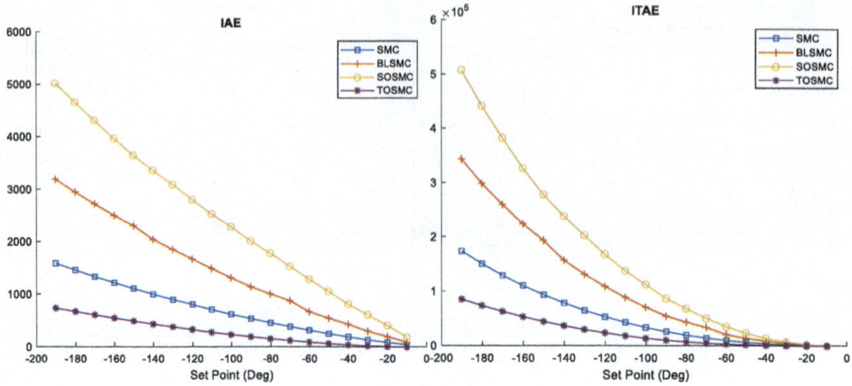

Fig. 6 IAE and ITAE of the presented controllers when varying set points at $a = 50$

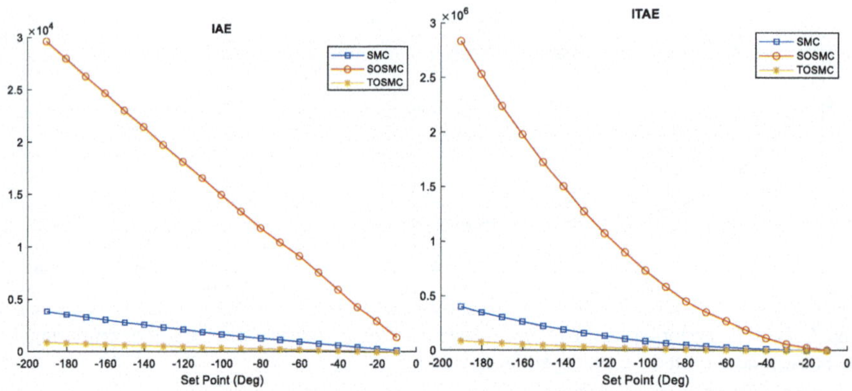

Fig. 7 IAE and ITAE of the presented controllers when varying set points at $a = 100$

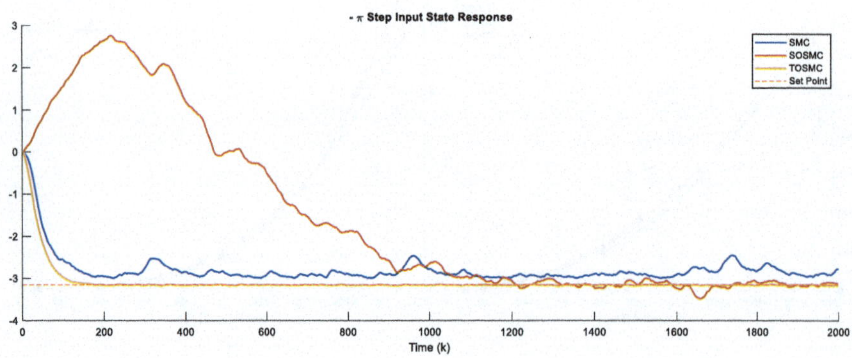

Fig. 8 Step input state response with disturbance $a = 100$

only the ankle and hip trajectories and state responses are presented below. As in Sect. 3.1, the controllers are evaluated over 2 s (or 200 samples at dt = 0.01 s), and a comparison of the IAE, ITAE, and ISE was conducted for the trajectory tracking problem. State response and controller inputs are both compared (Figs. 9, 10, 11, 12, 13, and 14).

As shown in Table 2, the SOSMC has the best trajectory tracking and performance in IAE and ITAE, which is a good indicator of accurate trajectory tracking. The BLSMC has a smooth control input ideal for actuators. Similarly, the SOSMC also has a smooth control input. However, unlike the BLSMC, the SOSMC has lower

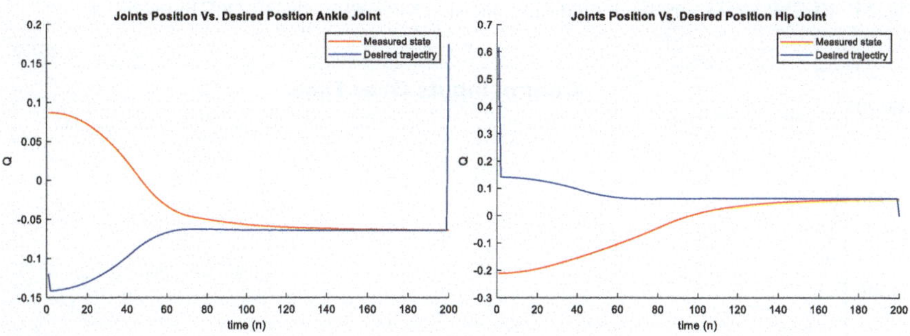

Fig. 9 SMC trajectory tracking during one step forward by the 10-DoF biped robot

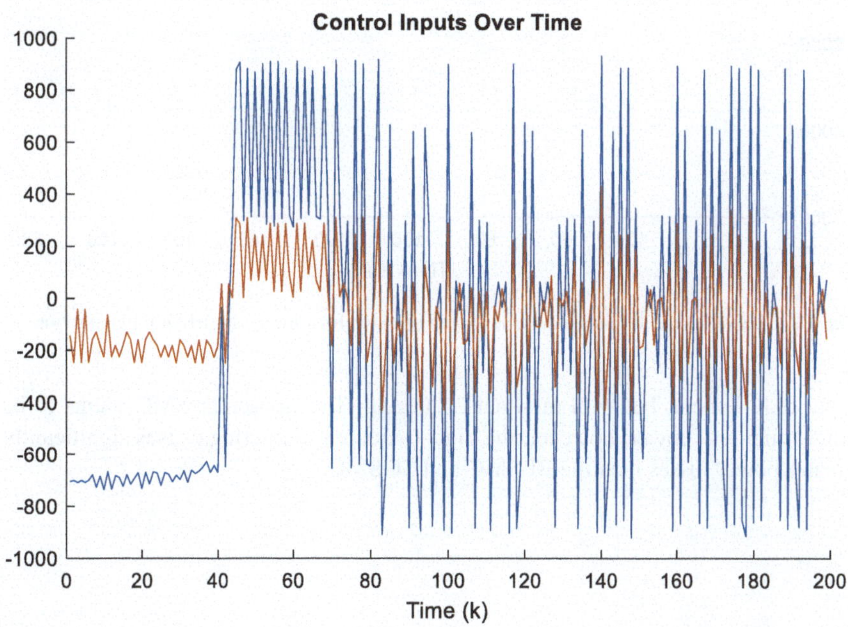

Fig. 10 SMC control inputs over time during one step forward by the 10-DoF biped robot

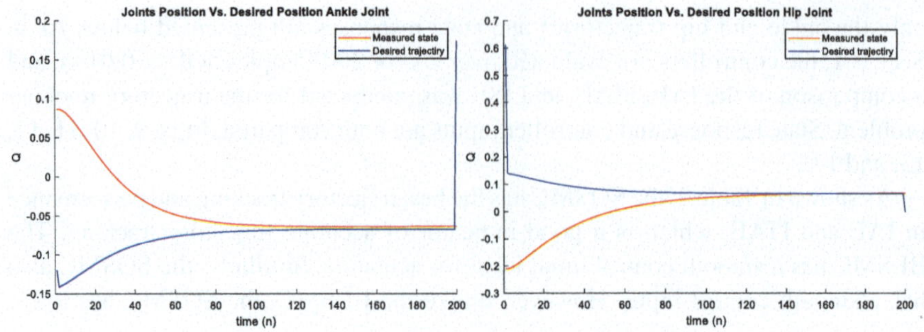

Fig. 11 BLSMC trajectory tracking during one step forward by the 10-DoF biped robot

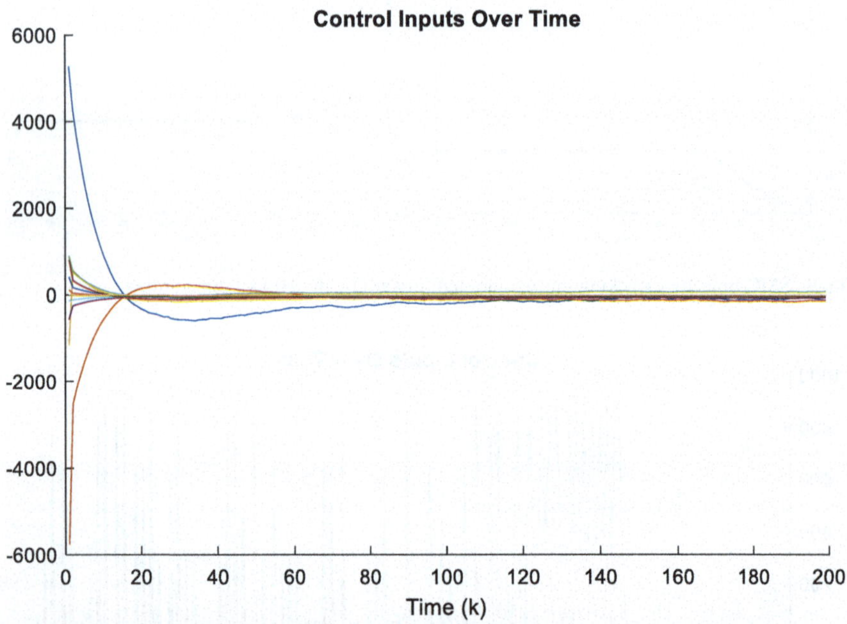

Fig. 12 BLSMC control inputs over time during one step forward by the 10-DoF biped robot

input requirement, but with minimal chattering. The first-order SMC seems to be consuming less power; however, the high-frequency chattering causes significantly greater power losses than the BLSMC and SOSMC.

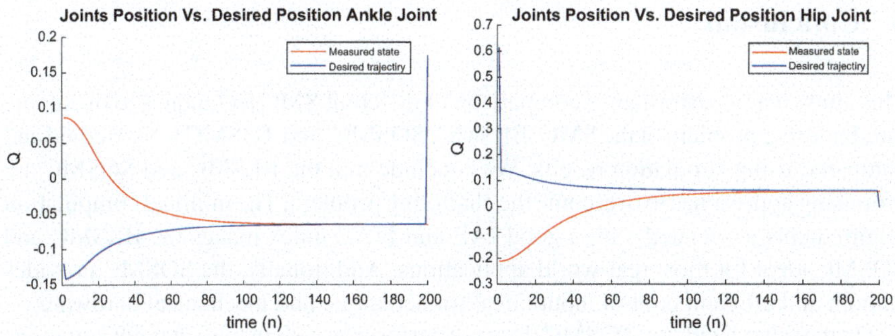

Fig. 13 SOSMC trajectory tracking during one step forward by the 10-DoF biped robot

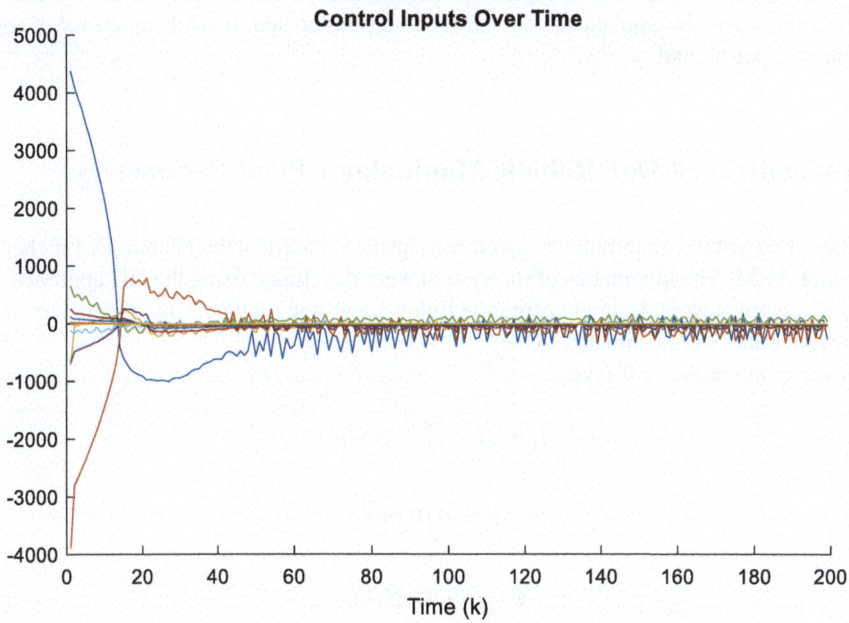

Fig. 14 SOSMC control inputs over time during one step forward by the 10-DoF biped robot

Table 2 Walking gait trajectory tracking performance comparison of the presented sliding mode controllers applied to a 10-DoF bipedal robot

Controller	Joint	IAE	ITAE	ISE
SMC	Ankle	5.7177	1.6920	53.4965
	Hip	13.2569	5.6371	196.7944
BLSMC	Ankle	4.4297	1.4546	25.5889
	Hip	7.5940	2.2791	103.1125
SOSMC	Ankle	3.7957	0.9144	29.7684
	Hip	7.5881	2.2583	104.8804

4 Conclusion

This study was a performance comparison of different SMC techniques to overcome the chattering problem of the SMC. BLSMC, SOSMC, and TOSMC were studied and compared using simulation results. We conclude that the BLSMC and SOSMC are promising approaches to overcome the chattering problem. The minimal computation requirements associated with a good IAE and ITAE index makes the BLSMC and SOSMC ideal for most real-world applications. Additionally, the SOSMC provides smooth and efficient control input, ideal for actuators, and mechanical hardware.

On the other hand, the TOSMC does not perform as well in low set point situations but was found to be ideal for high set points. Furthermore, the TOSMC is by far the most robust control technique explored in this study. In the future, the presented algorithms will be applied to the 4-DoF manipulator and 10-DoF biped robot to obtain experimental results.

Appendix A: 4-DoF Robotic Manipulator Plant Parameters

The 4-DoF robotic manipulator system was inspired based on the PhantomX Pincher Robot ARM. The kinematics of the system were developed using the DH approach. The dynamics were designed using the Euler–Lagrange method. The parameters of the dynamics were estimated to be:

$m_1 = m_2 = m_3 = 0.1(\text{kg})$..

$$l_1 = l_2 = l_3 = 0.1(m)$$

$$d = 0.01(m)$$

$$g = -9.81(\text{N/kg})$$

where m_i is the mass of the link i, l_i the link length of link i, d the z-axis distance offset between joint 1 and 2, and g is the gravity. In this model, inertia, friction and spring forces were not considered (Fig. 15).

Appendix B: 10-DoF Bipedal Robot Plant Parameters

The 10-DoF bipedal robot system was inspired based on the TonyPi biped robot. The kinematics of the system were developed using the DH approach. The dynamics were designed using the Euler–Lagrange method. Table 3 shows the estimated parameters values (Fig. 16).

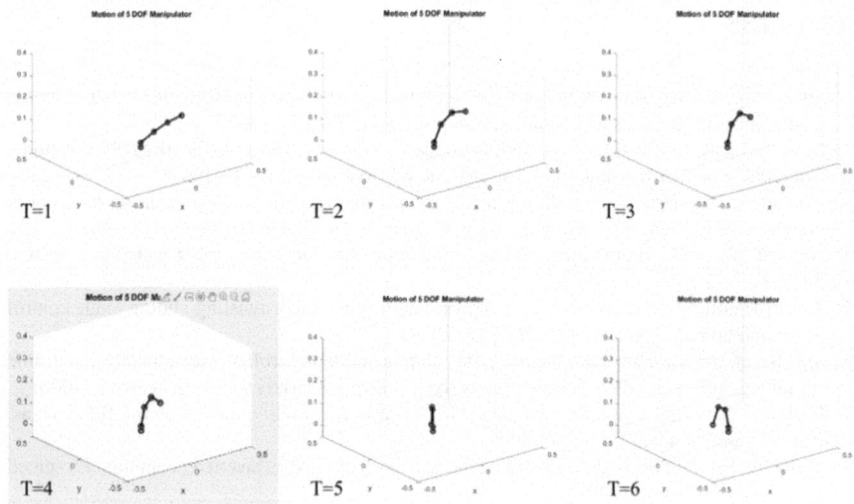

Fig. 15. 4-DoF robotic manipulator simulation platform on MATLAB

Table 3 Estimated parameters of 10-DoF bipedal robot based on the TonyPi biped robot

Link	Mass (kg)	Link Length (cm)	Inertia (kg cm^2)
1	0.1	4.5	2.025
2	0.1	6	3.600
3	0.1	6	3.600
4	0.1	4.5	2.025
5	1.5	6.5	63.375
6	0.1	4.5	2.025
7	0.1	6	3.600
8	0.1	6	3.600
9	0.1	4.5	2.025
10	0.05	1.5	0.1125

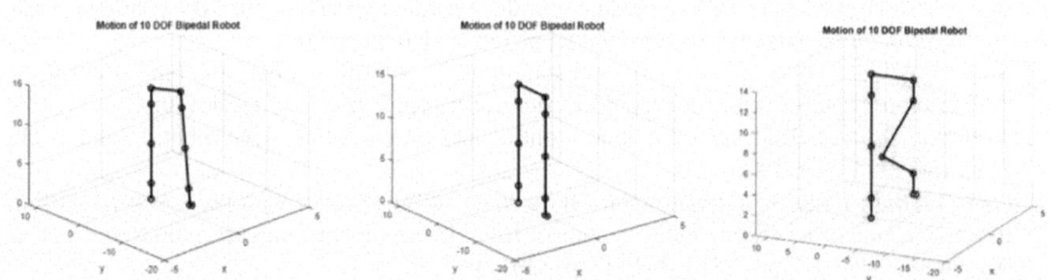

Fig. 16. 10-DoF biped robot simulation platform on MATLAB

References

1. Shen, W. et al.: A stable tele-robotic neurosurgical system based on SMC. IEEE International Conference on Robotics and Biomimetics. pp. 150–155 IEEE (2007)
2. Moosavian, S. et al.: Regulated sliding mode control of a biped robot. IEEE International Conference on Mechatronics and Automation. pp. 1547–1552 IEEE (2007)
3. Ahmed, S.: Invariant sets in sliding mode control theory with application to servo actuator system with friction. WSEAS Trans. Syst. Control. **8**(2), 33–45 (2013)
4. Nguyen, V., et al.: Hierarchical sliding mode algorithm for athlete robot walking. J. Robot. **2017**, 1–13 (2017)
5. Tayebi-Haghighi, S. et al.: Robust composite high-order super-twisting sliding mode control of robot manipulators. Robotics. **7**(1), 13 (2018)
6. Shi, S., et al.: Variable-gain second-order sliding mode controller with globally fixed-time stability guarantees. IEEE Trans. Circuits Syst. II Express Briefs **67**(8), 1414–1418 (2020)
7. Bartolini, G., et al.: Chattering avoidance by second-order sliding mode control. IEEE Trans. Autom. Control **43**(2), 241–246 (1998)
8. Mahjoub, S. et al.: Second-order sliding mode control applied to inverted pendulum. Presented at the , Sousse, Tunisia (2013)
9. Eker, İ: Second-order sliding mode control with experimental application. ISA Trans. **49**(3), 394–405 (2010)
10. JKD Power and Energy Solutions: Super-Twisting Controller, https://www.mathworks.com/matlabcentral/fileexchange/75146-super-twisting-controller, Last accessed 2020/09/02
11. Jedda, O. et al.: Second order sliding mode control for inverted pendulum. 12th International Multi-Conference on Systems, Signals & Devices. pp. 1–5 IEEE (2015)
12. Shi, S., et al.: Global high-order sliding mode controller design subject to mismatched terms: application to buck converter. IEEE Trans. Circuits Syst. I Regul. Pap. **66**(12), 4840–4849 (2019)
13. Erbatur, K., Kawamura, A.: Chattering elimination via fuzzy boundary layer tuning. Annual Conference of Industrial Electronics Society. pp. 2131–2136 IEEE (2002).
14. Capisani, L. et al.: Second order sliding mode motion control of rigid robot manipulators. IEEE Conference on Decision and Control. pp. 3691–3696 IEEE (2007)
15. Bekit, B. et al.: Sliding mode control for robot manipulators using time-varying switching gain and boundary layer. UKACC International Conference on Control. pp. 1198–1203 IET (1998)
16. Bartolini, G. et al.: Application of a second order sliding mode control to constrained manipulators. european control conference. pp. 3037–3042 IEEE (1997)
17. Arisoy, A. et al.: High Order Sliding Mode Control of a Space Robot Manipulator. International Conference on Recent Advances in Space Technologies. pp. 833–838 IEEE (2020)
18. Purwar, S.: Higher order sliding mode controller for robotic manipulator. IEEE International Symposium on Intelligent Control (ISIC). pp. 556–561 IEEE (2007)
19. Mu, X., Wu, Q.: Dynamic modeling and sliding mode control of a five-link biped during the double support phase. American Control Conference. pp. 2609–2614 IEEE (2004)
20. Tzafestas, S., et al.: Robust sliding-mode control applied to a 5-link biped robot. J. Intell. Rob. Syst. **15**(1), 67–133 (1996)
21. Oza, H. et al.: Finite time tracking of a fully actuated biped robot with pre-specified settling time: A second order sliding mode synthesis. IEEE International Conference on Robotics and Automation (ICRA). pp. 2570–2575 IEEE (2014)
22. Kljuno, E., Williams, R.: Humanoid Walking robot: modeling, inverse dynamics, and gain scheduling control. J. Robot. **2010**, 1–19 (2010)
23. Kim, M.: Euler-Lagrange tool package, https://www.mathworks.com/matlabcentral/fileexchange/49796-euler-lagrange-tool-package, last accessed 2020/09/02

Adaptive Backstepping-Based Non-singular Finite-Time Sliding Mode Controller for Suspension of Maglev Platforms

Nabanita Adhikary and Jobin Mathew

Abstract In this paper, an adaptive backstepping-based sliding mode controller is proposed for a magnetic levitation (maglev) platform. A novel non-singular finite-time sliding surface is proposed to reduce the steady-state error. The proposed controller is tested using numerical simulation in MATLAB/Simulink, and the results are compared with an adaptive backstepping-based sliding mode controller having a conventional non-singular terminal sliding surface. Backstepping is used to derive the sliding surface, and two adaptively tuned gains are included in the controller. The control law is obtained at the final step of backstepping where a power rate reaching law is used to obtain a faster convergence. Detailed stability analysis of the entire controller, including the smaller subsystems obtained via backstepping is also presented, and conditions based on the stability criteria are derived which can be used to design the controller parameters.

1 Introduction

Magnetic levitation technology has been in scientific knowledge since 1906 when Alfred Zehden filed a patent for "electric transport system using a traveling field motor" [11] and ever since it has been used in diverse fields including transportation such as magnetic levitation (maglev) trains [10]. Due to the advantages of less energy consumption, noiseless operation, and high riding quality, the research on maglev transportation has gained new traction [6, 10]. But maglev systems are inherently nonlinear unstable systems that make its controller design a challenging task [6].

Sliding mode control [7] (SMC) has proved to be one of the most effective robust controllers for nonlinear systems. It, however, suffers from certain drawbacks such as the requirement of the uncertainty matching conditions, chattering in the input

N. Adhikary (✉)
Department of Electrical Engineering, National Institute of Technology, Silchar, India
e-mail: nabanita@ee.nits.ac.in

J. Mathew
Department of Electrical Engineering, Assam Engineering College, Guwahati, India
e-mail: jobin.ele@aec.ac.in

J. Gu et al. (eds.), *Communication and Control for Robotic Systems*, Smart Innovation, Systems and Technologies 229, https://doi.org/10.1007/978-981-16-1777-5_5

due to the switching nature of the controller, and requirement of the uncertainty bounds [1]. But over the years, several modifications have been suggested [7, 8] that can deal with these drawbacks of SMC, increasing the applicability of the controller over a wide range of systems. Backstepping [5] is an adaptive recursive controller design method that has been successfully used in conjunction with SMC to tackle mismatched uncertainty. Backstepping is based on dividing the system into smaller first-order subsystems starting from the output and recursively defining a virtual control for each subsystem until the actual input is reached. When combined with SMC, the adaptively tuned gain can significantly reduce the control effort without compromising with the performance [2, 3], and the controller can be designed without the knowledge of the system uncertainty.

In this paper, an adaptive backstepping-based non-singular terminal sliding mode controller is proposed for the suspension control of a maglev transportation platform. Detailed stability analysis of the closed-loop system is presented that gives a series of conditions that can be used to calculate the range for the controller parameters. A non-singular terminal sliding surface [4] is produced using backstepping, which gives the closed-loop system finite-time convergence properties. The sliding surface is so designed that it is a function of the system error as well its integral and derivative, and this ensures that as the sliding surface reaches equilibrium in finite time, so does the system error.

The organization of the paper is as follows: Sect. 2 describes the dynamics of the levitating platform along with the challenges posed in controlling them. The controller design process is described in Sect. 3, and the stability analysis in Sect. 4. The numerical simulations are presented in Sect. 5, and finally, the conclusion is drawn in Sect. 6 along with the description of the future scope of the presented work.

2 Problem Definition

2.1 Dynamics of Maglev System

For the levitation control, the following three degrees of freedom are considered [9], viz. the linear motion along the z-axis, roll θ along the x-axis, and pitch ϕ along the y-axis as shown in Fig. 1. The platform is divided into four subsections A, B, C, and D where four electromagnets are placed for suspension. The vertical displacement of the platform is z_0 with $z_{\max}$ as the maximum allowable displacement, and this is measured at the Center Of Mass (COM) of the entire platform. The levitation heights for the four subsections at their corresponding COMs are given by z_A, z_B, z_C, z_D. The other physical attributes of the platform are: l_t and w_t are, respectively, the length and width of the platform, l_l and l_w are the distance of the four COMs from the x-axis and y-axis, respectively, and $l_d = \sqrt{l_w^2 + l_l^2}$. Considering only the suspension of the platform, the control forces acting on the system are $f = [f_A\ f_B\ f_C\ f_D]^T$. As

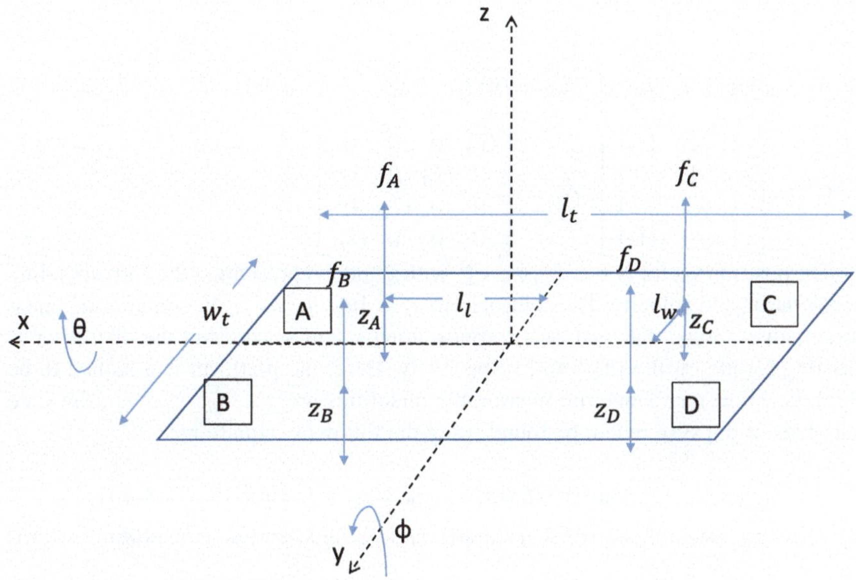

Fig. 1 Layout of the maglev platform

derived in [9], the magnetic levitation force created due to the electromagnet can be generally defined as

$$f_e(x_e, i_e) = \frac{1}{2}\left(\frac{\mathrm{d}P}{\mathrm{d}x_e}\right) N_{me}^2 i_e^2 = G_e i_e^2 \tag{1}$$

where x_e is the air-gap length, i_e is the current, P is the permeance and N_{me} is the number of turns of the electromagnet and $G_e = \frac{1}{2}\left(\frac{\mathrm{d}P}{\mathrm{d}x_e}\right) N_{me}^2$ is a time varying coefficient linking the force and the current. The subscript e can be replaced by A, B, C, D to represent the four corners of the maglev platform as depicted in Fig. 1.

The dynamics of the maglev platform [9] is as follows:

$$M_0\ddot{x}_0 + d_0 = u_c \tag{2}$$

where $M_0 = \begin{bmatrix} m_t & 0 & 0 & 0 \\ 0 & I_x & 0 & 0 \\ 0 & 0 & I_y & 0 \\ 0 & 0 & 0 & 1 \end{bmatrix}$, $d_0 = \begin{bmatrix} m_t g \\ 0 \\ 0 \\ 0 \end{bmatrix}$

$u_c = \Lambda_1 G_s f = G_g G_s u = \Lambda_1 G_s \Lambda_g (G_s^T \Lambda_N \Lambda_N G_s) u,$

$$u = [i_A^2\, i_B^2\, i_C^2\, i_D^2]^T,$$

$$\Lambda_1 = \text{diag}\{1, l_w, l_l, l_d\},\ \Lambda_N = 0.5I_4,$$

$$G_s = \begin{bmatrix} 1 & 1 & 1 & 1 \\ -1 & 1 & -1 & 1 \\ -1 & -1 & 1 & 1 \\ 1 & -1 & -1 & 1 \end{bmatrix}, \Lambda_g = \begin{bmatrix} G_A & 0 & 0 & 0 \\ 0 & G_B & 0 & 0 \\ 0 & 0 & G_C & 0 \\ 0 & 0 & 0 & G_D \end{bmatrix}.$$

The position vector is $x = [z_0\, \theta\, \phi\, \kappa]^T$ with $\dot{x}$ and $\ddot{x}$ representing the corresponding acceleration and velocity. The control term $u_c = [u_{F_z}\, u_{\tau_x}\, u_{\tau_y}\, u_{\kappa}]^T$ contains the linear force along z-axis, the rotational torques along x and y axes and the distorting or twisting torque on the platform, respectively. Here, the platform is assumed to be rigid, i.e., $\ddot{\kappa} = u_{\tau_\kappa}$. Since the measurable quantities are $z_A,\ z_B,\ z_C,\ z_D$, the state variables of the system can be found using the following equations:

$$\begin{aligned} z_A =& z_0 - l_w \sin(\theta) - l_l sin(\phi),\ z_B = z_0 + l_w \sin(\theta) - l_l \sin(\phi) \\ z_C =& z_0 - l_w \sin(\theta) + l_l \sin(\phi),\ z_B = z_0 + l_w \sin(\theta) + l_l \sin(\phi) \end{aligned} \tag{3}$$

Considering the current as the controlling element and including the effect of parameter variation as ΔM and Δd and adding a bounded external disturbance f_d to the system (2), a more practical representation can be derived as

$$(M + \Delta M)\ddot{x} + (d + \Delta d) + f_d = G_s u \tag{4}$$

$$M\ddot{x} + d + l_u = G_s u \tag{5}$$

where $M = G_g^{-1} M_0,\ d = G_g^{-1} d_0$ and $l_u = \Delta M\ddot{x} + \Delta d + f_d$ represents the lumped disturbance bounded by $||l_u|| < b_1$, where $b_1 > 0$ is a finite number.

2.2 *Challenges in Control*

The control of a maglev suspension platform is difficult owing to the following factors: (a) nonlinear relation between the control forces and the control current producing these forces, (b) the inherent open-loop instability of the system, (c) control currents belonging to non-negative regions, (d) the absence of damping elements and (e) the time-variant nature of the maglev dynamics [6]. This paper aims to develop a two-layered adaptive backstepping-based non-singular finite-time sliding mode controller (ABNTSMC) for the maglev suspension platform. The controller is developed keeping the following factors in focus

- Lower the overshoot as much as possible to any step disturbance to the system.
- The settling after the disturbance should be smooth and fast.
- Achieve a finite-time convergence.
- Obtain the desired performance without any excessive use of input energy.

3 Controller Design

The controller will be designed based on backstepping sliding mode methodology [1] so that it can take care of the system nonlinearity as much as possible. Moreover, adaptive tuning of the controller gain is performed in two consecutive steps of backstepping. The controller design procedure will be described considering a trajectory tracking problem so that it can be applied for regulation problems as well. The steps involved in the ABNTSMC design are elaborated in the following:

Step I
Considering the desired trajectory to be x_d, the system error is defined as follows:

$$x_{\text{error}} = x - x_d. \tag{6}$$

Now, the first regulatory variable of backstepping is defined as follows

$$z_1 = \int x_{\text{error}} \mathrm{d}t \tag{7}$$

$$\dot{z}_1 = x_{\text{error}} = x - x_d. \tag{8}$$

A Candidate Lyapunov Function (CLF) for (8) is defined as

$$V_1 = \frac{1}{2} z_1^T z_1 \tag{9}$$

$$\dot{V}_1 = z_1^T \dot{z}_1 = z_1^T (x - x_d). \tag{10}$$

Considering $x = \alpha_1$ to be the virtual control input for (10), α_1 is derived based on (10) as follows

$$\alpha_1 = -c_1 z_1 + x_d \tag{11}$$

where $c_1 \in \mathbb{R}^{4\times4}$ is the positive definite matrix of the user-defined controller gain.

Step II:
As the actual control is yet to be applied, it is clear that x and α_1 are not equa,l and hence, the next regulatory variable of the backstepping algorithm is defined as

$$z_2 = x - \alpha_1 = x - x_d + c_1 z_1 \tag{12}$$

$$\dot{z}_2 = \dot{x} - \dot{x}_d + c_1 \dot{z}_1 \tag{13}$$

and this reshapes the dynamics (8) as follows

$$\dot{z}_1 = -c_1 z_1 + z_2 \tag{14}$$

The CLF for (13) is defined as

$$V_2 = \frac{1}{2} z_2^T z_2 \quad (15)$$

$$\dot{V}_2 = z_2^T \dot{z}_2 = z_2^T (\dot{x} - \dot{x}_d + c_1 \dot{z}_1). \quad (16)$$

The virtual control for (13) is considered as $\dot{x} = \alpha_2$ and based on (16) α_2 is derived as follows

$$\alpha_2 = -\hat{k}_1 z_2 + \dot{x}_d - c_1 \dot{z}_1 \quad (17)$$

where $\hat{k}_1$ is the adaptively tuned controller gain that is determined by the following tuning law [2]:

$$\dot{\hat{k}}_1 = \Gamma_1 (|z_2| - \epsilon_1 \hat{k}_1) \quad (18)$$

where $\Gamma_1,\ \epsilon_1 > 0$ are user-defined parameters. From (17), the time derivative of α_2 is derived as

$$\dot{\alpha}_2 = -\hat{k}_1 \dot{z}_2 - \dot{\hat{k}}_1 z_2 + \ddot{x}_d - c_1 (\dot{x} - \dot{x}_d) \quad (19)$$

Step III:
Similar to the previous steps, a third regulatory variable is now defined as follows:

$$z_3 = \dot{x} - \alpha_2 \quad (20)$$

$$\dot{z}_3 = \ddot{x} - \dot{\alpha}_2 \quad (21)$$

With the introduction of z_3, dynamics of z_2 (13) can be rewritten as follows:

$$\dot{z}_2 = -\hat{k}_1 z_2 + z_3 \quad (22)$$

A non-singular terminal sliding surface is now defined as

$$s = z_2 + \beta z_3^{\delta} = 0 \quad (23)$$

where $\beta > 0$ and $1 < \delta < 2$ are design parameters. To avoid the problem of complex number formulation when $z_3 < 0$, the sliding surface can be modified as follows without losing any of its properties [1]

$$s = z_2 + \beta |z_3|^{\delta} \mathrm{sign}(z_3) = 0 \quad (24)$$

$$\dot{s} = \dot{z}_2 + \beta \delta |z_3|^{\delta - 1} \dot{z}_3 \quad (25)$$

The structure of the sliding surface allows it to reach the equilibrium in finite time via the properties of NTSM [1], and since the error integral is also a component of the surface, it can be ensured that in the absence of disturbances the system error should converge to zero in finite time. The following CLF, V_3 is now defined as follows

$$V_3 = \frac{1}{2}s^T s \tag{26}$$

$$\dot{V}_3 = s^T \dot{s} \tag{27}$$

$$= s^T(\dot{z}_2 + \beta\delta|z_3|^{\delta-1}\dot{z}_3)$$

$$= s^T[\dot{z}_2 + \beta\delta|z_3|^{\delta-1}(\ddot{x} - \dot{\alpha}_2)] \tag{28}$$

Based on (28) and the reaching law approach of SMC design [1], the control law is derived as follows

$$u = G_s^{-1}[d + M\dot{\alpha}_2 - (\beta\delta)^{-1}M|z_3|^{1-\delta}(\dot{z}_2 + \hat{k}_2|s|^{\rho}\mathrm{sign}(s) + Ws)] \tag{29}$$

where $W > 0$, $\rho > 0$ are control parameters and $\hat{k}_2$ is the controller gain that is tuned adaptively using the tuning law as given below

$$\dot{\hat{k}}_2 = \Gamma_2(|s| - \epsilon_2\hat{k}_2 + \gamma|z_3|) \tag{30}$$

where $\Gamma_2 > 0$, $\epsilon_2 > 0$ and $\gamma > 0$ are design parameters.

4 Stability Analysis

The system stability will be analyzed by checking the input to state stability (ISS) of the individual smaller subsystems obtained via backstepping, and the following lemmas are proposed for this purpose:

Lemma 1 *The subsystem (8) will be ISS with respect to z_2 as input if $\lambda_{\min}(c_1)|z_1| > |z_2|$.*

Proof The following CLF (V_1) is considered for analyzing the ISS of (13),

$$V_1 = \frac{1}{2}z_1^T z_1$$

$$\dot{V}_z = -z_1^T c_1 z_c + z_1^T z_2$$

$$\leq -\lambda_{\min}(c_1)z_1^T z_1 + |z_1|^T|z_2|$$

$$\leq -|z_1|^T(\lambda_{\min}(c_1)|z_1| - |z_2|) \tag{31}$$

Therefore, the subsystem will be ISS with z_2 as input if $\lambda_{\min}(c_1)|z_1| > |z_2|$.

Lemma 2 *The subsystem (13) will be ISS with respect to z_3 as input if the following condition is satisfied*

$$\lambda_{\min}(K_1)|z_2| \leq |z_3| \tag{32}$$

where $\lambda_{\min}()$ indicates the minimum eigenvalue, K_1 is the diagonal 4×4 *matrix with entries from $\hat{k}_1$.*

Proof The following CLF (V_z) is considered for analyzing the stability analysis of (13),

$$V_z = \frac{1}{2} z_2^T z_2 \tag{33}$$

$$\begin{aligned} \dot{V}_z =& - z_2^T \hat{K}_1 z_2 + z_2^T z_3 \\ \leq& - \lambda_{\min}(\hat{K}_1)(|z_2|^T |z_2| - |z_2|^T |z_3|) \\ \leq& - \lambda_{\min}(\hat{K}_1)|z_2|^T |z_2| \;\; if \;\; \lambda_{\min}(K_1)|z_2| \leq |z_3| \end{aligned} \tag{34}$$

Therefore, the subsystem will be ISS with z_3 as input if $\lambda_{\min}(K_1)|z_2| \leq |z_3|$.

Lemma 3 *The subsystem (13) including the adaptive law (18) will be ISS with respect to k_1 (k_1 is the arbitrary value to which $\hat{k}_1$ converges) as input if along with Lemma 1 the following two conditions are satisfied,*

$$\min \left[2\lambda_{\min}(\hat{K}_1), \lambda_{\min}(\epsilon_1)\lambda_{\min}(\Gamma_1) \right] > \frac{1}{2}\lambda_{\max}(\epsilon_1)|k_1|^T |k_1| \tag{35}$$

Proof The following CLF (V_k) is used for the proof

$$V_k = \frac{1}{2}(z_2^T z_2 + \tilde{k}_1^T \Gamma_1^{-1} \tilde{k}_1) \tag{36}$$

Using (18), (34) and Lemma 1, the derivative of V_k can be written as follows

$$\dot{V}_k \leq - \lambda_{\min}(\hat{K}_1)|z_2|^T |z_2| - \tilde{k}_1 \epsilon_1 \dot{\hat{k}}_1 \tag{37}$$

Since $\tilde{k}_1 = \hat{k}_1 - k_1$, it can be proved as shown in [4] that

$$\tilde{k}_1^T \epsilon_1 \hat{k}_1 \geq \frac{1}{2}(\tilde{k}_1^T \epsilon_1 \tilde{k}_1 - k_1^T \epsilon_1 k_1) \tag{38}$$

Using (37), (38) the following can be derived,

$$\dot{V}_k \leq -\lambda_{\min}(\hat{K}_1)|z_2|^T|z_2| - \frac{1}{2}(\tilde{k}_1^T\epsilon_1\tilde{k}_1 - k_1^T\epsilon_1 k_1) \tag{39}$$
$$\leq -\lambda_{\min}(\hat{K}_1)|z_2|^T|z_2| - \frac{1}{2}\tilde{k}_1^T\epsilon_1\tilde{k}_1 + -\frac{1}{2}k_1^T\epsilon_1 k_1$$
$$\leq -\lambda_{\min}(\hat{K}_1)|z_2|^T|z_2| - \frac{1}{2}\lambda_{\min}(\epsilon_1)(|\tilde{k}_1|^T|\tilde{k}_1|) + \frac{1}{2}\lambda_{\max}(\epsilon_1)|k_1|^T|k_1|$$
$$\leq -\lambda_{\min}(\hat{K}_1)|z_2|^T|z_2| - \frac{1}{2}\lambda_{\min}(\epsilon_1)\Gamma_1(|\tilde{k}_1|^T\Gamma_1^{-1}|\tilde{k}_1|) + \frac{1}{2}\lambda_{\max}(\epsilon_1)|k_1|^T|k_1|$$
$$\leq -\lambda_{\min}(\hat{K}_1)|z_2|^T|z_2| - \frac{1}{2}\lambda_{\min}(\epsilon_1)\lambda_{\min}(\Gamma_1)(|\tilde{k}_1|^T\Gamma_1^{-1}|\tilde{k}_1|) + \frac{1}{2}\lambda_{\max}(\epsilon_1)|k_1|^T|k_1|$$
$$\leq -\min\left[2\lambda_{\min}(\hat{K}_1), \lambda_{\min}(\epsilon_1)\lambda_{\min}(\Gamma_1)\right]V_k + \frac{1}{2}\lambda_{\max}(\epsilon_1)|k_1|^T|k_1|$$
$$\leq -\psi V_k + \eta_1 \tag{40}$$

where $\psi = \min\left[2\lambda_{\min}(\hat{K}_1), \lambda_{\min}(\epsilon_1)\lambda_{\min}(\Gamma_1)\right]$ and $\eta = \frac{1}{2}\lambda_{\max}(\epsilon_1)|k_1|^T|k_1|$. If the controller parameter are so chosen that $\eta/\psi < 1$, then for $V_k(0) > \eta/\psi$, the system will be stable.

Lemma 4 *The subsystem (25) will be ISS with respect to the lumped system uncertainty denoted by* $\xi_l = \beta\delta|z_3|^{\delta-1}M^{-1}l_u$ *as input for all nonzero s satisfying*

$$\hat{K}_1|s|^\rho > |\xi_l| \tag{41}$$

Proof In order to analyze the ISS of (25), the following CLF (V_s) is used,

$$V_s = \frac{1}{2}s^T s \tag{42}$$
$$\dot{V}_s = s^T(-\hat{K}_2|s|^\rho \text{sign}(s) - Ws - \xi_l) \tag{43}$$
$$= -\hat{k}_1^T|s|^{\rho+1} - s^T Ws - s^T\xi_l$$
$$\leq -\hat{k}_1^T|s|^{\rho+1} - |s|^T W|s| + |s|^T|\xi_l|)$$
$$\leq -|s|^T W|s| - |s|^T(\hat{K}_1|s|^\rho - |\xi_l|)$$
$$\leq -\lambda_{\min}(W)s^T s, \ \ if \ \ \hat{K}_1|s|^\rho > |\xi_l|$$
$$\leq -2\lambda_{\min}(W)V_s \tag{44}$$

Thus, $\dot{V}_s$ will be negative semi-definite if $\hat{K}_1|s|^\rho > |\xi_l|$.

Lemma 5 *The input to state stability of the complete closed-loop system with respect to* k_2 *(*k_2 *is the arbitrary value to which* $\hat{k}_2$ *converges) as the input will be achieved if Lemmas 1, 2, 3 are satisfied along with the following condition*

$$\min(\psi, 2\lambda_{\min}(W), \lambda_{\min}(\epsilon_2)\lambda_{\min}(\Gamma_2)) > \eta_1 + \eta_2 \tag{45}$$

Proof To prove this lemma, the following CLF V is considered,

$$V = V_k + V_s + \frac{1}{2}\tilde{k}_2^T \Gamma_2^{-1} \tilde{k}_2. \tag{46}$$

Using Lemmas 2 and 3 and (38), $\dot{V}$ can be derived as follows:

$$s\dot{V} = \dot{V}_k + \dot{V}_s + \frac{1}{2}\tilde{k}_2^T \Gamma_2^{-1} \dot{\hat{k}}_2 \tag{47}$$

$$\begin{aligned}
&\leq -\psi V_k + \eta_1 - 2\lambda_{\min}(W)V_s - \frac{1}{2}\tilde{k}_2^T \epsilon_2 \tilde{k}_2 + \frac{1}{2}k_2^T \epsilon_2 k_2 \\
&\leq -\psi V_k - 2\lambda_{\min}(W)V_s - \lambda_{\min}(\epsilon_2)\Gamma_2 \frac{1}{2}\tilde{k}_2^T \Gamma_2^{-1} \tilde{k}_2 + \eta_1 + \frac{1}{2}k_2^T \epsilon_2 k_2 \\
&\leq -\psi V_k - 2\lambda_{\min}(W)V_s - \lambda_{\min}(\epsilon_2)\lambda_{\min}(\Gamma_2)\frac{1}{2}\tilde{k}_2^T \Gamma_2^{-1} \tilde{k}_2 + \eta_1 + \eta_2 \\
&\leq -\Psi V + \eta_1 + \eta_2
\end{aligned} \tag{48}$$

where $\Psi = \min(\psi, 2\lambda_{\min}(W), \lambda_{\min}(\epsilon_2)\lambda_{\min}(\Gamma_2))$. Thus, the system stability can be ensured if $\Psi > \eta_1 + \eta_2$

The Lemmas 1 through 4 presented in this section provide a series of conditions that can be used for parameter selection for the controller assuming the output and the disturbances remain bounded.

5 Numerical Simulations

The proposed controller is simulated using MATLAB/Simulink for a maglev platform whose parameters are listed in Table 1.

In order to observe the effect of adding an extra adaptive tuning layer in the control law, a non-singular finite-time sliding mode controller [1] is also designed for the same system given as follows

$$\begin{aligned}
u = G_s^{-1}(&d + M\ddot{x}_d - c_1 M\dot{z}_1 - M(-cz_1 + \hat{k}(|s_c|)^\rho)\text{sign}(s_c) + W_c s_c) \\
&- (\beta\delta)^{-1} M(z_2)^{2-\delta})
\end{aligned} \tag{49}$$

Table 1 Maglev platform parameters

Platform	Electromagnet
$l_t = 0.88$ m, $w_t = 0.48$ m, $l_l = 0.44$ m, $l_w = 0.24$ m, $z0_{\max} = 5 \times 10^{-3}$ m, $m_0 = 100$ kg	$R = 6.4\ \Omega$, $L = 23.7$ mH, $G_e = 3.3 \times 10^{-5}/z_e^2$, where $e = A, B, C, D$

where $z_1 = x - x_d$, $z_2 = \dot{z}_1 + c_1 z_1$, $s_c = z_1 + \beta|z_2|^{\delta})\text{sign}(z_2)$ and $\hat{k}$ is the adaptively tuned gain given by the tuning law

$$\dot{\hat{k}} = \Gamma_c(|s_c| - \epsilon_c \hat{k}) \tag{50}$$

with the initial value of $\hat{k}$ taken as 5. This controller is termed as 'Controller 2' in the figures.

The parameters of the controller (49) are as follows $c = 100I$, $W_c = 3500I$,[1] $\beta = 2$, $\delta = 1.5$, $\Gamma_c = 106I$, $\epsilon_c = 10^{-5}I$.

The parameters of the proposed controller (29) are $c_1 = 100I$, $\Gamma_1 = 10I$, $\epsilon_1 = 0.001I$, $\Gamma_2 = 105I$, $\epsilon_2 = 0.001I$, $\gamma = 20$, $\beta = 2$, $\delta = 1.5$, $\rho = 0.3$, $W = 104I$. The initial condition for $\hat{k}_1$ and $\hat{k}_2$ are taken as 5 and 0, respectively.

For controller (49), similar parameters as the proposed controller were initially chosen; however, certain parameters had to be tuned since the initial values were yielding an unstable operation in the simulations. After using the conditions derived from the stability analysis, the parameters of both the controllers were tuned further via trial and error to yield the best results in terms of suspension error (Fig. 2).

Figure 3 shows the suspension height (z_0), Fig. 3b shows the angular displacements (θ, ϕ, α) and Fig. 3a shows the heights z_A, z_B, z_C, z_D of the platform in the presence of the a step disturbance given as 25% increment of the original platform weight introduced at $t = 6$ s.

The angular displacements are all maintained at 0° as shown in Fig. 3b. It can be clearly seen from 2 that the step disturbance in the mass of the platform leaves a steady-state error with controller (49), whereas the proposed controller (29) can tackle this disturbance and the platform height is maintained at the reference level without error. The proposed method has a slightly slower response time than the controller (49); however, this is compensated by the accuracy of the proposed method (29).

Figures 6 and 7 show the comparison of the forces acting on the four parts of the platform, and Figs. 4 and 5 show the corresponding currents in the electromagnets for both the controllers (29) and (49). The results are also summarized in Table 2 in terms of control effort required and the tracking accuracy in terms of suspension height. From Table 2, it is clear that both the controller uses a similar amount of control effort, but the proposed method (29) gives much better accuracy than the controller (49) and it also provides a finite-time convergence along with zero steady-state error even in the presence of disturbance for the numerical simulations. The proposed controller is also simulated with measurement noise and the same step disturbance in mass, and the resulting suspension height and angular positions are shown in Fig. 9 and the suspension heights of the four parts of the platform are shown in Fig. 8. Here, the initial conditions are chosen as (0.0077 m, 0 rad, 0.001 rad, 0 rad). For the four divisions of the platform, the forces developed are shown in Figs. 10 and 11 and the corresponding input currents are shown in Figs. 12 and 13. The adaptively tuned controller gains are shown in Fig. 14. As can be observed from the results even in the

[1] It was observed that for the same parameters of the system, increasing the value of Wc beyond 3500 results in unstable operation of the closed-loop system.

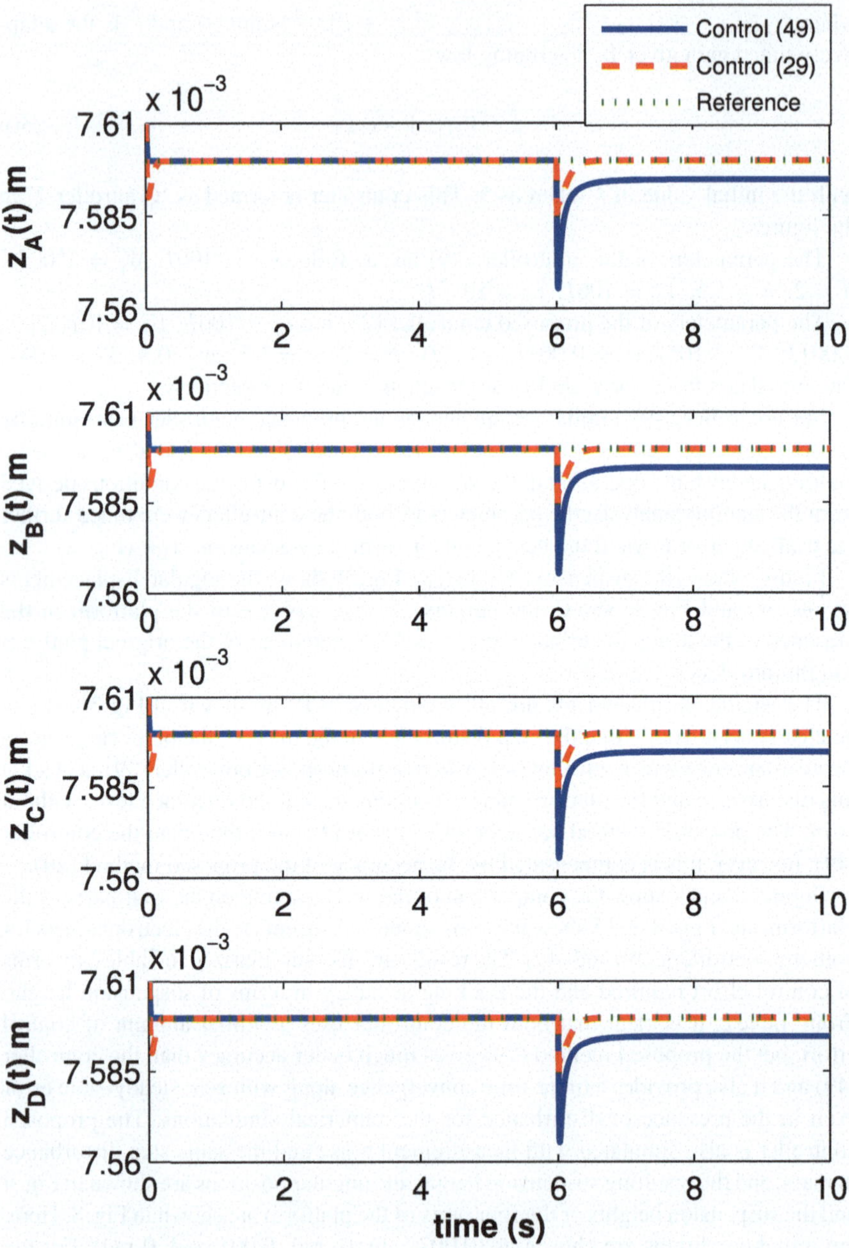

Fig. 2 Displacements z_A, z_B, z_C, z_D in the absence of measurement noise

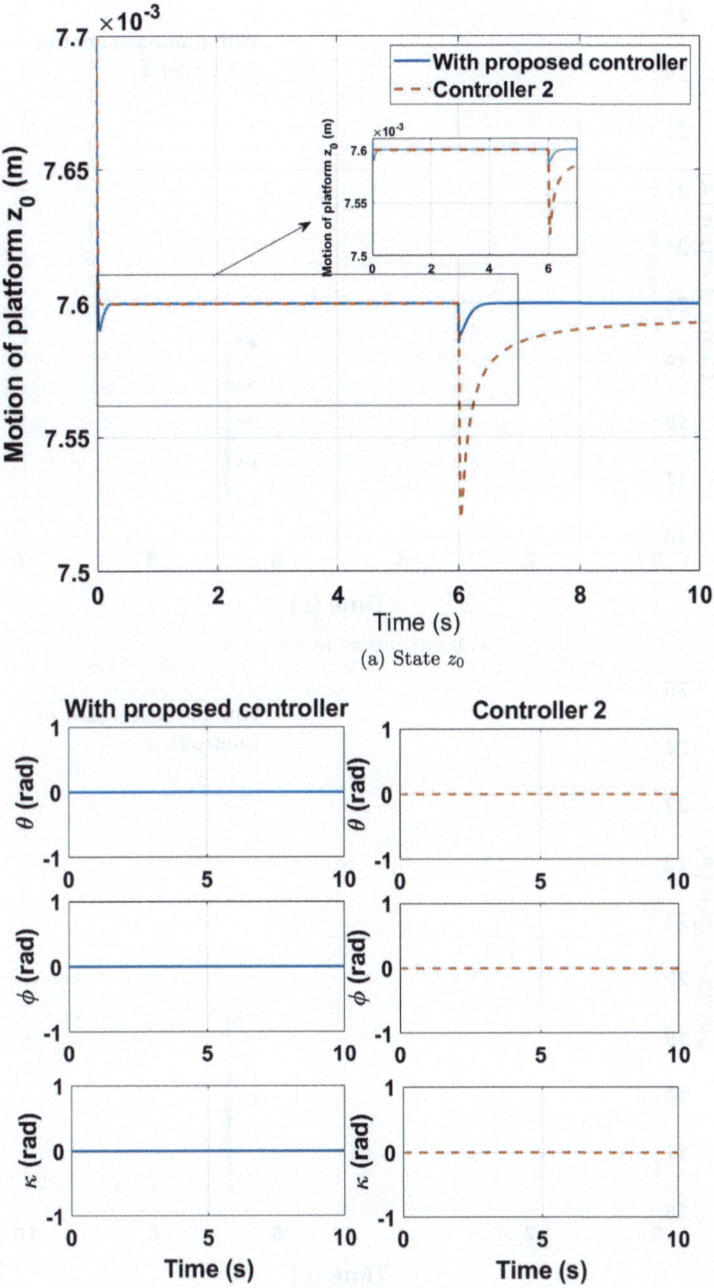

(a) State z_0

(b) States $\theta,\ \phi, \kappa$

Fig. 3 Motion of the platform in the absence of measurement noise

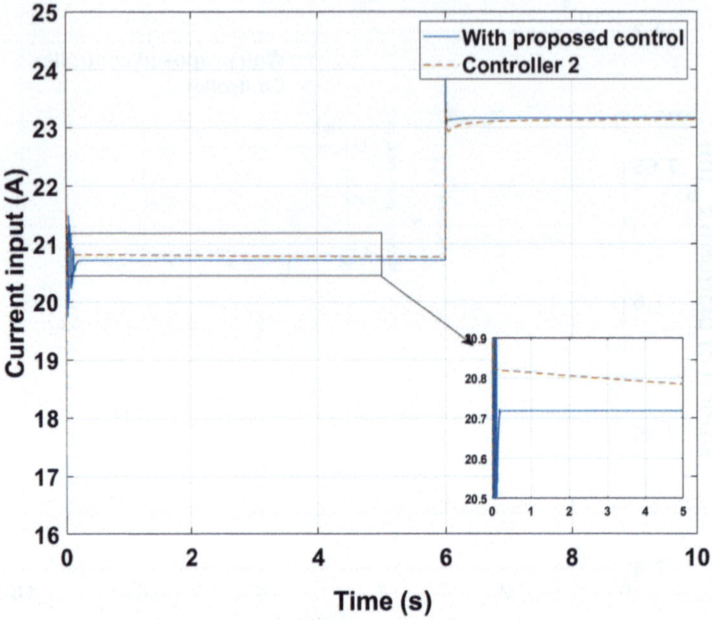

(a) Input current in section A

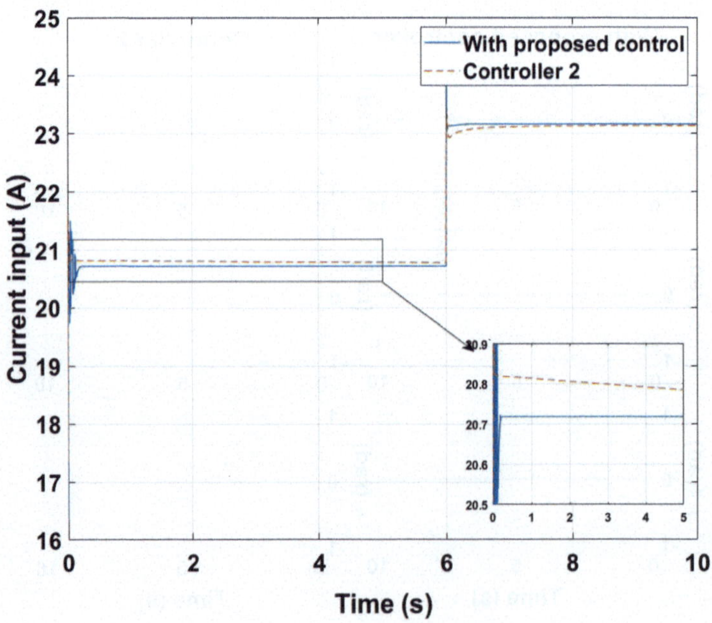

(b) Input current in section B

Fig. 4 Input currents in the absence of measurement noise for sections A and B

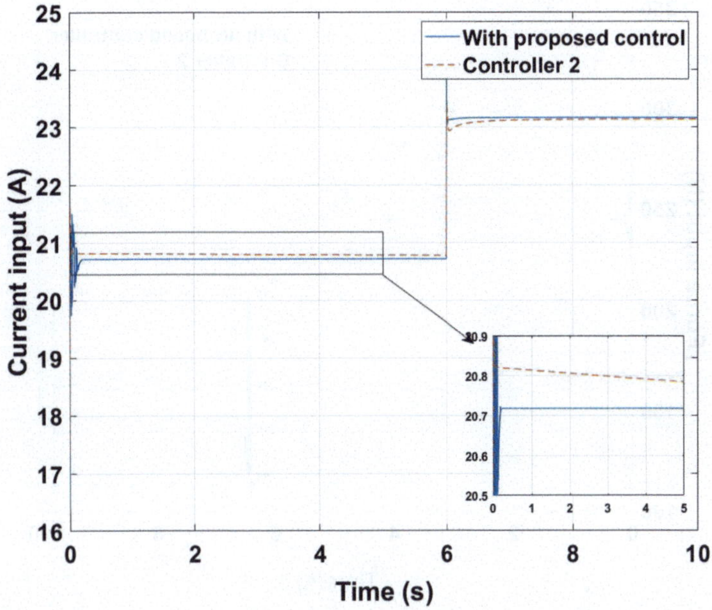

(a) Input current in section C

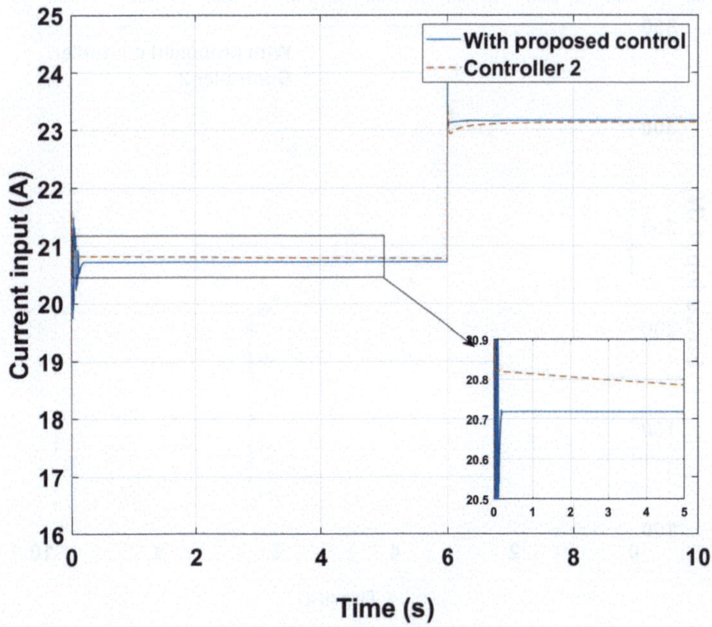

(b) Input current in section D

Fig. 5 Input currents in the absence of measurement noise for sections C and D

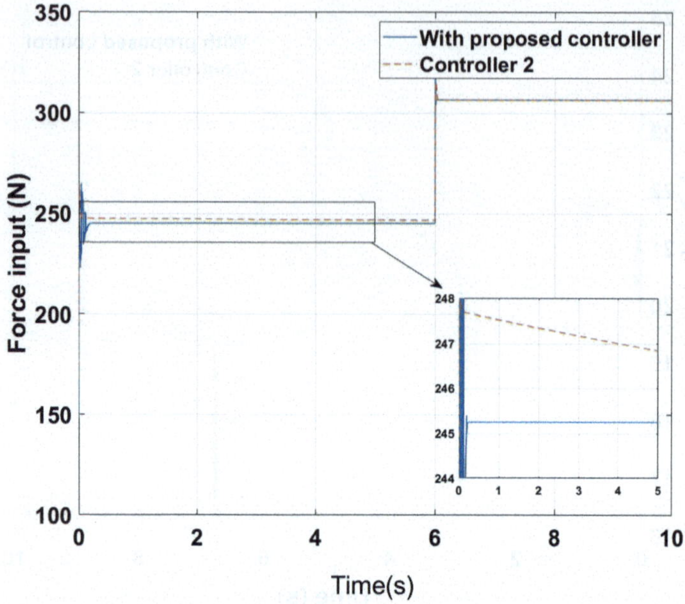

(a) Input force in section A

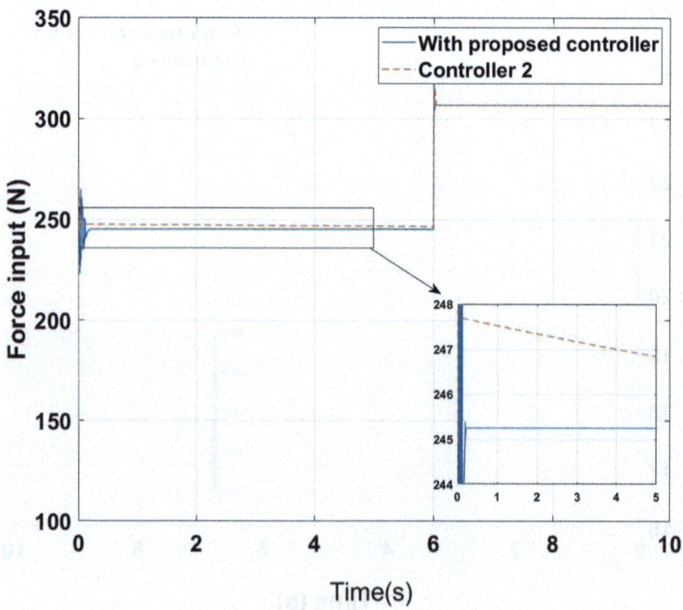

(b) Input force in section B

Fig. 6 Input currents in the absence of measurement noise for sections A and B

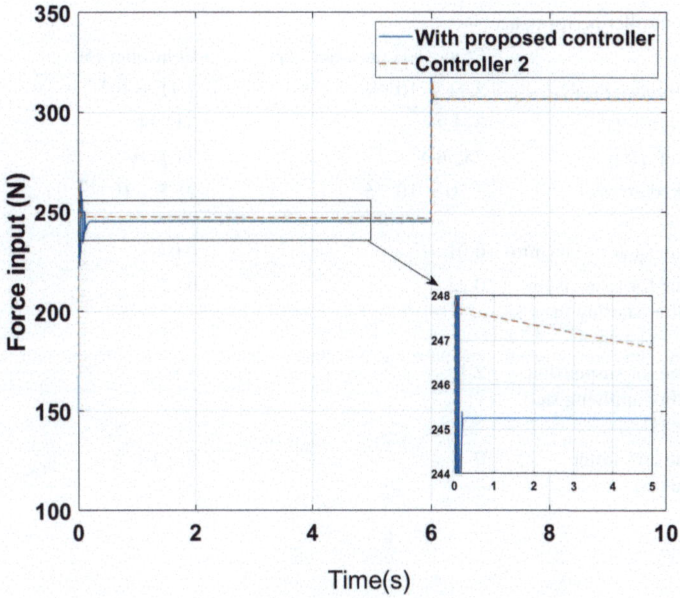

(a) Input force in section C

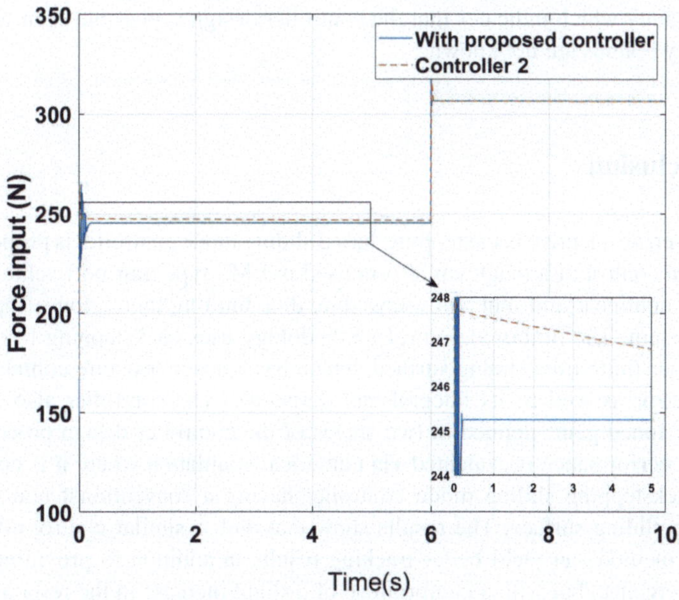

(b) Input force in section D

Fig. 7 Input forces in the absence of measurement noise for sections C and D

Table 2 Controller performance

	Proposed controller (29)	Controller (50)
2 norm of forces$\|\|f(t)\|\|$	2.71×10^4 N	2.71×10^4
Maximum of $\|f(t)\|$	374.91	342.34
Maximum of $\|i(t)\|$	25.56 A	24.71 A
Mean squared error of $x_{\text{error}_e}(t)$	7.50×10^{-12}	25.5×10^{-12}
Overshoot in z_0 at $t = 6$ (mm)	0.01	0.03
Settling time for suspension height z_0 after starting the system (s)	0.22	0.07
Settling time for suspension height z_0 after applying step disturbance (s)	2.1	1.8
Steady-state error after disturbance (m)	0	5×10^{-6}

presence of measurement noise and the step disturbance of the mass, the controller is able to maintain the suspension height at the reference level and can provide a finite convergence time. From Fig. 14, it is evident that the gain adjusts accordingly with the uncertainty which indicates that the controller design is possible even when the uncertainty bounds are not known.

6 Conclusion

In this paper, an adaptive backstepping-based sliding mode controller is proposed for suspension control of a maglev platform used in EMS type transport vehicles. The system is nonlinear and inherently unstable, thus limiting the region of operation for the system. The proposed control methodology uses backstepping to design a non-singular finite-time sliding surface, which has a novel structure containing the tracking error as well as its integral and derivative. The controller also has two adaptively tuned gains defined in two stages of the controller design process. The controller performance is evaluated via numerical simulation where it is compared with a backstepping sliding mode controller having a conventional non-singular finite-time sliding surface. The results show that with a similar control effort, the proposed method can yield better tracking results in addition to providing finite-time convergence, but with a compromise of a slight increase in the response time. As a future scope of this research, a thorough analysis on the possible bifurcation

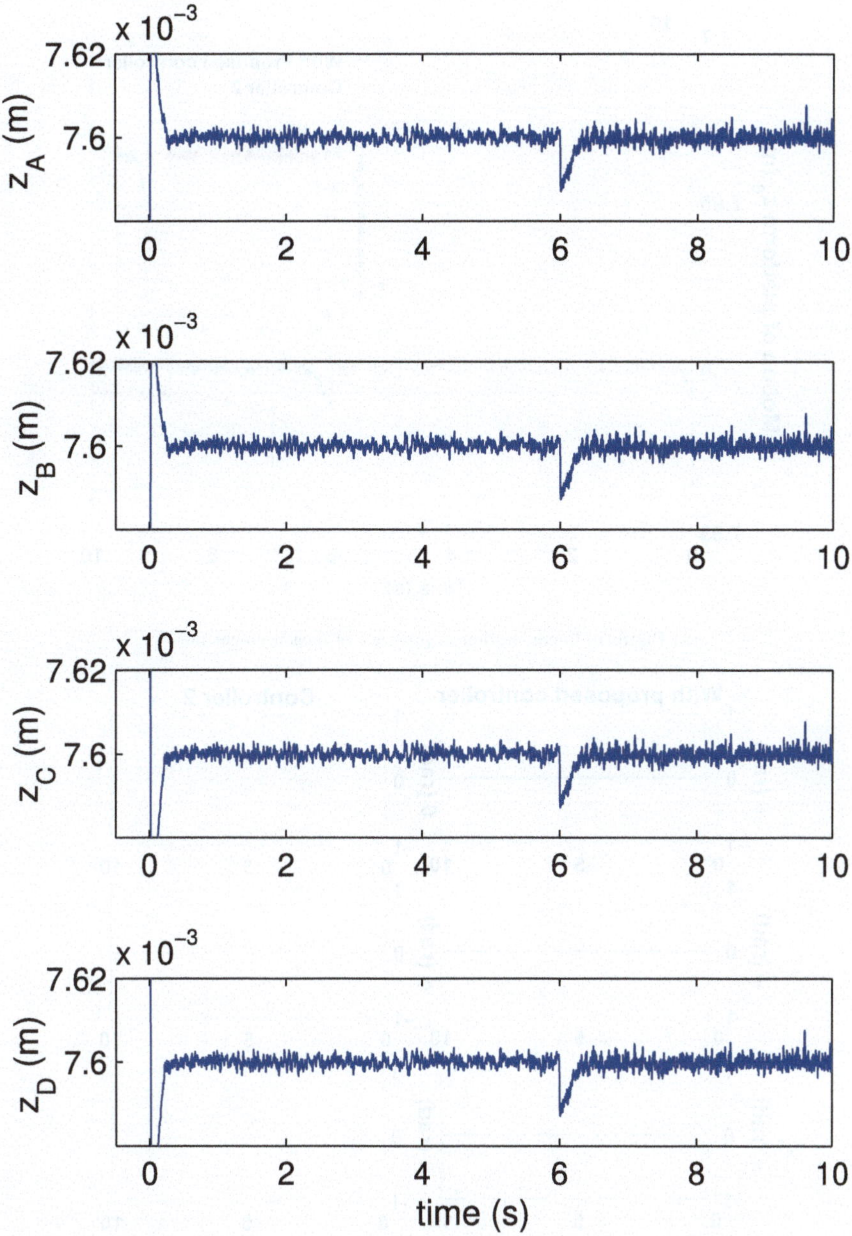

Fig. 8 Displacements z_A, z_B, z_C, z_D in the presence of measurement noise

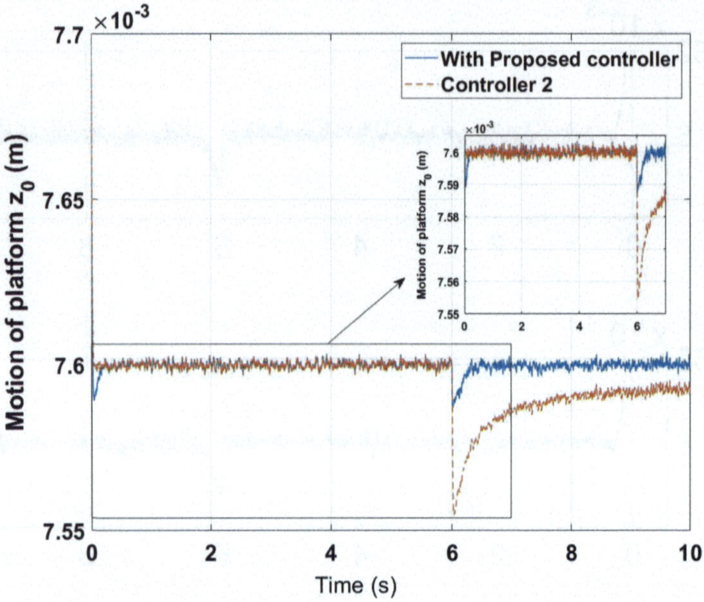

(a) Platform linear motion in presence of measurement noise

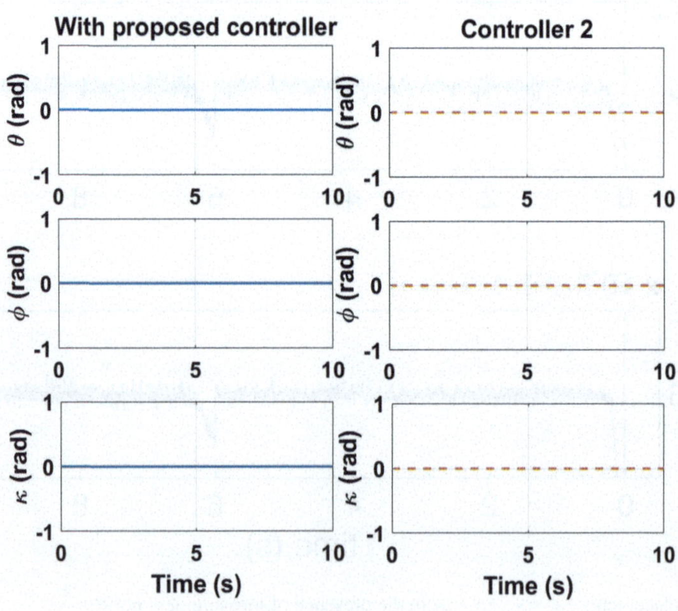

(b) Angular motions in presence of measurement noise

Fig. 9 States in the presence of measurement noise

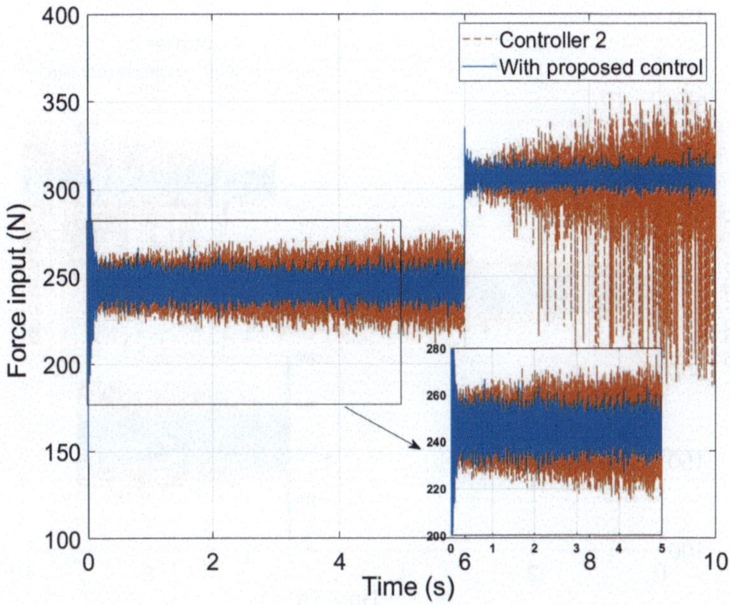

(a) Input force in section A

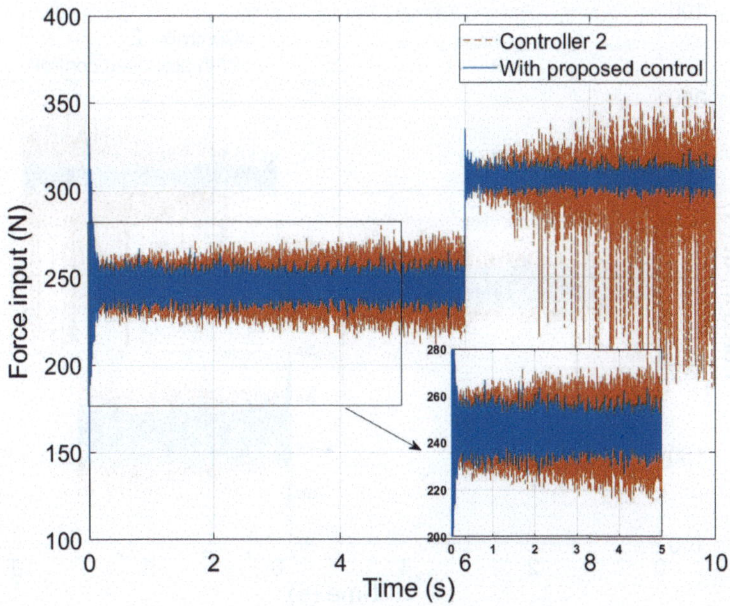

(b) Input force in section B

Fig. 10 Input forces in the presence of measurement noise

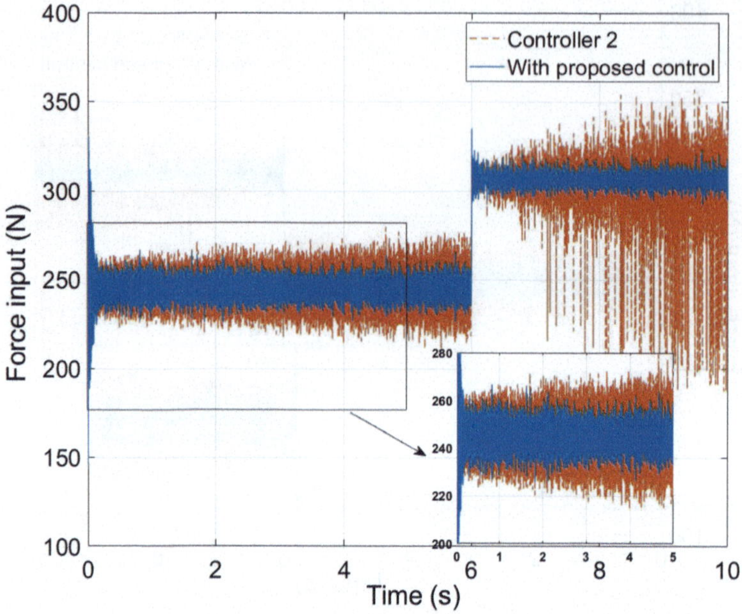

(a) Input force in section C

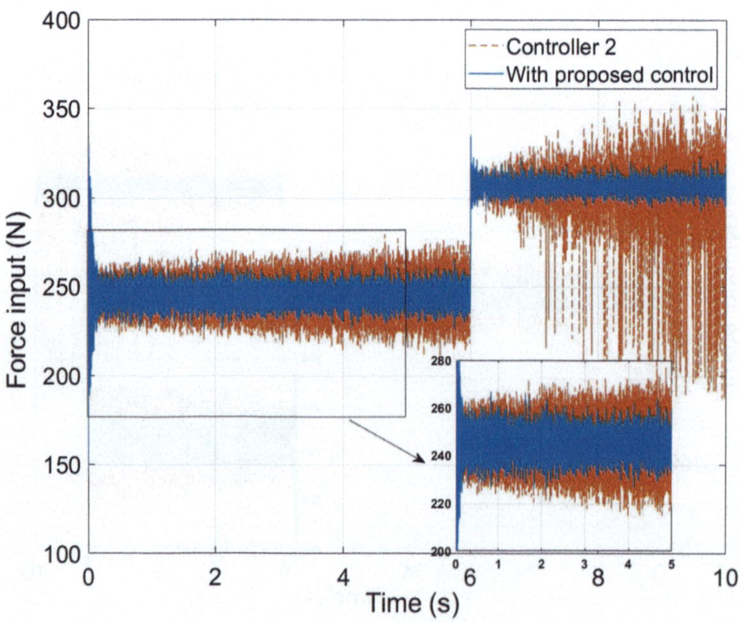

(b) Input force in section D

Fig. 11 Input forces in the presence of measurement noise

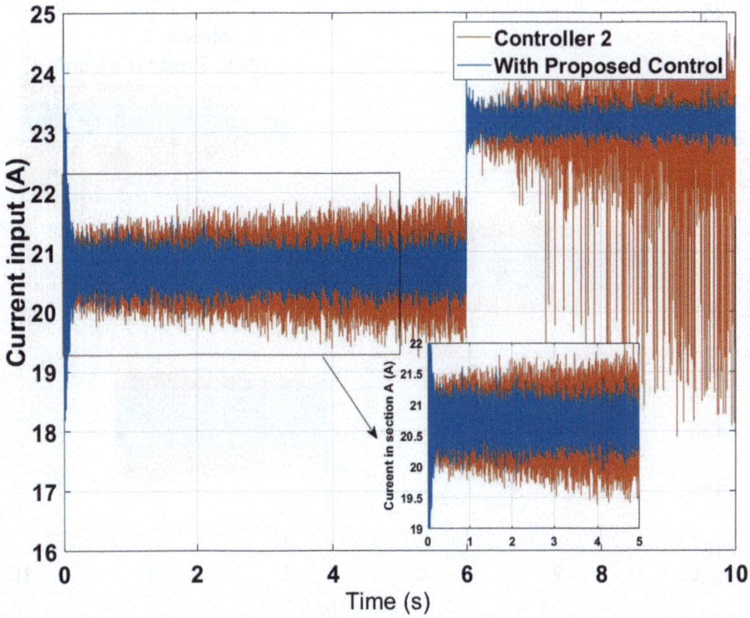

(a) Input current in section A

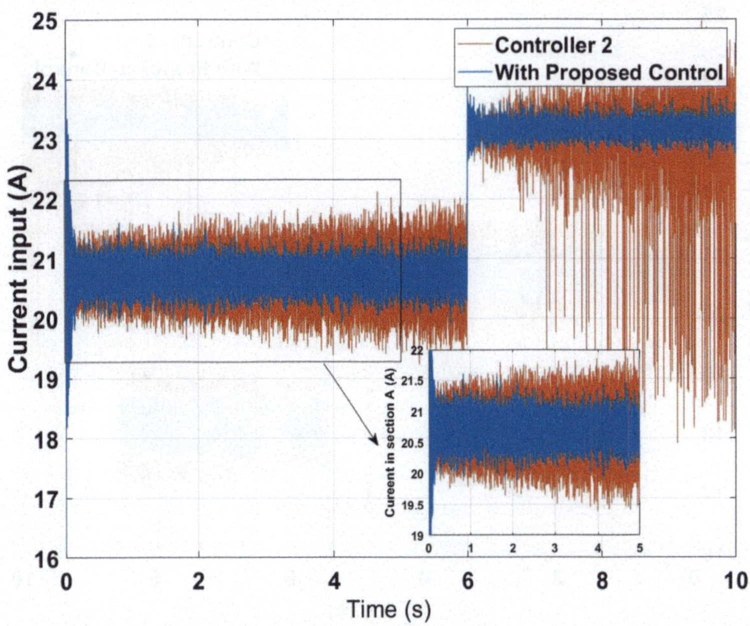

(b) Input current in section B

Fig. 12 Input currents in the presence of measurement noise

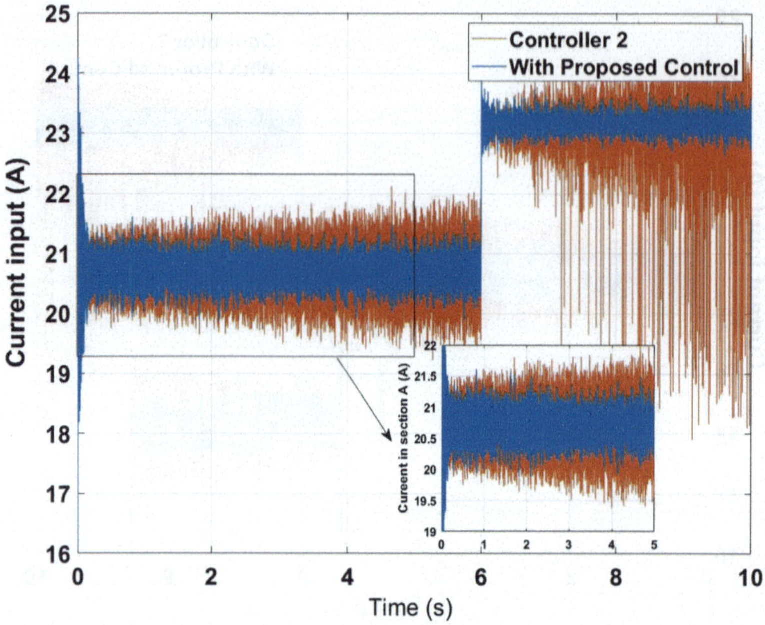

(a) Input current in section C

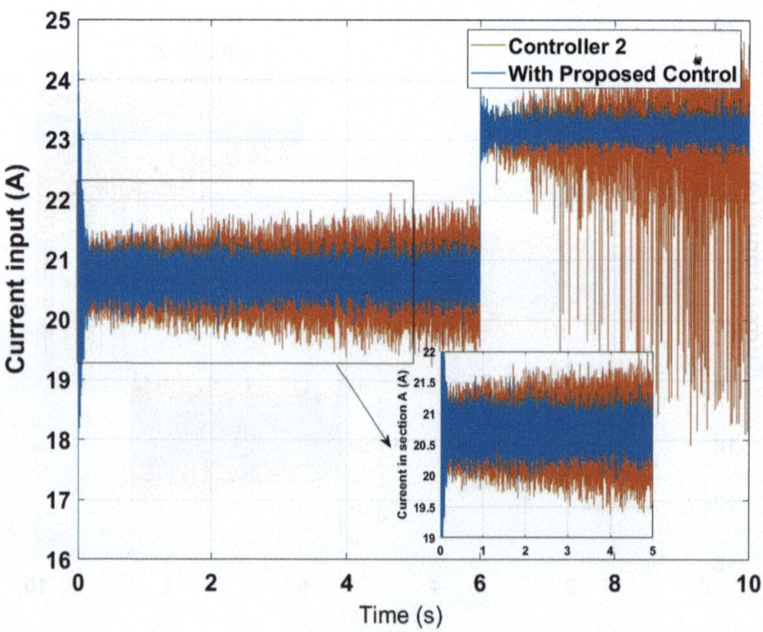

(b) Input current in section D

Fig. 13 Input currents in the presence of measurement noise

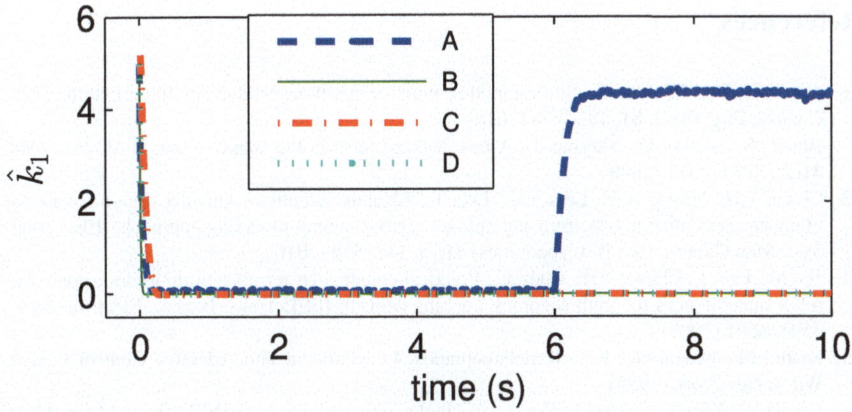

(a) $\hat{k}_1$

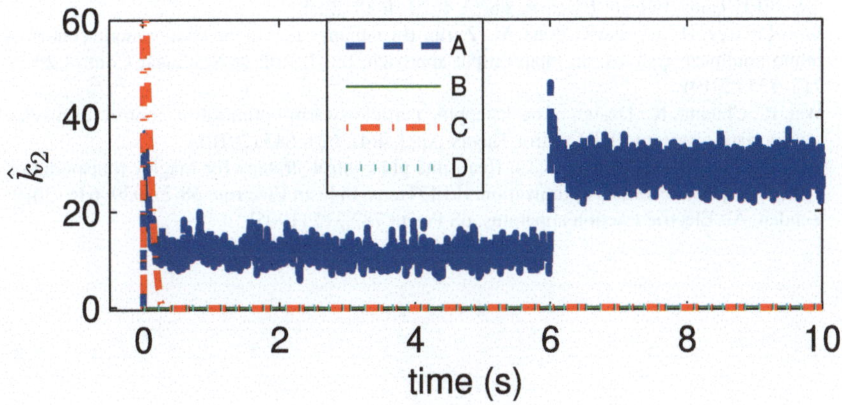

(b) $\hat{k}_2$

Fig. 14 Adaptively tuned controller gains $\hat{k}_1$ and $\hat{k}_2$ for the four corner electromagnets in A, B, C, D parts of Fig. 1

of the closed-loop system is also to be explored as that will not only provide an insight into the stability of the system but will also help in formulating more distinct criteria for parameter selection. Additionally, the effect of time delay and increasing the robustness of the proposed controller in the presence of time delay can also be explored.

References

1. Adhikary, N., Mahanta, C.: Sliding mode control of position commanded robot manipulators. Control. Eng. Pract. **81**, 183–198 (2018)
2. Blažič, S., Matko, D., Škrjanc, I.: A new leakage term in the adaptive law. IFAC Proc. Vol. **41**(2), 8913–8918 (2008)
3. Chien, Y.H., Wang, W.Y., Leu, Y.G., Lee, T.T.: Robust adaptive controller design for a class of uncertain nonlinear systems using online t–s fuzzy-neural modeling approach. IEEE Trans. Syst. Man Cybern. Part B (Cybernetics) **41**(2), 542–552 (2010)
4. Jin, M., Lee, J., Chang, P.H., Choi, C.: Practical nonsingular terminal sliding-mode control of robot manipulators for high-accuracy tracking control. IEEE Trans. Industr. Electron. **56**(9), 3593–3601 (2009)
5. Krstic, M., Kokotovic, P.V., Kanellakopoulos, I.: Nonlinear and Adaptive Control Design. Wiley, New York (1995)
6. Lee, H.W., Kim, K.C., Lee, J.: Review of Maglev train technologies. IEEE Trans. Magn. **42**(7), 1917–1925 (2006)
7. Ma, H., Li, Y., Xiong, Z.: Discrete-time sliding-mode control with enhanced power reaching law. IEEE Trans. Industr. Electron. **66**(6), 4629–4638 (2018)
8. Sira-Ramírez, H., Aguilar-Orduña, M., Zurita-Bustamante, E.: On the sliding mode control of mimo nonlinear systems: an input-output approach. Int. J. Robust Nonlinear Control **29**(3), 715–735 (2019)
9. Wai, R., Chuang, K.: Design of backstepping particle-swarm-optimisation control for maglev transportation system. IET Control Theory Appl. **4**(4), 625–645 (2010)
10. Wai, R.J., Lee, J.D., Chuang, K.L.: Real-time pid control strategy for maglev transportation system via particle swarm optimization. IEEE Trans. Industr. Electron. **58**(2), 629–646 (2010)
11. Zehden, A.: Electric traction apparatus. uS Patent 782,312 (1905)

Backstepping-Based Nonlinear Control of Underactuated 2-DoF Gyroscope for Robust Performance Analysis

Manish Patel and Bhanu Pratap

Abstract In this paper, a nonlinear controller has been proposed for a highly coupled 2-DoF underactuated mechanical gyroscope. The dynamics of the gyroscope have been derived using the Euler–Lagrange equation which is transformed into a strict feedback form. A backstepping-based sliding mode control (SMC) technique has been applied for the proposed controller design and the control laws obtained systematically. Thereafter, a switching law containing adaptive disturbance estimation term is derived for bounded input disturbance rejection which does not require knowledge of the upper bound of uncertainty. Lyapunov stability criteria with Barbashin–Krasvoskii theorem has been used to prove the asymptotic stability of the system. The effectiveness of the proposed controller has been investigated in a simulation environment with bounded matched uncertainty. A comparative study has been done for the proposed control scheme with a conventional SMC using the Monte Carlo simulation method. The proposed controller terminates the effect of disturbances considered with lesser control effort and exhibits improved tracking performance.

1 Introduction

The dynamics of a mechanical gyroscope is considered as one of the most interesting problems of mechanics [1]. Gyros find applications in attitude control of satellites [2], scanning and tracking in combat vehicles [3], submarine inertial auto-navigating [1], actuation of dual-arm space robots [4], underwater robots [5], anthropomorphic test device for two-wheeled self-balancing vehicles [6], and several more. The 3-DoF gyroscope considered herein is one of the benchmark nonlinear highly coupled systems to test new control strategies [7]. It has a disc mounted inside a blue gimbal which is in turn inside a red gimbal, these are placed inside a silver frame,

M. Patel · B. Pratap (✉)
Department of Electrical Engineering, National Institute of Technology Kurukshetra, Kurukshetra, Haryana 136119, India
e-mail: bhanu@nitkkr.ac.in

M. Patel
e-mail: manish_31904107@nitkkr.ac.in

J. Gu et al. (eds.), *Communication and Control for Robotic Systems*, Smart Innovation, Systems and Technologies 229, https://doi.org/10.1007/978-981-16-1777-5_6

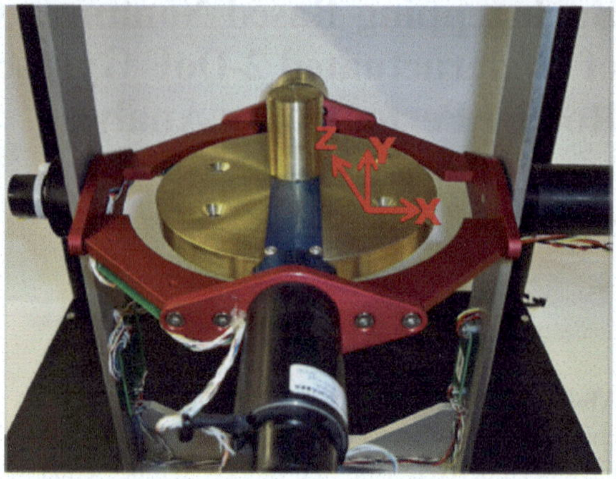

Fig. 1 Gyroscope

four DC motors each to rotate the disc, red and blue gimbal, and silver frame, and four high-resolution optical encoders to measure angular positions (Fig. 1).

Gyros are used in highly sophisticated systems so, to increase the tolerance for failure of actuators and to deal with space and weight constraints leads to an underactuated system. The underactuated system is a class of mechanical systems in which degrees of freedom available is more than actuators [8]. Quanser gyroscope can be used as an underactuated system by controlling the angular position of the red gimbal using DC motor at blue gimbal. In this paper, 2-DoF is considered by fixing the external silver frame.

Next, works related to the mechanical gyroscope and underactuated systems which has been done in the past decade are mentioned. Feedback linearization and adaptive control schemes have been applied to 2-DoF Quanser gyroscope for trajectory tracking with periodic matched disturbance in [7]; sliding mode and adaptive fuzzy control has been applied on Quanser Gyroscope for set-point tracking without any disturbance consideration in simulation [9]; active matched disturbance rejection for a 3-DoF gyroscope but results of internal dynamics are unavailable in [10]; sliding mode control applied to achieve tracking problem and periodic output disturbance rejection of a control moment gyroscope in [11]; backstepping-based nonlinear control applied for stabilizing pendubot near-equilibrium point [12]; nonlinear state feedback control of underactuated TORA system based on backstepping approach has been used to tackle with stabilization problems [13]; backstepping-based sliding mode control applied to accomplish trajectory tracking application for unmanned aerial vehicle [14]; integral backstepping-based sliding mode control has been used for swing-up and rejection of matched and mismatched uncertainties of cart pendulum system [8]; adaptive backstepping-based sliding mode controller has been proposed for 3D flight trajectory tracking and stabilization of an unmanned aerial vehicle [15]; adaptive backstepping control with finite time current observer has been proposed for output voltage regulation in buck converters [16], and neuro-adaptive backstepping control for tracking of angular velocity in buck converter-fed permanent magnet DC motor [17].

As shown in the literature, recent control problems for 2-DoF underactuated gyroscope have considered periodic uncertainty [7] and [11]. However, control design using a systematic approach like backstepping has not received much attention for 2-DoF underactuated gyroscope. The possibility of unpredictable interaction of unmodeled dynamics of 2-DoF underactuated gyroscope and uncertainty is also not mentioned in any of the recent works. In this document, trajectory tracking of the red gimbal using actuator at blue gimbal with bounded Gaussian-matched disturbance is considered. The system considered is underactuated with actuated shape variables, and Reza Olfati-Saber [18] proposed backstepping-based nonlinear control for underactuated systems with actuated shape variable and normal form in the strict feedback form. The normal form of this system is not in strict feedback form, so system equations are converted into strict feedback form using input–output linearization [19] but poses another challenge of controlling the internal dynamics of the system. Further, global asymptotic stability of the system can be ensured using the Barbashin–Krasvoskii theorem [19]. The internal dynamics of the system are also proved to be bounded for bounded reference tracking in this document. Due to complex dynamics of the gyroscope including nonlinear and coupling effects, the deviation of system variable from steady state at any instant of time the system may collapse because of unpredictable interaction between input uncertainty and unmodeled dynamics. Thus, the Monte Carlo simulation method with random bounded uncertainty is used while testing the robustness of the proposed control law and performance of the closed loop system.

In the remainder of the paper, the other sections are given as follows. The problem statement is formulated in Sect. 2, whereas the proposed controllers are designed in Sect. 3, and stability analysis is illustrated in Sect. 4. Section 5 presents the simulation studies, and the conclusion is given in Sect. 6 followed by the references.

2 Problem Formulation

2.1 Dynamical Model of 2-DoF Gyroscope

The dynamic model of the underactuated gyroscope can be derived using Euler–Lagrange's equation [20] given as,

$$\frac{\partial L}{\partial \psi} - \frac{\mathrm{d}}{\mathrm{d}t}\left(\frac{\partial L}{\partial \dot{\psi}}\right) = \tau \tag{1}$$

where $L = T - V$, T is the total kinetic energy and V is the total potential energy of the system, ψ is the vector of system variables and τ is the force.

There is no potential energy in the system and the kinetic energy of the system [1] is given as

$$T = \frac{1}{2}\left(J(\beta)\dot{\alpha}^2 + H\dot{\beta}^2 + C(\dot{\gamma} + \dot{\alpha}\sin\beta)^2\right) \tag{2}$$

where,

$$J(\beta) = A_2 + (A_1 + A)\cos^2\beta + C_1\sin^2\beta \tag{3}$$

$$H = B_1 + B \tag{4}$$

A_2, A_1, and A are moment of inertia of red gimbal, blue gimbal, and disc about x-axis, respectively, B_1 and B are moment of inertia of blue gimbal and disc about y-axis, respectively, and C_1 and C denote moment of inertia of blue gimbal and disc about z-axis, respectively. α, β and γ denote angular position of red gimbal, blue gimbal, and disc about x-axis, y-axis, and z-axis, respectively [1].

Using (1) and taking $\psi = \left[\alpha\ \beta\ \gamma\right]^T$, the system dynamics can be written in the following form

$$\ddot{\alpha} = -\frac{C}{J(\beta)}\left(\dot{\beta}\cos\beta\,(\dot{\gamma} + \dot{\alpha}\sin\beta)\right) - \frac{J'(\beta)}{J(\beta)}\dot{\alpha}\dot{\beta} - \frac{\sin\beta}{J(\beta)}\tau \tag{5}$$

$$\ddot{\beta} = \frac{C}{H}\left(\dot{\alpha}\dot{\gamma}\cos\beta + \dot{\alpha}^2\sin\beta\cos\beta\right) + \frac{J'(\beta)}{2H}\dot{\alpha}^2 + \frac{1}{H}\tau_1 \tag{6}$$

$$\ddot{\gamma} = \frac{C}{J(\beta)}\dot{\beta}\sin\beta\cos\beta\,(\dot{\gamma} + \dot{\alpha}\sin\beta) + \frac{J'(\beta)}{J(\beta)}\dot{\alpha}\dot{\beta}\sin\beta - \dot{\alpha}\dot{\beta}\cos\beta + \left(\frac{\sin^2\beta}{J(\beta)} + \frac{1}{C}\right)\tau \tag{7}$$

where τ and τ_1 are actuating torques at disc and blue gimbal, respectively, to achieve underactuation torque at red gimbal $\tau_2 = 0$.

2.2 Control Objective

For the underactuated mechanical gyroscope, it is desired to derive a control law τ_1 such that the angular position of red gimbal α tracks trajectory α^d, while the angular position of blue gimbal β remains bounded around origin. Also a controller is to be designed, so that the disc rotates at the desired constant angular velocity $\dot{\gamma}^d$.

Tracking error is defined as below

$$e = \alpha - \alpha^d \tag{8}$$

and a control law is to be derived, so that $e(t)$ as well as $\dot{e}(t)$ are asymptotically stable which can be written analytically as below
$e(t) \rightarrow 0$ as well as $\dot{e}(t) \rightarrow 0$ when $t \rightarrow \infty$.

It is assumed that α^d, $\dot{\alpha}^d$, $\ddot{\alpha}^d$ and $\dddot{\alpha}^d$ are available and bounded.

3 Control Law Design

3.1 *Proportional Controller*

In this section, a proportional control law τ is introduced to rotate the disc at the desired constant angular velocity $\dot{\gamma}^d$.

Linearizing (7) about $\dot{\gamma} = 0$ and keeping $\beta = 0$, $\dot{\beta} = 0$ to obtain differential equation as follows

$$\ddot{\gamma} = \frac{1}{C}\tau \tag{9}$$

Proportional controller with gain K_p mentioned below will force $\dot{\gamma}(t) \rightarrow \dot{\gamma^d}$ as $t \rightarrow \infty$.

$$\ddot{\gamma} = \frac{K_p}{C}\left(\dot{\gamma} - \dot{\gamma^d}\right) \tag{10}$$

Solution of above first-order linear differential equation is

$$\dot{\gamma}(t) = \dot{\gamma}^d\left(1 - \mathrm{e}^{-\frac{K_p}{C}t}\right) \tag{11}$$

It is evident from the above equation that with $K_p > 0$ $\dot{\gamma}(t) \rightarrow \dot{\gamma^d}$ as $t \rightarrow \infty$.

3.2 *Backstepping-Based Sliding Mode Controller*

Sliding mode control consists of two modes, and they are reaching mode and sliding mode. The sliding surface design should be done cautiously since phase trajectories on hitting the sliding mode region reaches the origin as per the dynamics of sliding surface [21].

Order of sliding surface depends on relative degree of output of the system [9] and from (5) comment on relative degree cannot be made since τ_1 is not appearing explicitly in (5). Hence, taking derivative of (5) w.r.t $'t'$ and considering the proportional controller mentioned in Sect. 3.1, reduced system dynamics are

$$\ddot{\alpha} = f_1\left(\psi, \dot{\psi}, \ddot{\psi}\right) + b_1\left(\psi, \dot{\psi}\right)\tau_1 \tag{12}$$

$$\ddot{\beta} = f_2\left(\psi, \dot{\psi}\right) + \frac{1}{H}\tau_1 \tag{13}$$

where

$$\begin{aligned} f_1\left(\psi, \dot{\psi}, \ddot{\psi}\right) = &-\frac{C}{J(\beta)}\left[f_2\cos(\beta)\left(\dot{\gamma} + \dot{\alpha}\sin\beta\right) + \dot{\beta}\cos\beta\left(\ddot{\gamma} + \ddot{\alpha}\sin\beta\right)\right] - \frac{J'(\beta)}{J(\beta)}f_2\dot{\alpha} \\ &+ \dot{\beta}^2\left(\dot{\alpha}\cos 2\beta - \dot{\gamma}\sin\beta\right) - \frac{J''(\beta)J'(\beta) - J'(\beta)^2}{J(\beta)^2}\dot{\alpha}\dot{\beta}^2 - \frac{J'(\beta)}{J(\beta)}\dot{\beta}\ddot{\alpha} \\ &- \frac{\left(J(\beta)\dot{\tau}\sin\beta + \tau\dot{\beta}\cos\beta - \tau\sin\beta\dot{\beta}J'(\beta)\right)}{J(\beta)^2} - \frac{f_1\dot{\beta}J'(\beta)}{J(\beta)} \\ b_1\left(\psi, \dot{\psi}\right) = &-\frac{C}{J(\beta)H}\left(\cos\beta\left(\dot{\gamma} + \dot{\alpha}\sin\beta\right) + J'(\beta)\dot{\alpha}\right) \\ f_2\left(\psi, \dot{\psi}\right) = &\frac{C}{H}\left(\dot{\alpha}\dot{\gamma}\cos\beta + \dot{\alpha}^2\sin\beta cos\beta\right) + \frac{J'(\beta)}{2H}\dot{\alpha}^2 \end{aligned}$$

From (12), it can be concluded that relative degree of the system is three.

The basic requirement to apply a backstepping-based sliding mode controller to any system is that the system should be in strict feedback form [18].

Taking state variables as

$$X = [\alpha, \dot{\alpha}, \ddot{\alpha}]^T \tag{14}$$

The system dynamics in (12), (13) can be written in strict feedback form as below

$$\dot{x_1} = x_2 \tag{15}$$

$$\dot{x_2} = x_3 \tag{16}$$

$$\dot{x_3} = f_1\left(\psi, \dot{\psi}, \ddot{\psi}\right) + b_1\left(\psi, \dot{\psi}\right)(\tau_1 + \tau_d) \tag{17}$$

leaving below internal dynamics of the system to be dealt separately.

$$\ddot{\beta} = f_2\left(\psi, \dot{\psi}\right) + \frac{1}{H}(\tau_1 + \tau_d)$$

where, τ_d is matched uncertainty.

***Step* 1**: Control variable and Lyapunov candidate in terms of control variable is chosen as below

$$e_1 = x_1 - x_1^d \tag{18}$$

$$V_1 = \frac{1}{2}e_1^2 \tag{19}$$

***Step* 2**: Time derivative of V_1 is $\dot{V_1} = e_1\dot{e_1}$ and $\dot{e_1} = x_2 - \dot{x_1}^d$, virtual control law $x_2 = -k_1e_1 + \dot{x_1}^d$ will stabilize the e_1 subsystem, so defining stabilizing function as $\rho_2 = -k_1e_1 + \dot{x_1}^d$ and error between the virtual control law and the stabilizing function is defined as

$$e_2 = x_2 - \rho_2 \tag{20}$$

$$\begin{aligned}\dot{V_1} &= e_1\left(x_2 - \dot{x_1}^d\right)\\ &= e_1\left(e_2 + \rho_2 - \dot{x_1}^d\right)\end{aligned}$$

$$\dot{V_1} = -k_1e_1^2 + e_1e_2 \tag{21}$$

Now, a new Lyapunov function is chosen for subsystem e_2 as follows

$$V_2 = V_1 + \frac{1}{2}e_2^2 \tag{22}$$

***Step* 3:** As done in earlier step, choosing stabilizing function $\rho_3 = -e_1 - k_1\dot{e_1} + \ddot{x1}^d - k_2e_2$ and error $e_3 = x_3 - \rho_3$, $\dot{V_2}$ will be as given below

$$\dot{V_2} = -k_1e_1^2 - k_2e_2^2 + e_2e_3 \tag{23}$$

$$\dot{e_3} = f_1\left(\psi, \dot{\psi}, \ddot{\psi}\right) + b_1\left(\psi, \dot{\psi}\right)(\tau_1 + \tau_d) + (1 + k_1k_2)\dot{e_1} + (k_1 + k_2)\ddot{e_1} - \dddot{x_1}^d \tag{24}$$

Derivative of e_3 contains input to the system, so now the sliding surface will be chosen as below

$$s = e_3 + c_1e_2 + c_2e_1 \tag{25}$$

order of sliding surface is $(n - 1)$ where (n) is relative degree, c_1 and c_2 are positive constants and can be chosen as per the required dynamics of the sliding surface and Lyapunov candidate for final step is as follows

$$V_3 = V_2 + \frac{1}{2}s^2 \tag{26}$$

***Step* 4:** Using (23), (24), (25), and $\dot{e_2} = \ddot{e_1} + k_1\dot{e_1}$ time derivative of V_3 is

$$\begin{aligned}\dot{V_3} = -k_1e_1^2 - k_2e_2^2 + e_2e_3 + s(f_1\left(\psi, \dot{\psi}, \ddot{\psi}\right) + b_1\left(\psi, \dot{\psi}\right)(\tau_1 + \tau_d)\\ +(k_1 + k_2 + c_1)\ddot{e_1} + (1 + k_1k_2 + c_1k_1 + c_2)\dot{e_1} - \dddot{x_1}^d)\end{aligned} \tag{27}$$

The constant plus proportional rate reaching law [21] given below will be used to design control law

$$\dot{s} = -Qsgn(s) - Ks \tag{28}$$

Using conventional sliding mode design law given in (28), control law is as follows

$$v = -\left(-\dddot{x}_1^d + (k_1 + k_2 + c_1)\ddot{e}_1 + (1 + k_1k_2 + c_1k_1 + c_2)\dot{e}_1 + Qsgn(s) + Ks\right) \tag{29}$$

and

$$\tau_1 = \frac{1}{b_1\left(\psi, \dot{\psi}\right)}(v - f_1\left(\psi, \dot{\psi}, \ddot{\psi}\right)) \tag{30}$$

control law (30) will achieve our control objective that is $e_1(t), \dot{e}_1(t) \to 0$ as $t \to \infty$ and can reject bounded continuous matched disturbances by choosing Q appropriately.

3.3 Adaptive Estimation of Matched Disturbance

For given matched disturbance τ_d, $\hat{d}$ estimates the disturbance and estimation error $\tilde{d}$ is defined as below

$$\tilde{d} = b_1\left(\psi, \dot{\psi}\right)\tau_d - \hat{d} \tag{31}$$

A Lyapunov candidate is chosen as below to deduce the adaptive estimation law

$$V_4 = V_3 + \frac{1}{2\gamma}\tilde{d}^2 \tag{32}$$

where γ is the adaptation gain.

Updating (29) with estimation term gives updated control law as below

$$v = -\left(-\dddot{x}_1^d + (k_1 + k_2 + c_1)\ddot{e}_1 + (1 + k_1k_2 + c_1k_1 + c_2)\dot{e}_1 + \hat{d} + Qsgn(s) + Ks\right) \tag{33}$$

Using (27), (30), (33) $\dot{V}_4$ is as follows

$$\begin{aligned}
\dot{V}_4 &= -k_1e_1^2 - k_2e_2^2 + e_2e_3 + s(b_1\left(\psi, \dot{\psi}\right)\tau_d - \hat{d} - Qsgn(s) - Ks) - \frac{1}{\gamma}\tilde{d}\dot{\hat{d}} \\
&= -k_1e_1^2 - k_2e_2^2 + e_2e_3 + s(\tilde{d} - Qsgn(s) - Ks) - \frac{1}{\gamma}\tilde{d}\dot{\hat{d}} \\
&= -k_1e_1^2 - k_2e_2^2 + e_2e_3 + s(-Qsgn(s) - Ks) - \frac{1}{\gamma}\tilde{d}(\dot{\hat{d}} - \gamma s)
\end{aligned}$$

Adaption law [22] given below and control law (33) achieves control objective by choosing Q, K and γ appropriately.

$$\dot{\hat{d}} = \gamma s \tag{34}$$

4 Stability Analysis

Assumption 1: It is assumed that the desired trajectory x_1^d, and its first derivative $\dot{x_1}^d$, second derivative $\ddot{x_1}^d$, and third derivative $\dddot{x_1}^d$ are available and bounded.
Assumption 2: $\bar{\tau_d}$ is the known upper bound for $b_1(\psi, \dot{\psi})\tau_d$, i.e., $\bar{\tau_d} > b_1(\psi, \dot{\psi})\tau_d$, where τ_d is the matched uncertainty.

4.1 Without Disturbance Estimation

$$V_3(e_1, e_2, s) = \frac{1}{2}e_1^2 + \frac{1}{2}e_2^2 + \frac{1}{2}s^2 \tag{35}$$

$$V_3(0, 0, 0) = 0 \text{ and } V_3(e_1, e_2, s) > 0, \ \forall\, e_1, e_2, s \neq 0 \tag{36}$$

$$||x|| \to \infty \Longrightarrow V_3(e_1, e_2, s) \to \infty \tag{37}$$

Lyapunov function V_3 given in (35) satisfies (36), (37), and if $\dot{V_3}(e_1, e_2, s) < 0$, $\forall\, e_1, e_2, s \neq 0$, then by invoking Barbashin–Krasovskii theorem, it can be inferred that $e_1(t), e_2(t), s(t)$ are globally asymptotically stable at origin.

Next, conditions will be established such that $\dot{V_3} < 0$, from (27), (29), (30) $\dot{V_3}$ can be written as

$$\begin{aligned}\dot{V_3} &= -k_1e_1^2 - k_2e_2^2 + e_2e_3 + s(b_1(\psi, \dot{\psi})\tau_d - Qsgn(s) - Ks)\\ &= -k_1e_1^2 - k_2e_2^2 + e_2e_3 - Ks^2 - s(Qsgn(s) - \bar{\tau_d})\\ &= -k_1e_1^2 - k_2e_2^2 + e_2e_3 - K(e_3 + c_1e_2 + c_2e_1)^2 - (Q||s||_1 - s\bar{\tau_d})\end{aligned}$$

Let $E = [e_1\ e_2\ e_3]^T$, putting $||s||_1\bar{\tau_d}$ in place of $s\bar{\tau_d}$ and since $||s||_1 \geq ||s||_p$ for any $p > 1$, $\dot{V_3}$ can be now written as

$$\dot{V_3} < -E^TPE - ||s||_1(Q - \bar{\tau_d}) \tag{38}$$

where, $P = \begin{bmatrix} k_1 + Kc_2^2 & Kc_1c_2 & Kc_2 \\ Kc_1c_2 & k_2 + Kc_1^2 & Kc_1 - \frac{1}{2} \\ Kc_2 & Kc_1 - \frac{1}{2} & K \end{bmatrix}$

Matrix P will be positive definite if its principal minors are positive definite and choosing Q such that $Q > \bar{\tau_d}$ will satisfy the following condition

$$\dot{V_3} < 0 \tag{39}$$

Hence, it can be concluded that with appropriate values of $c_1,\ c_2,\ k_1,\ k_2,\ K$ and Q control problem of tracking for the red gimbal subsystem can be stabilized globally asymptotically.

4.2 With Disturbance Estimation

$$V_4(e_1, e_2, s, \tilde{d}) = \frac{1}{2}e_1^2 + \frac{1}{2}e_2^2 + \frac{1}{2}s^2 + \frac{1}{2\gamma}\tilde{d}^2 \tag{40}$$

Conditions (36) and (37) are also satisfied by V_4, next conditions will be deduced, so that $\dot{V_4} < 0$, from (30), (32), (33) and (34)

$$\dot{V_4} = -k_1e_1^2 - k_2e_2^2 + e_2e_3 - K(e_3 + c_1e_2 + c_2e_1)^2 - Q||s||_1 \tag{41}$$

$$\dot{V_4} = -E^T PE - Q||s||_1 \tag{42}$$

Matrix E and P are same as that in the previous subsection and choosing c_1, c_2, k_1, k_2 and K such that P is a positive definite matrix will stabilize the error dynamics globally asymptotically. It can be observed that $Q > 0$ is a very mild condition to be followed unlike in earlier subsection which required exact knowledge of upper bound of matched disturbance. Here, knowledge of upper bound of disturbance is not required to decide control parameters.

4.3 Blue Gimbal Subsystem

Next stability analysis of the blue gimbal subsystem will be studied with the consideration that angular position of the red gimbal is tracking the desired trajectory α^d and the disc is rotating at the required angular velocity $\dot{\gamma}^d$.
From (5) $\dot{\beta}$ can be written as

$$\dot{\beta} = -\frac{J(\beta)\ddot{\alpha}^d}{C\dot{\gamma}^d \cos\beta + C\dot{\alpha}^d \sin\beta\cos\beta + J'(\beta)\dot{\alpha}^d} \tag{43}$$

Linearizing about $\beta = 0$, $\dot{\beta} = 0$

$$\dot{\beta} - p_1\ddot{\alpha}^d\dot{\alpha}^d\beta = p_2\ddot{\alpha}^d \tag{44}$$

where

$$p_1 = \frac{(A_2 + A_1 + A)(C + 2(C_1 - A_1 - A))}{(C\dot{\gamma}^d)^2}$$
$$p_2 = -\frac{A_2 + A_1 + A}{C\dot{\gamma}^d}$$

Linearized dynamics of the angular position of blue gimbal is first-order linear differential equation and can be solved using the integrating factor method.

Integrating factor = $\mathrm{e}^{-\phi(t)}$, where $\phi(t) = \int p_1\ddot{\alpha}^d\dot{\alpha}^{\mathrm{d}}\,\mathrm{d}t$

$$\beta(t) = \mathrm{e}^{\phi(t)}\int_0^t \mathrm{e}^{-\phi(t)}\,p_2\ddot{\alpha}^{\mathrm{d}}\,\mathrm{d}t \tag{45}$$

$\phi(t) = \frac{p_1}{2}((\dot{\alpha}^d)^2(t) - (\dot{\alpha}^d)^2(0))$ and approximating $\mathrm{e}^{-\phi(t)} \approx 1 - \phi(t)$

$$\begin{aligned}\beta(t) = \mathrm{e}^{\phi(t)}\left[p_2((\dot{\alpha}^d)(t) - (\dot{\alpha}^d)(0)) - \frac{p_1p_2}{6}((\dot{\alpha}^d)^3(t) - (\dot{\alpha}^d)^3(0))\right] \\ +\mathrm{e}^{\phi(t)}\left[\frac{p_1p_2}{2}(\dot{\alpha}^d)^2(0)((\dot{\alpha}^d)^2(t) - (\dot{\alpha}^d)^2(0))\right]\end{aligned} \tag{46}$$

It can be deduced from (46) that $\beta(t)$ will be bounded for bounded $\dot{\alpha}^d(t)$.

5 Simulation Results

In this section of the document, the simulation results of the backstepping-based sliding mode controller with disturbance estimation (DEBSMC) are compared with conventional backstepping-based sliding mode controller (CBSMC) with matched uncertainty using plant parameters (Table 1), and controller parameters (Table 2).

Profile of uncertainty is given in Fig. 3b which takes into account coulomb, static, viscous, and stribeck friction. Matched uncertainty is taken as below

$$\tau_d = -\frac{1}{\sigma\sqrt{2\pi}}\mathrm{e}^{-\frac{1}{2}(\frac{x-\mu}{\sigma})^2} \tag{47}$$

where $\mu = 0$, $\sigma = 2$ and $x = [-2.5, 2.5]$ is random gaussian variable.

Reference trajectory to be tracked is taken as:
$\alpha^d = (2 + 25\sin(2\pi f_1 t) + 3\cos(2\pi f_2 t))$(in deg.), where $f_1 = 0.05Hz$ and $f_2 = 0.025Hz$. $\dot{\gamma}^d$=34 rad/s.

To reduce chattering effect, $\tan h$ has been used instead of sgn [9].

Table 1 Moment of inertia of Quanser gyroscope

Subsystem of gyroscope	Moment of inertia (in kgm^2) about		
	x-axis	y-axis	z-axis
Red gimbal	0.00762058	0.02879104	0.02344554
Blue gimbal	0.00388552	0.00744682	0.00555680
Flywheel disc	0.00284584	0.00563843	0.00284582

Table 2 Controller parameters

Proportional controller constant:	$K_p = 1$
Sliding surface constants:	$c_1 = 4.2, c_2 = 9$
Switching control law constants	
For conventional BSMC:	$K = 1, Q = 130$
For DEBSMC:	$K = 1, Q = 1, \gamma = 20$
Backstepping control other constants:	$k_1 = 3, k_2 = 60$

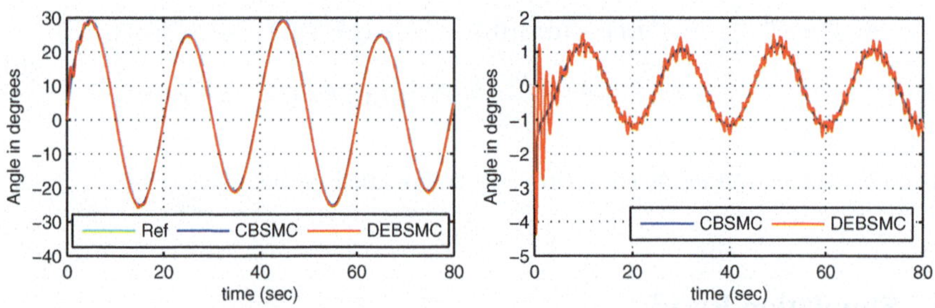

(a) (b)

Fig. 2 Angular position with matched uncertainty of the **a** red gimbal, **b** blue gimbal

From Fig. 2a, it can be said that the proposed controller has been able to reject the input uncertainty and able to track the desired trajectory, Fig. 2b shows that the angle β is bounded and the proportional controller is able to keep the disc rotating at the desired angular velocity which can be seen in Fig. 3a. Figure 4 shows that the control input is almost same in steady state for both the controllers but chattering in the proposed control law is less and can be observed in Fig. 5.

Graphically, it can be observed that both the controllers can stabilize the system asymptotically but information regarding the range of the tracking error and where the error is concentrated most of the time in steady state is absent, so to test and compare among the controllers, Monte Carlo simulation method is used, the system is simulated for 40,000 instants of time, and histogram of the tracking error is plotted. The control input RMS is also computed to get the exact value in steady state considering all 40,000 instants of time.

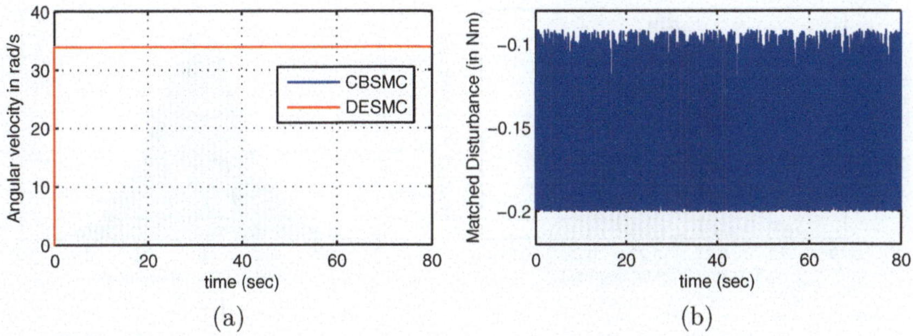

Fig. 3 **a** Angular velocity of the disc, **b** matched uncertainty profile

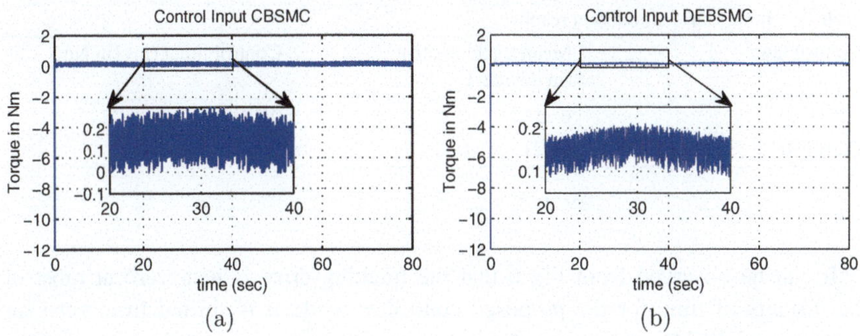

Fig. 4 Control input with matched uncertainty for **a** conventional BSMC, **b** BSMC with disturbance estimation

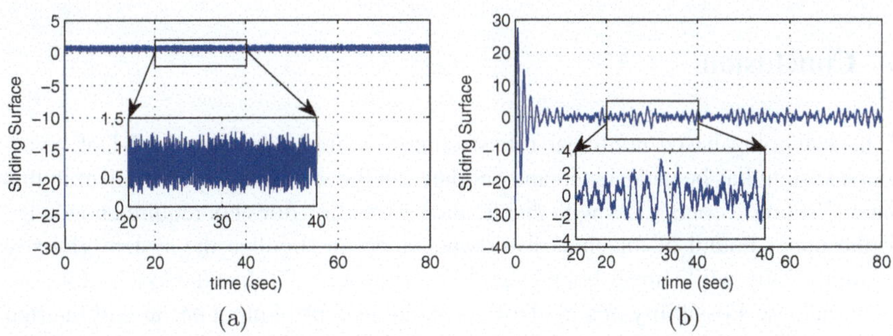

Fig. 5 Sliding surfaces with matched uncertainty for **a** conventional BSMC, **b** BSMC with disturbance estimation

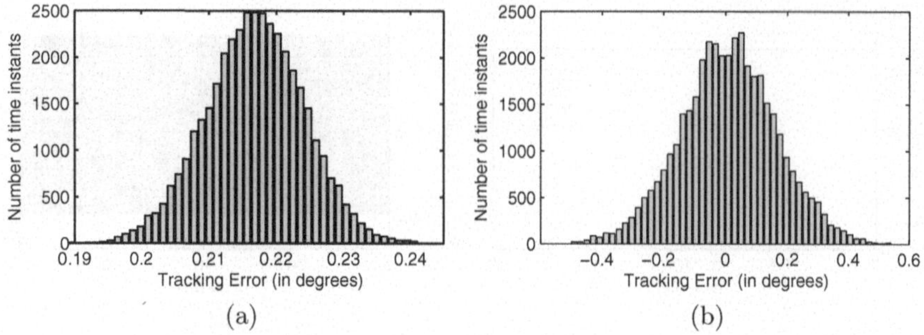

Fig. 6 Histogram of tracking error for **a** conventional BSMC, **b** BSMC with disturbance estimation

Table 3 Monte Carlo simulation results

Control law	Mean tracking error (in degrees)	Control input rms (in Nm)
CBSMC	0.2162	0.1717
DEBSMC	0.1215	0.1602

It can be observed from Fig. 6 that the tracking error is near zero at most of the instants of time for the proposed controller while it is shifted from zero for conventional BSMC. Also from Table 3, control input is marginally lesser for the proposed controller.

6 Conclusion

A backstepping-based nonlinear robust controller for underactuated 2-DoF gyroscope system has been designed and validated in the simulation environment in this paper. Proposed control law with disturbance estimation does not require knowledge of the upper bound of matched disturbance and can stabilize the system globally asymptotically while upper bound should be known to stabilize the system for conventional law. Possibility of singularity problem in control input because of internal dynamics, i.e., the angular position of blue gimbal, has been removed by keeping it bounded, and it is well documented in the stability section of this paper. Monte Carlo simulation method is used to compare the proposed control law with conventional law, which shows tracking error is closer to zero at most instants of time for the proposed controller than for conventional one with lesser control effort.

References

1. Polo, M.F.P., Molina, M.P.: Regular self-oscillating and chaotic behaviour of a pid controlled gimbal suspension gyro. Chaos Solitons Fract. **21**(5), 1057–1074 (2004)
2. Keshtkar, S., Moreno, J.A., Kojima, H., Uchiyama, K., Nohmi, M., Takaya, K.: Spherical gyroscopic moment stabilizer for attitude control of microsatellites. Acta Astronaut. **143**, 9–15 (2018)
3. Krzysztofik, I., Takosoglu, J., Koruba, Z.: Selected methods of control of the scanning and tracking gyroscope system mounted on a combat vehicle. Annu. Rev. Control. **44**, 173–182 (2017)
4. Jia, Y., Misra, A.K.: Robust trajectory tracking control of a dual-arm space robot actuated by control moment gyroscopes. Acta Astronaut. **137**, 287–301 (2017)
5. Yime, E., Moreno, H., Saltarén, R., Aracil, R.: Design of a cmg for underwater robots. In: OCEANS 2011 IEEE-Spain, pp. 1–6. IEEE (2011)
6. Yun, S.Y., Lee, W.S., Gwak, K.W.: Cmg-based anthropomorphic test device for human rider behavior reproduction for two-wheeled self-balancing personal mobility. Mechatronics **69** (2020)
7. Montoya-Cháirez, J., Santibáñez, V., Moreno-Valenzuela, J.: Adaptive control schemes applied to a control moment gyroscope of 2 degrees of freedom. Mechatronics **57**, 73–85 (2019)
8. Adhikary, N., Mahanta, C.: Integral backstepping sliding mode control for underactuated systems: swing-up and stabilization of the cart-pendulum system. ISA Trans. **52**(6), 870–880 (2013)
9. Mishra, A., Sinha, N.K.: Design and real time implementation of sliding mode and adaptive fuzzy control on quanser gyroscope. In: 2019 Fifth Indian Control Conference (ICC), pp. 312–317. IEEE (2019)
10. García, J.E.A., de Loza, A.F., Aguilar, L.T., Verdés, R.I.: Active disturbance rejection for a three degrees of freedom gyroscope. IFAC-PapersOnLine **51**(13), 372–377 (2018)
11. Toriumi, F.Y., Angélico, B.A.: Sliding mode control applied to a multivariate underactuated control moment gyroscope. In: 2019 IEEE 58th Conference on Decision and Control (CDC), pp. 4928–4933. IEEE (2019)
12. Rudra, S., Barai, R.K.: Design of block backstepping based nonlinear state feedback controller for pendubot. In: 2016 IEEE First International Conference on Control, Measurement and Instrumentation (CMI), pp. 1–5. IEEE (2016)
13. Rudra, S., Barai, R.K., Maitra, M., Mandal, D., Dam, S., Ghosh, S., Bhattacharyya, P., Dutta, A.: Design of nonlinear state feedback control law for underactuated tora system: a block backstepping approach. In: 2013 7th International Conference on Intelligent Systems and Control (ISCO), pp. 93–98. IEEE (2013)
14. Almakhles, D.J.: Robust backstepping sliding mode control for a quadrotor trajectory tracking application. IEEE Access **8**, 5515–5525 (2019)
15. Labbadi, M., Cherkaoui, M.: Robust adaptive backstepping fast terminal sliding mode controller for uncertain quadrotor uav. Aerosp. Sci. Technol. **93** (2019)
16. Nizami, T.K., Chakravarty, A., Mahanta, C.: Analysis and experimental investigation into a finite time current observer based adaptive backstepping control of buck converters. J. Franklin Inst. **355**(12), 4996–5017 (2018)
17. Nizami, T.K., Chakravarty, A., Mahanta, C.: Design and implementation of a neuro-adaptive backstepping controller for buck converter fed pmdc-motor. Control Eng. Pract. **58**, 78–87 (2017)
18. Olfati-Saber, R.: Nonlinear control of underactuated mechanical systems with application to robotics and aerospace vehicles. Ph.D. thesis, Massachusetts Institute of Technology (2001)
19. Khalil, H.K.: Nonlinear Systems. Macmilllan, London (1996)
20. Goldstein, H., Poole, C., Safko, J.: Classical mechanics (2002)

21. Hung, J.Y., Gao, W., Hung, J.C.: Variable structure control: a survey. IEEE Trans. Indust. Electron. **40**(1), 2–22 (1993)
22. Rudra, S., kumar Barai, R., Maitra, M., Mandal, D., Ghosh, S., Dam, S., Dutta, A., Bhattacharyya, P.: Stabilization of furuta pendulum: a backstepping based hierarchical sliding mode approach with disturbance estimation. In: 2013 7th International Conference on Intelligent Systems and Control (ISCO), pp. 99–105. IEEE (2013)

Communication and Control for Robotic Systems

Adaptive Backstepping Control of Multiple Mobile Robots Under Limited Communication; An Event-Triggered Approach

Sami Al Issa and Indrani Kar

Abstract Consensus problem of a leader–follower network of multiple mobile agents under limited communication is considered in this paper. Both kinematic and dynamic models are considered for mobile robot agents. Based on backstepping method, an adaptive consensus controller is designed to handle the uncertainties in each agent of the system. To deal with limited bandwidth restrictions, an event-triggered strategy is incorporated at the controller end resulting in reducing the communication burden. The event-triggered conditions are derived by ensuring the existence of an appropriate Lyapunov function. It is proved that under the proposed control scheme, all agents are practically converged to the leader, and all system signals are bounded. Simulation results show the effectiveness of proposed event-triggered scheme in terms of bandwidth utilization compared to traditional time-triggered control.

1 Introduction

Recent years have witnessed a compelling attention and extensive studies on multi-agent systems (See [1] for mobile robots, [2] for quad-rotors and [3] for smart sensors networks). Without the cooperation of these agents, many practical applications and complex control tasks would be difficult or even not possible to be achieved (e.g., large object manipulating [4], search and rescue in large-scale areas [5], security [6] and many others). Nevertheless, the communications between these agents are generally occurred periodically over a shared network which has a limited bandwidth

Supported in part by the Fellowship from Damascus University, Syrian Arab Republic.

S. Al Issa (✉) · I. Kar
Department of Electronics and Electrical Engineering, IIT Guwahati, Guwahati, India
e-mail: sami.issa@iitg.ac.in

I. Kar
e-mail: indranik@iitg.ac.in

S. Al Issa
Department of Computer Engineering and Automation, Damascus University, Damascus, Syria

J. Gu et al. (eds.), *Communication and Control for Robotic Systems*, Smart Innovation, Systems and Technologies 229, https://doi.org/10.1007/978-981-16-1777-5_7

channel. This periodic transmissions undoubtedly lead to inefficient utilization of network resources [7]. Besides, the uncertainties encountered in the agents may result in deterioration of system performance or even destabilize the overall closed-loop system [8].

Consensus problem is the major topic in multi-agent systems, and it is indeed challenging. In consensus, states of all agents converge to a common desired quantity. This states agreement can be achieved in two manners, namely leader–follower and leaderless scenario. Due to its wide applications and natural physical meaning, leader–follower scenario has been widely attempted by control theorists and practitioners [9]. In literature, consensus of linear multi-agent systems has been extensively investigated [10–12]. However, practical engineering systems are nonlinear in nature and affected by many factors that increase the complexity of control design and system analysis. Recently, some efforts have been devoted to investigate nonlinear multi-agent systems under some practical circumstances [13, 14]. In [13], Zhao et al. proposed an adaptive sliding mode control scheme based on backstepping for multi-agent systems. Trajectory tracking problem of multi-agent mobile robots was further studied under this control scheme. An adaptive fuzzy controller was designed for a class of strict feedback nonlinear multi-agent systems in [14]. However, all aforementioned studies are generally based on classical time-triggered implementation which is inefficient and inappropriate especially for practical situations wherein the shared bandwidth of network channel is limited.

In traditional time-triggered control, communications between agents occur periodically with no consideration for their behaviors. In other words, data are continuously transmitted over the network (even if it is not necessary) just because a fixed amount of time is elapsed, called sampling time (Δt). It is worth mentioning that this sampling time is commonly chosen by ensuring the system stability under worst conditions. Although control design and analysis under time-triggered implementation are simple and straightforward, it is evident that this periodic mechanism results in inefficient utilization of network resources. To handle this limitation, an alternative mechanism called event-triggered mechanism was proposed and proved to be more efficient [15]. In this scheme, communications over the network are subjected to triggering condition. In case this condition gets violated, data are to be updated and transmitted over the network in order to maintain the system performance and stability. On the other hand, no need for data update/transmission if the triggering condition is satisfied. In [16], event-triggered sliding mode control scheme was investigated for a class of nonlinear systems. Xing et al. designed an event-triggered adaptive control scheme for a class of strict feedback nonlinear uncertain systems based on fixed and relative threshold in [17]. They further proposed a switching strategy between these two thresholds that shows more saving in channel bandwidth utilization. In [18], we have derived the triggering condition based on Lyapunov analysis, and the proposed scheme shows more reduction in number of transmission over the network compared to [16, 17].

In the context of multi-agent mobile robot systems, it is observed that most studies in literature consider only the kinematic model of the robot [19–21]. Nevertheless, kinematic-based controllers are inadequate and do not consider the internal parameters of the robot. Besides, the uncertainties associated with the dynamic model may deteriorate the system performance or even results in instability. Hence, it is more appropriate to incorporate the dynamic model as well in order to design more accurate controllers [22]. In [23], trajectory tracking controller based on backstepping is developed for non-holonomic mobile robot considering both kinematic and dynamic models. To deal with parameter uncertainties in the dynamic model, an adaptive controller is designed in [24] under event-triggered implementation. It is to be mentioned that these studies consider the case of a single mobile robot controlled over a communication network. The results of these works are extended to multi-agent mobile robots in this paper.

Based on foregoing discussion, the consensus control problem for uncertain multiple mobile robots communicated over a limited bandwidth channel is attempted in this paper. Twofold design objective is accomplished under the proposed control scheme in terms of achieving the consensus with minimum communication. The main contributions are as follows: (1) An adaptive controller based on backstepping method is designed to handle the unknown parameters in the dynamic model of followers achieving the consensus tracking with the leader. (2) An event-triggered mechanism is incorporated to reduce the communication burden resulting in more saving in channel bandwidth. The Zeno behavior under the proposed event-triggered condition is proved to be excluded.

2 Problem Formulation

In this paper, a multi-mobile robots system is considered which consists of one leader and N follower agents. The kinematic and dynamic models of ith follower can be written in a strict feedback form as follows (See [24] for more details):

$$\begin{aligned} \dot{q}_i &= S_i(q_i)\nu_i, \\ \dot{\nu}_i &= \varphi_i(\nu_i)\theta_i + g_i\bar{u}_i \,. \end{aligned} \tag{1}$$

The first equation of (1) represents the kinematic model of the robot, where $q_i = [x_i\ y_i\ \theta_i]^T$ denotes its position and orientation and $\nu_i = [v_i\ w_i]^T$ represents its linear and angular velocities. The decoupling matrix S_i is given as

$$S_i(q) = \begin{bmatrix} \cos\theta_i & 0 \\ \sin\theta_i & 0 \\ 0 & 1 \end{bmatrix}.$$

Further, the dynamic model of the robot can be represented by the second equation of (1) where

$$\varphi_i(v_i) := \begin{bmatrix} w_i^2 & 0 \\ 0 & -v_i w_i \end{bmatrix}, \quad g_i := \begin{bmatrix} \frac{1}{m_i r_i} & \frac{1}{m_i r_i} \\ \frac{b_i}{d_i^2 m_i + I_i} & \frac{-b_i}{d_i^2 m_i + I_i} \end{bmatrix} \tag{2}$$

The vector $\theta_i = \begin{bmatrix} d_i & \frac{m_i d_i}{m_i d_i^2 + I_i} \end{bmatrix}^T$ represents the unknown parameters vector, and $u_i = \begin{bmatrix} \tau_r & \tau_l \end{bmatrix}^T$ is the control signal characterized by the torques on the right and left motors of the robot, respectively. In these expressions, r_i, m_i and I_i denote the radius of wheels, mass and inertia of the robot, respectively. The distances b_i and d_i are illustrated in Fig. 1.

Similarly, the mathematical model of the leader is considered as

$$\begin{aligned} \dot{q}_0 &= S_0(q_0)v_0, \\ \dot{v}_0 &= \varphi_0(v_0)\theta_0 + g_0 u_0 \,. \end{aligned} \tag{3}$$

Assumption 1 It is assumed that the desired trajectory and its time derivative are available only for the leader. Moreover, each agent needs to interact only with the leader and has its position and orientation.

Remark 1 It is to be noted from (1) that followers are not actuated by the actual control signal (u_i). Instead, they are actuated by $\bar{u}_i$. This is due to event-triggered implementation where $\bar{u}_i = u_i - e_i$. The measurement error signal (e_i) is utilized to design the event-triggered condition in Sect. 3.2.

The control problem at this juncture is stated as follows: designing an event-triggered adaptive control scheme for uncertain multi-mobile robots such that all agents eventually converge to the leader with minimum number of transmissions.

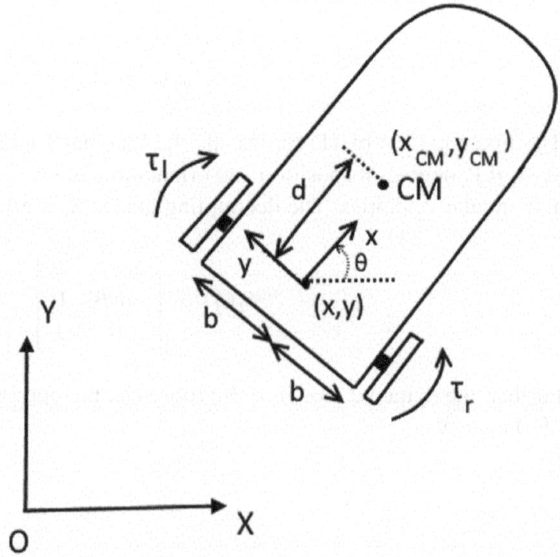

Fig. 1 Wheeled non-holonomic mobile robot

3 Control Design

Let us define the consensus error for each ith follower as

$$e_i^q = q_i - q_0, \tag{4}$$
$$e_i^\nu = S_i \nu_i - S_0 \nu_0 \,. \tag{5}$$

Let us also define $E_q = [(e_1^q)^T (e_2^q)^T \ldots (e_n^q)^T]^T$ and $E_\nu = [(e_1^\nu)^T (e_2^\nu)^T \ldots (e_n^\nu)^T]^T$. Then, the overall error dynamics can be written in a matrix form as follows:

$$\dot{E}_q = E_\nu, \tag{6}$$
$$\dot{E}_\nu = S_f \Phi_f \Theta_f + S_f G_f U_f - I S_0 \varphi_0 \theta_0 - I S_0 g_0 u_0 + \dot{S}_f V_f - I \dot{S}_0 \nu_0 \,. \tag{7}$$

where

$$S_f = \begin{bmatrix} S_1 & 0 & \ldots & 0 \\ 0 & S_2 & \ldots & 0 \\ \vdots & \vdots & & \vdots \\ 0 & 0 & \ldots & S_n \end{bmatrix}, \quad \Phi_f = \begin{bmatrix} \varphi_1 & 0 & \ldots & 0 \\ 0 & \varphi_2 & \ldots & 0 \\ \vdots & \vdots & & \vdots \\ 0 & 0 & \ldots & \varphi_n \end{bmatrix}, \quad G_f = \begin{bmatrix} G_1 & 0 & \ldots & 0 \\ 0 & G_2 & \ldots & 0 \\ \vdots & \vdots & & \vdots \\ 0 & 0 & \ldots & G_n \end{bmatrix},$$

and

$$\Theta_f = \begin{bmatrix} \theta_1 \\ \theta_2 \\ \vdots \\ \theta_n \end{bmatrix}, \quad U_f = \begin{bmatrix} u_1 \\ u_2 \\ \vdots \\ u_n \end{bmatrix}, \quad I = \begin{bmatrix} I_3 \\ I_3 \\ \vdots \\ I_3 \end{bmatrix}, \quad V_f = \begin{bmatrix} \nu_1 \\ \nu_2 \\ \vdots \\ \nu_n \end{bmatrix}.$$

Note that, $E_q \in R^{3n}$, $E_\nu \in R^{3n}$, $S_f \in R^{3n \times 2n}$, $\Phi_f \in R^{2n \times 2n}$ and $G_f \in R^{2n \times 2n}$. Moreover, the vectors Θ_f, U_f, $V_f \in R^{2n}$ and $I \in R^{3n \times 3}$ where $I_3 \in R^{3 \times 3}$ is the identity matrix.

3.1 Adaptive Backstepping Controller

In this subsection, the controller is designed based on backstepping method which is a powerful technique largely utilized in literature for stabilizing nonlinear systems. One of the advantages of this method is the systematic and standard construction of Lyapunov function for overall closed loop. Let us first define the error variables as

$$Z_1 = E_q, \tag{8}$$
$$Z_2 = E_\nu - \alpha \,. \tag{9}$$

Step 1 The dynamics of first error variable can be written as

$$\dot{Z}_1 = \dot{E}_q = E_v = Z_2 + \alpha \tag{10}$$

The Lyapunov function candidate is chosen at the first step as

$$V_1 = \frac{1}{2} Z_1^T Z_1, \tag{11}$$

whose derivative is

$$\dot{V}_1 = Z_1^T \dot{Z}_1 \tag{12}$$

Substituting (10) in (12), one gets

$$\dot{V}_1 = Z_1^T Z_2 + Z_1^T \alpha \tag{13}$$

The virtual control is now designed as

$$\alpha = -C_1 Z_1, \tag{14}$$

which renders the derivative of the Lyapunov function as

$$\dot{V}_1 = -Z_1^T C_1 Z_1 + Z_1^T Z_2 \tag{15}$$

It is to be noted that the negative definiteness of $\dot{V}_1$ and convergence of Z_1 are ensured if $Z_2 = 0$ which will be stabilize in next step.
Step 2 In this step, the Lyapunov function candidate is chosen as

$$V_2 = V_1 + \frac{1}{2} Z_2^T Z_2 + \frac{1}{2} \tilde{\Theta}_f^T \Gamma_1^{-1} \tilde{\Theta}_f + \text{trace}\left(\frac{1}{2} \tilde{G}_f^T \Gamma_2^{-1} \tilde{G}_f \right), \tag{16}$$

where $\tilde{\Theta}_f = \Theta_f - \hat{\Theta}_f$ and $\tilde{G}_f = G_f - \hat{G}_f$ are the estimation errors. The derivative of Lyapunov function candidate is given as

$$\dot{V}_2 = -Z_1^T C_1 Z_1 + Z_1^T Z_2 + Z_2^T \dot{Z}_2 - \tilde{\Theta}_f^T \Gamma^{-1} \dot{\hat{\Theta}}_f + \text{trace}\left(-\tilde{G}_f^T \Gamma_2^{-1} \dot{\hat{G}}_f \right), \tag{17}$$

where the dynamics of second error variable can be written as follows:

$$\begin{aligned} \dot{Z}_2 &= \dot{E}_v - \dot{\alpha} \\ &= S_f \Phi_f \Theta_f + S_f G_f U_f - I S_0 \varphi_0 \theta_0 - I S_0 g_0 u_0 + \dot{S}_f V_f - I \dot{S}_0 v_0 + C_1 \dot{Z}_1 \end{aligned} \tag{18}$$

Adding and subtracting $S_f \hat{G}_f U_f$ from (18), one gets

$$\begin{aligned}\dot{Z}_2 =& S_f\Phi_f\Theta_f + S_f G_f U_f - IS_0\varphi_0\theta_0 - IS_0 g_0 u_0 + \dot{S}_f V_f - I\dot{S}_0 v_0 + C_1\dot{Z}_1 \\ & + S_f\hat{G}_f U_f - S_f\hat{G}_f U_f \\ =& S_f\Phi_f\Theta_f + S_f\hat{G}_f U_f - IS_0\varphi_0\theta_0 - IS_0 g_0 u_0 + \dot{S}_f V_f - I\dot{S}_0 v_0 + C_1\dot{Z}_1 \\ & + S_f\tilde{G}_f U_f \end{aligned} \tag{19}$$

In order to render $\dot{V}_2$ negative definite, the actual control law is now designed as follows

$$\begin{aligned} S_f\hat{G}_f U_f =& - C_2Z_2 - Z_1 - C_1\dot{Z}_1 - S_f\Phi_f\hat{\Theta}_f + IS_0(\varphi_0\theta_0 - g_0u_0) \\ & - \dot{S}_f V_f + I\dot{S}_0 v_0 \end{aligned} \tag{20}$$

Since $S_f^T.S_f = I \in R^{2n\times 2n}$, we pre-multiply both sides of (20) by ${S_f}^T$ and get

$$\begin{aligned} \hat{G}_f U_f =& - S_f^T(C_2Z_2 + Z_1 + C_1\dot{Z}_1 + S_f\Phi_f\hat{\Theta}_f - IS_0(\varphi_0\theta_0 + g_0u_0) \\ & + \dot{S}_f V_f - I\dot{S}_0 v_0) \end{aligned} \tag{21}$$

The control law is now found as

$$\begin{aligned} U_f =& - \hat{G}_f^{-1} S_f^T(C_2Z_2 + Z_1 + C_1\dot{Z}_1 + S_f\Phi_f\hat{\Theta}_f - IS_0(\varphi_0\theta_0 + g_0u_0) \\ & + \dot{S}_f V_f - I\dot{S}_0 v_0) \end{aligned} \tag{22}$$

Substituting (22) in (19) and the result in (17), one gets

$$\begin{aligned} \dot{V}_2 =& - Z_1^T C_1 Z_1 - Z_2^T C_2 Z_2 + Z_2^T S_f\Phi_f\tilde{\Theta}_f + Z_2^T S_f\tilde{G}_f U_f \\ & - \tilde{\Theta}_f^T\Gamma^{-1}\dot{\hat{\Theta}}_f + \text{trace}\left(-\tilde{G}_f^T\Gamma_2^{-1}\dot{\hat{G}}_f\right) \\ =& -Z_1^T C_1 Z_1 - Z_2^T C_2 Z_2 + \tilde{\Theta}_f^T\left(\Phi_f^T S_f^T Z_2 - \Gamma^{-1}\dot{\hat{\Theta}}_f\right) \\ & + \text{trace}\left(\tilde{G}_f^T S_f^T Z_2 U_f^T - \tilde{G}_f^T\Gamma_2^{-1}\dot{\hat{G}}_f\right) \end{aligned} \tag{23}$$

With the adaptation law designed as

$$\dot{\hat{\Theta}}_f = \Gamma_1\left(\Phi_f^T S_f^T Z_2\right), \tag{24}$$

$$\dot{\hat{G}}_f = \Gamma_2\left(S_f^T Z_2 U_f^T\right), \tag{25}$$

the derivative of Lyapunov function can be written as

$$\dot{V}_2 = -Z_1^T C_1 Z_1 - Z_2^T C_2 Z_2 \tag{26}$$

3.2 Event-Triggered Mechanism

Instead of periodic control update, an event-triggered condition is placed now at the control end and checked periodically to determine when it is necessary to update the control signal. A new control input is transmitted over the network just in case of violation of triggering condition. Otherwise, the robot is actuated by the last transmitted control input. Undoubtedly, the event-triggered mechanism reduces the communication burden and results in more efficient utilization of network resources compared to traditional time-triggered control.

In light of the above discussion, the robots in this scenario are not actuated by the actual control signal (u_i). Instead, it is driven by $\bar{u}_i = u_i - e_i$ where e_i represents the measurement errors between the actual and event-triggered control signals. With integration of event-triggered mechanism, a new term related to measurement errors will appear in the derivative of Lyapunov function as follows:

$$\dot{V}_2 = -Z_1^T C_1 Z_1 - Z_2^T C_2 Z_2 - Z_2^T S_f \hat{G}_f E_f, \tag{27}$$

where $E_f = [e_1^T e_2^T \dots e_n^T]^T \in R^{2n}$. Defining $Z_1 = [z_{11}^T z_{12}^T \dots z_{1n}^T]^T$ and $Z_2 = [z_{21}^T z_{22}^T \dots z_{2n}^T]^T$, Eq. (27) can be written in a distributed form, and then the event-triggered condition for each follower can be obtained by ensuring $\dot{V}_{2i} \leq 0$ as follows.

$$\begin{aligned} \dot{V}_{2i} &= - z_{1i}^T C_{1i} z_{1i} - z_{2i}^T C_{2i} z_{2i} - z_2^T S_i \hat{g}_i e_i \\ &\leq -z_{1i}^T C_{1i} z_{1i} - z_{2i}^T C_{2i} z_{2i} + \|z_{2i}^T\| \|\hat{g}_i\| \|e_i\|, \end{aligned} \tag{28}$$

where $\|S_i\| = 1$. Applying Peter—Paul inequality on the third term, one gets

$$\dot{V}_{2i} \leq -z_{1i}^T C_{1i} z_{1i} - z_{2i}^T C_{2i} z_{2i} + \frac{1}{2\zeta} \|z_{2i}^T\|^2 \|\hat{g}_i\|^2 + \frac{\zeta \|e_i\|^2}{2} \tag{29}$$

where $\zeta > 0$ is a user-defined parameter. At this juncture, event-triggered condition that maintains the negative definiteness of $\dot{V}_{2i}$ can be formulated as

$$\|e_i\|^2 > \frac{2}{\zeta} (z_{1i}^T C_{1i} z_{1i} + z_{2i}^T C_{2i} z_{2i} - \frac{1}{2\zeta} \|z_{2i}^T\|^2 \|\hat{g}_i\|^2) \tag{30}$$

Theorem 1 *Under the action of control law* (22)*, adaptation law* (24) *and event-triggered condition* (30)*, all closed-loop signals are bounded and all followers converge to the leader with a bounded tracking error. Moreover, a minimum inter-event time* ($t_{k+1} - t_k$) *is ensured.*

Proof We choose a Lyapunov function candidate as $V = \frac{1}{2} {V_1}^2 + \frac{1}{2} {V_2}^2$. From Sect. 3, the derivative of Lyapunov function can be written as

$$\dot{V} = -Z^T\bar{C}Z + \frac{1}{2}\|E_f\|^2$$
$$\leq -\lambda_{\min}(\bar{C})V + K(E_f) \quad (31)$$

where $\bar{C} = \text{diag}(C_1, C_2, \frac{1}{2\zeta}\|G_f\|)$, and $K(E_f)$ is a class of κ function. Thus, the definition of input to state stability (ISS) with respect to the measurement errors $\|E_f\|$ as input is satisfied, and all signals are bounded under the proposed control scheme (See [25]). Further, the existence of a lower bound of inter-event time $(t_{k+1} - t_k)$ can be proven as follows [17]:

We have $e_i = u_i - \bar{u}_i$ that represents the measurement error between the actual (u_i) and event-triggered $(\bar{u}_i)$ control signals. We also have

$$\frac{\mathrm{d}}{\mathrm{d}t}|e_i| = \text{sign}(e_i)\dot{e}_i \leq |\dot{u}_i| \quad (32)$$

Moreover, we have $|\dot{u}_i| \leq \epsilon$ holds for any $\epsilon > 0$ as u_i is differentiable and bounded. Since $e_i(t_k) = 0$ and $\lim_{t \to t_{k+1}} e_i(t) = \eta$, we can ensure a lower bound of inter-event time $(t_{k+1} - t_k)$ which is found to be η/ϵ.

4 Simulation Results

To demonstrate the proposed control scheme, simulation study for multi-robot system consisting of three followers and one leader has been carried out in MATLAB for a total time of $T = 45$ s with sampling time $\mathrm{d}t = 0.02$.

The parameters of the robots are chosen as $m_i = 10\,\text{kg}$, $r_i = 0.1\,\text{m}$, $b_i = 0.5\,\text{m}$, $d_i(0) = 0.5\,\text{m}$ and $I_i(0) = 2.5\ \text{kgm}^2$. The uncertainties are defined as $d_i = d_i(0) + 0.2\sin(2t)$ and $I_i = I_i(0) + 0.2\sin(t)$. Circular trajectory is selected as reference with radius $R_r = 1$ m and angular speed $W_r = 0.5$ rad/s. The initial positions and orientations for the followers are set as $x_i = [0.2, -0.2, -0.2]$, $y_i = [0.6, 0.8, 1.3]$ and $\theta_i = [20, 20, 0]$. The controller parameters are tuned using trial and error method in such a way that minimizes the number of transmissions over the network in tandem with an acceptable tracking performance, and they are chosen as $C_{11} = \text{diag}(0.9, 1, 4)$, $C_{21} = \text{diag}(4, 4, 2)$, $C_{12} = \text{diag}(2, 2, 6)$, $C_{22} = \text{diag}(8, 8, 6)$, $C_{13} = \text{diag}(1, 0.4, 6)$, $C_{23} = \text{diag}(4, 4, 1)$. Other parameters are chosen as $\zeta_i = [0.5, 0.7, 0.9]$ and $\Gamma = 10$.

The trajectory tracking in X-Y plane for a leader–follower mobile robots is shown in Fig. 2. The performance and consensus errors in X, Y, Θ directions are shown in Fig. 3. It can be observed that all followers are able to synchronize to the leader in terms of outputs. Moreover, the consensus errors are bounded and converged to a small set around zero. The performance index is calculated in the time interval $[\frac{1}{2}T, T]$, i.e., after the transient period as

$$\text{Performance Index} = \max\ (|x_i - x_0|, |y_i - y_0|, |\theta_i - \theta_0|) \quad (33)$$

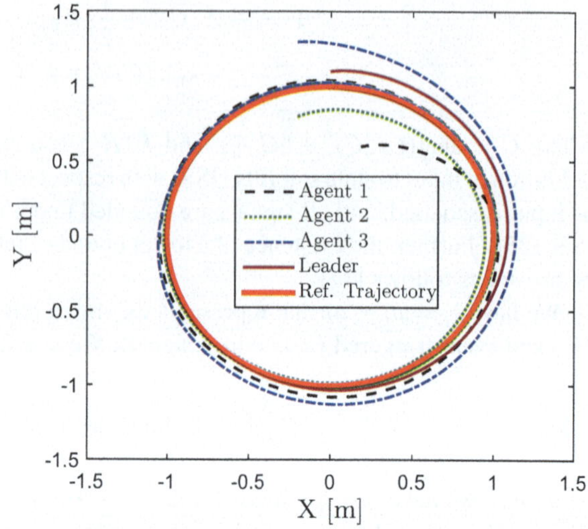

Fig. 2 Trajectory tracking in *X*-*Y* plane of leader–follower multi-agent systems

The time-triggered and event-triggered control signals for each agent are shown in Fig. 4 for right wheels and Fig. 5 for left wheels, respectively. The triggering instants when the designed condition is violated are illustrated for each follower in Fig. 6. At these time instants, the control signals are updated, and the motors are hence actuated by the actual signals. It is to be mentioned that the number of transmissions under classical time-triggered control is equal to $\frac{T}{\mathrm{d}t} = 2250$ since it is continuously occurred every $\mathrm{d}t$. It can be observed that the number of transmissions is dramatically reduced under the proposed event-triggered mechanism. Hence, more saving in the shared limited bandwidth is compared to time-triggered implementation. The bandwidth usage is calculated as

$$\text{Bandwidth Usage} = 100 \times \text{Number of updates} \times \frac{\mathrm{d}t}{T} \tag{34}$$

Furthermore, the inter-event time between two consecutive triggering instants ($t_k - t_{k-1}$) is shown in Fig. 7. The existence of a lower bound of this inter-event time is also proved theoretically in Sect. 3.2. The average and maximum values of inter-event time in addition to main simulation results are summarized in Table 1.

Table 1 Summarization of simulation results

	Number of updates	Bandwidth usage (%)	Performance index	Inter-event time (s)	
				Average	Maximum
Agent 1	257	11.42	0.028	0.17	1.18
Agent 2	207	9.2	0.022	0.22	1
Agent 3	301	13.38	0.025	0.15	0.78

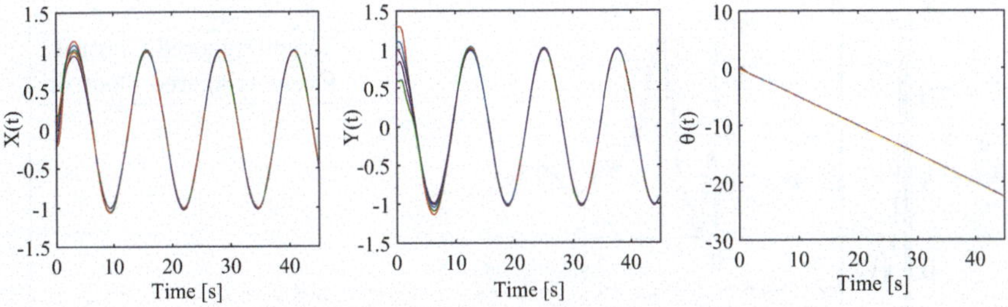

(a) Trajectory tracking in X,Y and θ directions.

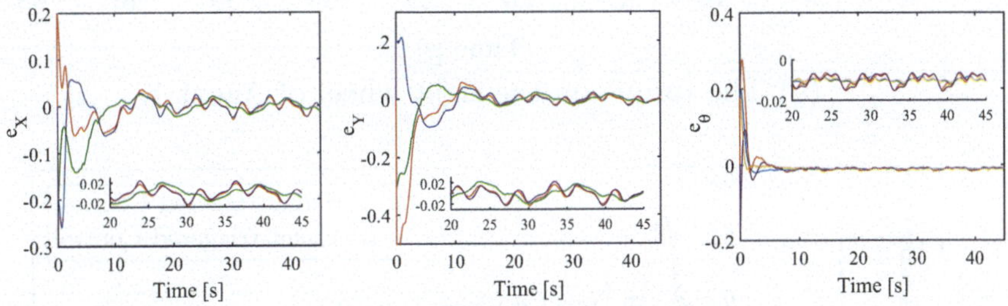

(b) Consensus errors in X,Y and θ directions.

Fig. 3 Performance and consensus error in X, Y and Θ directions

5 Conclusion

The consensus tracking control problem is investigated in this paper for a leader–follower network of multi-robot system under limited communication. For that, an adaptive backstepping controller is designed considering both kinematic and dynamic models of mobile robots. Further, event-triggered strategy based on Lyapunov analysis is incorporated resulting in more saving of channel bandwidth. It is observed from simulations that all agents are practically converged to the leader with less number of transmissions over the network. Future works include analyzing the effects of time delays and experimental studies on real mobile robots.

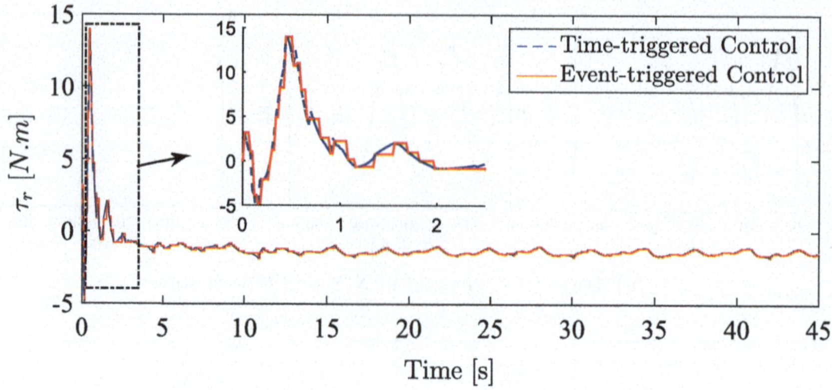

(a) The torque on the right wheel of Agent 1.

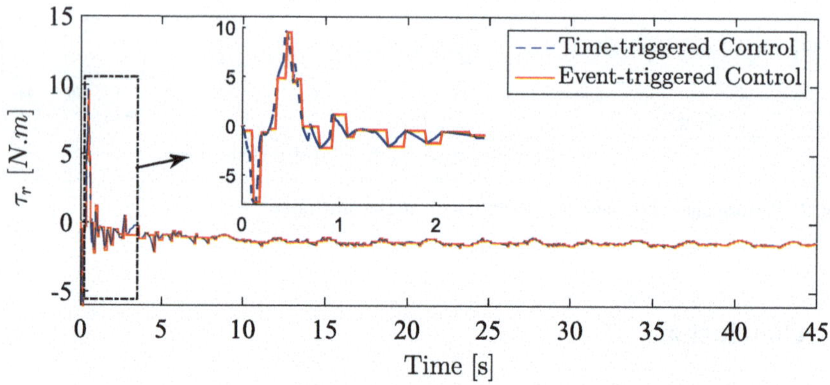

(b) The torque on the right wheel of Agent 2.

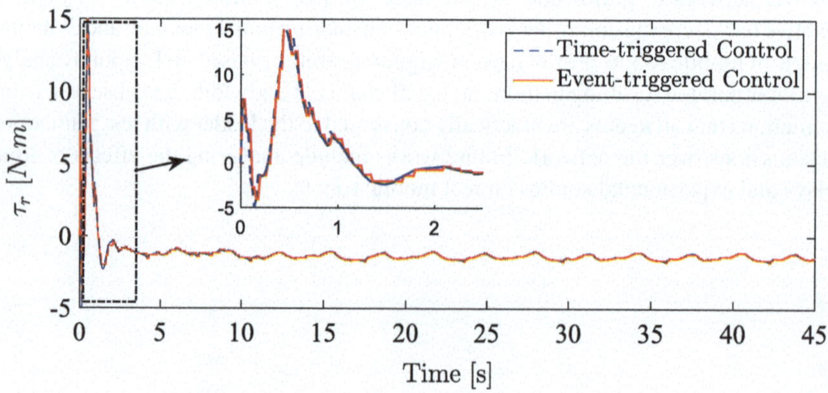

(c) The torque on the right wheel of Agent 3.

Fig. 4 Control signals of the right motor for each follower

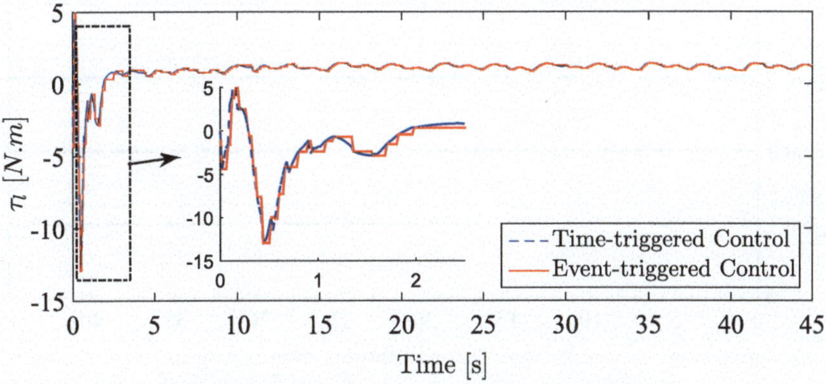

(a) The torque on the left wheel of Agent 1.

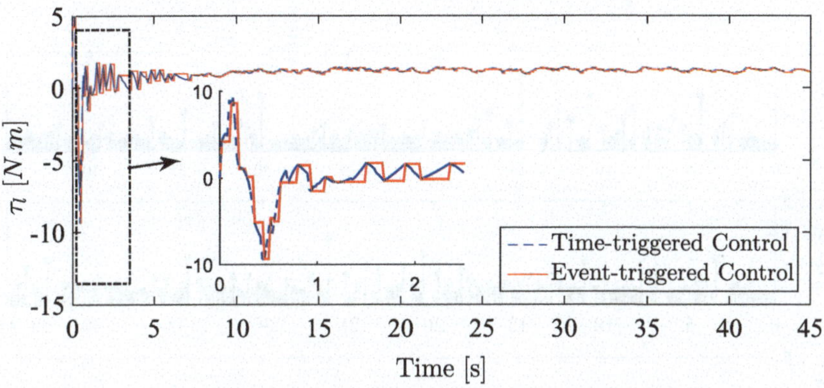

(b) The torque on the left wheel of Agent 2.

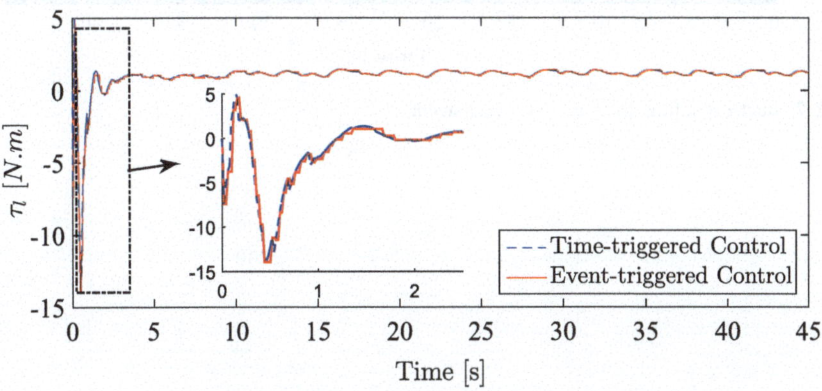

(c) The torque on the left wheel of Agent 3.

Fig. 5 Control signals of the left motor for each follower

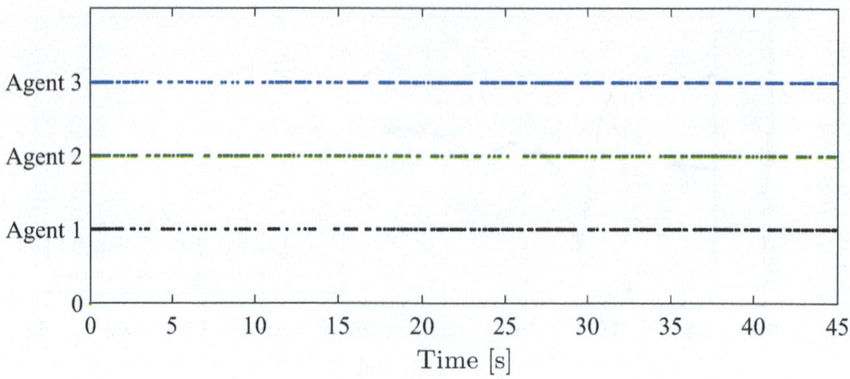

Fig. 6 Illustration of triggering events for each agent

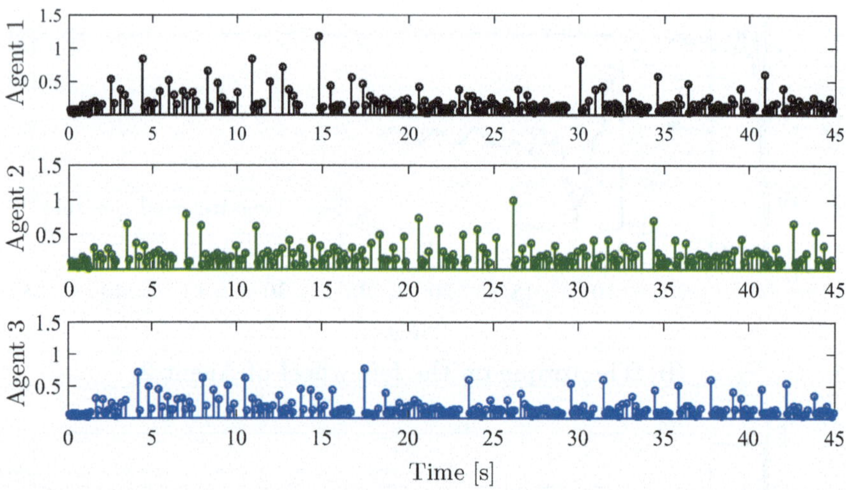

Fig. 7 Inter-event time $(t_k - t_{k-1})$ for each agent

References

1. Hsu, H.C.-H., Liu, A.: Multiagent-based multi-team formation control for mobile robots. J. Intell. Robot. Syst. **42**(4), 337–360 (2005)
2. Dong, X., Zhou, Y., Ren, Z., Zhong, Y.: Time-varying formation tracking for second-order multi-agent systems subjected to switching topologies with application to quadrotor formation flying. IEEE Trans. Indus. Electron. **64**(6), 5014–5024 (2016)
3. Lesser, V., Ortiz, C.L., Jr., Tambe, M.: Distributed Sensor Networks: A Multiagent Perspective, vol 9. Springer Science & Business Media (2012)
4. Fyler, D., Sullivan, B., Raptis, I.A.: Distributed object manipulation using a mobile multi-agent system. In: 2015 IEEE International Conference on Technologies for Practical Robot Applications (TePRA), pp. 1–6. IEEE (2015)
5. Kitano, H., Tadokoro, S., Noda, I., Matsubara, H., Takahashi, T., Shinjou, A., Shimada, S.: Robocup rescue: search and rescue in large-scale disasters as a domain for autonomous agents research. In: IEEE SMC'99 Conference Proceedings. 1999 IEEE International Conference on Systems, Man, and Cybernetics (Cat. No. 99CH37028), vol 6, pp. 739–743. IEEE (1999)
6. Tapia, D.I., De Paz, J.F., Rodriguez, S., Bajo, J., Corchado, J.M.: Multi-agent system for security control on industrial environments. Int. Trans. Syst. Sci. Appl. J. **4**(3), 222–226 (2008)
7. Ding, L., Han, Q.-L., Ge, X., Zhang, X.-M.: An overview of recent advances in event-triggered consensus of multiagent systems. IEEE Trans. Cybernet. **48**(4), 1110–1123 (2017)
8. Wang, J.: Distributed coordinated tracking control for a class of uncertain multiagent systems. IEEE Trans. Automat. Control **62**(7), 3423–3429 (2016)
9. Huang, J., Wang, W., Wen, C., Zhou, J., Li, G.: Distributed adaptive leader-follower and leaderless consensus control of a class of strict-feedback nonlinear systems: a unified approach. Automatica **118** (2020)
10. Liu, D., Yang, G.-H.: A dynamic event-triggered control approach to leader-following consensus for linear multiagent systems. IEEE Trans. Syst. Man Cybernet, Syst (2020)
11. Zhang, D., Feng, G.: A new switched system approach to leader-follower consensus of heterogeneous linear multiagent systems with dos attack. IEEE Trans. Syst. Man Cybernet, Syst (2019)
12. Cai, Y., Zhang, H., Zhang, J., He, Q.: Distributed bipartite leader-following consensus of linear multi-agent systems with input time delay based on event-triggered transmission mechanism. ISA Trans. **100**, 221–234 (2020)
13. Zhao, D., Zou, T., Li, S., Zhu, Q.: Adaptive s sliding mode control for leader-follower multi-agent systems. IET Control Theory Appl. **6**(8), 1109–1117 (2012)
14. Wang, W., Liang, H., Zhang, Y., Li, T.: Adaptive cooperative control for a class of nonlinear multi-agent systems with dead zone and input delay. Nonlinear Dyn. **96**(4), 2707–2719 (2019)
15. Tabuada, P.: Event-triggered real-time scheduling of stabilizing control tasks. IEEE Trans. Autom. Control **52**(9), 1680–1685 (2007)
16. Behera, A.K., Bandyopadhyay, B.: Event-triggered sliding mode control for a class of nonlinear systems. Int. J. Control **89**(9), 1916–1931 (2016)
17. Xing, L., Wen, C., Liu, Z., Hongye, S., Cai, J.: Event-triggered adaptive control for a class of uncertain nonlinear systems. IEEE Trans. Autom. Control **62**(4), 2071–2076 (2016)
18. Issa, S.A., Chakravarty, A., Kar, I.: Improved event-triggered adaptive control of non-linear uncertain networked systems. IET Control Theory Appl. **13**(13), 2146–2152 (2019)
19. Yang, Q., Fang, H., Cao, M., Chen, J.: Distributed trajectory tracking control for multiple nonholonomic mobile robots. IFAC-PapersOnLine **49**(4), 31–36 (2016)
20. Guinaldo, M., Fábregas, E., Farias, G., Dormido-Canto, S., Chaos, D., Sánchez, J., Dormido, S.: A mobile robots experimental environment with event-based wireless communication. Sensors **13**(7), 9396–9413 (2013)
21. Ayedi, D., Boujelben, M., Rekik, C.: A multiagent architecture for mobile robot navigation using hierarchical fuzzy and sliding mode controllers. Math. Probl. Eng. **2018** (2018)
22. Fierro, R., Lewis, F.L.: Control of a nonholomic mobile robot: Backstepping kinematics into dynamics. J. Robot. Syst. **14**(3), 149–163 (1997)

23. Al Issa, S., Sharma, M., Kar, I.: Event-triggered backstepping control scheme for networked mobile robots. In: Proceedings of the Advances in Robotics 2019, AIR 2019, New York. Association for Computing Machinery (2019)
24. Issa, S.A., Chakravarty, A., Kar, I.: Adaptive control of a networked mobile robot subject to parameter uncertainties and limited communications. In: 2019 Sixth Indian Control Conference (ICC), pp. 308–313 (2019)
25. Krstic, M., Kokotovic, P.V., Kanellakopoulos, I.: Nonlinear and Adaptive Control Design. Wiley, New York (1995)

Adaptive Robust Control of Tele-operated Master-Slave Manipulators with Communication Delay

Nabanita Adhikary, Rajeeb Dey, Muhammad Usman Asad, Jason Gu, Umar Farooq, and Rupak Dutta

Abstract In this paper, a robust controller for the bilateral teleoperation of a master-slave manipulator system is proposed. The controller is designed based on the sliding mode control method with an adaptively tuned gain to tackle the unknown uncertainty bounds of the system. A sliding surface, having a proportional integral derivative type structure, is proposed which is designed as a function of the tracking error between the master and the slave trajectory. The controller design constraints and parameter selection criteria are derived based on the analysis of the closed-loop system. The proposed method is validated using simulation performed for a master-slave system with each arm having two degrees of freedom (DoF) and a constant communication delay for position tracking of the system. The simulation shows that the controller can handle delay of up to 2.5 s while giving satisfactory tracking performance.

Supported by SPARC, MHRD.

N. Adhikary (✉) · R. Dey
Department of Electrical Engineering, National Institute of Technology Silchar, Silchar, India
e-mail: nabanita@ee.nits.ac.in

R. Dey
e-mail: rajeeb@ee.nits.ac.in

M. U. Asad · J. Gu · U. Farooq
Electrical & Computer Engineering, Dalhousie University, Halifax, Canada
e-mail: mh549096@dal.ca

J. Gu
e-mail: Jason.gu@dal.ca

U. Farooq
e-mail: umar.farooq@dal.ca

R. Dutta
Mathematics Department, NIT Agartala, Agartala, India

J. Gu et al. (eds.), *Communication and Control for Robotic Systems*, Smart Innovation, Systems and Technologies 229, https://doi.org/10.1007/978-981-16-1777-5_8

1 Introduction

In recent times due to the advancement of network communication and robotic systems, teleoperated robot manipulator systems have gained worldwide popularity in various applications such as remote surgery, robotic rehabilitation, disaster management, etc. In most of these applications, the manipulators are set up as master and slave devices whose motion and interaction forces require synchronization. To establish this synchronous motion, controllers are required in both the master and the slave devices. Such controller design is a challenging task since the robot manipulators are nonlinear systems, and in addition to this, the communication channel delay adds to the complexity of the problem.

Some recent works in the control of such bilateral system involves application of robust control [3], fuzzy logic [10], neural networks [13], adaptive control [12], etc. Among these methods, robust controllers, especially sliding mode controllers [4, 5], are widely used for such systems owing to their inherent robustness and capability to provide finite-time convergence.

In [7], an adaptive finite-time controller is proposed where a non-singular fast terminal sliding mode controller is used for the bilateral manipulation. The maximum delay bounds used in the paper is 0.7 s. In [9], a state observer-based sliding mode controller with finite-time convergence properties are used in presence of a time-varying delay with a maximum bound of 0.5 s, and their experimental results show the efficacy of the controller within this delay bound. In [14], Zhao et al. proposed an observer-based sliding mode controller with an integral sliding surface that can tackle delays up to 200 ms in an experimental setup and presence of uncertainties. In [8], Wang et al. have proposed an anti-jittering scheme with finite-time controller for a master-slave system with jittering delay. The literature shows that the application of sliding mode control can significantly improve system performance. However, tackling a larger delay is still an open problem, as the controller performance deteriorates with increasing delay.

In this paper, an adaptive sliding mode control law is proposed for a teleoperated dual manipulator system in presence of communication delay. It is assumed that both position and velocity information are available. The controller gain is tuned adaptively where the gain dynamics are proportional to the sliding variable. This allows the user to operate the system in presence of an unknown bound of uncertainty. Moreover, the gain adaptation eliminates the chattering in the control input. The rest of the paper is structured as follows: in Sect. 2, the manipulator model is introduced, and the objective of the controller is defined. In Sect. 3, the controller design and analysis are described. The simulation results are presented in Sect. 4, and the conclusion is given in Sect. 5.

2 Problem Formulation

2.1 System Dynamics

The dynamics of the n-DoF master-slave manipulator system are given as follows:

$$\boldsymbol{M}_m(\boldsymbol{q}_m)\ddot{\boldsymbol{q}}_m + \boldsymbol{C}_m(\boldsymbol{q}_m, \dot{\boldsymbol{q}}_m)\dot{\boldsymbol{q}}_m + \boldsymbol{G}_m(\boldsymbol{q}_m) = \boldsymbol{u}_m + \boldsymbol{\tau}_h + \boldsymbol{d}_m(\boldsymbol{q}_m, \dot{\boldsymbol{q}}_m) \tag{1a}$$

$$\boldsymbol{M}_s(\boldsymbol{q}_s)\ddot{\boldsymbol{q}}_s + \boldsymbol{C}_s(\boldsymbol{q}_s, \dot{\boldsymbol{q}}_s)\dot{\boldsymbol{q}}_s + \boldsymbol{G}_s(\boldsymbol{q}_s) = \boldsymbol{u}_s - \boldsymbol{\tau}_e + \boldsymbol{d}_s(\boldsymbol{q}_s, \dot{\boldsymbol{q}}_s) \tag{1b}$$

where m and s indicates master and the slave robots, respectively, $\boldsymbol{q}_i, \dot{\boldsymbol{q}}_i, \ddot{\boldsymbol{q}}_i \in \mathbb{R}^n$, $i = m(master), s(slave)$ are the angular position, velocity, and acceleration of the manipulator joints, respectively. $\boldsymbol{M}_i(\boldsymbol{q}_i) \in \mathbb{R}^{n\times n}$, $i = m, n$ is the system inertia matrix, $\boldsymbol{C}_i(\boldsymbol{q}_i, \dot{\boldsymbol{q}}_i \in \mathbb{R}^{n\times n}$, $i = m, n$ represents the effects Coriolis friction and the centrifugal forces, $\boldsymbol{G}_i(\boldsymbol{q}_i) \in \mathbb{R}^n$, $i = m, n$ is the effect of gravitational force, $u_i \in \mathbb{R}^n$, $i = m, n$ is the input torque in the manipulator joints, $\boldsymbol{\tau}_h \in \mathbb{R}^n$ is the torque exerted by the operator on the master device and $\boldsymbol{\tau}_e \in \mathbb{R}^n$ is the environmental interaction torque affecting the slave device and $\boldsymbol{d}_i(\boldsymbol{q}_i, \dot{\boldsymbol{q}}_i) \in \mathbb{R}^n$, $i = m, n$ are the lumped uncertainties affecting the devices. For the current work, it is assumed that there is no environmental or external interaction of the slave manipulator, i.e., $\boldsymbol{\tau}_e = \boldsymbol{0}_{n\times 1}$.

Both the master and slave manipulators are assumed to have the same structure, and the manipulator matrices are as follows [11]:

$$\boldsymbol{M}_i(\boldsymbol{q}_i) = \begin{bmatrix} (m_{1i} + m_{2i})l_{1i}^2 + m_{2i}l_{2i}^2 + 2m_{2i}l_{1i}l_{2i}\cos(q_{2i}) & m_{2i}l_{2i}^2 + m_{2i}l_{1i}l_{2i}\cos(q_{2i}) \\ m_{2i}l_{2i}^2 + m_{2i}l_{1i}l_{2i}\cos(q_{2i}) & m_{2i}l_{2i}^2 \end{bmatrix} \tag{2a}$$

$$\boldsymbol{C}_i(\boldsymbol{q}_i, \dot{\boldsymbol{q}}_i) = \begin{bmatrix} -2m_{2i}l_{1i}l_{2i}\dot{q}_{2i}\sin(q_{2i}) & -m_{2i}l_{1i}l_{2i}\dot{q}_{2i}\sin(q_{2i}) \\ m_{2i}l_{1i}l_{2i}\dot{q}_{1i}\sin(q_{2i}) & 0 \end{bmatrix} \tag{2b}$$

$$\boldsymbol{G}_i(\boldsymbol{q}_i) = \begin{bmatrix} (m_{1i} + m_{2i})gl_{1i}\sin(q_{1i}) + m_{2i}gl_{2i}\sin(q_{1i} + q_{2i}) \\ m_{2i}g\sin(q_{1i} + q_{2i}) \end{bmatrix} \tag{2c}$$

The manipulator Jacobian is

$$\boldsymbol{J}_i(\boldsymbol{q}_i) = \begin{bmatrix} -l_{1i}\sin(q_{1i}) - l_{2i}\sin(q_{1i} + q_{2i}) & -l_{2i}\sin(q_{1i} + q_{2i}) \\ l_{1i}\cos(q_{1i}) + l_{2i}\cos(q_{1i} + q_{2i}) & l_{2i}\cos(q_{1i} + q_{2i}) \end{bmatrix} \tag{3}$$

where $i = m(master), s(slave)$.

State Space Representation

The nominal form of the manipulator dynamics described in (1a) and (1b) without considering the unknown uncertainties can be rewritten as

$$\begin{aligned}\ddot{q}_m =& M_m^{-1}(q_m)[-C_m(q_m, \dot{q}_m)\dot{q}_m - G_m(q_m) + \tau_h + d_m(q_m, \dot{q}_m)] \\ &+ M_m^{-1}(q_m)u_m \end{aligned} \tag{4a}$$

$$\ddot{q}_s = M_s^{-1}(q_s)[-C_s(q_s, \dot{q}_s)\dot{q}_s - G_s(q_s) + d_s(q_s, \dot{q}_s)] + M_s^{-1}(q_s)u_s \tag{4b}$$

Considering the state variables for the system as $x_{1i} = q_i \in \mathbb{R}^n$, $x_{2i} = \dot{q}_i \in \mathbb{R}^n$, $i = m, s$ the dynamics for the master and the slave manipulator in the state space form can be written as follows:

$$\begin{aligned}\dot{x}_{1i} =& x_{2i} \\ \dot{x}_{2i} =& f_i(x_{1i}, x_{2i}) + g_i(x_{1i}, x_{2i})u_i \end{aligned} \tag{5}$$

where $i = m, s$, $x_i = [x_{1i}^T \; x_{2i}^T]^T \in \mathbb{R}^{2n}$ are the state variables and

$$\begin{aligned} f_m =& M_m^{-1}(q_m)[-C_m(q_m, \dot{q}_m)\dot{q}_m - G_m(q_m) + \tau_h] \in \mathbb{R}^n \\ g_m =& M_m^{-1}(q_m) \in \mathbb{R}^n \\ f_s =& M_s^{-1}(q_s)[-C_s(q_s, \dot{q}_s)\dot{q}_s - G_s(q_s)] \in \mathbb{R}^n \\ g_s =& M_s^{-1}(q_s) \in \mathbb{R}^n \end{aligned}$$

Some assumptions for the considered master-slave manipulator set are as follows:

Assumption 1 All the joints of both the manipulator are revolute.

Assumption 2 The parametric uncertainty and the external disturbances are bounded, i.e.,

$$|d_m(q_m, \dot{q}_m)| \leq \bar{d}_m \tag{7a}$$

$$|d_s(q_s, \dot{q}_s)| \leq \bar{d}_s \tag{7b}$$

where $\bar{d}_m$ and $\bar{d}_s$ are real valued and absolute value and inequality are taken element-wise.

Assumption 3 The external forces applied to the devices by the user are bounded, which means,

$$|\tau_h(t)| \leq \rho, \; \rho \in \mathbb{R}^n \tag{8}$$

Here, absolute value and inequality are taken element-wise.

Assumption 4 The manipulator joint position, velocity, and accelerations are bounded owing their the mechanical structures and the actuator limits, i.e.,

$$|\boldsymbol{q}_i(t)| \leq \bar{\boldsymbol{q}}_i, \quad \bar{\boldsymbol{q}}_i \in \mathbb{R}^n \tag{9a}$$

$$|\dot{\boldsymbol{q}}_i(t)| \leq \dot{\bar{\boldsymbol{q}}}_i, \quad \dot{\bar{\boldsymbol{q}}}_i \in \mathbb{R}^n \tag{9b}$$

$$|\ddot{\boldsymbol{q}}_i(t)| \leq \ddot{\bar{\boldsymbol{q}}}_i, \quad \ddot{\bar{\boldsymbol{q}}}_i \in \mathbb{R}^n \tag{9c}$$

Here, absolute value and inequality are taken element-wise.

Assumption 5 There are no interaction with the external environment by the slave manipulator.

Some important properties of the robot manipulator are [6]

Proposition 1 *The inertia matrix is bounded, symmetric, and positive definite, and due to assumption (1), the bounds can be expressed as follows:*

$$m_{0i}\boldsymbol{I}_n \leq |\boldsymbol{M}_i(\boldsymbol{q}_i)| \leq m_{1i}\boldsymbol{I}_n \tag{10}$$

$$\bar{m}_{0i}\boldsymbol{I}_n \leq |\boldsymbol{M}_i^{-1}(\boldsymbol{q}_i)| \leq \bar{m}_{1i}\boldsymbol{I}_n \tag{11}$$

$$\mu_{\min}||\boldsymbol{x}||^2 \leq \boldsymbol{x}^T\boldsymbol{M}_i(\boldsymbol{q}_i)\boldsymbol{x} \leq \mu_{\max}||\boldsymbol{x}||^2 \tag{12}$$

where $i = m, n$, $\boldsymbol{x} \in \mathbb{R}^n$ *is an arbitrary vector and* $0 < \mu_{\min} < \mu_{\max} \in \mathbb{R}$ *and* $\boldsymbol{I}_n$ *is an* $n \times n$ *identity matrix.*

Proposition 2 *The manipulator system is passive which means*

$$\boldsymbol{x}^T\left(\frac{1}{2}\dot{\boldsymbol{M}}_i(\boldsymbol{q}_i) - \boldsymbol{C}_i(\boldsymbol{q}_i, \dot{\boldsymbol{q}}_i)\right)\boldsymbol{x} = 0, \forall \boldsymbol{x} \neq 0, \boldsymbol{x} \in \mathbb{R}^n. \tag{13}$$

2.2 Objective

The purpose of this research is to establish a synchronized motion between the master and the slave manipulator when a motion is created in the master manipulator by applying an external force, and the slave manipulator has to follow the trajectory of the master manipulator.

The position tracking errors are defined as

$$\boldsymbol{e}_m(t) = \boldsymbol{q}_m(t) - \boldsymbol{q}_s(t-h) \tag{14a}$$

$$\boldsymbol{e}_s(t) = \boldsymbol{q}_s(t) - \boldsymbol{q}_m(t-h) \tag{14b}$$

The velocity tracking errors are defined as

$$\dot{\boldsymbol{e}}_m(t) = \dot{\boldsymbol{q}}_m(t) - \dot{q}_s(t-h) \tag{14c}$$

$$\dot{\boldsymbol{e}}_s(t) = \dot{\boldsymbol{q}}_s(t) - \dot{q}_m(t-h) \tag{14d}$$

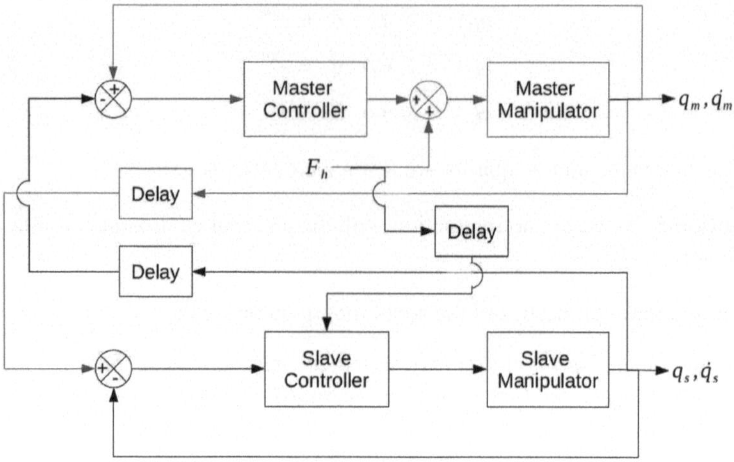

Fig. 1 Block diagram of the tele-operated bi-directional master-slave manipulator

The detailed block diagram of the teleoperated bi-directional master-slave manipulator is shown in Fig. 1. An external force $\boldsymbol{F}_h(t)$ will be applied to the master manipulator such that $\boldsymbol{\tau}_h(t) = \boldsymbol{J}_m^T(\boldsymbol{q}_m)\boldsymbol{F}_h(t)$ that will create a motion in the master manipulator. The movement of the master manipulator is transmitted to the slave manipulator which is expected to follow the motion of the master robot arm. It is assumed that there is no environmental interaction in the slave manipulator side, i.e., $\boldsymbol{\tau}_e(t) = \boldsymbol{0}$. The transmission channel between the master and the slave side is assumed to have communication delays. Thus, the objective is to design robust feedback controllers for both master and slave manipulators such that the motion created in the master manipulator by the external force is synchronously followed by the slave manipulator in presence of the communication delay, and the resulting system is stable. Thus, the objectives can be summarized as the following points:

1. The motion of the master will be directed by the external force $\boldsymbol{F}_h(t)$ applied by the operator, and this motion will be followed by the slave manipulator. Only the position and velocity information from the master side will be transmitted to the slave side.
2. The tracking errors $\dot{\boldsymbol{e}}_m(t)$ and $\dot{\boldsymbol{e}}_s(t)$ should converge to zero.

3 Controller Design

A sliding mode controller with adaptive gain and a sliding surface having the proportional integral derivative (PID) structure will be designed for the above-mentioned system. Due to the communication delay, at any time instant on both the master and slave sides, the delayed information of the other side is available. Hence on both sides, the tracking error will be considered as the difference between the current information and the delayed information from the other side.

3.1 Sliding Surface Design

The position tracking errors are defined as

$$\boldsymbol{e}_m(t) = \boldsymbol{q}_m(t) - \boldsymbol{q}_s(t-h) \tag{15a}$$

$$\boldsymbol{e}_s(t) = \boldsymbol{q}_s(t) - \boldsymbol{q}_m(t-h) \tag{15b}$$

The velocity tracking errors are defined as

$$\dot{\boldsymbol{e}}_m(t) = \dot{\boldsymbol{q}}_m(t) - \dot{q}_s(t-h) \tag{16a}$$

$$\dot{\boldsymbol{e}}_s(t) = \dot{\boldsymbol{q}}_s(t) - \dot{q}_m(t-h) \tag{16b}$$

On both sides, a variable $\sigma_i(t),\ i = m, s$ is defined as follows

$$\boldsymbol{\sigma}_i(t) = \dot{\boldsymbol{e}}_i(t) + \boldsymbol{C}_1 \boldsymbol{e}_i(t) \tag{17}$$

where $\boldsymbol{C}_1 = \text{diag}\{c_{1j}\},\ j = 1, \ldots, n$ is a user-defined positive definite matrix, $i = m, s$.

Thus, the sliding surface for the master manipulator is defined as

$$\boldsymbol{s}_m(t) = \dot{\boldsymbol{e}}_m(t) + \boldsymbol{C}_2 \int_0^t \boldsymbol{\sigma}_m(\boldsymbol{\theta}) \mathrm{d}\boldsymbol{\theta}. \tag{18}$$

where $\boldsymbol{C}_2 = \text{diag}\{c_{2j}\},\ j = 1, \ldots, n$ is a user-defined positive definite matrix and $\boldsymbol{A}$ is a diagonal matrix with diagonal elements ≤ 1, such that $||\boldsymbol{A}|| \leq 1$.

The sliding surface for the slave manipulator is defined as

$$\boldsymbol{s}_s(t) = \dot{\boldsymbol{e}}_s(t) + \boldsymbol{C}_2 \int_0^t \boldsymbol{\sigma}_s(\boldsymbol{\theta}) \mathrm{d}\boldsymbol{\theta}. \tag{19}$$

Analysis of the Sliding Mode Dynamics: The stable dynamics of the sliding surfaces, $\boldsymbol{s}_m(t) = \boldsymbol{0}$ and $\boldsymbol{s}_s(t) = \boldsymbol{0}$, will make sure that the tracking error for the master, and slave manipulator motions converges to zero, i.e., objective (2.2) is satisfied. Considering the sliding surface defined for the master manipulator in (18) and using (17), (15a) and (16a), the dynamics of the sliding surface can be defined as follows:

$$s_m(t) = 0 \tag{20}$$

$$\dot{e}_m(t) + C_2 e_m(t) + C_1 C_2 \int e_m(t)\mathrm{d}t = 0 \tag{21}$$

Considering $z_1 = \int e_m(t)\mathrm{d}t$, $z_2 = e_m(t)$, the dynamics of (21) can be written as follows:

$$\begin{aligned} \dot{z}_1 &= z_2 \\ \dot{z}_2 &= -C_2 z_2 - C_1 C_2 z_1 \end{aligned} \tag{22}$$

The dynamics represented by (22) can be rendered stable with the proper selection of controller parameters C_1 and C_2. Thus, it shows that when $s_m(t) = 0$ is stable, the surface dynamics will cause z_1 and z_2 to converge to zero, which again using (17) leads to σ_i to converge to zero.

3.2 Control Law

Sliding mode control laws for both the master and slave manipulator will be designed as follows:

$$u_i = u_{eq(i)} + u_{sw(i)} \tag{23}$$

where $i = m, s$. Here $u_{eq(i)}$ is the equivalent control which will maintain the sliding dynamics and $u_{sw(i)}$ is the switching control which brings the system dynamics to the sliding surface and compensates for any deviations occurring due to disturbances or change of references.

It is important to note that sensing the angular acceleration is expensive and taking derivative of the velocity signal to obtain the acceleration leads to a highly noisy signal, often unusable. Moreover, using such a noise corrupted and delayed acceleration signal in the control law can cause unnecessary complications in the control input such as the rise of the unaccounted high-frequency dynamics which can prove to be harmful to the system. Hence, while designing the control law, $\ddot{q}_s(t-h)$ is assumed to be unavailable and considered as uniformly zero.

The control law for the master manipulator is as follows:

$$u_m(t) = M_m(q_m)\left[-K_m \mathrm{sign}(s_m(t)) - W s_m(t) - C_2 \sigma_m(t)\right] \tag{24}$$
$$+ C_m(q_m, \dot{q}_m)\dot{q}_m + G_m(q_m) \tag{25}$$

where $K_m = \mathrm{diag}\{k_{mi}\}$ and $W_m = \mathrm{diag}\{w_{mi}\}$, $i = 1, \ldots, n$ are positive definite matrices of the controller gains.

One important point to note here is that, since the interaction force in the slave side is considered to be zero, hence there is no requirement of including any compensatory term in the master controller for the slave side interaction forces. However, there is an external force applied to the master manipulator. The movement of the master manipulator is commanded by this force, and hence, its inclusion in the slave manipulator controller is necessary for the proper depiction of the transmitted dynamics. Thus, the controller for the slave manipulator has the same structure as the master manipulator controller apart from the extra term for the applied force (delayed signal) in the master arm as follows:

$$\begin{aligned} \boldsymbol{u}_s(t) =& \boldsymbol{M}_s(\boldsymbol{q}_s)\left[-\boldsymbol{K}_s \text{sign}(\boldsymbol{s}_s(t)) - \boldsymbol{W}_s \boldsymbol{s}_s(t) - \boldsymbol{C}_2 \boldsymbol{\sigma}_s(t)\right] \\ &+ \boldsymbol{C}_s(\boldsymbol{q}_s, \dot{\boldsymbol{q}}_s)\dot{\boldsymbol{q}}_s + \boldsymbol{G}_s(\boldsymbol{q}_s) + \boldsymbol{\tau}_h(t-h) \end{aligned} \tag{26}$$

where $\boldsymbol{K}_s = \text{diag}\{k_{si}\}$ and $\boldsymbol{W}_s = \text{diag}\{w_{si}\}$, $i = 1, \ldots, n$ are positive definite matrices of the controller gains.

3.3 Closed-Loop System

The closed-loop systems during the reaching phase resulting from the application of the control laws (25) and (26) to (1a) and (1b), respectively, are as follows

$$\begin{aligned} \ddot{\boldsymbol{q}}_m(t) =& -\boldsymbol{K} \text{sign}(\boldsymbol{s}_m(t)) - \boldsymbol{W} \boldsymbol{s}_m(t) - \boldsymbol{C}_2 \boldsymbol{\sigma}_m(t) + \boldsymbol{M}_m^{-1}(\boldsymbol{q}_m)\boldsymbol{\tau}_h(t) \\ &+ \boldsymbol{M}_m^{-1}(\boldsymbol{q}_m)\boldsymbol{d}_m(\boldsymbol{q}_m, \dot{\boldsymbol{q}}_m) \end{aligned} \tag{27a}$$

$$\begin{aligned} \ddot{\boldsymbol{q}}_s(t) =& -\boldsymbol{K} \text{sign}(\boldsymbol{s}_s(t)) - \boldsymbol{W} \boldsymbol{s}_s(t) - \boldsymbol{C}_2 \boldsymbol{\sigma}_s(t) + \boldsymbol{M}_s^{-1}(\boldsymbol{q}_s)\boldsymbol{\tau}_h(t-h) \\ &+ \boldsymbol{M}_s^{-1}(\boldsymbol{q}_s)\boldsymbol{d}_s(\boldsymbol{q}_s, \dot{\boldsymbol{q}}_s) \end{aligned} \tag{27b}$$

Considering the following candidate Lyapunov Function

$$V_s = \frac{1}{2}(\boldsymbol{s}_m^T \boldsymbol{s}_m + \boldsymbol{s}_s^T \boldsymbol{s}_s) \tag{28}$$

$$\begin{aligned} \dot{V}_s =& \boldsymbol{s}_m^T\left[\ddot{\boldsymbol{q}}_m - \ddot{\boldsymbol{q}}_s(t-h) + \boldsymbol{C}_2\boldsymbol{\sigma}_m\right] + \boldsymbol{s}_s^T\left[\ddot{\boldsymbol{q}}_s - \ddot{\boldsymbol{q}}_m(t-h) + \boldsymbol{C}_2\boldsymbol{\sigma}_s\right] \\ =& \boldsymbol{s}_m^T\big[-\boldsymbol{K}_m \text{sign}(\boldsymbol{s}_m) - \boldsymbol{W}_m \boldsymbol{s}_m + \boldsymbol{M}_m^{-1}(\boldsymbol{q}_m)\boldsymbol{\tau}_h(t) - \ddot{\boldsymbol{q}}_s(t-h) \\ &+ \boldsymbol{M}_m^{-1}(\boldsymbol{q}_m)\boldsymbol{d}_m(\boldsymbol{q}_m, \dot{\boldsymbol{q}}_m)\big] + \boldsymbol{s}_s^T\big[-\boldsymbol{K}_s \text{sign}(\boldsymbol{s}_s) - \boldsymbol{W}_s \boldsymbol{s}_s \\ &+ \boldsymbol{M}_s^{-1}(\boldsymbol{q}_s)\boldsymbol{\tau}_h(t-h) - \ddot{\boldsymbol{q}}_m(t-h) + \boldsymbol{M}_s^{-1}(\boldsymbol{q}_s)\boldsymbol{d}_s(\boldsymbol{q}_s, \dot{\boldsymbol{q}}_s)\big] \\ =& -|\boldsymbol{s}_m|^T \boldsymbol{k}_m - \boldsymbol{s}_m^T \boldsymbol{W}_m \boldsymbol{s}_m + \boldsymbol{s}_m^T \boldsymbol{M}_m^{-1}(\boldsymbol{q}_m)\boldsymbol{\tau}_h(t) - \boldsymbol{s}_m^T \ddot{\boldsymbol{q}}_s(t-h) \\ &+ \boldsymbol{s}_m^T \boldsymbol{M}_m^{-1}(\boldsymbol{q}_m)\boldsymbol{d}_m(\boldsymbol{q}_m, \dot{\boldsymbol{q}}_m) - |\boldsymbol{s}_s|^T \boldsymbol{k}_s - \boldsymbol{s}_s^T \boldsymbol{W}_s \boldsymbol{s}_s \\ &+ \boldsymbol{s}_s^T \boldsymbol{M}_s^{-1}(\boldsymbol{q}_s)\boldsymbol{\tau}_h(t-h) + \boldsymbol{s}_s^T \boldsymbol{M}_s^{-1}(\boldsymbol{q}_s)\boldsymbol{d}_s(\boldsymbol{q}_s, \dot{\boldsymbol{q}}_s) \end{aligned} \tag{29}$$

where $\boldsymbol{k} = [k_{m1}\ k_{m2} \ldots k_{mn}]^T$ with k_{mi} being the diagonal elements of $\boldsymbol{K}_m$ and $\boldsymbol{k}_s = [k_{s1}\ \ k_{s2} \ldots k_{sn}]^T$ with k_{si} being the diagonal elements of $\boldsymbol{K}_s$ Since $\boldsymbol{M}_m^{-1}(\boldsymbol{q}_m)$, $\boldsymbol{M}_s^{-1}(\boldsymbol{q}_s)$, $\boldsymbol{\tau}_h$, $\ddot{\boldsymbol{q}}_m$ and $\ddot{\boldsymbol{q}}_s$ are bounded as per the assumptions (1)–(3)

the following inequalities can be derived:

$$s_m^T M_m^{-1}(q_m)\tau_h(t) \leq \bar{m}_{1m}|s_m|^T \rho \tag{30a}$$
$$s_m^T \ddot{q}_s(t-h) \leq |s_m|^T \bar{\ddot{q}}_s \tag{30b}$$
$$s_m^T M_m^{-1}(q_m)d_m(q_m, \dot{q}_m) \leq m_{1m}|s_m|^T \bar{d}_m \tag{30c}$$
$$s_s^T M_s^{-1}(q_s)\tau_h(t-h) \leq \bar{m}_{1s}|s_s|^T \rho \tag{30d}$$
$$s_s^T \ddot{q}_m(t-h) \leq |s_s|^T \bar{\ddot{q}}_m \tag{30e}$$
$$s_s^T M_s^{-1}(q_s)d_s(q_s, \dot{q}_s) \leq m_{1s}|s_s|^T \bar{d}_s \tag{30f}$$

Using (30) in (29), the following is derived

$$\begin{aligned}\dot{V}_s \leq & -|s_m|^T k_m - s_m^T W_m s_m + \bar{m}_{1m}|s_m|^T \rho + |s_m|^T \bar{\ddot{q}}_s + m_{1m}|s_m|^T \bar{d}_m \\ & -|s_s|^T k_s - s_s^T W_s s_s + \bar{m}_{1s}|s_s|^T \rho + |s_s|^T \bar{\ddot{q}}_m + m_{1s}|s_s|^T \bar{d}_s \\ \leq & -|s_m|^T (k_m - m_{1m}\rho - \bar{\ddot{q}}_s - m_{1m}\bar{d}_m) - s_m^T W_m s_m \\ & -|s_s|^T (k_s - m_{1s}\rho - \bar{\ddot{q}}_m - m_{1s}\bar{d}_s) - s_s^T W_s s_s\end{aligned} \tag{31}$$

Thus from (31), if the switching gains for the controller are large enough to satisfy

$$k_m > m_{1m}\rho + \bar{\ddot{q}}_s + m_{1m}\bar{d}_m \tag{32a}$$
$$k_s > m_{1s}\rho + \bar{\ddot{q}}_m + m_{1s}\bar{d}_s \tag{32b}$$

then, the time derivative of CLF V_s will satisfy the following

$$\dot{V}_s \leq -s_m^T W_m s_m - s_s^T W_s s_s \leq 0 \tag{32c}$$

which indicates that the sliding surface can be reached in finite time, and the state trajectories can be maintained there for the subsequent times. From (32), the robustness of the controller will increase with higher values of the switching controller gains $\boldsymbol{K}_i,\ i = m, s$, but this high value produces a significant amount of chattering in the control law. This is a compromise at which the robustness of the SMC is achieved.

3.4 Gain Adaptation

The robustness of the sliding mode controller lies in its gain value $\boldsymbol{K}_i$. However, this robustness is achieved at the expense of high chattering and thus higher energy utilization at the input. Although chattering reduction can be achieved by replacing the signum function of the sliding mode controller with a smoother approximation, it often causes deterioration in the controller performance. One very effective method of maintaining controller robustness while reducing chattering is the use of an adaptively tuned gain for the SMC which automatically reduces the controller gain once the states approach steady-state thus eliminating unnecessary use of input energy. The adaptive tuning of the gain with a leakage term [2] is defined as follows

$$\dot{\hat{\boldsymbol{k}}}(t) = \boldsymbol{\Gamma}(|\boldsymbol{s}_i(t)| - \boldsymbol{\epsilon}\hat{\boldsymbol{k}}(t)) \tag{33}$$

$$\hat{\boldsymbol{k}}(t) = \int_0^t \dot{\hat{\boldsymbol{k}}}(s)d(s) \tag{34}$$

where $i = m, s$ and $\hat{\boldsymbol{k}}(t) = [\hat{k}_j(t)],\ j = 1, \ldots, n$ and $\boldsymbol{K}$ in the control law is replaced with $\hat{\boldsymbol{K}}(t) = \text{diag}\{\hat{k}_j(t)\},\ j = 1, \ldots, n$. $\boldsymbol{\Gamma} = \text{diag}\{\gamma_j\}, \gamma_j > 0,\ j = 1, \ldots, n$ and $\boldsymbol{\epsilon} = \text{diag}\{\epsilon_j\}, \epsilon_j > 0,\ j = 1, \ldots, n$ are user defined parameters.

3.5 Analysis

A new CLF is now defined as follows in order to analyze the system with the adaptively tuned gain:

$$\begin{aligned} V &= V_s + \frac{1}{2}\sum_{i=m,s} \tilde{\boldsymbol{k}}_i^T \boldsymbol{\Gamma}^{-1}\tilde{\boldsymbol{k}}_i \\ &= \frac{1}{2}\sum_{i=m,s} \boldsymbol{s}_i^T \boldsymbol{s}_i + \frac{1}{2}\sum_{i=m,s} \tilde{\boldsymbol{k}}_i^T \boldsymbol{\Gamma}^{-1}\tilde{\boldsymbol{k}}_i \end{aligned} \tag{35}$$

where $\tilde{\boldsymbol{k}}_i = \hat{\boldsymbol{k}}_i - \boldsymbol{k}_i,\ i = m, s$ and $0 < \boldsymbol{k}_i < \infty,\ \boldsymbol{k}_i \in \mathbb{R}^n,\ i = m, s$ is the arbitrary value to which the adaptively tune gain converges to.

The following Lemma will be used in the further analysis

Lemma 1 *For real vectors* $\tilde{\boldsymbol{k}}_i, \hat{\boldsymbol{k}}, k_i > 0, i = m, s$ *and positive definite diagonal matrix* $\boldsymbol{\epsilon} \in \mathbb{R}^{n \times n}$, *if* $\tilde{\boldsymbol{k}}_i = \hat{\boldsymbol{k}}_i - k_i$ *then* $\tilde{\boldsymbol{k}}_i^T \boldsymbol{\epsilon}\hat{\boldsymbol{k}}_i \geq \frac{1}{2}\tilde{\boldsymbol{k}}_i^T \boldsymbol{\epsilon}\tilde{\boldsymbol{k}}_i - k_i^T \boldsymbol{\epsilon} k_i$

Proof The proof can be found in [1], page 111.

Taking time derivative of V, and using (32c), (33), and Lemma (1) , the following can be derived

$$\begin{aligned}
\dot{V} \leq & -\sum_{i=m,s} s_i^T W_i s_i + \sum_{i=m,s} (\tilde{k}_i^T \Gamma^{-1} \dot{\tilde{k}}_i) \\
\leq & -\sum_{i=m,s} s_i^T W_i s_i + \sum_{i=m,s} (\tilde{k}_i^T \Gamma^{-1} \Gamma(|s_i| - \epsilon \hat{k}_i)) \\
\leq & -\sum_{i=m,s} s_i^T W_i s_i + \sum_{i=m,s} (\tilde{k}_i^T |s_i| - \tilde{k}_i^T \epsilon \hat{k}_i)) \\
\leq & -\sum_{i=m,s} s_i^T W_i s_i + \frac{1}{2} \sum_{i=m,s} (\tilde{k}_i^T \epsilon \tilde{k}_i - k_i^T \epsilon k_i) \\
\leq & -\sum_{i=m,s} s_i^T W_i s_i - \frac{1}{2} \sum_{i=m,s} k_i^T \epsilon k_i + \frac{1}{2} \sum_{i=m,s} \tilde{k}_i^T \epsilon \tilde{k}_i \\
\leq & -\sum_{i=m,s} \lambda_{\min}(W_i) s_i^T s_i - \sum_{i=m,s} \lambda_{\min}(\Gamma \epsilon) k_i^T \Gamma^{-1} k_i + \frac{1}{2} \sum_{i=m,s} |\tilde{k}_i^T \epsilon \tilde{k}_i| \\
\leq & -\eta \left(\sum_{i=m,s} s_i^T s_i + \sum_{i=m,s} k_i^T \Gamma^{-1} k_i \right) + \frac{1}{2} \xi_k \\
\leq & -2\eta V + \frac{1}{2} \xi_k
\end{aligned} \tag{36}$$

where $\lambda_{\min}(\boldsymbol{A})$ indicates the minimum eigenvalue of matrix $\boldsymbol{A}$, $\eta = \min(\lambda_{\min}(\boldsymbol{W}_i), \lambda_{\min}(\boldsymbol{\Gamma}\boldsymbol{\epsilon}))$, $i = m, s$ and $\frac{1}{2}\xi_k = \frac{1}{2}\sum_{i=m,s} \tilde{k}_i^T \epsilon \tilde{k}_i$, $i = m, s$. Thus for $V(0) > \frac{\xi_k}{4\eta}$ and $\frac{\xi_k}{4\eta} < 1$, $V(t)$ will be a decreasing function, indicating the stability of the system.

4 Simulation Results

The proposed controllers (25), (26) are tested via simulation for the bilateral master-slave operation of two 2DoF manipulators having identical structures as given in (2).

A Cartesian force, as shown in Fig. 2, is applied to the end-effector of the master manipulator. The resultant joint motions are transmitted to the slave manipulator where the aim is to synchronize with the motion of the master manipulator. The parameters of the manipulators are as follows: $m_{1i} = m_{2i} = 1$ kg, $l_{1i} = l_{2i} = 1$ m, $g = 9.81$ kg m/s^2.

The controller gains used in the simulation are $C_1 = 2$, $C_2 = 0.001$. The parameters for the adaptive law for the gain are $\Gamma = 100$, $\epsilon = 0.1$, $W = 2$.

The controller can withstand a communication delay up to 2.5 s, beyond which its performance deteriorates with the given control parameters. Parametric uncertainty in the form of a 0.01 kg deviation of the nominal mass of the manipulator joints

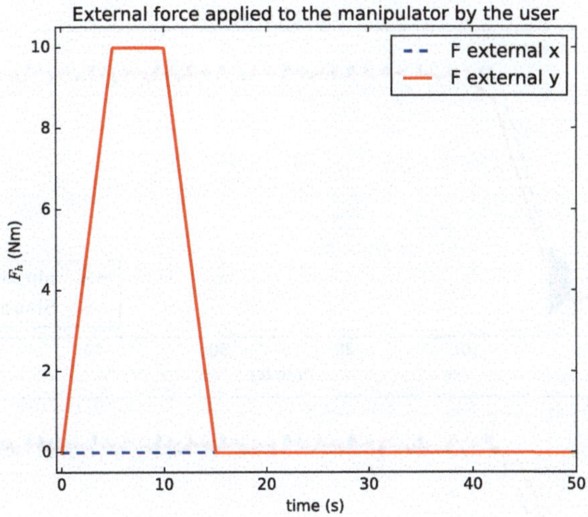

Fig. 2 External force applied by the user

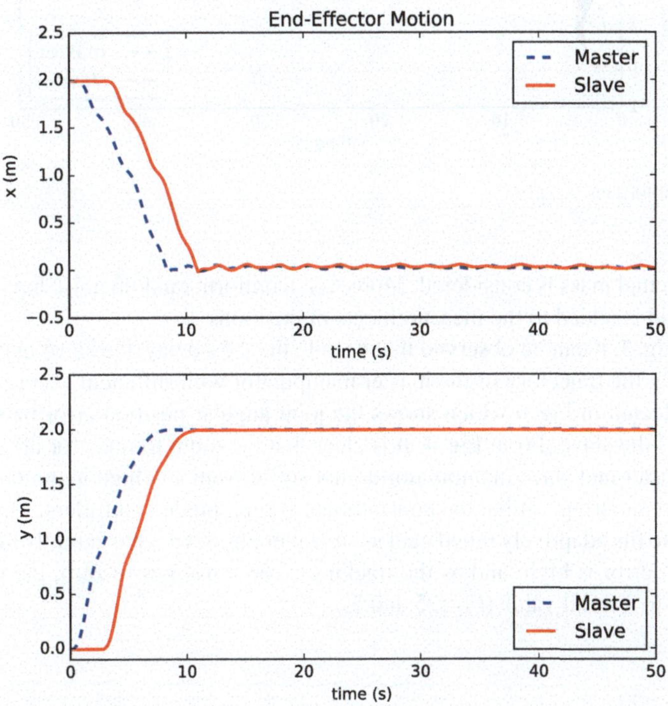

Fig. 3 End-effector trajectory

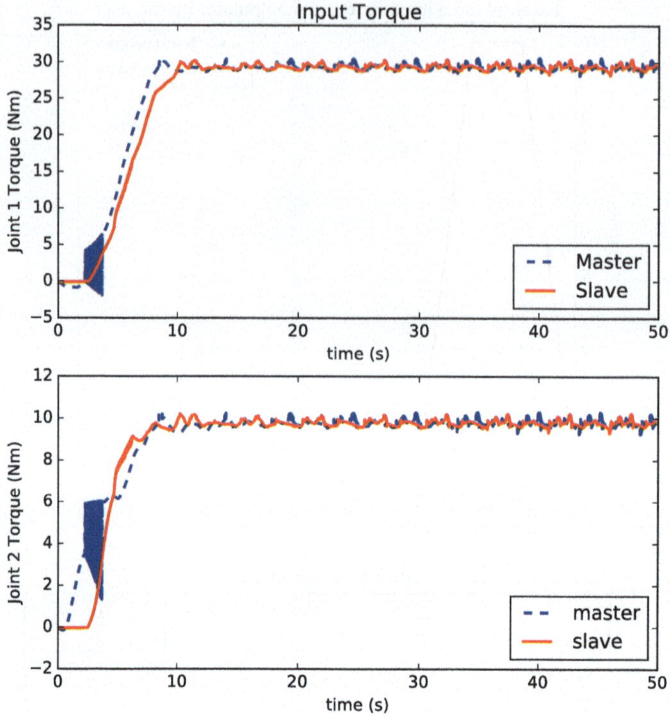

Fig. 4 Input torques

from the actual mass is considered. Moreover, a uniform random noise having limits ±0.0001rad is added to the measurements of positions.

From Fig. 3, it can be observed that despite the 2.5 s delay, the slave manipulator is following the trajectory of the master manipulator with sufficient accuracy which is also reflected in Fig. 6 which shows the joint angular position error between the master and the slave. From Fig. 4, it is clear that the input torques for all the joints in both master and slave manipulator do not suffer from any high magnitude, high-frequency chattering, unlike the conventional sliding mode controllers. This can be attributed to the adaptively tuned gain as shown in Fig. 8, which is initially high when the system error is high, and as the tracking error converges to zero, the gain also converges to a small value (Figs. 5 and 7).

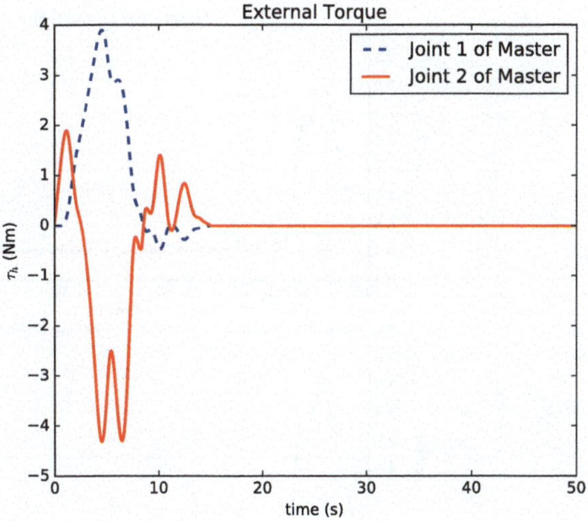

Fig. 5 External torques on master manipulator

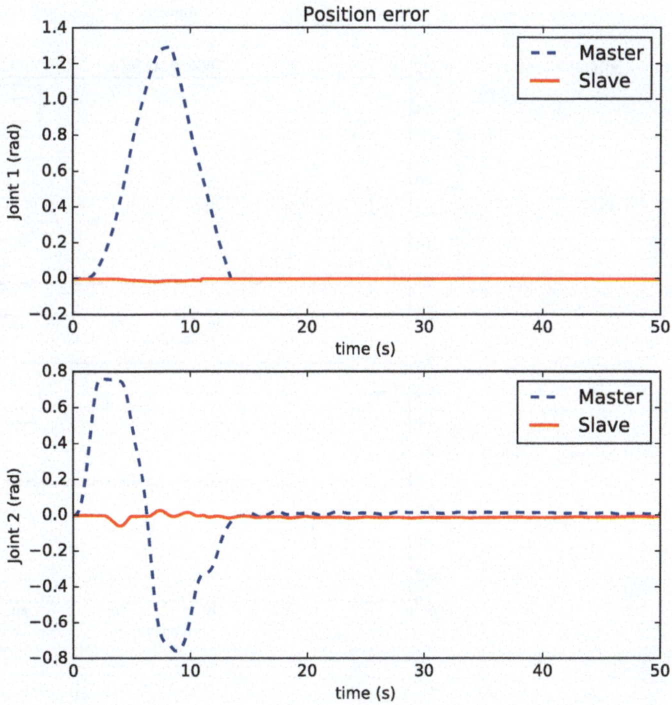

Fig. 6 Tracking error

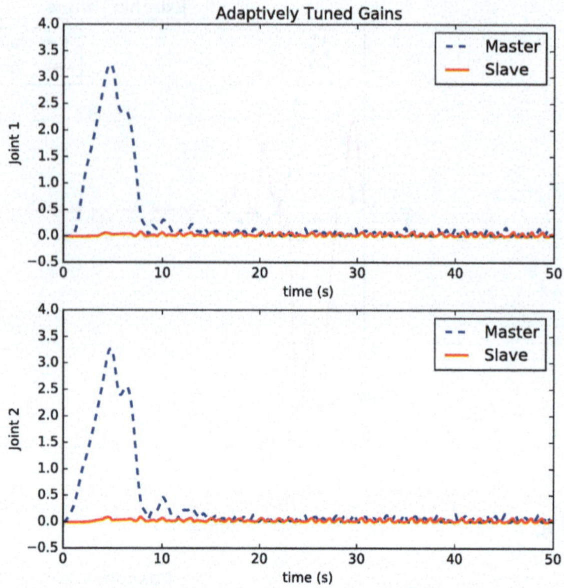

Fig. 7 Adaptively tuned gain K

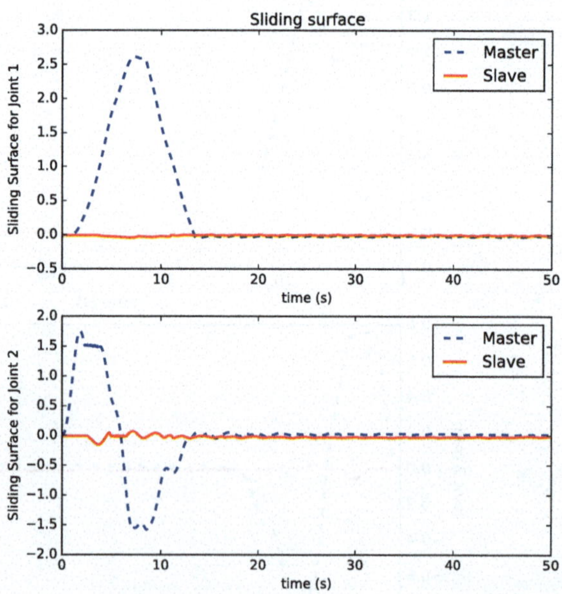

Fig. 8 Sliding surface

5 Conclusion

The proposed adaptive sliding mode controller for bilateral telemanipulation with a delay has a simple structure that is easy to implement. In presence of uncertainty and a delay of up to 2.5 s, the controller shows good performance without any requirements of re-tuning its parameters. The primary analysis regarding the stability of the controller shows that it is robust and can be used even when uncertainty bounds (parametric and external disturbances) are unknown. This can be attributed to the adaptive tuning of the switching gain of the controller, which increases with increasing tracking error and vice versa. Thus, under steady-state conditions, the gain remains at a very low value which prevents the unnecessary use of input energy and also prevents high amplitude and high-frequency chattering. Future work for the proposed method is to perform further analysis to gain detailed information regarding the delay tolerance of the system, and the same research can be extended to multilateral telemanipulation.

Acknowledgements This work has been supported by Scheme for Promotion of Academic and Research Collaboration (SPARC), MHRD, Government of India, no. SPARC/2018–2019/P1087/SL.

References

1. Adhikary, N.: Design of a few backstepping sliding mode based robust control techniques for robot manipulators. Ph.D. thesis (2017)
2. Blažič, S., Matko, D., Škrjanc, I.: A new leakage term in the adaptive law. IFAC Proc. Vol. **41**(2), 8913–8918 (2008). https://doi.org/10.3182/20080706-5-KR-1001.01506 (17th IFAC World Congress)
3. Cruz-Ortiz, D., Chairez, I., Poznyak, A.: Robust control of a master-slave manipulator under restricted task-space. In: Hernandez, E.E., Keshtkar, S., Valdez, S.I. (eds.) Industrial and Robotic Systems, pp. 240–248. Springer International Publishing, Cham (2020)
4. Huang, F., Zhang, W., Chen, Z., Tang, J., Song, W., Zhu, S.: Rbfnn-based adaptive sliding mode control design for nonlinear bilateral teleoperation system under time-varying delays. IEEE Access **7**, 11905–11912 (2019)
5. Moreau, R., Pham, M., Tavakoli, M., Le, M., Redarce, T.: Sliding-mode bilateral teleoperation control design for master-slave pneumatic servo systems. Control Eng. Pract. **20**(6), 584–597 (2012). https://doi.org/10.1016/j.conengprac.2012.02.003, http://www.sciencedirect.com/science/article/pii/S0967066112000275
6. Murray, R.M., Li, Z., Sastry, S.S., Sastry, S.S.: A Mathematical Introduction to Robotic Manipulation. CRC Press, Boca Raton (1994)
7. Wang, Z., Chen, Z., Liang, B., Zhang, B.: A novel adaptive finite time controller for bilateral teleoperation system. Acta Astronautica **144**, 263–270 (2018). https://doi.org/10.1016/j.actaastro.2017.12.046, http://www.sciencedirect.com/science/article/pii/S0094576517301224
8. Wang, Z., Chen, Z., Zhang, Y., Yu, X., Wang, X., Liang, B.: Adaptive finite-time control for bilateral teleoperation systems with jittering time delays. Int. J. Robust Nonlinear Control **29**(4), 1007–1030 (2019)
9. Yang, H., Liu, L., Wang, Y.: Observer-based sliding mode control for bilateral teleoperation with time-varying delays. Control Eng. Pract. **91** (2019). https://doi.org/10.1016/j.conengprac.2019.07.015, http://www.sciencedirect.com/science/article/pii/S0967066119301145

10. Yang, L., Chen, Y., Liu, Z., Chen, K., Zhang, Z.: Adaptive fuzzy control for teleoperation system with uncertain kinematics and dynamics. Int. J. Control Autom. Syst. **17**(5), 1158–1166 (2019)
11. Yang, X., Hua, C., Yan, J., Guan, X.: A new master-slave torque design for teleoperation system by t-s fuzzy approach. IEEE Trans. Control Syst. Technol. **23**(4), 1611–1619 (2015)
12. Zhai, D., Xia, Y.: Adaptive control for teleoperation system with varying time delays and input saturation constraints. IEEE Trans. Industr. Electron. **63**(11), 6921–6929 (2016)
13. Zhang, J., Liu, W., Gao, L., Li, L., Li, Z.: The master adaptive impedance control and slave adaptive neural network control in underwater manipulator uncertainty teleoperation. Ocean Eng. **165**, 465–479 (2018). https://doi.org/10.1016/j.oceaneng.2018.07.055, http://www.sciencedirect.com/science/article/pii/S0029801818313763
14. Zhao, L., Zhang, H., Yang, Y., Yang, H.: Integral sliding mode control of a bilateral teleoperation system based on extended state observers. Int. J. Control Autom. Syst. **15**(5), 2118–2125 (2017)

A Multi-Master Single-Slave Teleoperation System Through Composite State Convergence

Muhammad Usman Asad, Jason Gu, Umar Farooq, Rajeeb Dey, Nabanita Adhikary, Rupak Datta, and Chunqi Chang

Abstract This paper extends the transparent bilateral state convergence control architecture to multilateral state convergence control architecture. The case of multi-master-single-slave teleoperation system is considered to show the design procedure. In the proposed architecture, composite variables are transmitted across the communication channel instead of full systems' states. In addition, control gains are defined to consider masters-slave interactions. Through the method of state convergence, design conditions are derived to determine the control gains by allowing the tracking error to evolve as an autonomous system. To validate the findings, three-master-single-slave time-delayed system is simulated in MATLAB/QUARC/Simulink environment. It is found that the proposed architecture can establish communication between multiple systems to achieve position and force tracking.

M. U. Asad (✉) · J. Gu · U. Farooq
Electrical and Computer Engineering, Dalhousie University, Halifax, Canada
e-mail: mh549096@dal.ca

J. Gu
e-mail: Jason.gu@dal.ca

U. Farooq
e-mail: umar.farooq@dal.ca

R. Dey · N. Adhikary
Deparment of Electrical Engineering, NIT, Silchar, Assam, India
e-mail: rajeeb@ee.nits.ac.in

R. Datta
Department of Mathematics, National Institute of Technology Agartala, Agartala, India

C. Chang
School of Biomedical Engineering, Health Sciences Center, Shenzhen University, Shenzhen 518060, China
e-mail: cqchang@szu.edu.cn

J. Gu et al. (eds.), *Communication and Control for Robotic Systems*, Smart Innovation, Systems and Technologies 229, https://doi.org/10.1007/978-981-16-1777-5_9

1 Introduction

In a conventional teleoperation system, the human operator generates motion commands to the slave robot located at a distant location with unknown environmental conditions and forces present in its surroundings. Teleoperation has become more popular with the advancement in computational, network models and robotic technologies which in turn increased the number of applications ranging from space, nuclear, underwater exploration, military, and telesurgery [1]. It also increases the human capacity to do tasks in a hazard environment using the telerobotics technology which was almost impossible for humans to accomplish beforehand. Telesurgery is also another milestone of this technology which reduces the human limitation to provide medical care and surgery to remote locations and communities around the globe [2]. Extension of this telerobotics technology has also appeared to improve the lifestyle of disabled people by incorporating human control [3].

A telerobotic framework, a single master/single slave (SM/SS) teleoperation system as depicted in Fig. 1 is intended to transfer the environmental forces and related information present at a remote end to master operator which is also known as telepresence. The extent of telepresence could be graded as system transparency. In an ideal scenario, the position and velocities of the master and slave would be tracked, and the master operator will experience the same amount of forces present between the slave robot and its surrounding environment. Efforts to make the system completely transparent can make the system unstable. Several other factors which can includes modeling issues of master/slave robots and uncertain environmental conditions, and delays present in communication channel [4].

The first SM/SS is being proposed in the mid-1940s by Goertz. Until now, several researchers have suggested numerous state-of-art theories to deal with teleoperation/telerobotics as discussed in [2, 5, 6]. In the past few decades, apart from using a unilateral/bilateral approach, researchers have shifted their interest towards using a multilateral framework. It becomes rapidly popular due to the availability of multiple

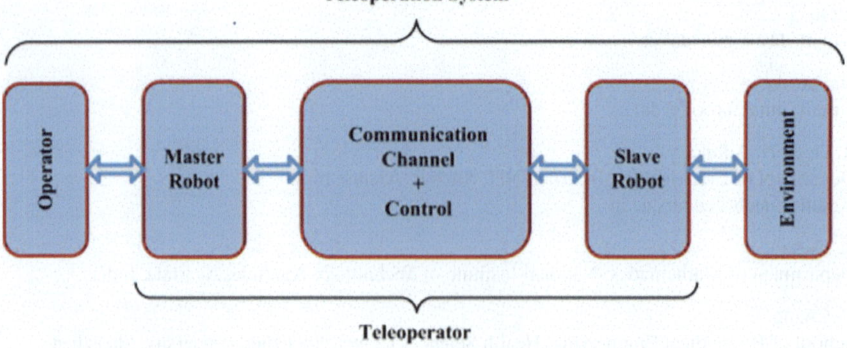

Fig. 1 A teleoperation system block diagram

collaborative scenarios among multiple master operator and multiple slave robots. It also becomes possible to incorporate human behavior [7] to improve the collaborative performance to do a certain task with more accuracy, efficacy, and dexterity [8].

Due to the complex framework of the multilateral teleoperation system, it is comparatively difficult to do stability analysis as compared to SM/SS system. Involvement of multi-robotic agents, unlimited topologies, and architectures can be implemented using a multilateral teleoperation system. To limit the variability in the multilateral framework, researchers have divided into four categories, as below:

1. Multi-Master/Single Slave (MM/SS), which permits multiple operators to control a single slave.
2. Single Master/Multi Slave (SM/MS), which allows us to control multiple slaves through a single master.
3. Multi-Master/Multi Slave (MS/MS), which enables collaboration of multiple operators and multiple slave robots to do a task.
4. Trilateral/Multilateral architecture refers to use a set of three human master and slave robots in total in this configuration [9].

The aim of the present paper is to explore the possibility of extending the transparent bilateral state convergence architecture to accommodate multiple systems. In addition, we want to keep the channel complexity at minimum when multiple systems are communicating. To achieve the former objective, multiple masters-slave interconnections are considered with new set of control gains while later objective is achieved by adopting composite variables from authors' earlier work [10, 11]. To determine the control-gains, method of state convergence is utilized which gives rise to several design conditions based on the evolution of error as an autonomous system. The proposal is validated through MATLAB simulation by considering a single degree-of-freedom tri-master-single-slave system.

This paper is organized as: Sect. 2 describes the architecture of proposed multilateral system. Section 3 presents the design methodology while simulation results are included in Sect. 4 followed by conclusions and references.

2 Architecture

Proposed MMSS architecture is shown in Fig. 2. Interactions between masters and slave are modeled by different control gains which will be determined by using the method of state convergence. Different parameters associated with the proposed architecture are defined below:

F_m^k: It represents the force exerted by the kth operator.
g_{s1}^k: It represents the influence of the kth operator's force on the slave system.
s_m^k: It represents the composite variable of kth master system and is formed by fusing the velocity and position signals of the respective master system.

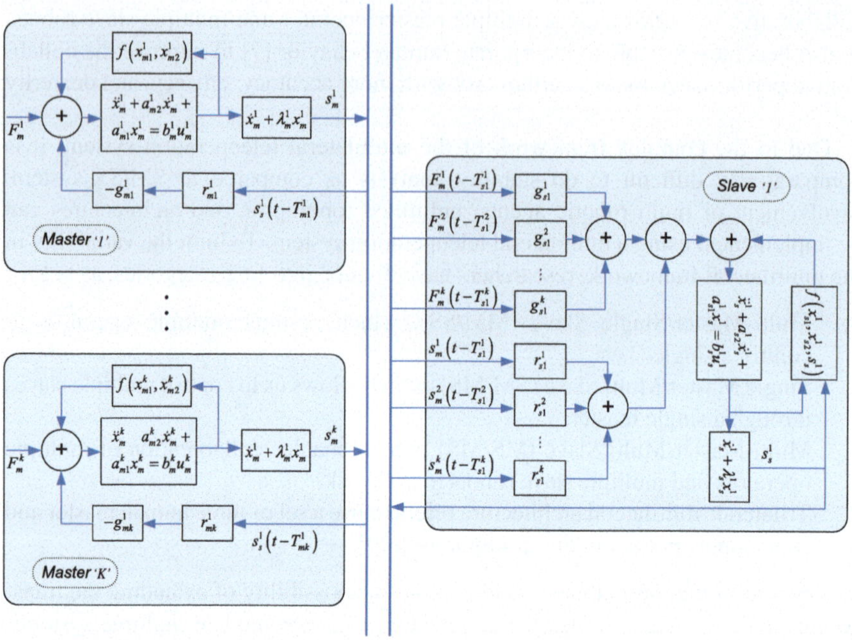

Fig. 2 Proposed MMSS architecture

s_s^1: It represents the composite variable of the slave system and is formed by fusing the velocity and position signals of the slave system.
r_{s1}^k: It models the effect of the motion of kth master system onto the slave system.
r_{mk}^1: It models the effect of the motion of slave system onto kth master system. It also carries environment information to the masters.
g_{mk}^1: It scales the environmental force as it is reflected the kth master system.
T_{mk}^1: It represents communication time delay from the slave to the kth master system.
T_{s1}^k: It represents communication time delay from the kth master system to the slave system.
k_s^1: It represents the stabilizing gain for the slave system.

Of these, $2k + 1$ parameters, $k_s^1, g_{s1}^k, r_{s1}^k$ are designed through state convergence methodology as detailed in next section.

3 Design Procedure

Consider a single degree of freedom MMSS teleoperation system:

$$
\begin{aligned}
\dot{x}_{m1}^{1} &= x_{m2}^{1} \\
\dot{x}_{m2}^{1} &= a_{m1}^{1}x_{m1}^{1} + a_{m2}^{1}x_{m2}^{1} + b_{m}^{1}u_{m}^{1} \\
&\vdots \\
\dot{x}_{m1}^{k} &= x_{m2}^{k} \\
\dot{x}_{m2}^{k} &= a_{m1}^{k}x_{m1}^{k} + a_{m2}^{k}x_{m2}^{k} + b_{m}^{k}u_{m}^{k} \\
\dot{x}_{s1}^{1} &= x_{s2}^{1} \\
\dot{x}_{s2}^{1} &= a_{s1}^{1}x_{s1}^{1} + a_{s2}^{1}x_{s2}^{1} + b_{s}^{1}u_{s}^{1}
\end{aligned} \tag{1}
$$

Let composite variables for the participant systems be defined as:

$$
\begin{aligned}
s_{m}^{1} &= x_{m2}^{1} + \lambda_{m1}x_{m1}^{1} \\
&\vdots \\
s_{m}^{k} &= x_{m2}^{k} + \lambda_{mk}x_{m1}^{k} \\
s_{s}^{1} &= x_{s2}^{1} + \lambda_{s1}x_{s1}^{1}
\end{aligned} \tag{2}
$$

We now introduce control inputs for the participant systems as:

$$
\begin{aligned}
u_{m}^{1} &= \frac{1}{b_{m}^{1}}\left(-a_{m1}^{1}x_{m1}^{1} - \overline{a_{m2}^{1} + \lambda_{m1}x_{m2}^{1}}\right) - g_{m1}^{1}r_{m1}^{1}s_{s}^{1}\left(t - T_{m1}^{1}\right) + F_{m}^{1} \\
&\vdots \\
u_{m}^{k} &= \frac{1}{b_{m}^{k}}\left(-a_{m1}^{k}x_{m1}^{k} - \overline{a_{m2}^{k} + \lambda_{mk}x_{m2}^{k}}\right) - g_{mk}^{1}r_{mk}^{1}s_{s}^{1}\left(t - T_{mk}^{1}\right) + F_{m}^{k} \\
u_{s}^{1} &= \frac{1}{b_{s}^{1}}\left(-a_{s1}^{1}x_{s1}^{1} - \overline{a_{s2}^{1} + \lambda_{s1}x_{s2}^{1}} + k_{s}^{1}s_{s}^{1}\right) + \sum_{j=1}^{k} r_{s1}^{j}s_{m}^{j}\left(t - T_{s1}^{j}\right) + \sum_{j=1}^{k} g_{s1}^{j}F_{m}^{j}\left(t - T_{s1}^{j}\right)
\end{aligned} \tag{3}
$$

The time-derivative of the composite variables (2) along with control inputs (3) yields the following closed loop composite systems:

$$
\begin{aligned}
\dot{s}_{m}^{1} &= -b_{m}^{1}g_{m1}^{1}r_{m1}^{1}s_{s}^{1}\left(t - T_{m1}^{1}\right) + b_{m}^{1}F_{m}^{1} \\
&\vdots \\
\dot{s}_{m}^{k} &= -b_{m}^{k}g_{mk}^{1}r_{mk}^{1}s_{s}^{1}\left(t - T_{mk}^{1}\right) + b_{m}^{k}F_{m}^{k} \\
\dot{s}_{s}^{1} &= k_{s}^{1}s_{s}^{1} + \sum_{j=1}^{k} b_{s}^{1}r_{s1}^{j}s_{m}^{j}\left(t - T_{s1}^{j}\right) + \sum_{j=1}^{k} b_{s}^{1}g_{s1}^{j}F_{m}^{j}\left(t - T_{s1}^{j}\right)
\end{aligned} \tag{4}
$$

By linearizing (4) and employing constant operators' force assumption, we obtain:

$$
\begin{aligned}
\dot{s}_m^1 &= -b_m^1 g_{m1}^1 r_{m1}^1 s_s^1 + b_m^1 g_{m1}^1 r_{m1}^1 T_{m1}^1 \dot{s}_s^1 + b_m^1 F_m^1 \\
&\vdots \\
\dot{s}_m^k &= -b_m^k g_{mk}^1 r_{mk}^1 s_s^1 + b_m^k g_{mk}^1 r_{mk}^1 T_{mk}^1 \dot{s}_s^1 + b_m^k F_m^k \\
\dot{s}_s^1 &= k_s^1 s_s^1 + \sum_{j=1}^{k} b_s^1 r_{s1}^j s_m^j - \sum_{j=1}^{k} b_s^1 r_{s1}^j T_{s1}^j \dot{s}_m^j + \sum_{j=1}^{k} b_s^1 g_{s1}^j F_m^j
\end{aligned} \tag{5}
$$

Further processing of (5) yields the following:

$$
\begin{aligned}
\dot{s}_m^1 &= -b_m^1 g_{m1}^1 r_{m1}^1 s_s^1 + b_m^1 g_{m1}^1 r_{m1}^1 T_{m1}^1 \dot{s}_s^1 + b_m^1 F_m^1 \\
&\vdots \\
\dot{s}_m^k &= -b_m^k g_{mk}^1 r_{mk}^1 s_s^1 + b_m^k g_{mk}^1 r_{mk}^1 T_{mk}^1 \dot{s}_s^1 + b_m^k F_m^k \\
\dot{s}_s^1 &= \frac{1}{\left(1 + \sum_{j=1}^{k} b_s^1 r_{s1}^j T_{s1}^j b_m^j g_{mj}^1 r_{mj}^1 T_{mj}^1\right)} \\
&\times \left(\left(k_s^1 + \sum_{j=1}^{k} b_s^1 r_{s1}^j T_{s1}^j b_m^j g_{mj}^1 r_{mj}^1\right) s_s^1 + \sum_{j=1}^{k} b_s^1 r_{s1}^j s_m^j + \sum_{j=1}^{k} \left(b_s^1 g_{s1}^j - b_s^1 r_{s1}^j T_{s1}^j b_m^j\right) F_m^j\right)
\end{aligned} \tag{6}
$$

Let us define composite error as:

$$
s_e = s_s^1 - \sum_{j=1}^{k} \alpha_{1j} s_m^j \tag{7}
$$

The time derivative of (7) along with (6) yields composite error dynamics as:

$$
\begin{aligned}
\dot{s}_e = &\left(k_s^1 + \left(b_s^1 r_{s1}^j T_{s1}^j + \alpha_{1j}\right) b_m^j g_{mj}^1 r_{mj}^1\right) s_s^1 + \sum_{j=1}^{k} b_s^1 r_{s1}^j s_m^j \\
&- \sum_{j=1}^{k} \left(b_s^1 r_{s1}^j T_{s1}^j + \alpha_{1j}\right)\left(b_m^j g_{mj}^1 r_{mj}^1 T_{mj}^1\right) \dot{s}_s^1 \\
&+ \sum_{j=1}^{k} \left(b_s^1 g_{s1}^j - \left(b_s^1 r_{s1}^j T_{s1}^j + \alpha_{1j}\right) b_m^j\right) F_m^j
\end{aligned} \tag{8}
$$

Further processing of (8) yields:

$$\dot{s}_e = \left(\begin{array}{l} k_s^1 + \left(b_s^1 r_{s1}^j T_{s1}^j + \alpha_{1j}\right) b_m^j g_{mj}^1 r_{mj}^1 \\ -\dfrac{\sum\limits_{j=1}^{k} \left(b_s^1 r_{s1}^j T_{s1}^j + \alpha_{1j}\right)\left(b_m^j g_{mj}^1 r_{mj}^1 T_{mj}^1\right)}{\left(1 + \sum\limits_{j=1}^{k} b_s^1 r_{s1}^j T_{s1}^j b_m^j g_{mj}^1 r_{mj}^1 T_{mj}^1\right)} \left(k_s^1 + \sum\limits_{j=1}^{k} b_s^1 r_{s1}^j T_{s1}^j b_m^j g_{mj}^1 r_{mj}^1\right) \end{array} \right) s_s^1$$
$$+ \sum_{j=1}^{k} b_s^1 r_{s1}^j \left(1 - \frac{\sum\limits_{j=1}^{k} \left(b_s^1 r_{s1}^j T_{s1}^j + \alpha_{1j}\right)\left(b_m^j g_{mj}^1 r_{mj}^1 T_{mj}^1\right)}{\left(1 + \sum\limits_{j=1}^{k} b_s^1 r_{s1}^j T_{s1}^j b_m^j g_{mj}^1 r_{mj}^1 T_{mj}^1\right)} \right) s_m^j$$
$$+ \sum_{j=1}^{k} \left(\begin{array}{l} \left(b_s^1 g_{s1}^j - \left(b_s^1 r_{s1}^j T_{s1}^j + \alpha_{1j}\right) b_m^j\right) \\ -\dfrac{\sum\limits_{j=1}^{k} \left(b_s^1 r_{s1}^j T_{s1}^j + \alpha_{1j}\right)\left(b_m^j g_{mj}^1 r_{mj}^1 T_{mj}^1\right)}{\left(1 + \sum\limits_{j=1}^{k} b_s^1 r_{s1}^j T_{s1}^j b_m^j g_{mj}^1 r_{mj}^1 T_{mj}^1\right)} \sum\limits_{j=1}^{k} \left(b_s^1 g_{s1}^j - b_s^1 r_{s1}^j T_{s1}^j b_m^j\right) \end{array} \right) F_m^j \tag{9}$$

Let us introduce the following assignments:

$$\begin{aligned} & k_s^1 + \left(b_s^1 r_{s1}^j T_{s1}^j + \alpha_{1j}\right) b_m^j g_{mj}^1 r_{mj}^1 \\ & \quad - \frac{\sum\limits_{j=1}^{k} \left(b_s^1 r_{s1}^j T_{s1}^j + \alpha_{1j}\right)\left(b_m^j g_{mj}^1 r_{mj}^1 T_{mj}^1\right)}{\left(1 + \sum\limits_{j=1}^{k} b_s^1 r_{s1}^j T_{s1}^j b_m^j g_{mj}^1 r_{mj}^1 T_{mj}^1\right)} \left(k_s^1 + \sum_{j=1}^{k} b_s^1 r_{s1}^j T_{s1}^j b_m^j g_{mj}^1 r_{mj}^1\right) \\ & \quad = -q \end{aligned} \tag{10}$$

$$b_s^1 r_{s1}^j \left(1 - \frac{\sum\limits_{j=1}^{k} \left(b_s^1 r_{s1}^j T_{s1}^j + \alpha_{1j}\right)\left(b_m^j g_{mj}^1 r_{mj}^1 T_{mj}^1\right)}{\left(1 + \sum\limits_{j=1}^{k} b_s^1 r_{s1}^j T_{s1}^j b_m^j g_{mj}^1 r_{mj}^1 T_{mj}^1\right)} \right) = \alpha_{1j} q \tag{11}$$

$$\frac{\sum\limits_{j=1}^{k} \left(b_s^1 r_{s1}^j T_{s1}^j + \alpha_{1j}\right)\left(b_m^j g_{mj}^1 r_{mj}^1 T_{mj}^1\right)}{\left(1 + \sum\limits_{j=1}^{k} b_s^1 r_{s1}^j T_{s1}^j b_m^j g_{mj}^1 r_{mj}^1 T_{mj}^1\right)} \sum_{j=1}^{k} \left(b_s^1 g_{s1}^j - b_s^1 r_{s1}^j T_{s1}^j b_m^j\right)$$

$$= b_s^1 g_{s1}^j - \left(b_s^1 r_{s1}^j T_{s1}^j + \alpha_{1j}\right) b_m^j \tag{12}$$

The above assignments allow the composite error to evolve as an autonomous system. Thus, convergence of composite error is ensured i.e. $s_s^1 - \sum_{j=1}^{k} \alpha_{1j} s_m^j \rightarrow 0$. Since, $\dot{x}_{s2}^1 + \lambda_{s1} x_{s2}^1 = \dot{s}_s^1 \rightarrow 0$, we have $x_{s1}^1 = \sum_{j=1}^{k} \alpha_{1j} x_{m1}^j$ in steady-state. In this way, position coordination is achieved. A similar analysis shows that force reflected to all the masters converges to the environmental force when applied forces are the same.

4 Simulation Results

To validate the proposed framework, 3MSS is adopted and tested in MATLAB Simulink environment. The following system parameters are assumed:

$$\begin{aligned}
&a_{m1}^k = 0, a_{m2}^k = -7.1429, b_m^k = 0.2656, g_{mk}^1 = 1 \\
&a_{s1}^1 = 0, a_{s2}^1 = -6.25, b_m^k = 0.2729 \\
&k_e^1 = 10, \alpha_{11} = 0.1, \alpha_{12} = 0.5, \alpha_{13} = 0.4 \\
&T_{m1}^1 = T_{s1}^1 = 0.1, T_{m2}^1 = T_{s1}^2 = 0.15, T_{m3}^1 = T_{s1}^3 = 0.2
\end{aligned} \tag{13}$$

The control gains are obtained as a solution of design conditions (10)–(12):

$$\begin{aligned}
&g_{s1}^1 = 0.7824, g_{s1}^2 = 4.5637, g_{s1}^3 = 4.1726 \\
&k_s^1 = -52.7717 \\
&r_{s1}^1 = 9.8187, r_{s1}^2 = 49.0936, r_{s1}^3 = 39.2749
\end{aligned} \tag{14}$$

Now, under constant operators' forces of $F_m^k = 0.25$ N, we simulate 3MSS system and the results are shown in Fig. 3. Slave is following the weighted motion of the master system while environment force is also reflected to the operators. It is also evident that transient phase of force reflection behavior is not appropriate while steady-state phase truly reflects the environment to the operators.

We have also performed semi-real-time experiments in MATLAB/QUARC/Simulink environment where one haptic device is used to drive three master systems. During the operation, force from the virtual slave environment is transmitted to the haptic device via the time-delayed communication channel. The recorded experimental results are displayed in Fig. 4. It is evident that slave is synchronized to the combined motion of the master systems while environment force is also being reflected to the masters. The results obtained from both the simulations and experiment validate the proposed extended architecture.

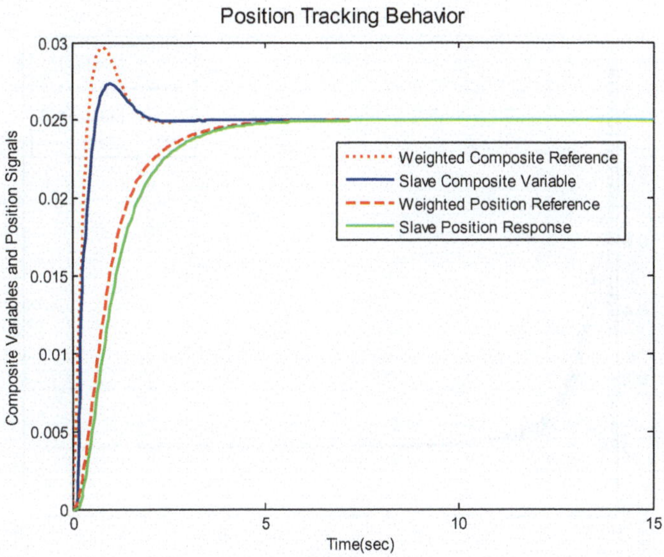

(a)

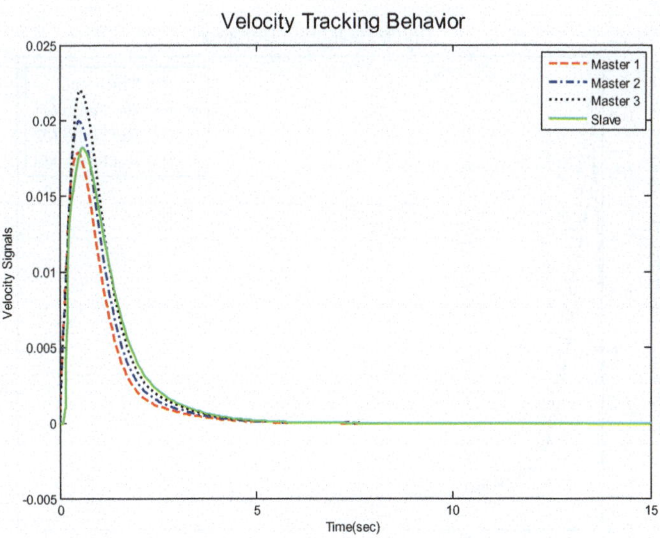

(b)

Fig. 3 Simulation results of 3MSS teleoperation system, **a** Position response, **b** velocity signals, **c** control inputs, **d** force reflection

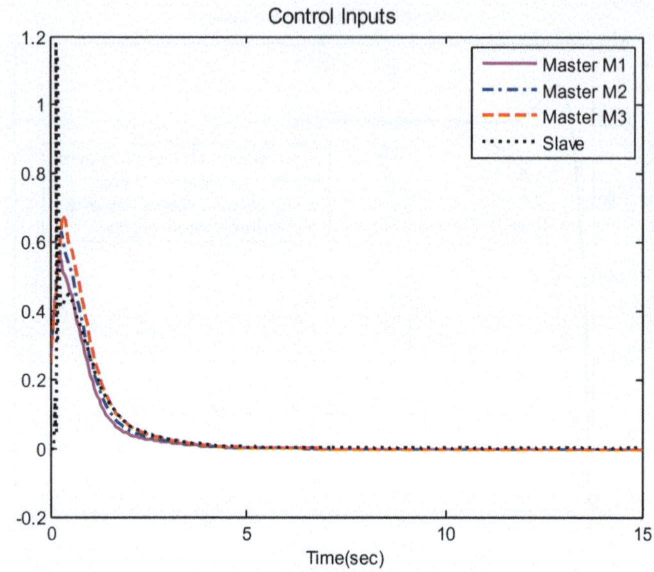

(c)

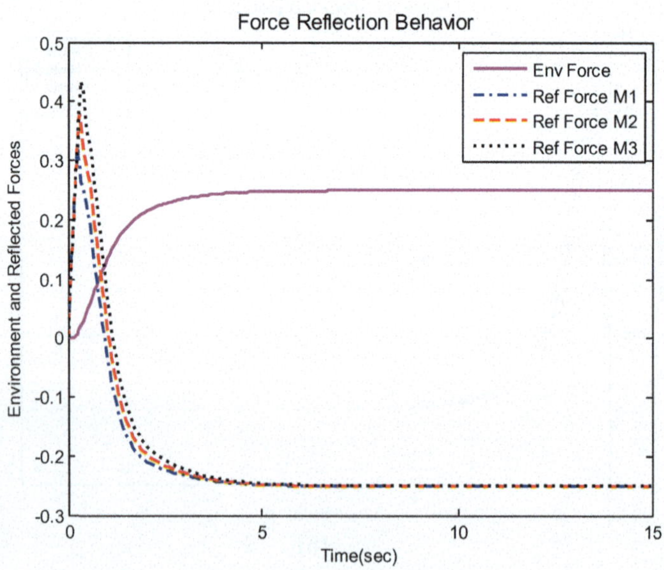

(d)

Fig. 3 (continued)

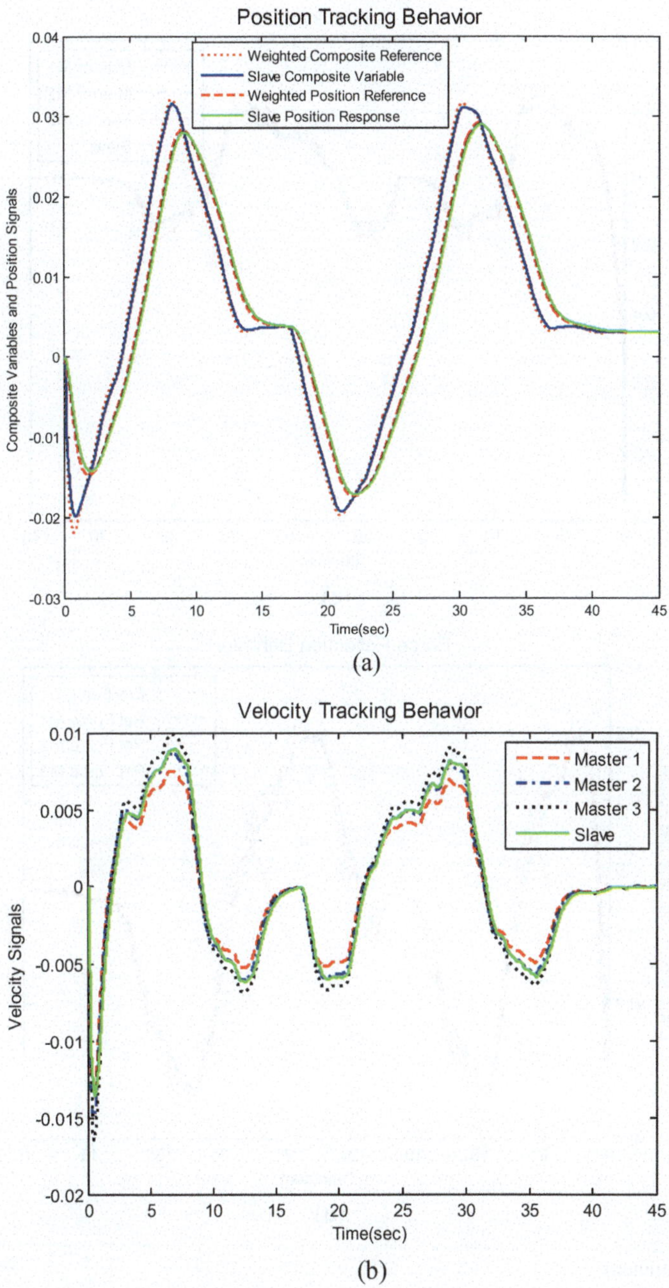

Fig. 4 Experimental results of 3MSS teleoperation system, **a** position response, **b** velocity signals, **c** control inputs, **d** force reflection

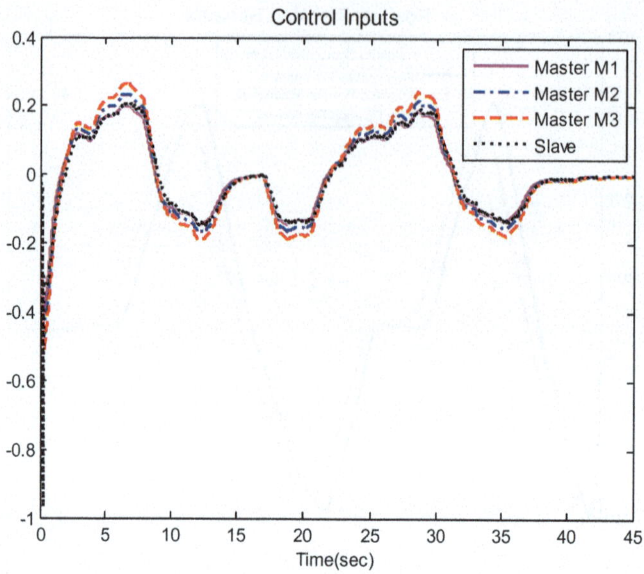

(c)

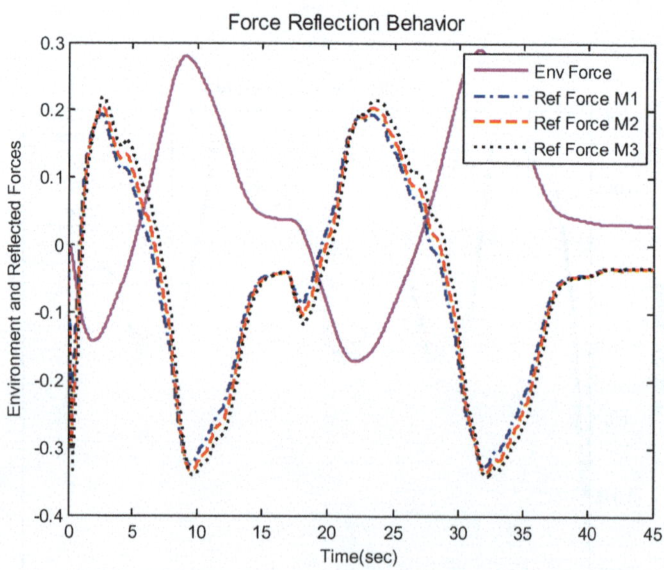

(d)

Fig. 4 (continued)

5 Conclusions

In this paper, design of a multilateral teleoperation system is presented by considering multiple masters and single slave system. The proposed architecture is extended version of transparent state convergence architecture which is developed earlier for bilateral control. Extension is realized by considering additional interactions and control gains. At the same time, concept of bilateral composite state convergence scheme is employed to reduce the control gains and to simplify the communication channel. Method of state convergence is finally utilized to derive the design conditions and control gains are determined as a solution of coupled equations. MATLAB simulations show that proposed architecture possesses position and force tracking ability. Future works involves robustifying the proposed architecture against parameter variations.

References

1. Ferre, M., Buss, M., Aracil, R., Melchiorri, C., Balaguer, C., (Eds.): Advances in Telerobotics. Springer-Verlag, Berlin (2007)
2. Hokayem, P., Spong, M.: Bilateral teleoperation: an historical survey. Automatica **42**(12), 2035–2057 (2006)
3. Yu, W.: Intelligent telerobotic assistance for enhancing manipulation capabilities of persons with disabilities. Ph.D. Thesis, University of South Florida (2004)
4. Malysz, P.: A kinematic control framework for asymmetric semi-autonomous teleoperation systems. Ph.D. Thesis, Department of Electrical & Computer engineering, McMaster University (2011)
5. Passenberg, C., Peer, A., Buss, M.: A survey of environment-, operator-, and task-adapted controllers for teleoperation systems. Mechatronics **20**(7), 787–801 (2010)
6. Niemeyer, G., Preusche, C., Stramigioli, S., Lee, D.: Telerobotics. In: Springer Handbook of Robotics, pp. 1085–1108. Springer, Berlin, Germany (2016)
7. Goldberg, K., et al.: Collaborative teleoperation via the internet. In: Proceedings of IEEE International Conference on Robotics and Automation, pp. 2019–2024 (2000)
8. Shahbazi, M., Atashzar, S.F., Talebi, H.A., Patel, R.V.: Novel cooperative teleoperation framework: Multi-master/single-slave system. IEEE/ASME Trans. Mechatron. **20**(4), 1668–1679 (2015)
9. Shahbazi, M., Atashzar, S.F., Patel, R.V.: A systematic review of multilateral teleoperation systems. IEEE Trans. Haptics **11**(3), (2018)
10. Farooq, U., Gu, J., El-Hawary, M., Abbas, G., Asad, M.U.: Fuzzy model based bilateral control design of nonlinear teleoperation system using method of state convergence. IEEE Access **4**, 4119–4135 (2016)
11. Asad, M.U., Farooq, U., Gu, J., Abbas, G., Liu, R., Balas, V.E.: A composite state convergence scheme for bilateral teleoperation system. IEEE/CAA J Autom. Sinica **6**(5), 1166–1178 (2019)

Event-Triggered Integral Sliding Mode Control for an Uncertain Euler–Lagrange System with Actuator Saturation

Krishanu Nath, Asifa Yesmin, and Manas Kumar Bera

Abstract This paper presents the design of integral sliding mode controller for an Euler–Lagrange system in event-triggered framework considering the actuator saturation. The event-triggered control strategy is used with an aim that the communication load will be reduced over the feedback link in networked control system by adapting a need-based information exchange policy. The integral sliding surface is designed with saturation function, whose finite time reachability is established with anti-saturated sliding mode controller. The closed loop system stability has been ensured by showing the convergence of state trajectories within any desired bound neighbourhood of the origin. The event-triggered sliding mode control is shown to be free from accumulation of events, i.e. Zeno free behaviour has been guaranteed. Finally, with these theoretical formulations, the numerical simulations have been shown to prove the effectiveness of the proposed control strategy.

1 Introduction

Euler–Lagrange (EL) system describes a wide class of physical systems of practical interest specially, robot manipulators, quad rotor helicopters, tethered satellite systems, and spacecraft attitude control systems. There has been a great effort by the control engineers to improve the performance of these class of nonlinear systems. One of the major practical issues which is frequently encountered during the operation of any controlled system is the input saturation. The performance of the system may be degraded or even instability may cause if it is ignored in the design. The

K. Nath · A. Yesmin · M. K. Bera (✉)
Department of Electronics and Instrumentation Engineering, NIT Silchar, Silchar, India
e-mail: manas@ei.nits.ac.in

K. Nath
e-mail: krishanu_rs@ei.nits.ac.in

A. Yesmin
e-mail: asifa_rs@ei.nits.ac.in

J. Gu et al. (eds.), *Communication and Control for Robotic Systems*, Smart Innovation, Systems and Technologies 229, https://doi.org/10.1007/978-981-16-1777-5_10

situation will be more complex if the system is affected by the external disturbances which is inevitable when the system is expected to work in harsh environment.

Several control strategies for EL systems have been investigated such as proportional–integral–derivative control, adaptive control, optimal control, and many more can be found in the literature [1–3]. The different control strategies of EL systems are still an evolving research area, and it has been widely investigated to achieve the desired performance of these special class of systems.

To cope up with the inherent nonlinear dynamics of EL systems with input saturation and external disturbances, the design of robust control strategy is the obvious choice. One of the most extensively used robust control mechanisms is the sliding mode control (SMC) which has been widely applied to control the EL system [4, 5]. SMC is a powerful nonlinear control technique that ensures the stability of the plant against matched uncertainties [6]. The conventional SMC has two motions in its response, first is the reaching phase and then is the sliding phase. The reaching phase is the motion of the state trajectory from its initial condition to the designed manifold, and the corresponding time required is called the reaching time. Once the state trajectory hits the manifold, the trajectory slides on the hyper-plane, and this phase is called the sliding phase. The robustness of the system with respect to the perturbations can be guaranteed only during the sliding phase. So, to improve the robustness of the system, integral SMC (ISMC) was proposed which can completely remove the reaching phase, and robustness is guaranteed from the initial time response [7, 8]. ISMC has been successfully applied in the EL system in [9–11]. Input saturation may cause overall performance degradation of the EL systems. So, anti-saturation control strategies are investigated extensively in the literatures. [12–14].

For the practical implementation of these control strategies, the use of digital computer is inevitable, and in classical digital control system, the sensing and actuation action are performed periodically in time. The revolutionary development of control and communication are leading to an era of networked and wireless control systems such as cyber-physical systems (CPSs) and networked control systems (NCSs). These systems are popular due to their numerous advantages, but at the same time, they pose challenges with limited computation, communication, and energy resources. The closed loop performance of the NCSs is greatly affected by the load of the communication network because the quality of service of the network is influenced by the induced delays or packet losses which degrades the system performance or even cause the instability of the control loop system. To efficiently use the available resources, to reduce the adverse effect of network induced imperfections, and to reduce the load of the communication network, it is desirable to develop a robust control strategy where the control action is updated when the system really needs attention. In such situation, classical time-triggered method to implement the control may lead to increase the communication burden and unnecessary energy consumption[15].

The control paradigm which considers designing of controllers with efficient usage of communication resources while preserving desirable closed loop dynamics is the event-triggering mechanism. The main idea of this design is to introduce a need-based sampling strategy which is governed by some predefined threshold [16, 17]. Essentially a stabilizing (event) condition is maintained such that the application of the control signal at certain time instants, known as the triggering instants, to the plant can ensure the desired stability. Eventually, event-based control strategy has become a powerful tool for the modern applications because the remarkable outcomes of this technique overweigh its counterpart. Considerable amount of research on event-triggered SMC has been done for both linear and nonlinear system [18–20]. The designs of event-triggered-based ISMC can be found in literatures as in [21–23]. But the event-triggered implementation of ISMC for the EL system considering the input saturation and external disturbances is still an open problem.

The aim of this paper is to redesign an ISMC technique [9] based on event-triggering mechanism for the EL system considering the input saturation. Here, the event-triggering condition is designed to guarantee the desired behaviour of the EL system.

The main contributions of this paper are as follows:

- The design of the ISMC law is proposed with actuator saturation in an event-triggered framework to guarantee the reduction in communication burden, system stability, and robustness of the closed loop system.
- An unique event condition is proposed which ensures the Zeno free control to stabilize the EL system in the presence of matched uncertainties and actuator saturation.
- The effect of event design parameter on the performance of the system is analysed.

Notation: $\mathbb{R}$ and $\mathbb{Z}$ represent the set of real numbers and integers, respectively. The absolute value is given as $|a|$ for the given $a \in \mathbb{R}$. Similarly, for any vector $a \in \mathbb{R}^n$, $\|a\|$ denotes the 2-norm of the vector. For a given matrix $Q \in \mathbb{R}^{m\times n}$, we write by $\|Q\|$ the induced 2-norm of this matrix. The symmetric matrix $\psi = \psi^T$ and $\psi > 0 (\geq 0)$ denotes positive (positive semi) definite matrix, i.e. $x\psi^\top x > 0(\geq 0)$ for all $x \in \mathbb{R}^n$.

2 Problem Statement

Consider a class of EL system dynamics can be represented by

$$M(q)\ddot{q}(t) + C(q,\dot{q})\dot{q}(t) + G(q) = u(t) + d(t) \tag{1}$$

where $q(t) \in \mathbb{R}^n$ represents the generalized coordinates, $\dot{q}(t) \in \mathbb{R}^n$ represents the generalized velocities, and $u(t) \in \mathbb{R}^n$ is the control input. The $M(q) \in \mathbb{R}^{n\times n}$ is the generalized inertia matrix, $C(q,\dot{q}) \in \mathbb{R}^{n\times n}$ represents the generalized centripetal Coriolis matrix, and $G(q) \in \mathbb{R}^n$ is the generalized gravitational vector. The disturbance $d(t)$ is assumed to be matched.

Some properties of the EL system (1) are stated below [24].

Property 1 *The inertia matrix is symmetric and positive definite, and also, it satisfies the inequality* $m_1 I \leq M(q) \leq m_2 I$ *for some positive* m_1 *and* m_2.

Property 2 *The centripetal Coriolis matrix satisfies inequality* $\|C(q, \dot{q})\| \leq c_m \|\dot{q}\|$ *for some* $c_m > 0$.

Property 3 *The gravitational and frictional vectors satisfy* $\|G(q)\| \leq g_m$ *for some* $g_m > 0$.

Property 4 *The matrix* $\dot{M}(q) - 2C(q, \dot{q})$ *is skew symmetric, i.e.* $z^\top((q) - 2C(q, \dot{q}))z = 0$ *for* $z \in \mathbb{R}^n$.

The following assumptions are true for the system described by (1).

Assumption 1 The disturbance signal $d(t)$ is bounded by constant d_m $\forall\, t > 0$[25].

Assumption 2 All the system states measurements are directly available for use, and also, the domain of interest is $[q^\top \dot{q}^\top]^\top \in \Omega \in \mathbb{R}^{2n}$.

Assumption 3 The actuator has constraint to have maximum input of u_{max}[9].

Assumption 4 The EL system (1) is connected to a cyber-physical space, and the control action is generated from a central controlling unit through a wireless transmission. The communication network is assumed to be ideal, i.e. it has no delays or packet loss.

Note: *To reduce the complexity in the expressions, the argument of the variables of* $M(q)$, $C(q, \dot{q})$, and $G(q)$ *has not been used in the later part of the paper.*

The system (1) is considered to have uncertainty in the form of external disturbance. The aim here is to design a robust sliding mode control such that the states $q(t)$ and $\dot{q}(t)$ of the system (1) can be stabilized at the equilibrium, i.e. $q(t),\ \dot{q}(t) \to 0$ as $t \to \infty$ in presence of the uncertainty. As the system and the controller are in a cyber-physical space, connected through a wireless communication channel, the traditional sampled-data control is likely to cause greater number of control updates. A control law is designed which is implemented with need-based sampling, known as event-triggering such that the information exchange is minimum which reduces the communication load by reducing the control updates keeping the system performance consistent. The controller design also needs to take care of the actuator saturation which is not only theoretically challenging but practically imperative.

One such robust approach is addressed by using ISMC design [9] where the sliding surface is defined as

$$s(t) = \dot{q}(t) + \int_0^t M^{-1}\Big(C\dot{q}(t) + k_1 \mathrm{sat}(\lambda_1 q(t)) + k_2 \mathrm{sat}(\lambda_2 \dot{q}(t))\Big)dt \qquad (2)$$

and the continuous-time control law that drives the system trajectories to the manifold is given by

$$u(t) = -k_1 \mathrm{sat}(\lambda_1 q(t)) - k_2 \mathrm{sat}(\lambda_2 \dot{q}(t)) - k_3 \mathrm{sgn}(s(t)) \tag{3}$$

where $k_1 > 0, k_2 > 0, k_3 > d_m, \lambda_1 > 0$ and $\lambda_2 > 0$. This control law is sufficient to make $s = 0$ in finite time which guarantees the occurrence of ideal sliding motion.

The saturation function is defined as

$$\mathrm{sat}(x) = \begin{cases} \mathrm{sgn}(x), & \text{if } |x| > 1 \\ x, & \text{otherwise.} \end{cases} \tag{4}$$

The time triggered implementation of control law (3) requires periodic sensing and actuation, which increases the burned on the shared communication network and also on the shared central processing unit. The frequency of information exchange and the number of control updates can be reduced if a need-based sampling called the event-triggered paradigm can be introduced. In this mechanism, an *event condition* is continuously monitored, such that whenever it is violated, the sampling occurs, and the control is updated. The time between two consecutive triggering instance is known as *inter-event execution time* and is denoted by

$$\Delta T_k := t_{k+1} - t_k \tag{5}$$

where t_k and t_{k+1} are consecutive sampling instances. To establish the stability of the event-triggered mechanism, it is important to show that the minimum inter-event execution time is positive. Unless no triggering instant is generated, the plant runs in open-loop mode under the action of previous control input which is applied through a zero-order hold device. The block diagram of closed loop plant with the event-triggered scheme in the feedback path is shown in Fig. 1. One important feature of the event-triggered SMC is that the ideal sliding motion cannot be achieved due to the aperiodic control effort. However, the state trajectories remain confined near the designed hyper-plane/manifold.

The next section showcases the design of event-based integral sliding mode controller for the system (1) where the control law (3) is redesigned because the same gain condition which is evaluated for periodic sampling is not sufficient to stabilize the system.

3 Main Results

To extend the continuous-time control in event-triggered paradigm, the control law has to be formulated in terms of the sampled sliding variable $s(t_k)$ and the sampled states $q(t_k)$ and $\dot{q}(t_k)$. The control law (3) can be rewritten for event-triggered implementation as

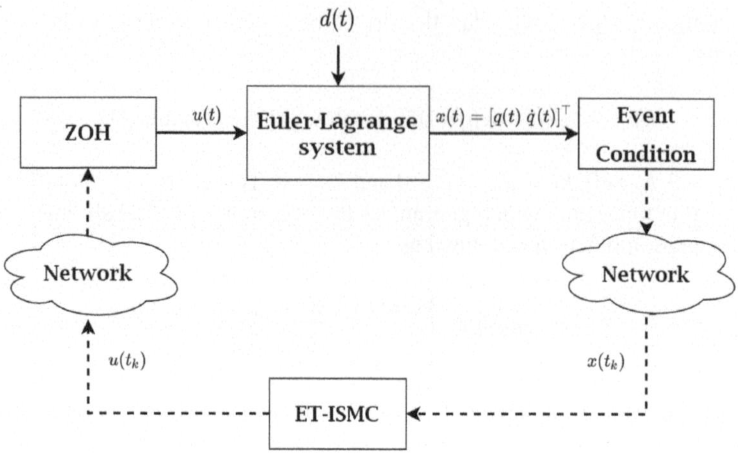

Fig. 1 Block diagram of the event-triggered control configuration

$$u(t) = -k_1 \mathrm{sat}(\lambda_1 q(t_k)) - k_2 \mathrm{sat}(\lambda_2 \dot{q}(t_k)) - k_3 \mathrm{sgn}(s(t_k)) \tag{6}$$

where $\lambda_1 = \frac{1}{k_1}$, $\lambda_2 = \frac{1}{k_2}$. The choice of gains $k_1,\ k_2$, and k_3 is introduced later. The triggering condition that governs the aperiodic sampling is given by

$$\varepsilon(t) \le \alpha \tag{7}$$

where

$$\varepsilon(t) = \|q(t) - q(t_k)\| + \|\dot{q}(t) - \dot{q}(t_k)\| \tag{8}$$

and $\alpha > 0$ is the event design parameter. The choice of this design parameter is crucial as this decides the bound of system states, sliding variable, and the minimum inter-event execution time which is established in theorems later. Next we introduce the dynamics of the sliding variable and notion of practical sliding mode which will be required later.

The dynamics of the sliding variable can be obtained by differentiating (2) with respect to time given by

$$\dot{s}(t) = \ddot{q}(t) + M^{-1}\Big(C\dot{q}(t) + k_1 \mathrm{sat}(\lambda_1 q(t)) - k_2 \mathrm{sat}(\lambda_2 \dot{q}(t))\Big). \tag{9}$$

Definition 1 Utkin [6] The condition which can ensures the finite time reaching of the sliding variable onto the sliding manifold is called the reachability condition which is satisfied iff

$$s^{\top}(t)\dot{s}(t) \le \bar{\eta}\|s(t)\| \tag{10}$$

where $\bar{\eta}$ is a positive scalar.

Definition 2 Bandyopadhyay and Behera [20] A system is said to have practical sliding motion if the system state trajectories remain in the vicinity of the manifold such that the sliding variable s remains bounded as

$$\|s\| \leq \rho \tag{11}$$

where ρ is a positive scalar.

With these preliminaries, next we present the following theorem.

Theorem 1 *Consider the EL system* (1) *with the sliding variable* (2) *and the event-based control law* (6) *with the gains*

$$\begin{aligned} k_1 &> 0 \\ k_2 &> \beta \\ k_3 &> g_m + d_m + \alpha + \eta \end{aligned} \tag{12}$$

where β is a positive scalar defined in Appendix, η is also a positive scalar, and α is the event design parameter satisfying the event condition (7)*. Then, the sliding variable is contained in the invariant set*

$$\mathcal{B}_s = \left\{ q(t) \in \Omega : \|s(t)\| \leq \alpha + \Xi \right\} \tag{13}$$

where $\Xi = \left\| \int_{t_k}^{t} M^{-1} \left(C\dot{q}(t) + k_1 sat(\lambda_1 q(t)) + k_2 sat(\lambda_2 \dot{q}(t)) \right) dt \right\|$. *Also, the states $q(t)$ and $\dot{q}(t)$ are contained in the set $\mathcal{B}_q$ and $\mathcal{B}_{\dot{q}}$ given by*

$$\mathcal{B}_q = \left\{ q(t) \in \Omega : q(t) \leq \alpha \right\} \tag{14}$$

$$\mathcal{B}_{\dot{q}} = \left\{ q(t) \in \Omega : \dot{q}(t) \leq max\left(\alpha, \beta \right) \right\} \tag{15}$$

Proof The proof has two parts, first to establish the existence of practical sliding mode and then verify the boundedness of the system states.

Existence of practical sliding motion:

Consider a trivial Lyapunov candidate function in the time interval $t \in [t_k, t_{k+1})$

$$V_1 = \frac{1}{2} s^{\top}(t) s(t).$$

Obtaining its time derivative along the dynamics (1), we get

$$\dot{V}_1 = s^\top(t)\Big[\ddot{q}(t) + M^{-1}\Big(C\dot{q}(t) + k_1\mathrm{sat}(\lambda_1 q(t)) + k_2\mathrm{sat}(\lambda_2\dot{q}(t))\Big)\Big]$$
$$= s^\top(t)M^{-1}\Big[-C\dot{q}(t) - G + u(t) + d(t) + \Big(C\dot{q}(t) + k_1\mathrm{sat}(\lambda_1 q(t))$$
$$+ k_2\mathrm{sat}(\lambda_2\dot{q}(t))\Big)\Big].$$

Substituting the control (6) in the above equation and simplifying it, we get

$$\dot{V}_1 = s^\top(t)M^{-1}\Big[-G + d(t) + k_1\mathrm{sat}(\lambda_1 q(t)) - k_1\mathrm{sat}(\lambda_1 q(t_k))$$
$$+ k_2\mathrm{sat}(\lambda_2\dot{q}(t)) - k_2\mathrm{sat}(\lambda_2\dot{q}(t_k)) - k_3\mathrm{sgn}(s(t_k))\Big)\Big]. \tag{16}$$

The proof has to be continued for the two cases

- **Case 1: None of the sliding surfaces reach their respective manifold.** When none of the sliding variables have reached their respective manifold, it can be inferred that $\mathrm{sgn}(s(t)) = \mathrm{sgn}(s(t_k))$. Thus, applying the properties of norm, (16) can be converted to

$$\dot{V}_1 \leq \|s(t)\|\|M^{-1}\|\Big[\|G\| + \|d(t)\| + k_1\|\mathrm{sat}(\lambda_1 q(t)) - \mathrm{sat}(\lambda_1 q(t_k))\|$$
$$+ k_2\|\mathrm{sat}(\lambda_2\dot{q}(t)) - \mathrm{sat}(\lambda_2\dot{q}(t_k))\|\Big] - k_3\|M^{-1}\|\|s(t)\|.$$

Using the fact that the Lipchitz constant of sat(.) function is $L = 1$, which is applied to the third and fourth term in the above inequality along with the substitution of maximum bounds of G and $d(t)$ as per the Assumption 1 and the Property 3, we have

$$\dot{V}_1 \leq \|s(t)\|\|M^{-1}\|\Big[g_m + d_m + k_1\lambda_1\|q(t) - q(t_k)\|$$
$$+ k_2\lambda_2\|\dot{q}(t) - \dot{q}(t_k)\|\Big] - k_3\|M^{-1}\|\|s(t)\|.$$

Putting the values $\lambda_1 = \frac{1}{k_1}$ and $\lambda_2 = \frac{1}{k_2}$ as per the design consideration and using the event condition (7), the above inequality results in

$$\dot{V}_1 \leq \|s(t)\| \|M^{-1}\| \Big[g_m + d_m + \|q(t) - q(t_k)\| + \|\dot{q}(t) - \dot{q}(t_k)\| \Big] - k_3 \|M^{-1}\| \|s(t)\|$$

$$\dot{V}_1 \leq \|s(t)\| \|M^{-1}\| \Big[g_m + d_m + \alpha - k_3 \Big].$$

Using the gain condition (12), we finally arrive at

$$\dot{V}_1 \leq -\eta_1 \|s\| \|M^{-1}\|. \tag{17}$$

As the mass matrix M is bounded by a positive scalar as stated in Property 1, thus the reachability condition is satisfied which ensures finite time dynamics of the sliding variable.

- **Case 2: All trajectories have reached their respective manifold.**
  Once all the trajectories hit the manifold, it can be shown that they are contained in the set $\mathcal{B}_s$. As the relation (7) holds, this restricts the trajectory of $s(t)$ to be within a band, which can be easily calculated by the maximum deviation of the $s(t)$ in the interval $[t_k, t_{k+1})$.

$$\begin{aligned}
\|s(t) - s(t_k)\| =& \|\dot{q}(t) - \dot{q}(t_k) + \int_0^t M^{-1} \Big(C\dot{q}(t) + k_1 \mathrm{sat}(\lambda_1 q(t)) \\
&+ k_2 \mathrm{sat}(\lambda_2 \dot{q}(t)) \Big) dt - \int_0^{t_k} M^{-1} \Big(C\dot{q}(t) \\
&+ k_1 \mathrm{sat}(\lambda_1 q(t)) + k_2 \mathrm{sat}(\lambda_2 \dot{q}(t)) \Big) dt \| \\
\leq& \|\dot{q}(t) - \dot{q}(t_k)\| + \left\| \int_{t_k}^t M^{-1} \Big(C\dot{q}(t) + k_1 \mathrm{sat}(\lambda_1 q(t)) \right. \\
&\left. + k_2 \mathrm{sat}(\lambda_2 \dot{q}(t)) \Big) dt \right\| \\
\leq& \alpha + \Xi.
\end{aligned} \tag{18}$$

Thus, due to event condition (7), once the sliding variable reaches $\mathcal{B}_s$, then it is always contained in this invariant set which establishes the practical sliding mode. Next we need to affirm the stability of the states of the EL system.

Boundedness of system states:

To start with we first show that the set $\mathcal{B}_q$ and $\mathcal{B}_{\dot{q}}$ are attractive for which we choose an Lyapunov candidate function

$$V_2 = \sum_{i=1}^{n} \int_{0}^{q_i(t)} k_1 \text{sat}(\lambda_1 \gamma) d\gamma + \frac{1}{2} \dot{q}^\top(t) M \dot{q}(t) \tag{19}$$

where k_1 is positive as in (12), γ is a dummy variable, and $q_i(t)$ are the generalized coordinates where $i \in [1, n]$. The time derivative is given as

$$\begin{aligned}\dot{V}_2 &= \sum_{i=1}^{n} k_1 \text{sat}(\lambda_1 q_i(t)) \dot{q}_i(t) + \dot{q}^\top(t) M \ddot{q}(t) + \frac{1}{2} \dot{q}^\top(t) M \dot{q}(t) \\ &= \dot{q}^\top(t) k_1 \text{sat}(\lambda_1 q(t)) + \dot{q}^\top(t) M \ddot{q}(t) + \frac{1}{2} \dot{q}^\top(t) M \dot{q}(t)\end{aligned}$$

Substituting the value of $\ddot{q}(t)$ from (9) in the above equation and using the property 4

$$\begin{aligned}\dot{V}_2 =& \dot{q}^\top(t) k_1 \text{sat}(\lambda_1 q(t)) + \dot{q}^\top(t) M \Big[\dot{s}(t) - M^{-1} \Big(C\dot{q}(t) + k_1 \text{sat}(\lambda_1 q(t)) \\ &+ k_2 \text{sat}(\lambda_2 \dot{q}(t)) \Big) \Big] + \frac{1}{2} \dot{q}^\top(t) M \dot{q}(t) \\ =& \dot{q}^\top(t) M \dot{s}(t) - k_2 \dot{q}^\top(t) \text{sat}(\lambda_2 \dot{q}(t))\end{aligned} \tag{20}$$

Now, the proof is proceeded in two parts

- **Case 1**: $\|\lambda_2 \dot{q}(t)\| > 1$:
  For $\|\lambda_2 \dot{q}(t)\| > 1$ we have $\text{sat}(\lambda_2 \dot{q}(t)) = \text{sgn}(\lambda_2 \dot{q}(t))$ which replaced in (20) and applying properties of norm

$$\dot{V}_2 \leq \|\dot{q}(t)\| \|M\dot{s}(t)\| - k_2 \|\dot{q}(t)\|.$$

  Substituting the maximum value of $\|M\dot{s}(t)\|$ from (35) which is calculated in Appendix, the above expression we have

$$\begin{aligned}\dot{V}_2 \leq& \beta \|\dot{q}(t)\| - k_2 \|\dot{q}(t)\| \\ =& - \|\dot{q}(t)\| (k_2 - \beta).\end{aligned} \tag{21}$$

  Due to the gain condition (12), it can be assured that $\dot{V}_2$ is negative definite causing the state trajectories to converge towards the equilibrium. Thus, by Lasalle's invariance principal, [26] both q and $\dot{q}$ are asymptotically stable and reach the origin.
- **Case 2**: $\|\lambda_2 \dot{q}(t)\| < 1$:
  For $\|\lambda_2 \dot{q}(t)\| \leq 1$, we have $\text{sat}(\lambda_2 \dot{q}(t)) = \lambda_2 \dot{q}(t)$ which replaced in (20), and applying properties of norm, we have

$$\begin{aligned}\dot{V}_2 \leq & m_2\beta\|\dot{q}(t)\| - \|\dot{q}(t)\|^2 \\ = & -\|\dot{q}(t)\|(\|\dot{q}(t)\| - \beta)\end{aligned} \tag{22}$$

Hence, as long as $\|\dot{q}(t)\| > \beta$, $\dot{V}_2$ is negative definite causing the origin to be attractive for the states of the EL system.

Thus, by Lasalle's invariance principal, [26] $q(t)$ is asymptotically stable and reaches the origin. Now, once the trajectories hit the equilibrium, due to (7), the maximum growth of $q(t)$ can be calculated as

$$\|q(t) - q(t_k)\| \leq \alpha. \tag{23}$$

This ensure that $q(t)$ remains in the invariant set $\mathcal{B}_q$ given by (14). Also, due to (7), $\dot{q}(t)$ can grow maximum by

$$\|\dot{q}(t) - \dot{q}(t_k)\| \leq \alpha. \tag{24}$$

Hence, due to (22) and the above expression, $\dot{q}(t)$ is contained in $\mathcal{B}_{\dot{q}(t)}$ given by (15). This completes the proof.

Thus, the stability of sliding mode dynamics is established. The next proposition ensures that the control is always bounded.

Proposition 1 *Consider the EL system* (1) *with the sliding surface* (2) *and the event-based control* (6). *The control u is always bounded and satisfies the following*

$$\|u(t)\| \leq \sqrt{n}(k_1 + k_2 + k_3). \tag{25}$$

Proof The proof can be done by applying norm on (6) we get

$$\begin{aligned}\|u(t)\| = & \| - k_1\text{sat}(\lambda_1 q(t_k)) - k_2\text{sat}(\lambda_2 \dot{q}(t_k)) - k_3\text{sgn}(s(t_k))\| \\ \leq & k_1\|\text{sat}(\lambda_1 q(t_k))\| + k_2\|\text{sat}(\lambda_1 \dot{q}(t_k))\| + k_3\|\text{sgn}(s(t_k))\| \\ \leq & \sqrt{n}(k_1 + k_2 + k_3)\end{aligned} \tag{26}$$

This completes the proof.

This satisfies the constrained actuator saturation design. Now, to ensure a Zeno free solution [16] of the event-triggered control, we finally establish that the inter-event execution time (ΔT_k) is nonzero positive value.

Theorem 2 *Consider the system* (1) *with the sliding variable* (2) *and the control* (6) *satisfying the event condition* (7), *then the inter-event execution time satisfies the following*

$$\Delta T_k > \frac{\alpha}{\rho_1(\alpha) + \mu} \tag{27}$$

where

$$\rho_1(\alpha) = \frac{\alpha + c_m\alpha^2}{m_2}$$
$$\mu = \frac{1}{m_2}\left(\sqrt{n}(k_1 + k_2 + k_3) + d_m + g_m\right) \tag{28}$$

Proof To start with the proof, we first find the maximum growth of ε in an interval $t \in [t_k, t_{k+1})$ and then establish the fact that minimum inter-event time is always greater than this growth time. So, differentiating ε, we get

$$\begin{aligned}\frac{d}{dt}\varepsilon(t) \leq& \|\frac{d}{dt}\varepsilon(t)\| \\ =& \|\dot{q}(t)\| + \|\ddot{q}(t)\|.\end{aligned}$$

Substituting $\ddot{q}(t)$ from (1) in the above inequality

$$\begin{aligned}\frac{d}{dt}\varepsilon(t) \leq& \|\dot{q}(t)\| + \|M^{-1}(-C\dot{q}(t) - G + u(t) + d(t))\| \\ \leq& \|\dot{q}(t)\| + \|M^{-1}\|\Big(\|C\|\|\dot{q}(t)\| + \|G\| + \|u(t)\| + \|d(t)\|\Big).\end{aligned} \tag{29}$$

Substituting the respective bounds from the Property 1-3, Assumption 1, (15) and (25) in the above equation

$$\begin{aligned}\frac{d}{dt}\varepsilon(t) \leq& \|\dot{q}(t)\| + \frac{1}{m_2}\left(c_m\|\dot{q}(t)\|^2 + g_m + d_m + \sqrt{n}(k_1 + k_2 + k_3)\right) \\ \leq& \alpha + \frac{1}{m_2}\left(c_m\alpha^2 + \sqrt{n}(k_1 + k_2 + k_3) + d_m + g_m\right) \\ =& \frac{\alpha + c_m\alpha^2}{m_2} + \frac{1}{m_2}\left(\sqrt{n}(k_1 + k_2 + k_3) + d_m + g_m\right).\end{aligned}$$

Again substituting (28) in the above inequality for simplification and the integrating both sides of the same inequality with the fact that $\varepsilon(t_k) = 0$ as initial condition and at $\varepsilon(t_{k+1}) = \alpha$ in subsequent steps, we have

$$\begin{aligned}\frac{d}{dt}\varepsilon(t) \leq& \rho_1(\alpha) + \mu \\ \|\varepsilon\| \leq& \int_{t_k}^{t_{k+1}} [\rho_1(\alpha) + \mu]dt \\ \alpha \leq& [\rho_1(\alpha) + \mu](t_{k+1} - t_k) \\ \alpha \leq& [\rho_1(\alpha) + \mu]\Delta T_k\end{aligned} \tag{30}$$

Finally, rearranging the terms we arrive at (27). Thus, we ensure that the inter-event execution time is greater than zero which completes the proof.

Thus, the proofs for the stability of the EL system and the event-triggered control have been established. Next, we present the simulation results.

4 Simulation

To verify the performance of event-triggered ISM controller, simulations are performed for a robotic manipulator system [22] described by

$$M_1\ddot{q} + C_1\dot{q} + G_1 = u_1 + d_1 \tag{31}$$

where

$$M_1 = \begin{bmatrix} p_1 + 2p_3\cos(q_2) & p_2 + p_3\cos(q_2) \\ p_2 + p_3\cos(q_2) & p_2 \end{bmatrix}$$
$$C_1 = \begin{bmatrix} -p_3\sin(q_2)\dot{q}_2 & -p_3\sin(q_2)(\dot{q}_1 + \dot{q}_2) \\ p_3\sin(q_2)\dot{q}_1 & 0 \end{bmatrix}$$
$$G_1 = \begin{bmatrix} g_1 & 0 \\ 0 & g_2 \end{bmatrix}.$$

The manipulator is assumed to have an initial condition of $[0.45,\ 0.5,\ 1,\ 1]^\top$ with the parameters given as $p_1 = 3.473\text{kgm}^2$, $p_2 = 0.196\text{kgm}^2$, $p_3 = 0.242\text{kgm}^2$, $g_1 = 2.3$ Nm and $g_2 = 1.1$ Nm. The two link manipulator is subjected to a sinusoidal disturbance of $[0.1\sin(3t)\ 0.1\sin(4t)]^\top$. Now, for the design of controller, the gains were calculated satisfying the gain condition (12) as $k_1 = 1,\ k_2 = 4,\ k_3 = 2,\ \lambda_1 = 1$ and $\lambda_2 = 0.25$. To get a comparative visualization, the simulation has been performed with the event parameters α as 0 and 0.2 ($\alpha = 0$ is considered as time-triggered). The two values of α have been chosen to give the reader a better understanding between time-triggered and event-triggered implementation. The simulations are performed for stabilization of (31), and the results of time-triggered implementation are shown in Figs. 2, 3, 4 and 5 and that of event-triggered implementation are shown in Figs. 6, 7, 8, 9 and 10.

The evolution of the sliding variable with time-triggered control action is shown in Fig. 2 where a finite time reaching is achieved. The trajectories of the system states, i.e. the respective arm position and velocities of the manipulator (31) with time sampled control, are shown in Figs. 3 and 4 which are asymptotically stable. The plot of the control effort with time-triggered sampling is displayed in Fig. 5. In Fig. 6, the trajectories of the sliding variable which are driven by event sampled control are shown. It can be observed that the sliding variables hit their respective manifolds in finite time. In comparison with the ideal sliding motion which is achieved using time-triggered sampling, the event-triggered implementation achieves a practical

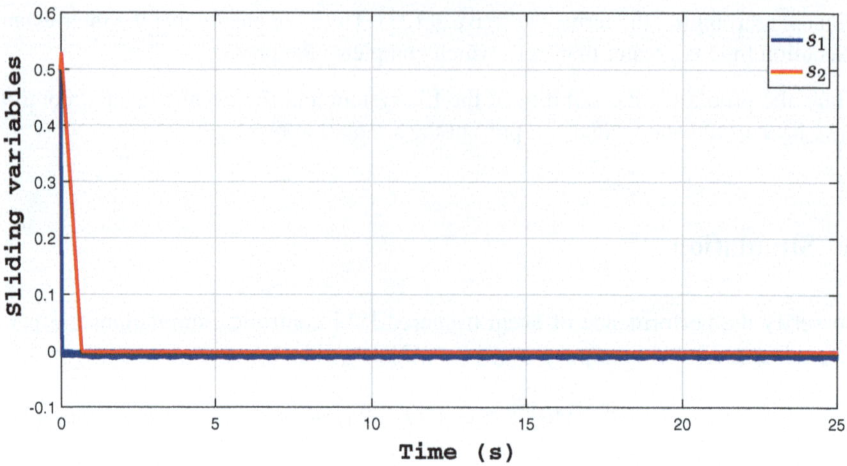

Fig. 2 Evolution of sliding variables with time-triggered SMC

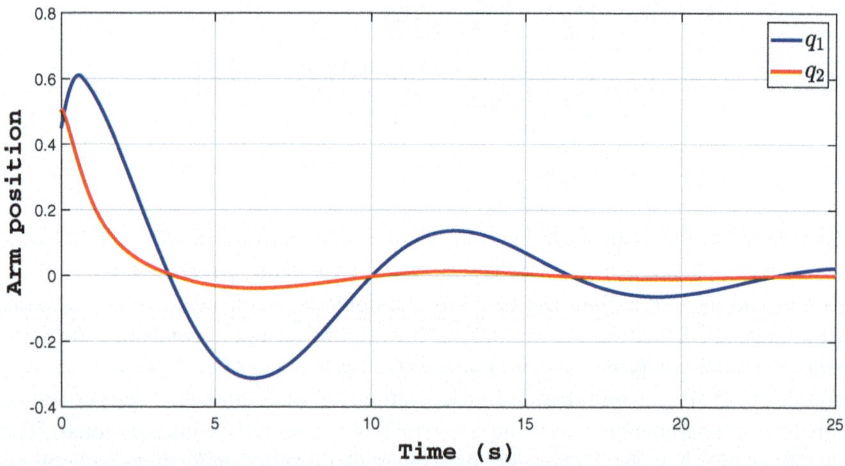

Fig. 3 Evolution of robot link angles with time-triggered SMC

sliding motion. The sliding variables are ultimately bounded within a desired band defined by (13). The evolution of the arm position and velocities with event-triggered implementation is shown in Figs. 7 and 8. It can be noticed that in contrast to an asymptotic behaviour, the system states with event sampled control do not settle to the origin, rather they are confined in a set defined by (14) and (15). The evolution of control signal with event sampled state information is displayed in Fig. 9 which is bounded as per (25). The inter-event execution time is shown in Fig. 10 which is always positive, and the control is free from accumulation.

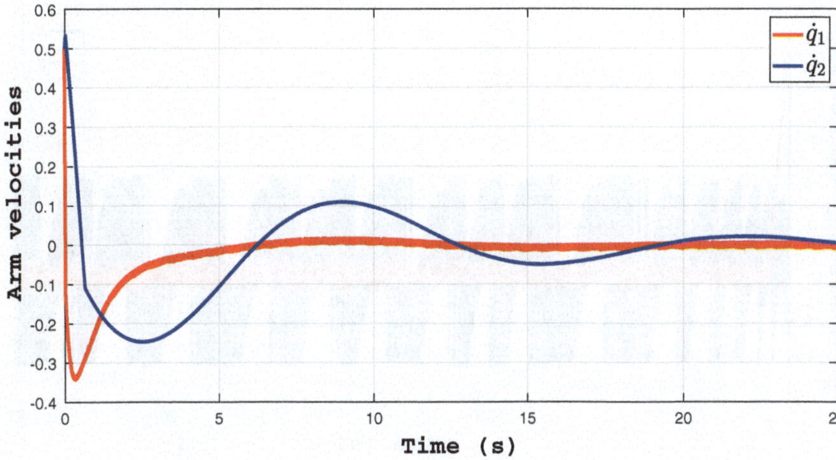

Fig. 4 Evolution of robot link angular velocities with time-triggered SMC

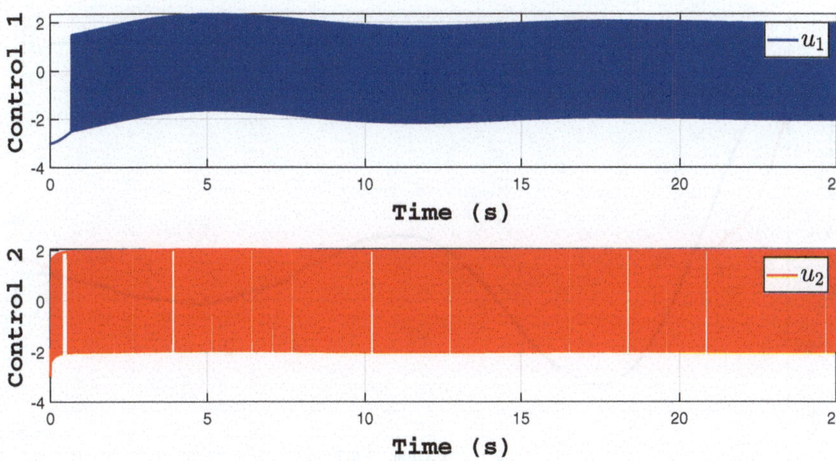

Fig. 5 Evolution of control signal with time-triggered SMC

To have an in-depth understanding of the event-based control strategy, we would like to introduce a few performance indices [27] as in Table 1. These performance indices give a measure how the event parameter α affects the system performance and to establish the effectiveness of the event-triggered control strategy with respect to time- triggered case. The event design parameter α is varied in steps from 0 to 0.3, simulations are performed, and the performance indices are tabulated in Table 2. The performance indices for time-triggered implementation ($\alpha = 0$) show that the total number of control updates with a sampling period of 0.001 s is 25000. The rms value of sliding variable is 0.054, and average control effort is 2.863 which are almost equivalent to that of event-triggered case with $\alpha = 0.01$, but the control updates are

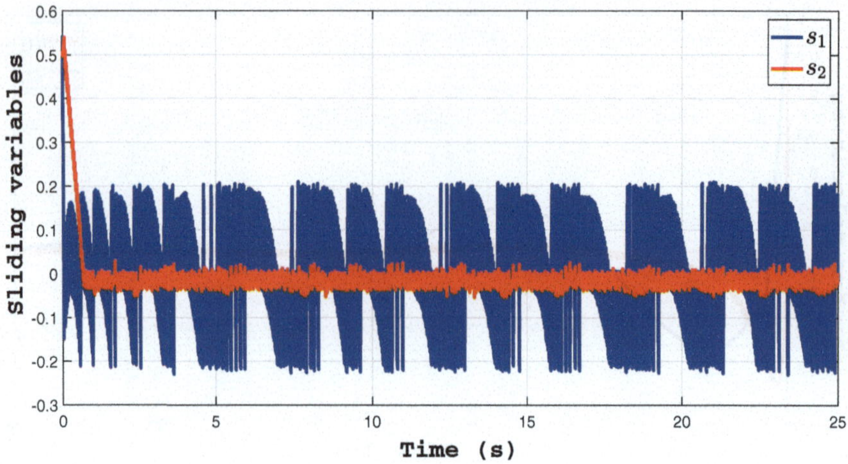

Fig. 6 Evolution of sliding variables with event-triggered SMC

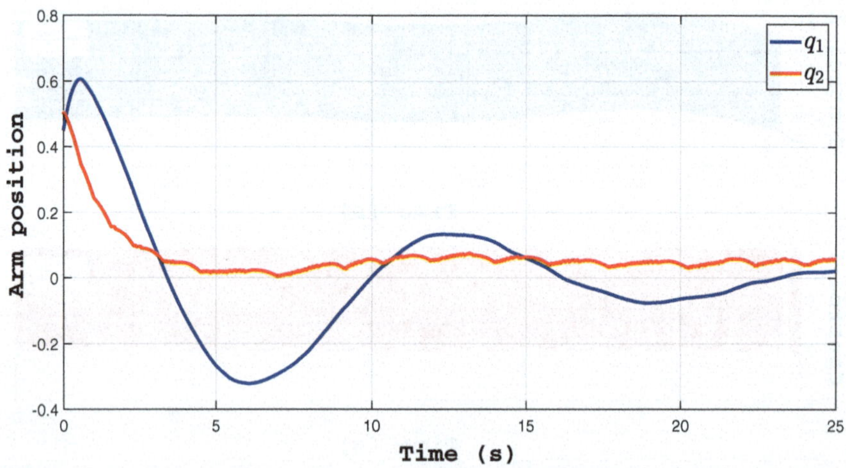

Fig. 7 Evolution of robot link angles with event-triggered SMC

reduced to almost 42%. This clearly shows the effectiveness of need-based sampling which not only reduces the control updates but also can keep the performance satisfactory, which is as per to the time-triggered implementation. Further as α increases, the number of updates reduces which reduces average time between sampled instances. Though the average control effort and rms value of the sliding variable increase, there is a large amount of reduction in the control updates which shows that a trade-off exists in design. The choice of α plays the most important role in the design of event-triggered ISMC which needs to be addressed depending on the application.

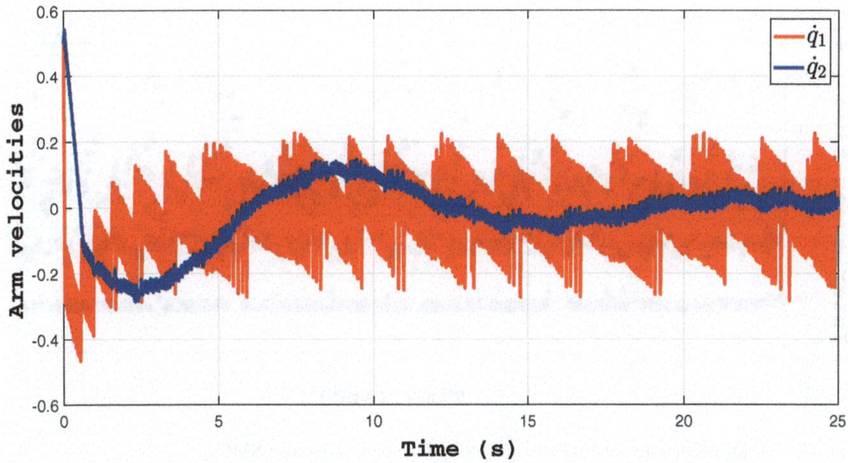

Fig. 8 Evolution of robot link angular velocities with event-triggered SMC

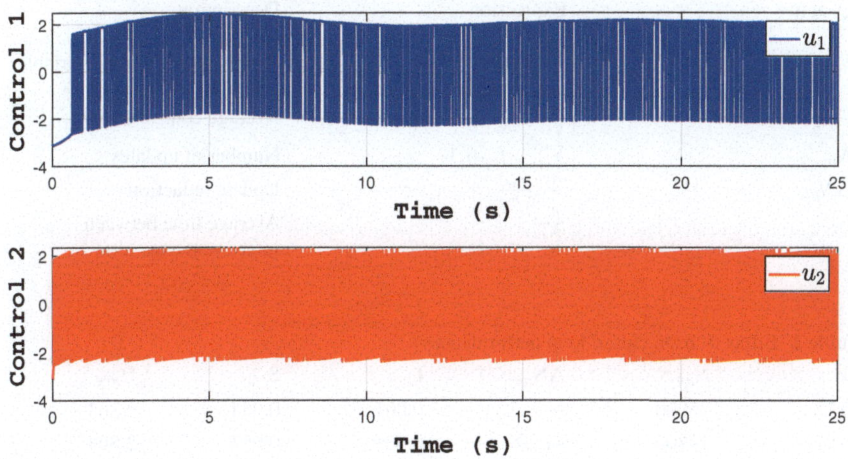

Fig. 9 Evolution of control signal with event-triggered SMC

5 Conclusion

In this paper, the design of event-based ISMC with input saturation has been proposed for uncertain EL system. A unique event-triggering mechanism has been designed for the EL system to realize the reduced control computation and minimum resource usage. The integral sliding surface is considered to achieve the reachability of the sliding variable time with anti-saturated control which achieves an overall practical stability. The minimum inter-event execution time has been derived to guarantee that the control is free from accumulation. Finally, the simulation results for the EL system have been shown to validate the effectiveness of the proposed control strategy.

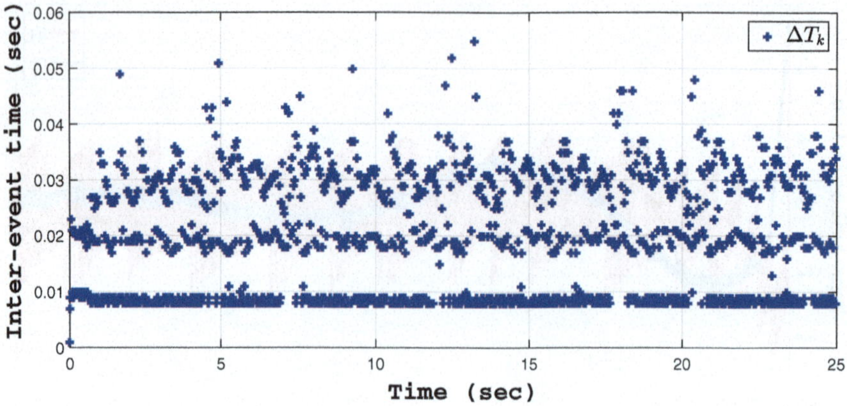

Fig. 10 Evolution of inter-event execution time for event-triggered SMC

Table 1 Performance Indices

Quantity	Expression	Description
S_{rms}	$\sqrt{\frac{\sum_{i=1}^{N_s} s^2(t_i)}{N_s}}$	RMS value of sliding variable
U_{avg}	$\sqrt{\frac{\sum_{i=1}^{N_s} u^2(t_i)}{N_s}}$	Average control effort
N_{up}	$\sum_{i=1}^{N_s} f_{up}(t_i)$	Number of updates
ΔN_{up}	$\frac{N_s - N_{up}}{N_s}$	Update reduction
t_{avg}	$\frac{T_{sim}}{N_{up}}$	Average time between inter-event execution

Table 2 Effect of α on closed loop performances

α	N_{up}	$\Delta\mathbf{N}_{up}(\%)$	$\mathbf{t}_{avg}$	S_{rms}	U_{avg}
0	25000	–	0.001	0.054	2.863
0.01	14706	41.20	0.0017	0.054	2.864
0.05	5442	78.23	0.0046	0.061	2.876
0.1	2538	89.85	0.010	0.076	2.889
0.2	1234	95.06	0.020	0.124	2.913
0.3	754	96.98	0.033	0.187	2.930

6 Appendix

Bound of $\|M\dot{s}(t)\|$:

Consider the EL system (1), which can be rewritten as

$$M\ddot{q}(t) = -C\dot{q}(t) - G + u(t) + d(t). \tag{32}$$

Recall the dynamics of the sliding variable

$$\dot{s}(t) = \ddot{q}(t) + M^{-1}\Big(C\dot{q}(t) + k_1\text{sat}(\lambda_1 q(t)) + k_2\text{sat}(\lambda_2\dot{q}(t))\Big). \tag{33}$$

Premultiplying M on both sides of the above equation we get

$$M\dot{s}(t) = M\ddot{q}(t) + C\dot{q}(t) + k_1\text{sat}(\lambda_1 q(t)) + k_2\text{sat}(\lambda_2\dot{q}(t)).$$

Substituting (32) in the above equation leads to

$$M\dot{s}(t) = -C\dot{q}(t) - G + u(t) + d(t) + C\dot{q}(t) + k_1\text{sat}(\lambda_1 q(t)) + k_2\text{sat}(\lambda_2\dot{q}(t)).$$

Substituting the event-based control law (6) in the above equation

$$\begin{aligned} M\dot{s}(t) =& d(t) - G + k_1[\text{sat}(\lambda_1 q(t)) - \text{sat}(\lambda_1 q(t_k))] \\ &+ k_2[\text{sat}(\lambda_2\dot{q}(t)) - \text{sat}(\lambda_2\dot{q}(t_k))] - k_3\text{sgn}(s(t_k)). \end{aligned}$$

Now, applying the properties of norms

$$\begin{aligned} \|M\dot{s}(t)\| \leq& \|d(t)\| + \|G\| + k_1\|\text{sat}(\lambda_1 q(t)) - \text{sat}(\lambda_1 q(t_k))\| \\ &+ k_2\|\text{sat}(\lambda_2\dot{q}(t)) - \text{sat}(\lambda_2\dot{q}(t_k))\| + k_3. \end{aligned}$$

Using the Lipchitz property of sat(.) function and substituting the values of λ_1 and λ_2, the above inequality results in

$$\|M\dot{s}(t)\| \leq \|d(t)\| + \|G\| + \|q(t) - q(t_k))\| + \|\dot{q}(t) - \dot{q}(t_k))\| + k_3.$$

Using the inequality (7), Property 3, and Assumption 1, finally, we arrive at

$$\begin{aligned} \|M\dot{s}(t)\| \leq& d_m + g_m + \alpha + k_3 \\ \|M\dot{s}(t)\| \leq& \beta \end{aligned} \tag{34}$$

where

$$\beta = d_m + g_m + \alpha + k_3. \tag{35}$$

References

1. Verastegui-Galván, J., Hernández-Guzmán, V.M., Orrante-Sakanassi, J.: Pid position regulation in one-degree-of-freedom euler–lagrange systems actuated by a pmsm. Int. J. Cont. **91**(2), 285–296 (2018). (Online). Available https://doi.org/10.1080/00207179.2016.1278272
2. Patre, P.M., MacKunis, W., Dupree, K., Dixon, W.E.: Modular adaptive control of uncertain Euler Lagrange systems with additive disturbances. IEEE Trans. Autom. Control **56**, 155–160 (2011)
3. Dupree, K., Patre, P.M., Wilcox, Z.D., Dixon, W.E.: Asymptotic optimal control of uncertain nonlinear Euler-Lagrange systems. Automatica **47**, 99–107 (2011)
4. Yang, Y., Hua, C., Li, J., Guan, X.-P.: Robust adaptive uniform exact tracking control for uncertain Euler-Lagrange system. Int. J. Control **90**, 2711–2720 (2017)
5. Sun, T., Cheng, L., Wang, W., Pan, Y.: Semiglobal exponential control of Euler-Lagrange systems using a sliding-mode disturbance observer. Automatica **112** (2020)
6. Utkin, V.I.: Sliding modes in control and optimization. Springer, Berlin Heidelberg (1992)
7. Utkin, V., Shi, J.: Integral sliding mode in systems operating under uncertainty conditions. In: Proceedings of 35th IEEE Conference on Decision and Control, vol. 4, pp. 4591–4596 (1996)
8. Lee, J.-H., Allaire, P.E., Tao, G., Zhang, X.: Integral sliding-mode control of a magnetically suspended balance beam: analysis, simulation, and experiment. IEEE/ASME Trans. Mech. **6**(3), 338–346 (2001)
9. Guo, Y., Huang, B., Li, A., Wang, C.-Q.: Integral sliding mode control for Euler-Lagrange systems with input saturation. Int. J. Robust Nonlinear Control **29**, 1088–1100 (2018)
10. Lee, J.Y., Chang, P.H., Jin, M.: Adaptive integral sliding mode control with time-delay estimation for robot manipulators. IEEE Trans. Indus. Electron. **64**, 6796–6804 (2017)
11. Tho, H.D., Tasaki, R., Kaneshige, A., Miyoshi, T., Terashima, K.: Robust sliding mode control with integral sliding surface of an underactuated rotary hook system. In: 2017 IEEE International Conference on Advanced Intelligent Mechatronics (AIM), pp. 998–1003 (2017)
12. Hong, Y., Yao, B.: A globally stable high-performance adaptive robust control algorithm with input saturation for precision motion control of linear motor drive systems. IEEE/ASME Trans. Mech. **12**(2), 198–207 (2007)
13. Gayaka, S., Lu, L., Yao, B.: Global stabilization of a chain of integrators with input saturation and disturbances: a new approach. Automatica **48**(7), 1389–1396 (2012)
14. Gayaka, S., Yao, B.: Adaptive robust actuator fault-tolerant control in presence of input saturation. In: Proceedings of the 2011 American Control Conference. IEEE, pp. 3766–3771 (2011)
15. Yesmin, A., Bera, M.K.: Design of event-based sliding mode controller with logarithmic quantized state measurement and delayed control update. In: ISA Trans. (2020). (Online). Available: http://www.sciencedirect.com/science/article/pii/S0019057820302871
16. Tabuada, P.: Event-triggered real-time scheduling of stabilizing control tasks. IEEE Trans. Autom. Control **52**(9), 1680–1685 (2007)
17. Heemels, W., Johansson, K.H., Tabuada, P.: An introduction to event-triggered and self-triggered control. In: 2012 IEEE 51st IEEE Conference on Decision and Control (CDC). IEEE, pp. 3270–3285 (2012)
18. Behera, A.K., Bandyopadhyay, B.: Robust sliding mode control: an event-triggering approach. IEEE Trans. Circ. Syst. II Express Briefs **64**(2), 146–150 (2016)
19. Yesmin, A., Bera, M.K.: Design of event-triggered sliding mode controller based on reaching law with time varying event generation approach. Euro. J. Control **48**, 30–41 (2019)
20. Bandyopadhyay, B., Behera, A.K.: Event-triggered sliding mode control a new approach to control system design. Springer (2018)
21. Nair, R.R., Behera, L., Kumar, S.: Event-triggered finite-time integral sliding mode controller for consensus-based formation of multirobot systems with disturbances. IEEE Trans. Control Syst. Technol. **27**(1), 39–47 (2017)

22. Kumari, K., Behera, A.K., Bandyopadhyay, B.: Event-triggered sliding mode-based tracking control for uncertain euler-lagrange systems. IET Control Theor. Appl. **12**(9), 1228–1235 (2018)
23. Yesmin, A., Bera, M.K.: Event-triggered integral sliding mode control. In: TENCON 2019-IEEE Region 10 Conference (TENCON). IEEE, pp. 1387–1392 (2019)
24. Spong, M.W., Vidyasagar, M.: Robot dynamics and control. Wiely, India (2008)
25. Shtessel, Y., Edwards, C., Fridman, L., Levant, A.: Sliding mode control and observation. Springer, New York (2014)
26. Khalil, H.K.: Nonlinear control systems. Prentice Hall (1995)
27. Cucuzzella, M., Incremona, G.P., Ferrara, A.: Event-triggered sliding mode control algorithms for a class of uncertain nonlinear systems: experimental assessment. Am. Control Conf. (ACC) **2016**, 6549–6554 (2016)

AKIRA—A Voice Based Virtual Assistant with Authentication

Vishnu Sai Bhonsle, Sailaja Thota, and Surekha Thota

Abstract Virtual voice assistance executes the natural language commands and reciprocates its response using synthesized voice. Adoption of voice-based assistance is increasing at a greater speed. Google assistant, Alexa from Amazon, Siri from Apple, Cortana from Microsoft are the leading voice assistants. Various surveys have been conducted to review the popularity and correctness of these voice assistance and results displayed that these virtual assistances were leading by gaining more popularity especially among the youngsters. Studies have also highlighted that these devices have a history of gaining unauthorized access to the device and caused damage to the owner. The current paper is an attempt to address the security aspect by categorizing the commands to either permission or permission-less. The permission commands have to undergo an authentication procedure to ensure high security and privacy of the owner.

1 Introduction

In the twentieth century, CEO's and Managing Director's appointed personal secretaries for assisting them in their daily business tasks. They were given oral instructions to perform various tasks like scheduling meetings, read and send emails, book tickets, and appointments, search for phone numbers and place calls. In the very beginning of twenty-first century, people use digital organizers for setting reminders, storing phone numbers and email ids, but here the manual feeding of information is required.

In the decade of 2010s, with the wide use of Natural Language Processing (NLP), Human–Computer Interaction (HCI), and Artificial Intelligence (AI) algorithms, virtual assistance came into existence. Virtual assistance/voice assistance not only execute commands but also reciprocate its response using synthesized voice [1, 2].

V. S. Bhonsle · S. Thota (✉) · S. Thota
REVA University, Bangalore, Karnataka, India
e-mail: sailaja.thota@reva.edu.in

S. Thota
e-mail: surekha.thota@reva.edu.in

J. Gu et al. (eds.), *Communication and Control for Robotic Systems*, Smart Innovation, Systems and Technologies 229, https://doi.org/10.1007/978-981-16-1777-5_11

Voice assistant technology is a reality now which was thought of as science fiction sometime back. These voice assistants are also termed as Intelligent Personal Assistants and Speech-based Natural User Interfaces (NUI) from the academic perspective.

1.1 Different Names for Voice Assistant

The alternative names for voice assistant are [3]:

- Intelligent Virtual Assistance (IVA)
- Speech-based Natural User Interface (NUI) (academic perspective)
- Digital Assistant
- Voice Activated Personal Assistant
- Voice-Enabled Assistant
- Conversational Agent
- Smart Assistant
- Personal Digital Assistant (PDA)
- Virtual Personal Assistant
- Intelligent Personal Assistant
- Mobile Assistant.

The most popular companies like Amazon, Apple, Google, and Microsoft are implementing the technologies like text-to-speech, automatic speech recognition, and dialog management to improvise their dialogue systems (Fig. 1).

The most familiar voice assistants are Alexa from Amazon, Siri from Apple, Google Assistant, and Cortana from Microsoft [5, 6]. Few others voice-based PDAs are Samsung's S Voice, Nuance's Dragon, Facebook's M Messenger. A particular

Fig. 1 Various voice assistants from Amazon, Apple, Google, and Microsoft [4]

salutation from the user's voice for each of the agent will activate the system and responds to the user command appropriately. The user's command is interpreted by a server and suitable response is generated in the interaction system [5]. Below we highlight some of the important voice assistants.

Amazon Alexa. Alexa is an IVA working in association with smart speaker "Echo" which is an independent voice-controlled system which searches for the desired music albums and plays, controls other devices at home, gets the required data from the internet, plays the news articles from various online newspapers, creates a to-do list and reminds, updates on the current weather, traffic situation and other such tasks. Alexa responds to some commands like "Alexa, find me a Hyderabadi biryani restaurant nearby.", "Alexa, fix an appointment with dentist". "Alexa, what is my schedule today?", "Alexa, set an alarm after one hour" and "Alexa, How is the traffic in my commute to office?"

Amazon Fire Phone, Kindle Fire, and FireTV are some of the devices that can work in integration with Alexa [6].

Google Assistant. Google Assistant is IVA designed to work with voice-activated speaker called Google Home. Google assistant recommends the user based on the available data predicting user's need. Google Assistant responds to the commands like: "Ok Google, play Titanic song", "Ok Google, direct me to nearby Bawarchi", "Ok Google, Let me know Electric Bulb was invented by whom?" "Ok Google, Send a birthday wish to Amit." [6].

Microsoft Cortana. Working on windows powered devices of both PCs and mobile is the uniqueness of Microsoft Cortana. Cortana can recognize natural voice and serve the purpose similar to the above-mentioned devices. Cortana responds to some of the commands like: "What's the today's weather forecast?", "Place a call to Raghu", "Tell me the value of 10 cm in inches", "Photosynthesis means?" [6, 7].

Apple Siri. Siri is a voice-activated PDA for Apple devices like iPhone. Siri is the earliest in smart personal assistants which was released in the year 2010. Siri is currently undergoing a design change to work with new devices. Commands that are for Siri include: "Display my photos from Scotland last year", "Tell me schedule of movies in star movies tomorrow?", "Message Rahul, 'Where to meet for lunch?", "Read my last WhatsApp message" [1, 6].

Though each of the smart agents are unique in their features, yet the basic function remains the same. Current personal assistants are much different from previous voice-activated technologies in terms of responding to many complicated questions as the latest assistants are constantly connected on the Internet to provide appropriate information or respond according to user's needs [5].

Depending on the application and its complexity, different approaches and techniques are followed by each individual product. Google Assistant and Alexa improvised by using deep learning architecture [8]. Dialogue system of Cortana is based on Microsoft Azure Machine Learning Studio. Facebook used machine-learning algorithms with contextual memory for M-Messenger.

Everyday lot of research is being carried out to develop a more efficient and accurate Virtual Assistant. But still, there are several instances where false positives and true negatives are identified in the test results and there is a lot of scope for improvement. When Human–computer interaction is blended with efficient AI algorithms, it provides an improvised virtual Voice Assistant. In this paper, we would like to propose a model that ensures security of the user.

2 Literature Survey

The methods applied in creation of a virtual assistant are different among the different products available in the market. One product may be good in quality of voice synthesis and the other may provide an accurate results. Study of various Voice Assistants proved that there is no single voice assistant which could cater to all the requirements. Hence, organizations are trying to improve by refining their existing products [9].

As per [10] the Voice Assistants are prone to security threats, hence the literature survey carried out in this paper is divided into two sections, the first section surveys the various features of popular products and the second section surveys about security issues in Voice Assistants.

2.1 Survey on Various Voice Assistants

Client's awareness about the technology and power of the devices that can be controlled via voice is still in the initial stages. However, it is gaining momentum among the young users of the technology which is creating a demand for the future of such technologies. Many organizations would adopt and implement this technology in their future products [8].

From different surveys about the various factors on the voice assistants, it is observed that in 2017, audience were most engaged with Alexa which was used at least once per day. In 2018, survey on voice recognition and correctness of the result, Google assistant was in the forefront with 58.8% whereas Siri responded at 41.64% and Cortana at 35% accuracy rate. In 2019, survey was conducted on the most used voice assistants in the US, Apple Siri and Google assistant stood at 36% each, Amazon Alexa stood at 25%, and Microsoft Cortana at 19% usage [11, 12].

The details like release date, manufacturer of various popular PDAs mentioned in Table 1 and survey results on different voice assistants in 2019 [12] shown in Fig. 2.

Virtual assistance technology provides features like playing music, controlling various smart devices like speakers and lights that are connected to internet, tracking flight schedules, and answering various queries by searching the knowledge base through search engines. Although computers have superior processing and computational capability, they are still inferior to humans in terms of intelligence as they

Table 1 Details of the PDAs from different companies

IVA/PDA	Manufacturer	Compatible devices	Date of release [5]	Number of language supported
Google Assistant	Google	Android and iOS contraptions	2016	4
Alexa	Amazon	Amazon's Resound, Fire, savvy speakers and headphones, brilliant watches, and advanced cell devices including TVs, radios, coolers, lights, and indoor controllers	2014	3
Cortana	Microsoft	Windows 10, Xbox One, iOS, Android, Skype, Cyanogen, Windows Mixed Reality devices	2013	8
Siri	Apple	Apple devices like iPad, iPod Touch, iPhone, Home Pod, Mac, Apple TV, Apple Watch	2010	20

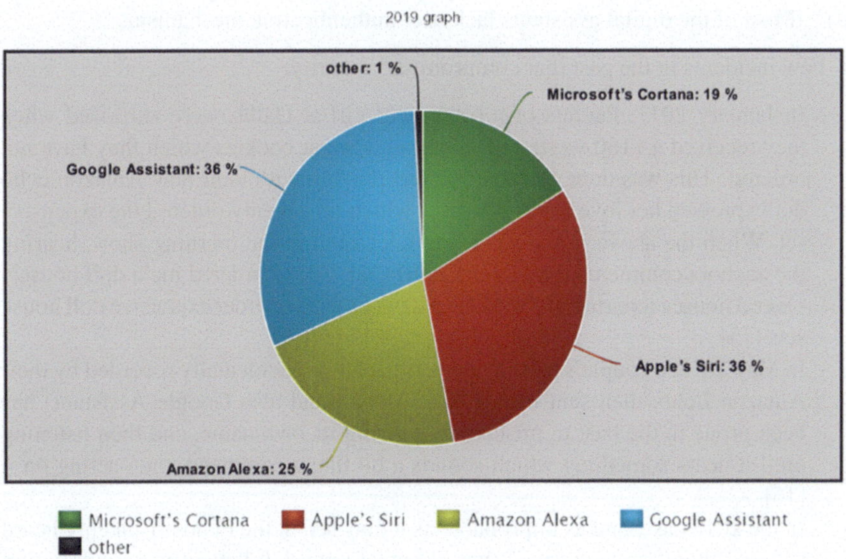

Fig. 2 Survey on different voice assistants—2019 [12]

are still in its initial stages of context-based learning. It was observed in the past that there were cases where Alexa was hacked [5].

Multi-modal dialogue techniques are used in organization's research to design upcoming generation dialogue systems. The current commercially available systems take inputs from multiple sources. For instance, latest infotainment systems in the automobiles take voice-based commands as well as touch screen inputs for navigation, turn on radio, climate control, answering a phone call [13].

With the advent of new technologies like machine learning, AI, Internet of Things (IOT), new features are being added constantly and the cost is maintained at a minimum level. University of Southern California is working on Semio, a research project based on voice recognition and body language for humanoid robots. This will help the amateur users to interact with the robot through natural language communication [14].

2.2 *Survey on Security Aspects in Voice Assistants*

Three vulnerability points while using Voice Assistants had discussed in [10].

(a) Digital assistants use voice recognition as its main technology, hence it suffers from Eavesdropping even it is not used
(b) Washington University had found that surfing attacks are possible
(c) (i.e., extracting the features of smartphone and data using remote control)
(d) Most of the digital assistants lack user authentication mechanism.

Few incidents in the past that compromised security.

1. In January 2017, Parents of a 6-year-old girl at Dallas were surprised when they received a $160 worth doll house and Sugar cookies which they have not ordered. This was done through the girl playing with their new Amazon echo dot expressed her love for doll houses which mistakenly ordered the expensive set. When the above news was being telecasted in the morning show, hearing the anchors comments that the little girl said "Alexa ordered me a doll house", several houses reported that their own devices tried to order expensive doll house sets [15].
2. In May 2018, a couple's private conversation was accidentally recorded by their Amazon Echo, then sent to a friend. Alexa, (and also Google Assistant) has been prone in the past to mistakenly hearing its own name, and then listening until it hears something which sounds a bit like a command, then acting on it [16].
3. In the 2018 case, and as improbable as it may seem, the Echo mistakenly heard "Alexa," then "send message," then a contact's name, followed by a confirmation word like "right." [16].
4. Use of voice assistance arbitrarily pose dangerous situations to the user such as it may share real-time video or location invading the privacy of the user.

Realtime enabling and disabling of the access of the voice assistant to several applications can minimize the damage caused [17, 18].

Retaining privacy and reconfirming before the execution of highly secured commands is of utmost importance. Hence this paper proposes a solution by introducing an authentication procedure to ensure security and privacy of the user along with his data.

3 Methodology

In this paper, we propose a lightweight voice-based Virtual Assistant—AKIRA. AKIRA takes voice input; process the request and responds to the user with a synthesized voice along with required action. Although Microsoft Cortana, Alexa, Siri etc. supports virtual voice assistance, it's a proprietary software which is specific to Microsoft, Amazon, and Apple products. Here we build AKIRA so that it can be easily deployed and used by any Operating system. The high-level architecture of AKIRA is as shown in Fig. 3.

The architecture is divided into four layers of processing.

1. **Speech Recognition**: The virtual assistance should be capable of accepting the verbal request and act accordingly. We have many speech-to-text converter API's available in the market. The most popular speech recognition API's are from Amazon, Google, IBM, Microsoft and so on. AKIRA uses the popular Google's speech recognition API that recognizes the verbal commands issued by the user.
2. **AKIRA Module**: AKIRA module maintains a knowledge base of predefined commands which are mapped to the corresponding API's that performs the required activity. The user commands provided to the Google's speech recognition is converted into the commands that are issued to the AKIRA Module. The

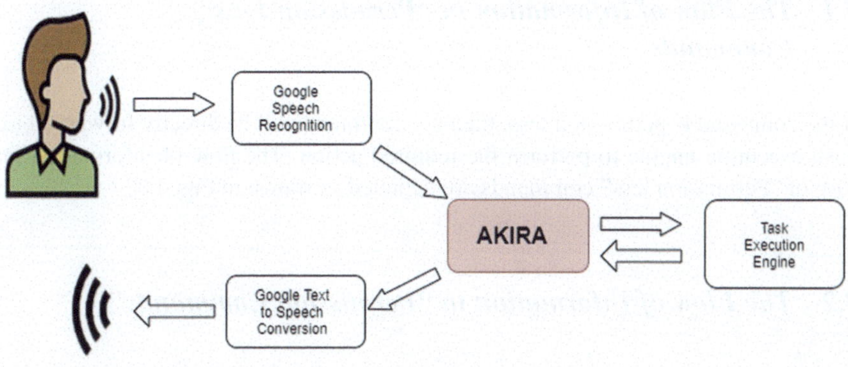

Fig. 3 High-level architecture diagram of AKIRA

Table 2 Classification of security labels for commands

Command	API/Software	Security label
Calculate strings into formula	Wolframalpha	Permission less
Playing music	Play sound	Permission less
Answering question	Wikipedia	Permission less
Searching the web engine	Web browser	Permission less
Shutdown of system	Operating system	Permission
File edit	Operating system	Permission
E-commerce shopping	Web browser	Permission

AKIRA module invokes the corresponding API to perform the required action. First two columns in Table 2 provide some examples of the commands and the API's that are invoked.

We have seen in the literature review that there are instances where Virtual Assistant can be hacked, hence authentication is required for the commands that may delete or edit a file or that debits the amount from the account (like online shopping). Hence we would like to design an authentication mechanism that controls the usage of Virtual Assistant. Using AKIRA, we can also stop people who mimic voices and provide unauthenticated commands.

As part of authentication mechanism, a security label is assigned for each command. The security label classified the command as "permission" or "permission less" as depicted in Table 2. If the commands are classified as "permission less", it specifies that the action has no or less impact if the command is executed by an unauthenticated user. If the command is classified as "permission", it signifies that the user needs to have a permission before he executes the command.

3.1 The Flow of Information in "Permission Less Commands"

If the command is permission less, then the command will be directly forwarded to Task execution engine to perform the required action. The flow of information in case of "Permission less" commands are depicted as shown in Fig. 4.

3.2 The Flow of Information in "Permission Commands"

If the command is classified as "permission", the system demands more authentication and high security. Hence AKIRA proposes an authentication process before

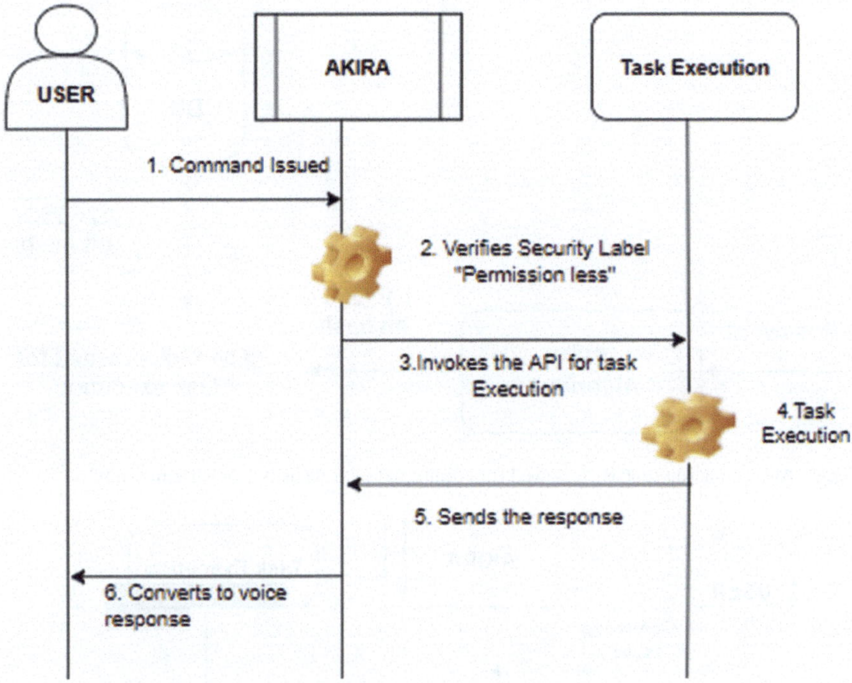

Fig. 4 Processing a command whose security label is "Permission less"

it actually executes the commands with security label "Permission". On receiving the request with Permission Label, AKIRA prompts the user for the authentication password. The voice-based authentication password will be converted to hash using SHA256 algorithm. The generated 256-bit hash is mapped with the hash of the actual password stored in AKIRA. If it matches then only the request is processed. Upon password verification, the corresponding command is executed by task execution engine. Figure 5 depicts the authentication process of the user and Fig. 6 depicts the overall process flow of commands having "Permission".

3.3 Task Execution Engine

Based on the Commands invoked, the appropriate Application Program Interface will be invoked and the required task will be executed. Few API's are depicted in Table 2.

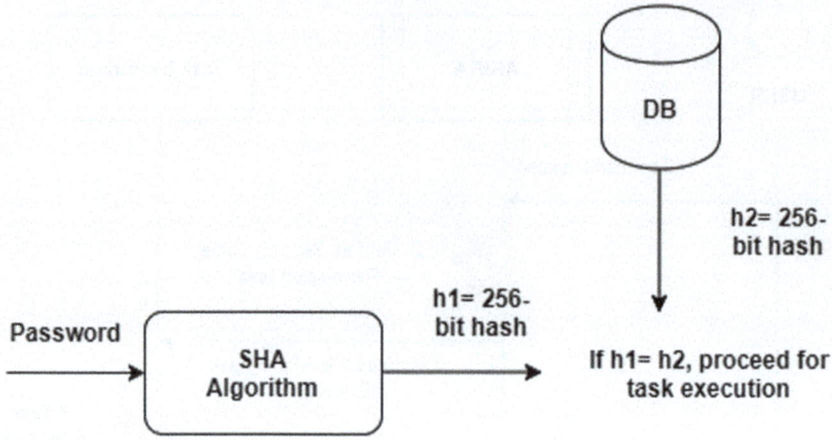

Fig. 5 Authentication process followed for commands with security label "Permission"

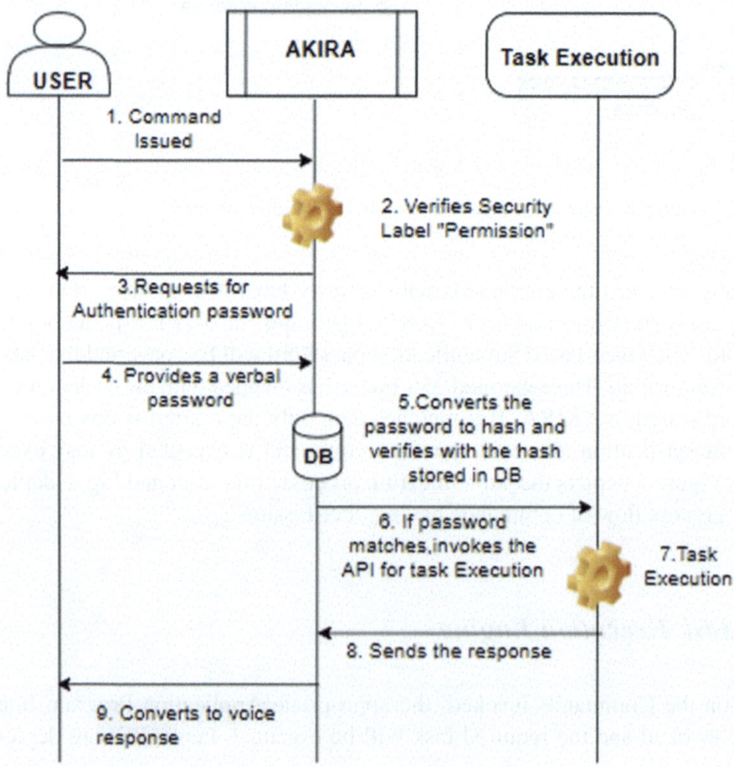

Fig. 6 Processing a command whose security label is "Permission"

3.4 *Google Text to Speech*

On task execution, the execution results are converted to synthesized voice and conveyed to the user.

4 Results

AKIRA is the voice-based Virtual Assistant built for laptops. AKIRA is built using Python technologies and tested on Intel(R) Core(TM) i5-6200U CPU @ 2.30 GHz, 8 GB RAM, × 64-based processor with Windows operating system. A comparative study of AKIRA had been carried out with the main features specified in [19]. We noticed that AKIRA even provides the basic functionalities of Greeting, Email Management, and Social Media. As of today, it does not support Social Gaming. In addition, our proposed model performs a command-level authentication mechanism that ensures monetary transactions and transactions that include confidential data are executed only by authentication.

4.1 *Invoking AKIRA*

Whenever the user utters the word "AKIRA", it will be invoked in listening mode to accept the commands issued by the user. The listening mode of AKIRA is as shown in Fig. 7.

4.2 *Issuing Command*

The commands issued by the user can of either permissioned or permission-less. To execute the permissioned commands, initially the password is set by the user and is stored in the form of hash in a key file as shown in Fig. 8.

4.2.1 Scenario 1—Executing Permission-Less Commands

When the user executes permission-less commands like browsing the search engine, the command can be executed without the need of authentication key. As an example if the user is searching for the keyword python, AKIRA opens the browser without asking password as shown in Fig. 9.

Fig. 7 Listening mode of AKIRA

Fig. 8 Hash of authentication password stored in key file

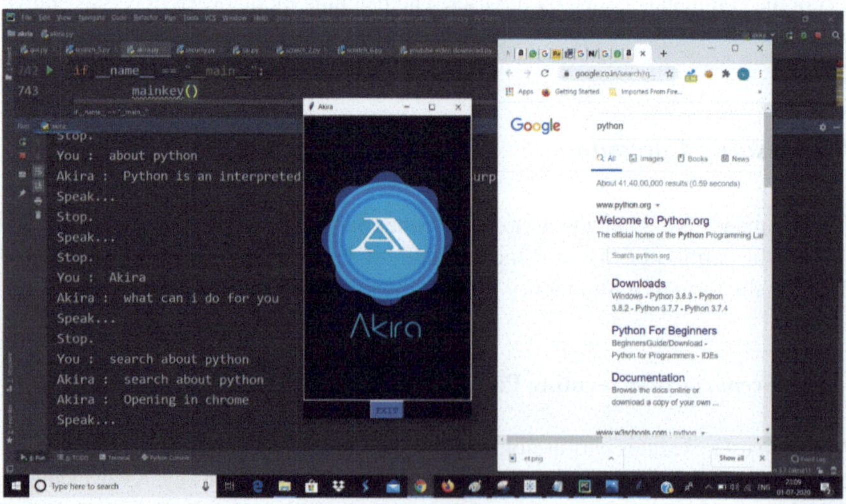

Fig. 9 Execution of permission-less commands

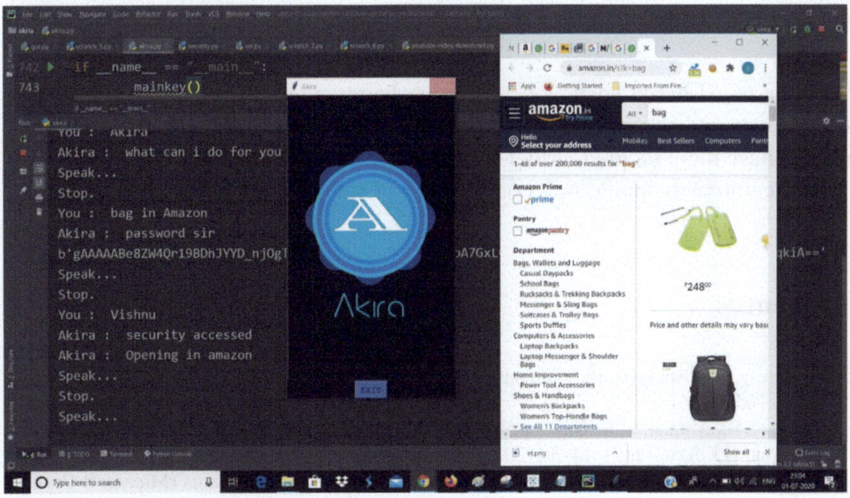

Fig. 10 Execution of permission commands

4.2.2 Scenario 2—Executing Permission Commands

When the user executes permission commands like opening an e-commerce website or deleting a file, these commands can be executed only on verification of authentication key. If the user is searching for bags in Amazon as shown in Fig. 10, AKIRA requests the password from the user and then verifies with the hash that is stored in Key file. Upon verification, it will open bags in Amazon website.

Thus, our project adds an additional layer of security while executing commands through Virtual Assistants.

5 Conclusion

It had become very intuitive and convenient for the users to use voice-based assistants to achieve the required result through natural language-based commands. Although various voice-based assistants are currently available in the market, our focus was to build a virtual assistant that categorizes the commands issued as permissioned and permission-less. The permission-less commands execute without any authentication mechanism whereas permissioned commands need to follow the authentication procedure before it executes. This kind of authentication ensures that the transaction is safe without any unauthorized use and prohibiting the intruder to gain access on the system.

6 Future Scope

Adoption of AI-based voice assistances are being increased day by day. However, the accuracy and security features of these products have to be still improvised. In order to have more secured virtual assistance, the authentication password of highly secured commands should be a combination of voice identification and password.

References

1. Bahcecia, O.: Analysis and Comparison of Intelligent Personal Assistants. instructure.com. Retrieved (2016)
2. Shakeel, B., Ahmad, M.S.: Siri–Apple's personal assistant: a review. Int. J. Comput. Sci. Mobile Comput. (IJCSMC) **6**(7), 44–48 (2017)
3. Kepuska, V., Bohouta, G.: Next-generation of virtual personal assistants (Microsoft Cortana, Apple Siri, Amazon Alexa And Google Home). In: 2018 IEEE 8th Annual Computing and Communication Workshop and Conference (CCWC). IEEE (2018)
4. G-BEEZ: Best voice assistant | Top 10 smart assistants for Phone, Laptop & IoT device, October 30 (2019)
5. Hoy, M.B.: Alexa, Siri, Cortana, and more: an introduction to voice assistants. Med. Ref. Serv. Q. **37**(1), 81–88 (2018)
6. López, G., Quesada, L., Guerrero, L.A.: Alexa vs. Siri vs. Cortana vs. Google Assistant: a comparison of speech-based natural user interfaces. In: International Conference on Applied Human Factors and Ergonomics. Springer, Cham (2017). https://www.greatbuyz.com/blog/best-voice-and-smart-assistant/
7. Berdasco, A., et al.: User experience comparison of intelligent personal assistants: Alexa, Google Assistant, Siri and Cortana. Multidiscip. Digital Publ. Inst. Proc. **31**(1), (2019)
8. Tulshan, A.S., Dhage, S.N.: Survey on virtual assistant: Google Assistant, Siri, Cortana, Alexa. In: International Symposium on Signal Processing and Intelligent Recognition Systems. Springer, Singapore (2018)
9. Polyakov, E.V., Mazhanov, M.S., Rolich, A.Y., Voskov, L.S., Kachalova, M.V., Polyakov, S.V.: Investigation and development of the intelligent voice assistant for the internet of things using machine learning. In: 2018 Moscow Workshop on Electronic and Networking Technologies (MWENT), pp. 1–5. Moscow (2018). https://doi.org/10.1109/MWENT.2018.8337236
10. THE KIUWAN TEAM: Cybersecurity: how safe are voice assistants. https://www.kiuwan.com/blog/safety-voice-assistants/#:~:text=Voice%20assistants%20carry%20the%20risk,kind%20of%20information%20you%20share. 9 Jun (2020)
11. Sterling, G.: Survey: Alexa the most frequently used assistant, Cortana seen as most accurate. https://searchengineland.com/survey-alexa-frequently-used-assistant-cortana-seen-accurate-269052. 9 Feb (2017)
12. Giret, L.: Microsoft releases voice assistant study, 19% of respondents using Cortana. https://www.onmsft.com/news/microsoft-releases-voice-assistant-study-19-of-respondents-using-cortana. 25 April (2019)
13. Pieraccini, R., et al.: A multimodal conversational interface for a concept vehicle. New School Psychol. Bull. **1**(1), 9–24 (2003)
14. Mead, R.: Semio: Developing a cloud-based platform for multimodal conversational AI in social robotics. In: International Conference on Consumer Electronics (ICCE). IEEE (2017)
15. Chung, H., et al.: Alexa, can I trust you? Computer **50**(9), 100–104 (2017)
16. Chokshi, N.: Is Alexa listening? Amazon Echo sent out recording of couple's conversation. https://www.nytimes.com/2018/05/25/business/amazon-alexa-conversation-shared-echo.html. May 25 (2018)

17. Alepis, E., Patsakis, C.: Monkey says, monkey does: security and privacy on voice assistants. IEEE Access **5**, 17841–17851 (2017). https://doi.org/10.1109/ACCESS.2017.2747626
18. Agarwal, Y., Rai, R., Chaubey, P.K.: BRAIN—THE A.I. (personal voice assistant). Int. Res. J. Eng. Technol. (IRJET) 5379–5382, (2020). e-ISSN: 2395-0056
19. Reis, A., et al.: Using intelligent personal assistants to assist the elderlies an evaluation of Amazon Alexa, Google Assistant, Microsoft Cortana, and Apple Siri. In: 2nd International Conference on Technology and Innovation in Sports, Health and Wellbeing (TISHW), pp. 1–5. Thessaloniki (2018). https://doi.org/10.1109/TISHW.2018.8559503

Survey on Internet of Things Based Intelligent Wireless Sensor Network for Fire Detection System in Building

Arpita Mukherjee, Saikat Kumar Shome, and Partha Bhattacharjee

Abstract Disaster caused by fire in various residential, commercial and industrial places is a major concern as it may result in huge damage of infrastructure as well as human life. Thus, an early detection of fire and notify the appropriate authority for prompt extinguishing to protect valuable lives and properties is a very important task. A real-time automatic intelligent fire detection system integrated with wireless sensor network (WSN), artificial intelligent (AI), and internet of things (IoT) and can solve this problem. In this paper, a literature survey on wireless sensor network capabilities through use of different intelligent algorithm using IoT aimed at fire detection has been presented. A schematic block diagram of IoT-based intelligent WSN for fire detection system (FDS) is also proposed for real-time automatic early detection of fire and disaster management.

1 Introduction

The disaster caused due to fire is still a serious problem and its occurrence is random moreover, the major characteristics of fire are it extends exponentially with time. Therefore, early detection of fire is critical for avoiding a major accident. Industry, residential and commercial building, hospitals, etc. are at risk to fire which may cause harm to its occupants and severe damage to property [1]. In India, about 25,000 persons on an average die every year due to fire and related cases among them 66% are female. As per the statistics of the National Crime Records Bureau (NCRB), about 5.9% (23,281) of the total reported deaths were due to fire among natural and unnatural causes during the year 2012 [2]. The fire accident data of India reveal that nearly 3 lakh people died in fire accidents between 2001 and 2014, i.e. average of 59 deaths per day. Every year the estimated loss of property due to fire is more than Rs. 1000 cores in India [3]. Thus It is very important to address it to save

A. Mukherjee · S. K. Shome (✉) · P. Bhattacharjee
CSIR-Central Mechanical Engineering Research Institute, Ministry of Science and Technology, Government of India, Durgapur 713209, India
e-mail: saikatkshome@cmeri.res.in

J. Gu et al. (eds.), *Communication and Control for Robotic Systems*, Smart Innovation, Systems and Technologies 229, https://doi.org/10.1007/978-981-16-1777-5_12

unparalleled human lives as well as to prevent damages critical civil infrastructure using IoT and AI along with wireless sensor network.

The occurrence of fire in residential, commercial, and building is one of the major concerns of builders, designers, and residents. Generally, an individual sensor is used for detecting fire which cannot detect the level of fire and notify the emergency response units. Another challenge in designing the fire detection system (FDS) is the detection and identification of the actual fire event in presence of different types of noise. Most of the commercially available fire detectors unable to discriminate between the combustion generated by an actual fire event, and the combustion generated by non-threatening event like cigarette smoke, automobile exhaust, etc., and generates false alarm. To solve this problem, we require an intelligent early fire detection and prediction, with real-time monitoring cyber-physical system that would not only detect the fire by using integrated multi-sensors and sensor fusion technology, but also inform the appropriate authorities, e.g., fire department, local police station, and ambulance services, instantaneously for rapid extinguishing of fire and take necessary step to protect valuable lives and properties. In this paper a survey on modern wireless sensor network capabilities through use of machine learning, Artificial Intelligence (AI) using IoT for fire detection has been presented and also a schematic of the real-time intelligent fire detection system (FDS) for disaster management is proposed. The remaining paper is arranged as follows. A survey on application of WSN in fire detection is presented in Sect. 2, use of IoT in fire detection is presented in Sect. 3, a survey on a different multisensory intelligent algorithm to increase the efficiency of the FDS has been discussed in Sect. 4. The schematic block diagram of the proposed FDS is presented in Sect. 5 and the conclusion is given in Sect. 6.

2 Application of Wireless Sensor Network for Detection of Fire Event

The wireless sensor network (WSN) is widely used for fire monitoring application in various fields like residential areas [4], buildings [5, 6], forest [7–9] rural areas [10], wild lands [11, 12] for the protection of human and animal life, infrastructure, land, etc. fire detection and monitoring purpose which is a low cost and reliable solution for fire detection. The author in [13] proposed a low-cost, power-efficient, reduced size WSN for fire detection using smoke, flame, humidity, temperature, and light intensity sensors also validated the approached in terms of accuracy and occurrence of false alarm by experimentation. In [6, 14, 15], the WSN for fire detection is implemented using multiple combination of cameras and sensors which also provides the visualization of the area under monitoring, but it requires continuous human monitoring. The author in [16] proposed a WSN model with an enhanced accuracy for fire detection using only low cost temperature sensors. This work [16] has been extended in [17] extended by adding more types of sensors like humidity, pressure and position sensors, etc. In [18], the author proposed a WSN consist of sensor nodes (SNs) to sense temperature of air, relative humidity, speed and direction of the wind, CO and CO_2 levels, for early fire detection and provided the data to central nodes which transmit the data to the control center. However, the above mentioned WSN lacked intelligence and provided less efficiency, thus different intelligent algorithm for fire detection is discussed in the next section.

3 Fire Detection Algorithm

Nowadays the multisensory fire detection algorithm is a vital research area in designing an intelligent and smart building. Several work have been done to develop an efficient fire detection algorithm. In the paper [19–21], multiple fuzzy logic approaches, e.g., neighbor-based fuzzy logic algorithms, fuzzy logic II, tree classifier has been used for designing improved fire detection system. In [22] an adaptive fusion method of the sensor data for detection of fire in a firefighting robot is proposed. The author in [23] have used neural network for development of a multisensory fire detection algorithm with fusion of a smoke sensor and a temperature sensor data. A real-time FDS is presented in [24], where the fire-detection algorithm is derived using the spatial, temporal, and spectral properties of the fire from color video input. The author in paper [25] applied genetic algorithms for automatic FDS where online identification of stochastic signal models was presented for measured fire signals. The acquired sensor data may contain error which can affect the correct output decision of fire detection system and can result in some dangerous accident. So, various methods have been proposed for detection and isolation of erroneous sensor data to minimize the error of the FDS. In [26, 27] the fire event and sensor failure was detected by adaptive fusion method of two temperatures, two smoke, and two flame sensors data. Other proposed methods to detect and isolate the sensory errors are the generalized likelihood ratio method [28], parity space method [29], measurement test method [30], pseudo nodes method [31] and extended Kalman filtering (EKF) method [32, 33]

4 Use of Internet of Things in Building Fire Detection System

Internet of Things (IoT) provides a powerful platform to fire detection system. The IoT offers a promising solution in many critical problems and it has been widely studied for its capability for connecting various devices and transmitting the data in large-scale systems. The paper [34] presented a survey on modern research and development of IoT technology in fire-related industries. The author in [35] presented an

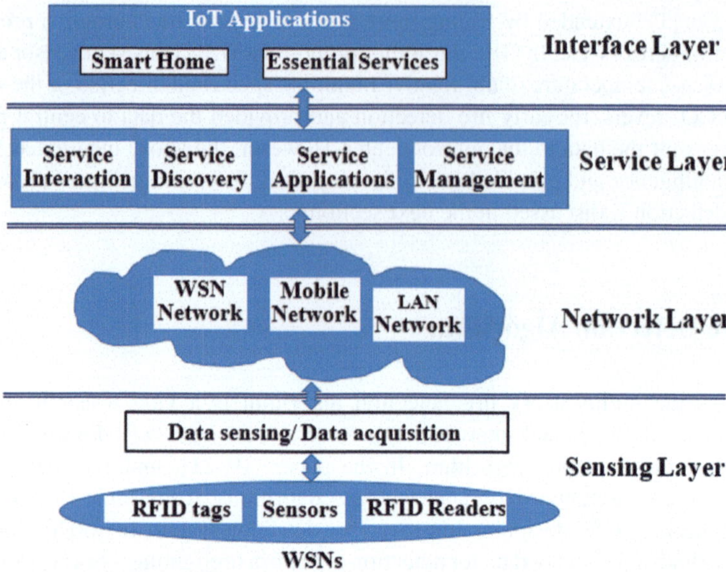

Fig. 1 SOA architecture of fire IoT

Android-supported scalable, cloud-based home automation system. The paper [36] proposed home automation system using Raspberry Pi as a web-based sensor node. The author in [37] analyzed the rescue operation supported by computer visualization. The paper [38] analyzed the disaster management scenario using IoT and Device-free Passive Localization. The objective of IoT in FDS is to connect different fire-related things over the networks which is normally supported by Service Oriented Architecture (SOA) as shown in Fig. 1, where the occurrence of fire is monitored by interaction between the four layers.

Sensing Layer: The function of sensing layer is to sense the environment for monitoring any fire incident with help of WSN node consist of sensors, radio-frequency identification tagged device.

Network Layer: This layer connects the different network together like sensor network, mobile network. This layer aggregate and share all the information related to fire incident and also solve the problem of heterogeneous network issues i.e. wireless and mobile.

Service Layer: The function of this layer is to address the issues related with services, e.g., exchange of information, and storage, management, process and communication of data. This layer consists of service interaction, service discovery, service applications, and service management module. Service interaction module supports the communication between all the connected things. The objects which required services are found out by Service discovery among several connected objects. Service application aim to provide all the services essential for fire IoT. Service management module determines the validity and evaluates the data supplied by all other services.

Interface Layer: The function of the interface layer is to simplify the management of different interconnected things by supporting universal protocol for interacting large number of devices. These devices do not follow uniform standards and protocols as made by dissimilar manufacturers.

5 Schematic of IoT-Based Intelligent WSN for Fire Detection System

The schematic block diagram of a proposed IoT-based intelligent WNS for FDS is shown in Fig. 2. The first unit consists sensor network of several sensors node which collect the information about different environmental conditions e.g. temperature, humidity, smoke, presence of CO and CO_2 gas along with image using different types of sensors and perform some local processing. The acquired sensor data is then sent to the base station i.e. the second unit where some initial processing is performed and it sends alert locally in case of any unusual event. Then the processed sensor data is sent to the cloud server through gateway for further analysis. These multi-sensor cloud data can be accessed by the user and real-time analysis is performed using different intelligent algorithm to avoid any error. In case of detection of fire event an immediate alert system will be activated and informed the concerned authority for further action to mitigate any damage due to occurrence of fire.

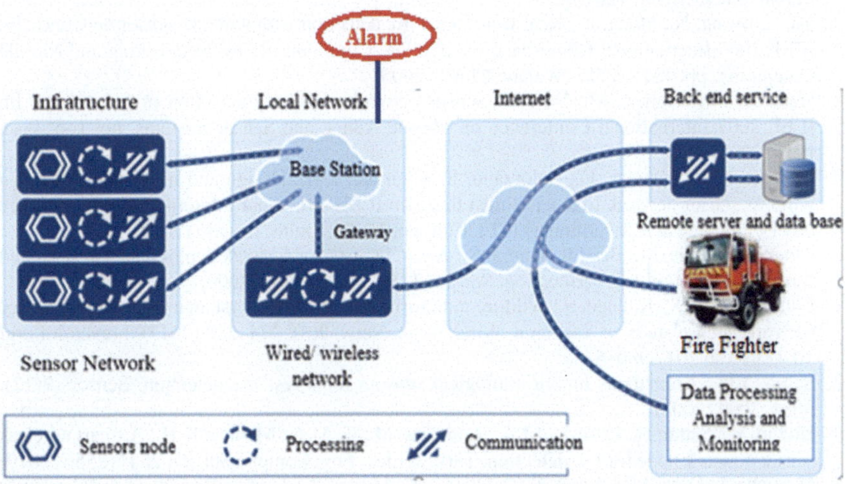

Fig. 2 Schematic block diagram of IoT-based intelligent WSN for fire detection system

6 Conclusion

Automatic, early, and accurate fire detection system is an extremely important task to avoid loss of valuable life and infrastructure due to occurrence of fire. Thus suitable integration of state of art technologies like IoT, intelligent algorithms, WSN can make an accurate fire detection system for real-time monitoring and disaster management. In this paper, a survey on different technology for fire detection system has been presented. Also, a schematic of IoT-based intelligent WSN for fire detection system has been proposed.

References

1. Schmidt-Rohr, K.: Why combustions are always exothermic, yielding about 418 kJ per mole of O_2. J. Chem. Educ. **92**(12), 2094–2099 (2015)
2. Fire Incidents from 2001–2014. National Crime Records Bureau (NCRB)
3. Tiwari, S., Bandopadhaya, S.: IoT based fire alarm and monitoring system. Int. J. Innov. Adv. Comput. Sci. IJIACS **6**(9), (2017)
4. Derbel, F.: Reliable wireless communication for fire detection systems in commercial and residential areas. In: IEEE 2003 Wireless Communications and Networking, pp. 654–659. New Orleans, LA, USA (2003)
5. Huang, Q., Cox, R.F., Shaurette, M., Wang, J.: Intelligent building hazard detection using wireless sensor network and machine learning techniques. In: International Conference on Computing in Civil Engineering, pp. 485–492. Clearwater Beach, Florida, USA (2012)
6. Zeng, Y., Sreenan, C.J., Sitanayah, L.: A real-time and robust routing protocol for building fire emergency applications using wireless sensor networks. In: IEEE 2010 8th International Conference on Pervasive Computing and Communications Workshops, pp. 358–363. Mannheim, Germany (2010)
7. Yu, L., Wang, N., Meng, X.: Real-time forest fire detection with wireless sensor networks. In: IEEE 2005 International Conference on Wireless Communications, Networking and Mobile Computing, pp. 1214–1217. Wuhan, China (2005)
8. Hefeeda, M., Bagheri, M.: Wireless sensor networks for early detection of forest fires. In: IEEE 2007 International Conference on Mobile Adhoc and Sensor Systems, pp. 1–6. Pisa, Italy (2007)
9. Cantuña, J.G., Bastidas, D., Solórzano, S., Clairand, J.M.: Design and implementation of a wireless sensor network to detect forest fires. In: IEEE 2017 Fourth International Conference on eDemocracy & eGovernment (ICEDEG), pp. 15–21. Quito, Ecuado (2017)
10. Lloret, J., Garcia, M., Bri, D., Sendra, S.: A wireless sensor network deployment for rural and forest fire detection and verification. Sensors **9**(11), 8722–8747 (2009)
11. Lutakamale, A.S., Kaijage, S.: Wildfire monitoring and detection system using wireless sensor network: a case study of Tanzania. Wirel. Sens. Netw. **9**(8), 274–289 (2017). https://doi.org/10.4236/wsn.2017.98015
12. Sahin, Y.G.: Animals as mobile biological sensors for forest fire detection. Sensors **7**(12), 3084–3099 (2007)
13. Khalid, W., Sattar, A., Qureshi, M.A., Amin, A., Malik, M.A., Memon, K.H.: A smart wireless sensor network node for fire detection. Turk. J. Elec. Eng. Comput. Sci. **27**, 2541–2556 (2019)
14. Martinez-de Dios, J.R., Arrue, B.C., Ollero, A., Merino, L., Gómez-Rodríguez, F.: Computer vision techniques for forest fire perception. Image Vis. Comput. **26**(4), 550–562 (2008)
15. Ko, A., Lee, N., Sham, R., So, C., Kwok, S.: Intelligent wireless sensor network for wildfire detection. WIT Trans. Ecol. Environ. **158**, 137–148 (2012). https://doi.org/10.2495/FIVA120121

16. Alkhatib, A.A.: Smart and low cost technique for forest fire detection using wireless sensor network. Int. J. Comput. Appl. **81**(11), 12–18 (2013)
17. Bouabdellah, K., Noureddine, H., Larbi, S.: Using wireless sensor networks for reliable forest fires detection. Procedia Comput. Sci. **19**, 794–801 (2013). https://doi.org/10.1016/j.procs.2013.06.104
18. Molina-Pico, A., Cuesta-Frau, D., Araujo, A., Alejandre, J., Rozas, A.: Forest monitoring and wildland early fire detection by a hierarchical wireless sensor network. J. Sens. **2016**, 1–8 (2016)
19. Maksimović, M., Vujović, V., Perišić, B., Milošević, V.: Developing a fuzzy logic based system for monitoring and early detection of residential fire based on thermistor sensors. Comput. Sci. Inf. Syst. **12**(1), 63–89 (2015)
20. Kapitanova, K., Son, S.H., Kang, K.D.: Event detection in wireless sensor networks—Can fuzzy values be accurate? In: International Conference on Ad Hoc Networks (ADHOCNETS), pp. 168–184. Victoria, BC, Canada (2010)
21. Wang, B., Zhuang, A., Sun, H., Li, T., Sun, X.: An improved spatial-based fuzzy logic event detecting algorithm for wireless sensor networks. Int. J. u- e-Serv. Sci. Technol. **8**(4), 265–278 (2015)
22. Su, K.L.: Automatic fire detection system using adaptive fusion algorithm for fire fighting robot. In: 2006 IEEE International Conference on Systems, Man and Cybernetics, pp. 966–971. Taipei (2006)
23. Xihuai, W., Jianmei, X., Minzhong, B.: A ship fire alarm system based on fuzzy neural network. In: Proceedings of the 3rd World Congress on intelligent Control and Automation, vol. 3, pp. 1734–1736 (2000)
24. Healey, O., Slater, D., Lin, T., Drda, B., Goedeke A.D.: A system for real-time fire detection. In: Proceedings of IEEE Computer Society Conference on Computer Vision and Pattern Recognition, pp. 605–606 (1993)
25. Neubauer, A.: Genetic algorithms in automatic fire detection technology. In: Second International Conference on Genetic Algorithms in Engineering Systems: Innovations and Applications, pp. 180–185 (1997)
26. Luo, R.C., Su, K.L., Tsai, K.H.: Fire detection and isolation for intelligent building system using adaptive sensory fusion method. In: Proceedings of the IEEE International Conference on Robotics and Automation, pp.1777–1781 (2002)
27. Luo, R.C., Su, K.L., Tsai, K.H.: Intelligent security robot eire detection system using adaptive sensory fusion method. In: The IEEE International Conference on Industrial Electronics Society, (IECON 2002), pp. 2663–2668 (2002)
28. Narasimhan, S., Mah, R.S.H.: Generalized likelihood ratio method for large error identification. AIChE J. **33**(9), 1514–1521 (1987)
29. Ragot, J., Aitouche, A., Kratz, F., Maquin, D.: Detection and localization of gross errors in instrument using parity space technique. Int. J. Miner. Process. **31**, 281–299 (1991)
30. Mah, R.S.H., Jamhane, A.C.: Detection of large errors in process data. AIChE J. **28**(5), 828–831 (1982)
31. Ragot, J., Maquin, D., Darouach: Analysis of generalized bilinear systems application to diagnosis. In: Proceeding of IMACS-IFAC Symposium (1991)
32. Del Gobbo, D.: Sensor Failure Detection and Identification using Extended Kalman Filtering (1998)
33. Del Gobbo, D., Napolitano, M., Famouri, P., Innocenti, M.: Experimental application of extended Kalman filtering for sensor validation. IEEE Trans. Control Syst. Technol. **9**(2), 376–380 (2001)
34. Vijayalakshmi, S.R., Muruganand, S.: A survey of Internet of Things in fire detection and fire industries. In: 2017 International Conference on I-SMAC (IoT in Social, Mobile, Analytics and Cloud) (I-SMAC), pp. 703–707. Palladam (2017)
35. Korkmaz, I., et al.: A cloud based and Android supported scalable home automation system. Comput. Electr. Eng. **43**, 112–128 (2015)

36. Vujovic, V., et al.: Raspberry Pi as a sensor web node for home automation. Comput. Electr. Eng. **44**, 153–171 (2015)
37. Morin, M., et al.: Computer-supported visualization of rescue operations. Saf. Sci. **35**, 3–27 (2010)
38. Deak, G., et al.: IoTs (Internet of Things) and DfPL (Device-free Passive Localisation) in a disaster management scenario. Simul. Model. Pract. Theory **35**, 86–96 (2013)

Computation Paradigm for Robotic Systems

Nonlinear State Estimation Using Adaptive Gaussian Filters with One-Step Randomly Delayed Measurements

Poluri Sri Mannarayana and Aritro Dey

Abstract This paper presents two adaptive sigma point Gaussian filters based on unscented transformation (UT) rule and cubature rule for one-step randomly delayed measurements when process noise covariance remains unknown due to parametric uncertainty. The proposed filter is designed to give satisfactory estimation performance by carrying out auto tuning of unknown element of process noise covariance. Efficacy of the adaptive Gaussian filters for one-step randomly delayed measurements is demonstrated by comparing root mean square error (RMSE), average RMSE (ARMSE), and percentage of track loss with corresponding non-adaptive filters. Monte Carlo (MC) simulations demonstrated the effectiveness and consistency of the proposed filters.

1 Introduction

Since R. E. Kalman introduced Kalman filter [1] in 1960, usage of linear and nonlinear estimation filters has become popular in applications like aerospace, weather prediction, underwater targeting, economics, ocean atmosphere, and state of charge estimation of lithium ion batteries [2–9]. Based on the standard Kalman filter [1], various linear and nonlinear variants of Kalman filters have been developed. Kalman filter gives an optimal solution for the linear systems; however, suboptimal solution is obtained from the nonlinear filters which are still evolving and drawing the attention of the researchers. Estimation algorithms can be divided into mainly two categories: point estimation and density estimation. Point estimation methods include extended Kalman filter (EKF), unscented Kalman filter (UKF) and its variants, cubature Kalman filter (CKF) and its variants, and cubature quadrature Kalman filter (CQKF) [10–15], which are the most widely used methods, and density estimation methods include particle filter [16].

P. S. Mannarayana (✉) · A. Dey
Department of Electrical Engineering, National Institute of Technology Durgapur, Durgapur 713209, West Bengal, India
e-mail: psm.19ee1101@phd.nitdgp.ac.in

J. Gu et al. (eds.), *Communication and Control for Robotic Systems*, Smart Innovation, Systems and Technologies 229, https://doi.org/10.1007/978-981-16-1777-5_13

Initially, EKF [10, 11] has been employed for the state estimation for nonlinear systems, mostly in aircraft tracking. Drawbacks of EKF method are: Taylor series limits the accuracy of linearization, if dimension of system increases, then the size of the Jacobian matrix increases, which eventually needs more memory, and differentiation of strong nonlinear systems is difficult. Due to the above-mentioned drawbacks, EKF fails to track the trajectories several times.

When it comes to deterministic sample point filters, unscented Kalman filter (UKF) was introduced at first, where UKF [12] is completely free from Jacobian matrices; however, accuracy of UKF depends on the tuning parameters of the algorithm. New deterministic sample point filter termed as cubature Kalman filter (CKF) was developed in [13], which overcomes the numerical instability of UKF and free of tuning of algorithm. Cubature rule has been employed to solve multi-dimensional intractable integrals, where intractable integrals are divided into surface integral and line integral. Surface integrals are solved using third degree spherical cubature rule and line integrals using first-order Gauss–Laguerre rule of integration. Later, Bhaumik et al. [14] proposed cubature quadrature Kalman filter (CQKF) based on cubature quadrature (CQ) rule with higher order Gauss–Laguerre rule of integration and shown its enhanced accuracy with increased computational cost compared to CKF and subsequently developed square root CQKF in [15]. Nonlinear estimation methods mentioned above are being improved in different fields.

In all above-mentioned literature, it is assumed that measurement is readily available at every instant, i.e. traditional filters developed based on assumption that there is no delay in the measurement. However, in practice, it is not always true because there are certain fields like underwater target tracking, target tracking in a sensor network, and there is a chance of delayed measurements. In that case, traditional filters give large estimation error. So, there is a need to modify traditional algorithms when there is delay in the measurements.

In the literature, randomly delayed measurement is not widely reported. Nevertheless, few authors [17–19] have addressed this problem for linear systems and for nonlinear systems [20–25]. The problem of delayed measurements for linear systems is introduced in [17] and then continued in [18, 19]. For nonlinear systems, in [20], Carazo and Perez developed EKF and UKF when measurements delayed by maximum one step and later extended it for two-step randomly delayed measurements in [22]. In [2], Wang et al. proposed CKF with one-step randomly delayed measurements and explained comprehensively. Not only measurements will be delayed but also there is a possibility of measurements will be lost at any instant. In that case, filters must be modified to avoid large estimation error. This issue has been identified and addressed in [23–25].

Filtering algorithms mentioned so far are assumed to have correct knowledge of both process and measurement noise covariance. Unavailability of either process or measurement noise cannot be eliminated by choosing some random values as it degrades the performance of the filter or may lead to filter divergence. This paper proposes nonlinear filtering algorithm for one-step randomly delayed measurements when process noise covariance remains unknown to the practitioner in face of parametric uncertainty. Usually, noise covariance is selected based on trial and error method during offline tuning of filters. However, during parametric uncertainty, choice of suitable value of process noise covariance is a challenging task. The present workers have proposed in this paper an adaptive version of nonlinear filters for one-step random delay in measurements which can automatically adjust the process noise covariance in face of unknown parameter variation. For solving multi-dimensional nonlinear intractable integrals, cubature rule has been preferred in this paper. There are mainly two approaches for the adaptation for the Kalman filtering as mentioned in [26]. Among them, maximum likelihood estimation method (MLE) [27–31] has been considered in this paper. In MLE, probability density function of measurement is expressed in terms of innovation sequence, and proposed formulation ensures positive definiteness.

In this work, mainly adaptive cubature Kalman filter and adaptive unscented Kalman filter are extended to solve nonlinear problem for one-step randomly delayed measurements when process noise covariance remains unknown due to parametric uncertainty and are abbreviated as ACKFRD and AUKFRD, respectively. Following sections are arranged as follows: Sect. 2 derives the problem formulation under Gaussian framework of filtering for one-step randomly delayed measurements. Algorithm steps of ACKFRD are given in Sect. 3, and same algorithm can extend for AUKFRD with corresponding sigma points and weights. Simulation procedure and results are enumerated in Sect. 4, followed by conclusion in Sect. 5.

2 Problem Formulation

A discrete stochastic nonlinear system can be represented as

$$\varkappa_k = \mathcal{F}_{k-1}(\varkappa_{k-1}) + w_{k-1}, \tag{1}$$

$$z_k = \mathcal{G}_k(\varkappa_k) + v_k, \tag{2}$$

where $\varkappa_k \in \Re^n$ and $z_k \in \Re^p$ are column vectors of the state and ideal measurement at any instant k, i.e. $k = \{1, 2, 3, \ldots, N\}$. Whereas $\mathcal{F}$ and $\mathcal{G}$ are the known nonlinear functions of state and measurement, respectively. w_k is the process noise, and v_k is the measurement noise. w_k and v_k are uncorrelated and are assumed to be zero mean white Gaussian noise such that $\mathbb{E}\left[w_k w_l^T\right] = \boldsymbol{Q}_k \delta_{kl}$ and $\mathbb{E}\left[v_k v^T\right] = \boldsymbol{R}_k \delta_{\mathrm{kl}}$, where δ_{kl} is a Kronecker delta function.

As stated previously, in this paper, measurements are considered to be maximum one-step randomly delayed, i.e. the arriving measurements may belong to that specific instant or previous instant. Then, measurement equation can be represented with the help of Bernoulli random variables as

$$y_k = (1 - \Theta_k) z_k + \alpha_k z_{k-1}, \ \ k > 1; \ \text{and} \ y_1 = z_1, \tag{3}$$

Θ_k, w_k, and v_k are mutually independent. For $k > 1$, the sequence of Bernoulli random variables takes independent values of either 0 or 1 with the probability

$$\begin{aligned} P(\Theta_k = 1) &= \mathbb{E}[\Theta_k] = p_k, \\ P(\Theta_k = 0) &= 1 - \mathbb{E}[\Theta_k] = 1 - p_k, \\ \mathbb{E}\big[(\Theta_k - p_k)^2\big] &= (1 - p_k)p_k. \end{aligned} \tag{4}$$

p_k represents the probability of delay in measurement at k^{th} instant. From Eq. (4), if $\Theta_k = 1$, then the measurement is delayed by one step with probability p_k such that $y_k = z_{k-1}$; otherwise, if $\Theta_k = 0$, then true measurements are arrived (no delay in measurement) with the probability $1 - p_k$ and $y_k = z_k$.

The main objective is to design adaptive Gaussian filter with $\boldsymbol{Q}$ adaptation algorithm for auto tuning of the elements of $\boldsymbol{Q}$ for the nonlinear system. Our objective is to determine posterior pdf $P(\varkappa_{k+1}|\boldsymbol{Y}_{k+1})$, where $\boldsymbol{Y}_{k+1} = \{\boldsymbol{Y}_j\}$ with $j = \{1, 2, 3, \ldots, k+1\}$ represents the set of randomly delayed measurements with unknown process noise covariance ($\boldsymbol{Q}$).

Certain approximations should be made before we proceed for the posteriori pdf, $P(\varkappa_{k+1}|y_{k+1})$. As $\mathcal{F}$ and $\mathcal{G}$ are nonlinear, for Gaussian noise posteriori pdf $P(\varkappa_{k+1}|y_{k+1})$ does not necessarily follow Gaussian distribution. However, it is assumed to follow Gaussian distribution with mean $\hat{\varkappa}_{k+1|k+1}$ and corresponding covariance of $\boldsymbol{\Sigma}^{\varkappa\varkappa}_{k+1|k+1}$.

2.1 Gaussian Framework of Filtering with One-Step Randomly Delayed Measurements

Modified delayed measurement equation can be written using Eqs. (2) and (3) as

$$y_{k+1} = (1 - \alpha_{k+1})(\mathcal{G}_{k+1}(\varkappa_{k+1}) + v_{k+1}) + \alpha_{k+1}(\mathcal{G}_k(\varkappa_k) + v_k). \tag{5}$$

from Eq. (5), it is understood that in order to estimate the states $P(\varkappa_{k+1}|\boldsymbol{Y}_{k+1})$, $P(v_{k+1}|\boldsymbol{Y}_{k+1})$ has to be determined. So, state vector should be augmented with the measurement noise and the augmented state vector $\varkappa^a_{k+1} = \left[\varkappa^T_{k+1}\ v^T_{k+1}\right]^T$.

Certain assumptions are as follows:

(i) The pdf $P(\varkappa_{k+1}|\boldsymbol{Y}_{k+1})$ is Gaussian with mean $\hat{\varkappa}_{k+1|k+1}$ and covariance $\boldsymbol{\Sigma}^{\varkappa\varkappa}_{k+1|k+1}$,

(ii) The pdf $P(v_{k+1}|\boldsymbol{Y}_{k+1})$ is Gaussian with mean $\hat{v}_{k+1|k+1}$ and covariance $\boldsymbol{\Sigma}^{vv}_{k+1|k+1}$,

(iii) Predictive pdf of $\varkappa_{k+1}$ is Gaussian, i.e. $P(\varkappa_{k+1}|\boldsymbol{Y}_k) = \aleph(\varkappa_{k+1}; \hat{\varkappa}_{k+1}, \boldsymbol{\Sigma}^{\varkappa\varkappa}_{k+1|k})$,

(iv) Predictive pdf of z_{k+1} is Gaussian, i.e. $P(y_{k+1}|\boldsymbol{Y}_k) = \aleph\Big(y_{k+1}; \hat{y}_{k+1}, \boldsymbol{\Sigma}^{yy}_{k+1|k}\Big)$,

If the pdf of $P(\varkappa_{k+1}|\boldsymbol{Y}_{k+1})$ and $P(v_{k+1}|\boldsymbol{Y}_{k+1})$ is assumed to be Gaussian, then the augmented state also becomes Gaussian, i.e.

$$P\left(\varkappa^a_{k+1}|\boldsymbol{Y}_{k+1}\right) = \aleph\left(\varkappa^a_{k+1}; \widehat{\varkappa}^a_{k+1|k+1}, \underset{k+1|k+1}{\overset{a}{\boldsymbol{\Sigma}}}\right) \tag{6}$$

where $\aleph(\boldsymbol{x}; \boldsymbol{\mu}, \boldsymbol{\Sigma})$ represents the Gaussian distribution of variable $\boldsymbol{x}$ with mean $\boldsymbol{\mu}$ and of covariance $\boldsymbol{\Sigma}$, and

$$\widehat{\varkappa}^a_{k+1|k+1} = \begin{bmatrix} \widehat{\varkappa}_{k+1|k+1} \\ \widehat{v}_{k+1|k+1} \end{bmatrix} \text{ and } \underset{k+1|k+1}{\overset{a}{\boldsymbol{\Sigma}}} = \begin{bmatrix} \underset{k+1|k+1}{\overset{\varkappa\varkappa}{\boldsymbol{\Sigma}}} & \underset{k+1|k+1}{\overset{\varkappa v}{\boldsymbol{\Sigma}}} \\ \left(\underset{k+1|k+1}{\overset{\varkappa v}{\boldsymbol{\Sigma}}}\right)^T & \underset{k+1|k+1}{\overset{vv}{\boldsymbol{\Sigma}}} \end{bmatrix}, \tag{7}$$

where $\boldsymbol{\Sigma}^{\varkappa v}$ is the cross covariance between the posteriori state and measurement noise.

Some of the parameters are as follows

$$\begin{cases} \underset{k+1|k}{\overset{\varkappa y}{\boldsymbol{\Sigma}}} = \mathbb{E}\left[\widetilde{\varkappa}_{k+1|k}\widetilde{y}^{\boldsymbol{T}}_{k+1|k}|\boldsymbol{Y}_k\right], \widetilde{z}_{k+1|k} = \mathbb{E}\left[z_{k+1}|\boldsymbol{Y}_k\right], \\ \underset{k+1|k}{\overset{\varkappa z}{\boldsymbol{\Sigma}}} = \mathbb{E}\left[\widetilde{\varkappa}_{k+1|k}\widetilde{z}^{\boldsymbol{T}}_{k+1|k}|\boldsymbol{Y}_k\right], \underset{k|k}{\overset{zz}{\boldsymbol{\Sigma}}} = \mathbb{E}\left[\widetilde{z}_{k|k}\widetilde{z}^{\boldsymbol{T}}_{k|k}|\boldsymbol{Y}_k\right], \\ \underset{k+1,k|k}{\overset{\varkappa z}{\boldsymbol{\Sigma}}} = \mathbb{E}\left[\widetilde{\varkappa}_{k+1|k}\widetilde{z}^{\boldsymbol{T}}_{k|k}|\boldsymbol{Y}_k\right], \widetilde{z}_{k|k} = \mathbb{E}[z_k|\boldsymbol{Y}_k] \\ \underset{k+1|k}{\overset{vy}{\boldsymbol{\Sigma}}} = \mathbb{E}\left[\widetilde{v}_{k+1}\widetilde{y}^{\boldsymbol{T}}_{k+1|k}|\boldsymbol{Y}_k\right], \underset{k+1|k}{\overset{zz}{\boldsymbol{\Sigma}}} = \mathbb{E}\left[\widetilde{z}_{k+1|k}\widetilde{z}^{\boldsymbol{T}}_{k+1|k}|\boldsymbol{Y}_k\right] \end{cases} \tag{8}$$

2.1.1 Measurement Noise Estimation

Priori and posteriori pdf of measurements are given by

$$\widetilde{z}_{k+1|k} = \int \mathcal{G}(\varkappa_{k+1})\aleph\left(\varkappa_{k+1}; \widehat{\varkappa}_{k+1}, \underset{k+1|k}{\overset{\varkappa\varkappa}{\boldsymbol{\Sigma}}}\right)\mathrm{d}\varkappa_{k+1}, \tag{9}$$

$$\begin{aligned} \underset{k+1|k}{\overset{zz}{\boldsymbol{\Sigma}}} = &\int \mathcal{G}(\varkappa_{k+1})\mathcal{G}^T(\varkappa_{k+1})\aleph\left(\varkappa_{k+1}; \widehat{\varkappa}_{k+1}, \underset{k+1|k}{\overset{\varkappa\varkappa}{\boldsymbol{\Sigma}}}\right)d\varkappa_{k+1} \\ &- z_{k+1|k}z^{\boldsymbol{T}}_{k+1|k} + \boldsymbol{R}_{k+1}, \end{aligned} \tag{10}$$

$$\widehat{z}_{k|k} = \int [\mathcal{G}(\varkappa_k) + v_k]\aleph\left(\varkappa^a_k; \widehat{\varkappa}^a_{k|k}, \underset{k|k}{\overset{a}{\boldsymbol{\Sigma}}}\right)\mathrm{d}\varkappa^a_k, \tag{11}$$

$$\underset{k|k}{\overset{zz}{\boldsymbol{\Sigma}}} = \int [\mathcal{G}(\varkappa_k) + v_k][\mathcal{G}(\varkappa_k) + v_k]^T\aleph\left(\varkappa^a_k; \widehat{\varkappa}^a_{k|k}, \underset{k|k}{\overset{a}{\boldsymbol{\Sigma}}}\right)d\varkappa^a_k - z_{k|k}z^T_{k|k}. \tag{12}$$

Priori mean and covariance of measurement is given by

$$y_{k+1|k} = (1 - p_{k+1})\widehat{z}_{k+1|k} + p_{k+1}\widehat{z}_{k|k} \tag{13}$$

$$\underset{k+1|k}{\overset{yy}{\Sigma}} = (1 - p_{k+1}) \underset{k+1|k}{\overset{zz}{\Sigma}} + p_{k+1} \underset{k|k}{\overset{zz}{\Sigma}} + (1 - p_{k+1}) p_{k+1} \left(\widehat{z}_{k+1|k} - \widehat{z}_{k|k}\right)\left(\widehat{z}_{k+1|k} - \widehat{z}_{k|k}\right)^T. \tag{14}$$

Kalman gain of noise can be calculated from Eq. (15)

$$\boldsymbol{K}_k^v = \underset{k+1|k}{\overset{vy}{\Sigma}} \left(\underset{k+1|k}{\overset{yy}{\Sigma}} \right)^{-1} \tag{15}$$

Cross covariance $\boldsymbol{\Sigma}_{k+1|k}^{vy}$ is given by

$$\underset{k+1|k}{\overset{vy}{\Sigma}} = (1 - p_{k+1})\boldsymbol{R}_{k+1}. \tag{16}$$

Finally, posterior mean and covariance of measurement noise can be obtained as

$$\widehat{v}_{k+1|k+1} = \boldsymbol{K}_k^v \left(y_{k+1} - \widehat{y}_{k+1|\mathrm{k}} \right) \tag{17}$$

$$\underset{k+1|k+1}{\overset{vv}{\Sigma}} = \boldsymbol{R}_{k+1} - \boldsymbol{K}_k^v \underset{k+1|k}{\overset{vy}{\Sigma}} \left(\boldsymbol{K}_k^v\right)^T \tag{18}$$

2.1.2 State Estimation

Prior mean and covariance of state can be written as

$$\widehat{\varkappa}_{k+1|k} = \int \mathcal{F}(\varkappa_k) \aleph\left(\varkappa_k; \widehat{\varkappa}_{k|\mathrm{k}}, \underset{k|k}{\overset{\varkappa\varkappa}{\Sigma}} \right) d\varkappa_k, \tag{19}$$

$$\underset{k+1|k}{\overset{\varkappa\varkappa}{\Sigma}} = \int \mathcal{F}(\varkappa_k)\mathcal{F}^T(\varkappa_k) \aleph\left(\varkappa_k; \widehat{\varkappa}_{k|\mathrm{k}}, \underset{k|k}{\overset{\varkappa\varkappa}{\Sigma}} \right) d\varkappa_k - \widehat{\varkappa}_{k+1|\mathrm{k}} \widehat{\varkappa}_{k+1|k}^T + \mathbf{Q}_k. \tag{20}$$

Cross covariances can be written as

$$\underset{k+1,k|k}{\overset{\varkappa z}{\Sigma}} = \int \mathcal{F}(\varkappa_k)[\mathcal{G}(\varkappa_k) + v_k]^T \aleph\left(\varkappa_k^a; \widehat{\varkappa}_{k|k}^a, \underset{k|k}{\overset{a}{\Sigma}} \right) d\varkappa_k^a - \widehat{\varkappa}_{k+1|\mathrm{k}} \widehat{\varkappa}_{k|k}^T, \tag{21}$$

$$\underset{k+1|k}{\overset{\varkappa z}{\Sigma}} = \int \varkappa_{k+1}\left[\mathcal{G}_{k+1}(\varkappa_{k+1})\right]^T \aleph\left(\varkappa_{k+1}; \widehat{\varkappa}_{k+1|\mathrm{k}}, \underset{k+1|k}{\overset{\varkappa\varkappa}{\Sigma}} \right) d\varkappa_{k+1}$$

$$- \widehat{\varkappa}_{k+1|k} \widehat{z}^T_{k+1|k}. \tag{22}$$

Kalman gain corresponding to state is given by

$$\boldsymbol{K}_k^{\varkappa} = \overset{\varkappa y}{\underset{k+1|k}{\Sigma}} \left(\overset{yy}{\underset{k+1|k}{\Sigma}} \right)^{-1} \tag{23}$$

where

$$\overset{\varkappa y}{\underset{k+1|k}{\Sigma}} = (1 - p_{k+1}) \overset{\varkappa z}{\underset{k+1|k}{\Sigma}} + p_{k+1} \overset{\varkappa z}{\underset{k+1,k|k}{\Sigma}} \tag{24}$$

Finally, the posteriori estimate of the state and error covariance can be obtained as

$$\widehat{\varkappa}_{k+1|k+1} = \widehat{\varkappa}_{k+1|k} + \boldsymbol{K}_k^{\varkappa} \left(y_{k+1} - \widehat{y}_{k+1|k} \right), \tag{25}$$

$$\overset{\varkappa\varkappa}{\underset{k+1|k+1}{\Sigma}} = \overset{\varkappa\varkappa}{\underset{k+1|k}{\Sigma}} - \boldsymbol{K}_k^{\varkappa} \overset{yy}{\underset{k+1|k}{\Sigma}} \left(\boldsymbol{K}_k^{\varkappa}\right)^T. \tag{26}$$

Note-1: If measurements are non-delayed, i.e. when $\Theta_k = 0$, above-mentioned delayed measurement equations become non-delayed measurement equations, and the formulated problem reduces to conventional non-delayed estimation problem.
Note-2: The intractable Gaussian integrals expressed in Eqs. (9)–(12) and Eqs. (19)–(22) are numerically approximated using cubature rule of integration.
Note-3: Proofs for the equations in Eq. (8) can found from [21].

2.1.3 Cubature Rule of Integration

. I Arasaratnam and Haykin, Simon developed cubature Kalman filter in 2009. CKF utilizes cubature rule to approximate the integrals of the form *nonlinear function* $\times$ *Gaussian*. On the basis of Bayesian filter, CKF uses 2n points with corresponding weights to capture the mean and covariance of the states.

Consider an integral of the form

$$I(f) = \int_D f(x) w(x) \mathrm{d}x \tag{27}$$

where $f(\cdot)$ is known nonlinear function, $D \in \Re^n$ is the region of integration, and $w(\cdot)$ is the known weighting function.

According to [13], Eq. (27) can be written as

$$I(f) = \int_D f(x)\aleph(\varkappa; 0, \mathrm{I})dx \approx \sum_{i=1}^{m} f(\xi_i)\omega_i \tag{28a}$$

where $m = 1, 2, 3\ldots, 2n$. $\xi_i = \sqrt{n}[U]_i$. U is an identity matrix, and suffix 'i' indicates the ith column of U. $\aleph(\varkappa; 0, \mathrm{I})$ represents the standard Gaussian distribution, and n represents the state vector dimension.

2.1.4 Unscented Transformation

The integral mentioned in (27) can be solved approximately with the help of sigma points by using unscented transformation (UT) [12]. Consider a random variable $\boldsymbol{x}$ of dimension D has mean $\overline{\boldsymbol{x}}$ and covariance $\boldsymbol{\Sigma}$. Let $\boldsymbol{y} = f(\boldsymbol{x})$, to know the statistics of $\boldsymbol{y}$, first we need to have set of sigma points ς_i and corresponding weights ω_i as follows:

$$\begin{aligned}
\varsigma_i &= \overline{\boldsymbol{x}} \quad i = 0 \\
\varsigma_i &= \overline{\boldsymbol{x}} + \left(\sqrt{(D+\lambda)\boldsymbol{\Sigma}}\right)_i \quad i = 1, \ldots, D \\
\varsigma_i &= \overline{\boldsymbol{x}} - \left(\sqrt{(D+\lambda)\boldsymbol{\Sigma}}\right)_{i-D} \quad i = D+1, \ldots, 2D \\
\omega_i^m &= \frac{\lambda}{(D+\lambda)} \quad = 0 \\
\omega_i^c &= \omega_i^m + 1 - \alpha^2 + \beta \quad i = 0 \\
\omega_i^c &= \omega_i^m = \frac{1}{\{2(D+\lambda)\}} \quad i = 1, \ldots, 2D
\end{aligned} \tag{28b}$$

where $\lambda = \alpha^2(D+k) - D$ is a scaling parameter, α denotes the spread of sigma points around mean and $k = 3 - D$. β is an another scaling factor which incorporates the prior knowledge of $\boldsymbol{x}$. Subscript i indicates to select i^{th} column of the square root matrix.

3 Adaptive Cubature Kalman Filter Algorithm for One-Step Randomly Delayed Measurements

Step I Initialization

- $\varkappa^a_{0|0} = \begin{bmatrix} \hat{\varkappa}_{0|0} \\ 0 \end{bmatrix}$, $\boldsymbol{\Sigma}^a_0 = \begin{bmatrix} \boldsymbol{\Sigma}^{\varkappa\varkappa}_{0|0} & 0 \\ 0 & \boldsymbol{R} \end{bmatrix}$, $\widehat{\boldsymbol{Q}}_0$.
- Construct two sets of cubature points and weights. One with $2(n+\rho)$ number of cubature points $\boldsymbol{\zeta}_j(j = 1, 2, 3, \ldots, 2(n+\rho))$ and corresponding weights $\boldsymbol{\omega}_j$ and another set of cubature points $\boldsymbol{\zeta}_i(i = 1, 2, 3, \ldots, 2n)$ with $2n$ number and

corresponding eights ω_i. Here, ρ indicates the dimension of the measurement, and p indicates probability of delay in the measurement.

Step II Calculation of cubature points and prediction

$$\underset{k|k}{\overset{a}{\Sigma}} = S^a_{k|k}\left(S^a_{k|k}\right)^T. \tag{29}$$

- Construct cubature points

$$\mathcal{E}_{j,k|k} = \left[\left(\mathcal{E}^{\varkappa}_{j,k|k}\right)^T \left(\mathcal{E}^{v}_{j,k|k}\right)^T\right]^T = \widehat{\varkappa}^a_{k|k} + S^a_{k|k}\zeta_j, \tag{30}$$

where $j = 1, 2, 3, \ldots, 2(n+\rho)$.

- Evaluation of predicted mean and covariance

$$\widehat{\varkappa}_{k+1|k} = \sum_{j=1}^{2(n+\rho)} \mathcal{F}_k\left(\mathcal{E}^{\varkappa}_{j,k|k}\right)\omega_i, \tag{31}$$

$$\underset{k+1|k}{\overset{\varkappa\varkappa}{\Sigma}} = \sum_{j=1}^{2(n+\rho)} \mathcal{F}_k\left(\mathcal{E}^{\varkappa}_{j,k|k}\right)\mathcal{F}_k\left(\mathcal{E}^{\varkappa}_{j,k|k}\right)^T\omega_i - \widehat{\varkappa}_{k+1|k}\widehat{\varkappa}^T_{k+1|k} + \widehat{Q}_{k-1}. \tag{32}$$

Step III Estimation of measurement noise statistics

Main objective of this step is to compute for $v_{k+1|k+1}$ and $\Sigma^{vv}_{k+1|k+1}$.

- Factorize error covariance in Eq. (32)

$$\underset{k+1|k}{\overset{\varkappa\varkappa}{\Sigma}} = S^{\varkappa}_{k+1|k}\left(S^{\varkappa}_{k+1|k}\right)^T. \tag{33}$$

- Generation of cubature points

$$\mathcal{E}^{\varkappa}_{i,k+1|k} = \widehat{\varkappa}_{k+1|k} + S^{\varkappa}_{k+1|k}\zeta_i. \tag{34}$$

- Calculation of the statistics of non-delayed measurement

$$\widehat{z}_{k+1|k} = \sum_{i=1}^{2n} \mathcal{G}_k\left(\varkappa_{i,k+1|\mathrm{k}}\right)\omega_i, \tag{35}$$

$$\underset{k+1|k}{\overset{zz}{\Sigma}} = \sum_{i=1}^{2n} \mathcal{G}_k\left(\mathcal{E}^{\varkappa}_{i,k+1|k}\right)\mathcal{G}_k\left(\mathcal{E}^{\varkappa}_{i,k+1|k}\right)^T\omega_i - \widehat{z}_{k+1|k}\widehat{z}^T_{k+1|k} + R_{k+1}, \tag{36}$$

$$\hat{z}_{k|k} = \sum_{j=1}^{2(n+\rho)} [\mathcal{G}_k(\mathcal{E}^{\varkappa}_{j,k|k}) + \mathcal{E}^{v}_{j,k|k}]\omega_j, \tag{37}$$

$$\overset{zz}{\underset{k|k}{\Sigma}} = \sum_{j=1}^{2(n+\rho)} [\mathcal{G}_k(\mathcal{E}^{\varkappa}_{j,k|k}) + \mathcal{E}^{v}_{j,k|k}][\mathcal{G}_k(\mathcal{E}^{\varkappa}_{j,k|k}) + \mathcal{E}^{v}_{j,k|k}]^T \omega_j - \hat{z}_{k|k}\hat{z}^T_{k|k}. \tag{38}$$

- Calculation of the statistics of delayed measurement

$$\hat{y}_{k+1|k} = (1 - p_{k+1})\hat{z}_{k+1|k} + p_{k+1}\hat{z}_{k|k} \tag{39}$$

$$\overset{yy}{\underset{k+1|k}{\Sigma}} = (1 - p_{k+1}) \overset{zz}{\underset{k+1|k}{\Sigma}} + p_{k+1} \overset{zz}{\underset{k|k}{\Sigma}} + (1 - p_{k+1})p_{k+1} \times \left(\hat{z}_{k+1|k} - \hat{z}_{k|k}\right)\left(\hat{z}_{k+1|k} - \hat{z}_{k|k}\right)^T. \tag{40}$$

- Kalman gain with respect to measurement noise

$$\boldsymbol{K}^v_k = \boldsymbol{\Sigma}^{vy}_{k+1|k}\left(\boldsymbol{\Sigma}^{yy}_{k+1|k}\right)^{-1}, \tag{41}$$

where $\boldsymbol{\Sigma}^{vy}_{k+1|k} = (1 - p_{k+1})\boldsymbol{R}_{k+1.}$.

Finally,

$$\hat{v}_{k+1|k+1} = \boldsymbol{K}^v_k\left(y_{k+1} - \hat{y}_{k+1|k}\right), \tag{42}$$

$$\overset{vv}{\underset{k+1|k+1}{\Sigma}} = \boldsymbol{R}_{k+1} - \boldsymbol{K}^v_k \overset{vy}{\underset{k+1|k}{\Sigma}} \left(\boldsymbol{K}^v_k\right)^T. \tag{43}$$

Step IV State estimation

Main objective of this step is to compute for $\hat{\varkappa}_{k+1|k+1}$ and $\boldsymbol{\Sigma}^{\varkappa\varkappa}_{k+1|k+1}$.

- Cross covariance between the state and measurements is given by

$$\overset{\varkappa z}{\underset{k+1,k|k}{\Sigma}} = \sum_{j=1}^{2(n+\rho)} \mathcal{F}_k(\mathcal{E}^{\varkappa}_{j,k|k})[\mathcal{G}_k(\mathcal{E}^{\varkappa}_{j,k|k}) + \mathcal{E}^{\varkappa}_{j,k|k}]^T \omega_j - \hat{\varkappa}_{k+1|k}\hat{z}^T_{k|k}, \tag{44}$$

$$\overset{\varkappa z}{\underset{k+1,k|k}{\Sigma}} = \sum_{i=1}^{2\boldsymbol{n}} \mathcal{E}^{\varkappa}_{i,k+1|k}\mathcal{G}_k\left(\mathcal{E}^{v}_{i,k+1|k}\right)^T \omega_i - \hat{\varkappa}_{k+1|k}\hat{z}^T_{k+1|k}, \tag{45}$$

$$\overset{\varkappa y}{\underset{k+1|k}{\Sigma}} = (1 - p_{k+1}) \overset{\varkappa z}{\underset{k+1|k}{\Sigma}} + p_{k+1} \overset{\varkappa z}{\underset{k+1,k|k}{\Sigma}}, \tag{46}$$

- Kalman gain for the state estimation can be computed as

$$\boldsymbol{K}_k^{\varkappa} = \underset{k+1|k}{\overset{\varkappa y}{\Sigma}} \left(\underset{k+1|k}{\overset{yy}{\Sigma}} \right)^{-1}, \tag{47}$$

- Updating state and state error covariance

$$\widehat{\varkappa}_{k+1|k+1} = \widehat{\varkappa}_{k+1|k} + \boldsymbol{K}_k^{\varkappa}\left(y_{k+1} - \widehat{y}_{k+1|k} \right), \tag{48}$$

$$\underset{k+1|k+1}{\overset{\varkappa\varkappa}{\Sigma}} = \underset{k+1|k}{\overset{\varkappa\varkappa}{\Sigma}} - \boldsymbol{K}_k^{\varkappa} \underset{k+1|k}{\overset{yy}{\Sigma}} \left(\boldsymbol{K}_k^{\varkappa} \right)^T. \tag{49}$$

- Cross covariance between state and measurement noise can be written as

$$\underset{k+1|k+1}{\overset{v\varkappa}{\Sigma}} = -\boldsymbol{K}_k^{\varkappa} \underset{k+1|k}{\overset{yy}{\Sigma}} \left(\boldsymbol{K}_k^{v} \right)^T. \tag{50}$$

Step V Q Adaptation steps

Innovation covariance sequence can be described as

$$\widehat{\boldsymbol{C}}_k^{\boldsymbol{\nu}} = \frac{1}{L} \sum_{j=k-L+1}^{k} \boldsymbol{\nu}(j)\boldsymbol{\nu}^T(j), \tag{51}$$

where L is the epoch length and $\boldsymbol{\nu}_k = y_k - \widehat{y}_{k|k-1}$. For partial adaptation of $\boldsymbol{Q}$, we have to select the corresponding elements of the Kalman gain $\left(\boldsymbol{K}_k^{\varkappa}\right)$ and termed as K_k^a. In this paper, as the turn rate (5th state) is to be adapted, so K_k^a will be the fifth row of $\boldsymbol{K}_k^{\varkappa}$.

Then, adapted noise covariance can be written as

$$\widehat{\boldsymbol{Q}}_k^a = \mathrm{K}_k^a \widehat{\boldsymbol{C}}_k^{\boldsymbol{\nu}} \left(\mathrm{K}_k^a\right)^T. \tag{52}$$

As the process noise covariance of the proper states (i.e. first four states) is known, we don't have to adapt the elements of Q belongs to proper states $\left(\widehat{\boldsymbol{Q}}_k^a\right)$ can be calculated. Then, the overall process noise covariance with partially adapted covariance can be presented as

$$\widehat{\boldsymbol{Q}}_k = \begin{bmatrix} \boldsymbol{Q}_k^x & 0 \\ 0 & \widehat{\boldsymbol{Q}}_k^a \end{bmatrix}, \tag{53}$$

Initially, up to $k < L$, $\widehat{\boldsymbol{Q}}_k^a$ is not getting updated, i.e. $\widehat{\boldsymbol{Q}}_k^a = \widehat{\boldsymbol{Q}}_{k-1}^a$. As and when $k > L$, $\widehat{\boldsymbol{Q}}_k^a$ will be updated according to Eq. (52). Considerable small window length would be better to tune adaptive parameters and which also depends on problem.

4 Simulation Procedure, Results, and Discussions

Case Study

Performance of the proposed adaptive Gaussian filters has been demonstrated using well-known bearing only tracking (BOT) problem, where aircraft is executing manoeuvering turn with unknown constant turn rate and is being sensed by two radars which are placed at different locations. As the turn rate of the aircraft is considered to be unknown, kinematic model of the system becomes strong nonlinear. As the aircraft is executing unknown turn rate, sometimes radars may lose the track of the aircraft. Kinematic mode of an aircraft is given below.

Process Model

An aircraft executing unknown turn rate can be represented [32] as

$$\boldsymbol{\varepsilon}_{k+1} = \begin{bmatrix} 1 & \frac{\sin(\Omega_k\tau)}{\Omega_k} & 0 & -\frac{1-\cos(\Omega_k\tau)}{\Omega_k} & 0 \\ 0 & \cos(\Omega_k\tau) & 0 & -\sin(\Omega_k\tau) & 0 \\ 0 & \frac{1-\cos(\Omega_k\tau)}{\Omega_k} & 1 & \frac{\sin(\Omega_k\tau)}{\Omega_k} & 0 \\ 0 & \sin(\Omega_k\tau) & 0 & \cos(\Omega_k\tau) & 0 \\ 0 & 0 & 0 & 0 & 1 \end{bmatrix} \boldsymbol{\varepsilon}_k + \boldsymbol{w}_k, \tag{54}$$

where Eq. (54) is state transition equation (or process equation). $\boldsymbol{w}_k$ represents white Gaussian noise which is included in order to compensate for the unmodelled dynamics of the system. Ω is the unknown turn rate. Where state vector $\boldsymbol{\varepsilon} = [x\dot{x}y\dot{y}\Omega]^T$. x and $\dot{x}$ indicates position and velocity in x direction. y and $\dot{y}$ indicates position and velocity in y direction. And τ indicates time interval between two instances, i.e. sampling time, which is 1 s.

Process noise covariance can be calculated from

$$\boldsymbol{Q} = \text{diag}([\boldsymbol{N}_1\boldsymbol{N}_2 N_3]),$$

where $\boldsymbol{N}_1 = \boldsymbol{N}_2 = q_1\begin{bmatrix} \frac{\tau^3}{3} & \frac{\tau^2}{2} \\ \frac{\tau^2}{2} & \tau \end{bmatrix}$, where $q_1 = 0.1$, and $N_3 = q_2\tau$, where as $q_2 = 1.745 * 10^{-3}$.

Note that turn rate has been modelled as a random walk model with a high covariance $\boldsymbol{Q}$ (true value). This includes unknown variation in turn rate. Readers have to note that only covariance of turn rate is unknown. However, in order to generate true trajectories, covariance of turn rate is chosen as N_3.

Measurement Model

Aircraft is being tracked with the help of two passive radars using angle measurements which are placed at different locations.

Measurement equation can be represented as

$$\varnothing_{k+1}^{\sigma} = \tan^{-1}\left(\frac{y_{k+1} - y^{\text{ref}}}{x_{k+1} - x^{\text{ref}}}\right) + v_{k+1}^{\sigma}, \sigma = 1, 2 \tag{55}$$

σ indicates radar number. Locations of radar are: $y^1 = 10^4$ m, $y^2 = 10^4$ m, $x^1 = -10^4$ m, and $x^2 = 10^4$ m. Measurement noise covariance of each radar is as follows: $R_1 = 30\text{m rad}^2$ and $R_2 = 40\text{m rad}^2$. Sampling time is considered as $\tau = 1$ s.

4.1 Simulation Procedure

To validate the performance of the proposed filters, 5000 Monte Carlo simulations are carried. True trajectories are generated with initial state as $\boldsymbol{\varepsilon}_{0|0} = \left[1000\text{m}\, 300\text{ms}^{-1}\, 1000\text{m}\, 0\text{ms}^{-1} - 3\text{rad s}^{-1}\right]^T$. For the filters, the initial estimates are generated randomly with error covariance, $\boldsymbol{\Sigma}_{0|0}^{\varepsilon\varepsilon} = \text{diag}([100\text{ m}\, 10\text{ ms}^{-1}\, 100\text{ m}\, 10\text{ ms}^{-1}\; 100 * 1e^{-6} rad\, s^{-1})]$. Each MC run is carried out for 100 s with different noise sequences.

Only covariance of turn rate is unknown so which is adapted partially. In process covariance matrix ($\boldsymbol{Q}$) except turn rate element (which is fifth element) others are known constants. True trajectories are generated with the correct knowledge of the process noise covariance $(\boldsymbol{Q}_{\text{true}})$. However, for the filter, $\boldsymbol{Q}_{\text{true}}$ is absent due to unknown parameter variations. The proposed adaptive filter automatically tunes it with the initial choice of $\boldsymbol{Q}$ $(\boldsymbol{Q}_{\text{filter}})$ made by the designer with substantial error, i.e. $\boldsymbol{Q}_{\text{filter}} = 50 * \boldsymbol{Q}_{\text{true}}$.

As turn rate is unknown to the designer and radars are placed at different locations, there is a possibility of track loss which means measurement partially loses its observability. So, track loss condition also observed in this case study, which happens when the line of sight (LOS) does not intersect each other. Track loss is identified by taking the divergence condition as presented by Eq. (56)

$$\left\| \left(P_{x,k}^{i} - \hat{P}_{x,k}^{i}\right)^2 + \left(P_{y,k}^{i} - \hat{P}_{y,k}^{i}\right)^2 \right\|_{\infty} \geq 800\text{ m} \tag{56}$$

if the above condition satisfies, then filter will fail to track the trajectory.

Performance Index

Superiority of the proposed filter is validated by comparing root mean square error (RMSE) [21] and average RMSE (ARMSE). RMSE can be calculated from the below equation

$$\text{RMSE}_k = \sqrt{\frac{1}{5000}\sum_{i=1}^{5000}\left(\left(P_{x,k}^{i} - \hat{P}_{x,k}^{i}\right)^2 + \left(P_{y,k}^{i} - \hat{P}_{y,k}^{i}\right)^2\right)^2} \tag{57}$$

where RMSE_k represents RMSE at kth instant. $\left(P^i_{x,k}, \hat{P}^i_{x,k}\right)$ is the true and estimated position in x direction, and $\left(P^i_{y,k}, \hat{P}^i_{y,k}\right)$ is the true and estimated velocity in and y direction. Similarly, RMSE of velocity and turn rate can be calculated from Eq. (57).

4.2 Simulation Results and Discussion

In this subsection, significant findings are enumerated. For validation, 5000 Monte Carlo simulation runs have been carried out. We have considered different probabilities for the random delay in measurement, i.e. $p = 0.2, 0.4, 0.5, 0.6,$ and 0.8. As explained earlier, there is a possibility of track loss, while making RMSE calculations, track loss cases are excluded. RMSE comparison is shown in figures for probability $p = 0.5$, and ARMSE comparison and percentage track loss for probability $p = 0.2, 0.4, 0.5, 0.6,$ and 0.8 are represented in Table 1. For the simulation purpose, unknown covariance of turn rate for adaptive filters is initialized with $\boldsymbol{Q}_{\text{filter}} = 50 * \boldsymbol{Q}_{\text{true}}$. Window length of $L = 10$ is chosen.

NAUKFRD, NACKFRD, AUKFRD, and ACKFRD are used for simulation. It has been observed that NAUKFRD and AUKFRD are facing numerical instability. Most of the literature [12] on UKF suggested to take small value for α, but it does not work satisfactory for all nonlinear applications as has been observed in this case study. In this paper, authors have carried out rigorous tuning of the scaling parameter α, and it is set to $\alpha = 3$ for which satisfactory performance has been recorded. This indicates that a careful tuning of parameters is necessary for UKF. As ARMSE difference is small between ACKFRD and AUKFRD, ACKFRD is completely free from such tuning, so it is suggested to prefer ACKFRD over AUKFRD in practical applications.

Figure 1 shows RMSE comparison of position with NAUKFRD, NACKFRD, AUKFRD, and ACKFRD. Velocity RMSE comparison is depicted in Fig. 2, and unknown turn rate RMSE comparison is presented by Fig. 3. ARMSE comparison of position, velocity, and turn rate, and percentage track loss are given in Table.1. From Fig. 4. we can see that proposed filter is able to track the true value of unknown covariance, i.e. $\boldsymbol{Q}(5, 5)$.

It can be observed from Figs. 2, 3, 4 and Table.1, when knowledge of the covariance of turn rate is unknown, ARMSE is more. Whereas, when it comes to adaptive Gaussian filters, RMSE of position, velocity, and turn rate are significantly less compared to non-adaptive Gaussian filters. MATLAB codes are executed in a computer with Intel(R) core (TM) i5-2450 M CPU @2.50GHZ, 8 GB RAM, and 64-bit operating system.

Table 1 Average RMSE and percentage track loss comparison

Probability (p)	Filter	Average RMSE of			Number of divergences out of 5000 MC runs	Percentage (%) of track loss
		Position (m)	Velocity (ms^{-1})	Turn rate ($\mathrm{rad\ s}^{-1}$)		
0.2	NAUKFRD	172.9770	77.0556	0.0719	3941	78.82
	NACKFRD	182.5396	78.1004	0.0713	3953	79.06
	AUKFRD	153.2510	46.4400	0.0381	482	09.64
	ACKFRD	155.4763	47.1512	0.0383	506	10.12
0.4	NAUKFRD	196.7879	84.8579	0.0730	4004	80.08
	NACKFRD	204.9192	86.3122	0.0726	4126	82.52
	AUKFRD	170.4540	50.6111	0.0392	635	12.70
	ACKFRD	173.4118	51.5683	0.0396	662	13.24
0.5	NAUKFRD	204.6035	87.8164	0.0735	4048	80.96
	NACKFRD	212.8410	89.5879	0.0735	4150	83.00
	AUKFRD	175.7871	51.8702	0.0395	685	13.70
	ACKFRD	178.4096	52.8139	0.0398	735	14.70
0.6	NAUKFRD	211.4647	89.9604	0.0735	4102	82.04
	NACKFRD	217.5769	91.6763	0.0736	4219	84.38
	AUKFRD	178.5446	52.2950	0.0394	665	13.30
	ACKFRD	181.2232	53.1803	0.0397	720	14.40
0.8	NAUKFRD	213.3889	92.2890	0.0741	4204	84.08
	NACKFRD	221.7596	94.3851	0.0739	4348	86.96
	AUKFRD	175.8062	51.4222	0.0386	631	12.62
	ACKFRD	178.8446	52.2468	0.0388	666	13.32

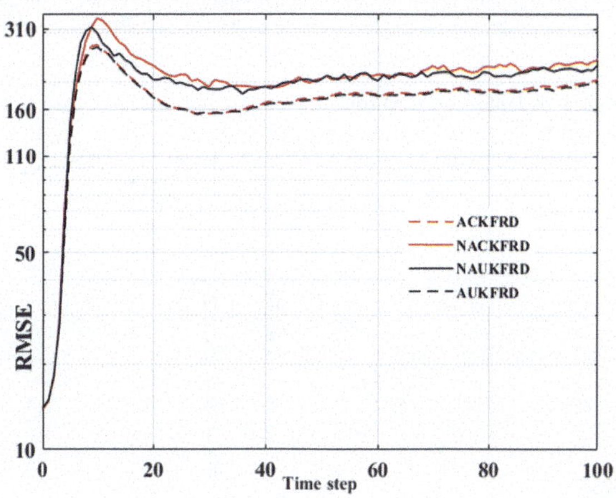

Fig. 1 RMSE of position for $p = 0.5$

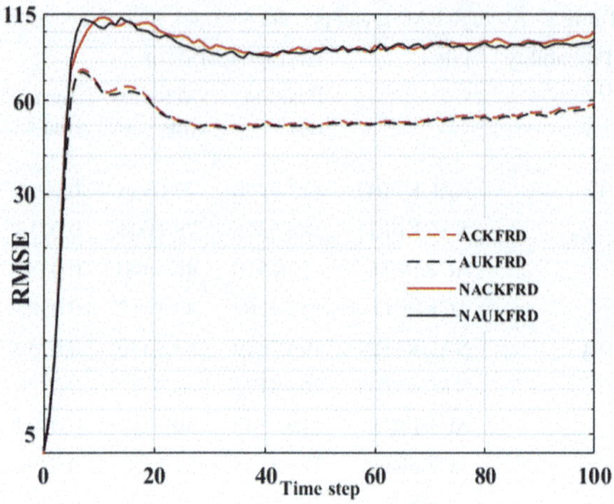

Fig. 2 RMSE of velocity for $p = 0.5$

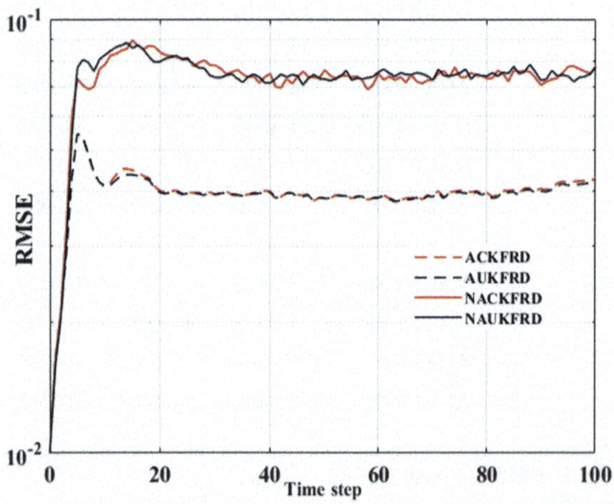

Fig. 3 RMSE of turn rate for $p = 0.5$

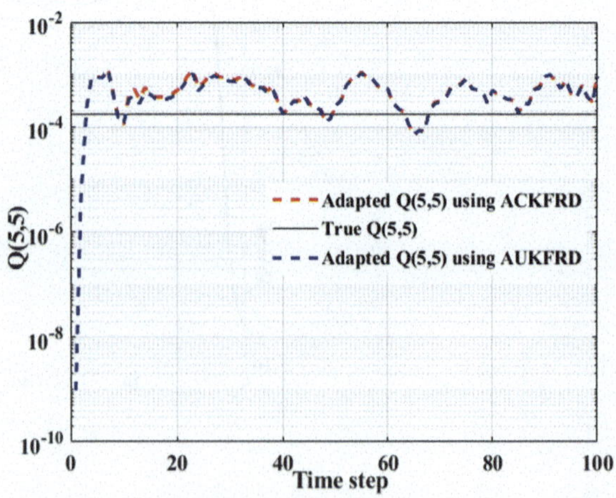

Fig. 4 Adapted $Q(5, 5)$ for $p = 0.5$

5 Conclusion

Adaptive cubature Kalman filer and adaptive unscented Kalman filter are proposed for one-step randomly delayed measurements in this paper for nonlinear state estimation when measurements are randomly delayed by maximum one step, and process noise covariance remains unknown due to parametric uncertainty. Superiority of the proposed ACKFRD and AUKFRD over NAUKFRD and NACKFRD is demonstrated with the help of Monte Carlo (MC) simulations by taking well-known target tracking problem. Comparison of root mean square error and average RMSE demonstrate the superiority and effectiveness of proposed adaptive filters over their non-adaptive part, which is verified from percentage of track loss. In addition to this, it is observed that the proposed adaptive unscented Kalman filter requires substantial tuning to avoid singularities unlike its closed competitor, adaptive cubature Kalman filer. Based on the findings as iterated above, adaptive cubature Kalman filer with one-step randomly delated measurement is promoted for nonlinear state estimation.

References

1. Kalman, R.E.: A new approach to linear filtering and prediction problems. J. Basic Eng. **82**(1), 35–45 (1960)
2. Song, Y., Nuske, S., Scherer, S.: A multi-sensor fusion MAV state estimation from long-range stereo, IMU, GPS and Barometric Sensors. Sensors **17**(1), 11 (2017)
3. Mu, X., Chen, J., Zhou, Z., Leng, Z., Fan, L.: Accurate initial state estimation in a monocular visual-inertial SLAM system. Sensors **18**(2), 506 (2018)
4. Yu, Q., Xiong, R., Lin, C., Shen, W., Deng, J.: Lithium-ion battery parameters and state-of-charge joint estimation based on H-infinity and unscented Kalman filters. IEEE Trans. Veh. Technol. **66**(10), 8693–8701 (2017)
5. Chen, T., Yse, F., Ling, K.V., Chen, X.: Distributed state estimation using a modified partitioned moving horizon strategy for power systems. Sensors **17**(10), 2310 (2017)
6. Dunik, J., Straka, O., Simandl, M., Blasch, E.: Random-point-based filters: analysis and comparison in target tracking. IEEE Trans. Aerosp. Electron. Syst. **51**(2), 1403–1421 (2015)
7. Zhao, X., Li, J., Yan, X., Ji, S.: Robust adaptive cubature Kalman filter and its application to ultra-tightly coupled SINS/GPS navigation system. Sensors **18**(7), 2352 (2018)
8. Katsikas, S.K., Lerosb, A.K., Lainiotis, D.G.: Underwater tracking of a maneuvering target using time delay measurements. Signal Process. **41**(1), 17–29 (1995)
9. Ikoma, N., Ichimura, N., Higuchi, T., Maeda, H.: Maneuvering target tracking by using particle filter. In: Proceedings, IFSA World Congress and 20th NAFIPS International Conference, vol. 4, pp. 2223–2228 (2001)
10. Bar-Shalom, Y., Li, X.R., Kirubarajan, T.: Estimation with application to tracking and navigation. Wiley-Interscience Publishing, New York (2001)
11. Anderson, B.D.O., Moore, J.B.: Optimal Filtering. Dover Publication, INC., Mineola, New York (2005)
12. He, W., Williard, N., Chen, C., Pecht, M.: State of charge estimation for electric vehicle batteries using unscented Kalman filtering. Microelectron. Reliab. **53**(6), 840–847 (2013)
13. Haykin, S., Arasaratnam, I.: Cubature kalman filters. IEEE Trans. Autom. Control **54**(6), 1254–1269 (2009)
14. Bhaumik, S., Swati: Cubature quadrature Kalman filter. IET Signal Process. **7**(7), 1–9 (2013)

15. Bhaumik, S., Swati: Square-root cubature-quadrature Kalman filter. Asian J. Control **16**(2), 617–622 (2014)
16. Gustafsson, F., et al.: Particle filters for positioning, navigation, and tracking. IEEE Trans. Signal Process. **50**(2), 425–437 (2002)
17. Ray, A., Liou, L.W., Shen, J.H.: State estimation using randomly delayed measurements. J. Dyn. Sys. Meas. Control **115**(1), 19–26 (1993)
18. Hsiao, F.H., Pan, S.T.: Robust Kalman filter synthesis for uncertain multiple time-delay stochastic systems. J. Dyn. Sys. Meas. Control **118**(4), 803–808 (1996)
19. Challa, S., Evans, R.J., Wang, X.: A Bayesian solution and its approximations to out-of-sequence measurement problems. Information Fusion **4**(3), 185–199 (2003)
20. Carazo, A.H., Perez, J.L.: Extended and unscented filtering algorithms using one-step randomly delayed observations. Appl. Math. Comp. **190**(2), 1375–1393 (2007)
21. Wang, X., Liang, Y., Pan, Q., Zhao, C.: Gaussian filter for nonlinear systems with one-step randomly delayed measurements. Automatica **49**(4), 976–986 (2013)
22. Carazo, A.H., Perez, J.L.: Unscented filtering algorithm using two-step randomly delayed observations in nonlinear systems. Appl. Math. Model. **33**(9), 3705–3717 (2009)
23. Singh, A.K., Date, P., Bhaumik, S.: A modified Bayesian filter for randomly delayed measurements. IEEE Trans. Autom. Control **62**(1), 419–424 (2016)
24. Wang, Z., Ho, D.W.C., Liu, X.: Variance-constrained filtering for uncertain stochastic systems with missing measurements. IEEE Trans. Autom. Control **48**(7), 1254–1258 (2003)
25. Zhang, Y., et al.: Particle filter with one-step randomly delayed measurements and unknown latency probability. Int. J. Syst. Sci. **47**(1), 209–221 (2016)
26. Mohamed, A.H., Schwarz, K.P.: Adaptive Kalman filtering for INS/GPS. J. Geodesy **73**(4), 193–203 (1999)
27. Zhou, W., Liu, L.: Adaptive cubature Kalman filter based on the expectation-maximization algorithm. IEEE Access **7**, 158198–158206 (2019)
28. Patra, N., Sadhu, S., Ghoshal, T.K.: Adaptive state estimation for tracking of civilian aircraft. IET Sci. Meas. Technol. **12**(6), 777–784 (2018)
29. Qiu, Z., Qian, H., Wang, G.: Adaptive robust cubature Kalman filtering for satellite attitude estimation. Chin. J. Aeronaut. **31**(4), 806–819 (2018)
30. Dey, A., Sadhu, S., Ghoshal, T.K.: Adaptive Gauss Hermite filter for parameter varying nonlinear systems. In: IEEE International Conference on Signal Processing and Communications (SPCOM), pp. 1–5 (2014)
31. Maybeck, P.S.: Stochastic Models, Estimation, and Control, 1st edn. Academic, New York (1982)
32. Dey, A., Sadhu, S., Ghoshal, T.K.: Adaptive nonlinear information filters for multiple sensor fusion. In: Informatics in Control, Automation and Robotics 12th International Conference, ICINCO 2015 Colmar, France, July 21–23, 2015, pp. 371–390. Springer, Cham (2016)

Extended Kalman Filter-based Attitude Estimation using Magnetometer- and Sun Sensor-Aided MEMS Gyros

Jiljo K. Moncy and Kesavabrahmaji Karuturi

Abstract A novel idea is being proposed to reuse the spent upper stages of launch vehicles to conduct micro-gravity experiments. The spent stages, which remain in orbit, acts as a small spacecraft with limited power and space to meet the requirements. In order to achieve the mission objectives, attitude of the spacecraft must be determined. A new system comprising of rate class micro-electro-mechanical system (MEMS) gyroscopes, three-axis magnetometers, and Sun sensors is to be used. The paper focuses on the design of an attitude estimation algorithm using the MEMS gyros as the primary sensor. Secondary sensors are to be used to aid the gyro measurements. Algorithm using inertial navigation system (INS) error state-based extended Kalman filter (EKF) is studied in this work. The paper includes detailed derivation of the algorithm and reference models for the secondary sensors. Performance of the algorithm during magnetometer failure and eclipse is also quantified.

1 Introduction

A recent trend in space technology is to use small spacecraft as a platform for micro-gravity experiments. For the successful conduct of these experiments, the attitude of the stage is to be precisely determined. As these spacecraft are of micro size, power as well as space availability for the navigation system is limited. Thus, MEMS sensors are proposed to be used for the attitude estimation.

Since the rate class MEMS gyros are to be used for the estimation, stand-alone INS cannot meet the mission accuracy requirements. In order to meet the requirements, magnetometer and Sun sensor measurements are to be fused to the gyro measurements using an optimal filter. Sun sensor measures the direction of the Sun with respect to the vehicle, and the magnetometer measures the geomagnetic field vector, whereas the gyros provide the instantaneous angular rates. The magnetometer and

J. K. Moncy (✉)
Vikram Sarabhai Space Centre, Thiruvananthapuram, India
e-mail: jkmoncy@alumni.caltech.edu

K. Karuturi
ISRO Inertial Systems Unit, Thiruvananthapuram, India

J. Gu et al. (eds.), *Communication and Control for Robotic Systems*, Smart Innovation, Systems and Technologies 229, https://doi.org/10.1007/978-981-16-1777-5_14

Sun sensor measurements may be unavailable for a period of time due to sensor malfunctions or due to phenomena like eclipse. As at least two distinct reference vectors are required to estimate orientation, the effect of such occurrences on attitude estimation is non-trivial. Such scenarios also have to be studied to characterize the navigation system.

2 Literature Review

Several solutions for the attitude estimation of spacecraft are available in the literature. In [3], the authors introduce an attitude estimation algorithm for a mini-satellite, QSAT, using a combination of weighted least squares and linearized-Kalman filter estimation methods .The weighted least squares method is used to produce a single attitude-angle measurement using the Sun sensor and magnetometer measurement. The linearized Kalman filter is used to fuse this combined measurement with the gyro rates. However, a major drawback of such a scheme is that the secondary measurement used in measurement update of the Kalman filter is no longer uncorrelated, which adversely affects the optimality and accuracy of the filter. In [4], estimation schemes during an eclipse are discussed. During an eclipse, the attitude should be estimated with the gyro rates and magnetometer. Linearized Kalman filter and extended Kalman filters are proposed in this paper too. This paper uses international geomagnetic reference field (IGRF) model and VSOP87 model for the reference sensor measurements.

The authors of [7] presents the results of using the magnetometer-aided navigation (MAGNAV) onboard WIRE spacecraft. The MAGNAV algorithm is similar to the algorithm discussed in this paper but varies in the states estimated. MAGNAV algorithm is a blending of an EKF with the pseudo-linear Kalman filter. Along with attitude, it also estimates the orbital parameters. The algorithm also employs Sun sensor to improve the accuracy. The performance of this algorithm has been flight proven even during eclipses. Convergence of the algorithm has also been verified after every restart of the spacecraft.

A deterministic attitude estimation method should be used to obtain the initial estimate to be provided to the filter. The authors of [1] propose a technique to estimate the attitude using the magnetometer and Sun sensor measurements. They claim that the technique can attain the attitude with an accuracy of less than 3° about roll and pitch axis and less than 2° about yaw axis, with comparable complexity as that of the tri-axial attitude determination (TRIAD) algorithm.

In order to compute the residuals of the magnetometer and Sun sensor measurements, accurate modeling of the geomagnetic field and Sun vector are required. In this study, the IGRF model is used for the geomagnetic field, whereas VSOP87 model is used for the Sun vector calculations. The IGRF model requires the Global Navigation Satellite System (GNSS) position as well as the time in Universal Time Coordinated (UTC) for the computation, while VSOP87 requires the Julian date in centuries.

2.1 Earth's Magnetic Field: IGRF Model

IGRF [6] is a mathematical model maintained by the International Association of Geomagnetism and Aeronomy (IAGA), describing the internal part of the Earth's magnetic field between epochs 1900 A.D. and the present. An update is released every five years since 1965. The model used in the study is the 12th generation IGRF model, referred as IGRF-12.

The mathematical formulation of the IGRF model includes the modelling of the magnetic scalar potential as a function of Schmidt quasi-normalized associated Legendre polynomial, geocentric latitude and longitude, radial distance from center of Earth, and also the time. The model is shown in Eq. (1).

$$V(r,\theta,\phi,t) = a\sum_{n=1}^{N}\sum_{m=0}^{n}\frac{a}{r}^{n+1}\Big[g_n^m(t)\cos(m\phi) + h_n^m(t)\sin(m\phi)\Big]P_n^m(\cos(\theta)) \tag{1}$$

where $a = 6371.2$ km is the geomagnetic conventional Earth's reference spheroid radius, θ is the geocentric co-latitude, ϕ is the longitude, and r is the radial distance. $P_n^m(\cos\theta)$ is the Schmidt quasi-normalized associated Legendre function.

The coefficients h_n^m and g_n^m are provided at five year interval epochs. The values are assumed to be linear within the five year interval. For the last five years between 2015 and 2020, the rates of the coefficients are directly provided as those values are the predicted secular variations. The data provided for the years after epoch 2005 is of precision 0.01 nT for DGRF and 0.1 nT for IGRF. The maximum degree of the IGRF-12 is 13. Reducing the order of the model results in a coarser estimate of the geomagnetic field.

The geocentric components of the geomagnetic field in the local frame (X-North, Y-East and Z-Down) are obtained by using Eq. (2).

$$\vec{B} = \begin{bmatrix} B_x \\ B_y \\ B_z \end{bmatrix} = \begin{bmatrix} \frac{1}{r}\frac{\partial V}{\partial \theta} \\ \frac{1}{r\sin\theta}\frac{\partial V}{\partial \phi} \\ \frac{\partial V}{\partial r} \end{bmatrix} \tag{2}$$

The Schmidt quasi-normalized associated Legendre function used in the IGRF model is given in Eq. (3).

$$P_n^m(x) = \begin{cases} \frac{1}{2^n n!}\frac{d^n}{dx^n}(x^2-1)^n, \; m = 0 \\ \sqrt{\frac{2(n-m)!}{(n+m)!}}(1-x^2)^m\left(\frac{1}{2^n n!}\right)\frac{d^{n+m}}{dx^{n+m}}(x^2-1)^n, \; m > 0 \end{cases} \tag{3}$$

The iterative algorithm used to compute the derivative of associated Legendre polynomial is given in Eq. (4).

$$\dot{P}_n^m(\cos\theta) = \begin{cases} 0, \text{ when } n, m = 0 \\ \cos(\theta), \text{ when } n, m = 1 \\ \dfrac{2n-1}{\sqrt{n^2-m^2}}\left[\cos\theta\,\dot{P}_{n-1}^{n-1} + -\sin\theta\,P_{n-1}^{m}\right] \\ \quad -\sqrt{\dfrac{(n-1)^2-m^2}{n^2-m^2}}\,\dot{P}_{n-2}^{m}, \text{ when } n \neq m \\ \sqrt{1-\dfrac{1}{2n}}\left[\sin\theta\,\dot{P}_{n-1}^{n-1} + \cos\theta\,P_{n-1}^{n-1}\right], \text{ when } n = m \end{cases} \tag{4}$$

2.2 Sun Vector: VSOP87 Model

Many algorithms are available to compute the reference Sun vector. However, most of them are computationally heavy. An approximated model VSOP87 [8] is used to compute the Sun vector in ECI (J2000) frame. The Sun vector with respect to the ECI origin is assumed to be parallel to the one with respect to the origin of the vehicle body. Thus, the Sun vector in body frame is obtained by transforming the reference Sun vector using the direction cosine matrix alone.

Sun vector in ECI frame is computed using Eqs. (5)–(9).

$$\vec{s} = \left[s\cos\lambda_e,\, s\cos\epsilon\sin\lambda_e,\, s\sin\epsilon\sin\lambda_e\right] \tag{5}$$

$$\epsilon = 23.4^\circ - 0.0130 \quad T_{\text{TDB}} \tag{6}$$

$$\lambda_e = \lambda_M + 1.91^\circ\sin(M_{\text{Sun}}) + 0.0199^\circ\sin(2M_{\text{Sun}}) \tag{7}$$

$$\lambda_M = 280.460^\circ + 36000.770 \quad T_{\text{UTI}} \tag{8}$$

$$M_{\text{Sun}} = 357.527^\circ + 35999.05^\circ \quad T_{\text{TDB}} \tag{9}$$

where,
$\vec{s}$: Sun vector
ϵ: Obliquity of ecliptic
λ_e: ecliptic Earth's longitude
λ_M: Mean longitude of Sun
M_{Sun}: Mean anomaly of Sun
T_{UTI}: Julian centuries of universal time
T_{TDB}: Julian centuries of barycentric dynamical time

3 Estimation Algorithm

A Kalman filter is used to estimate the internal states from noisy measurements using a linear system model. The measurement noise and process noise are assumed to be zero mean Gaussian process for the estimation. But, when the system is nonlinear, the Kalman filter is applied after linearizing the system at the point of estimation. This scheme is known as extended Kalman filter (EKF).

3.1 Kinematic Model

The navigation frame is chosen as the ECI (J2000) frame. The angular rate of the navigation frame with respect to the body frame is given by Eq. (10). ω_{ib}^{b} is provided by the gyroscope. p, q, r are the body rates relative to the inertial.

$$\omega_{bi}^{\mathrm{b}} = -\omega_{ib}^{\mathrm{b}} \tag{10}$$

$$\omega_{ib}^{\mathrm{b}} = [p \quad q \quad r]^{\mathrm{T}} \tag{11}$$

$$\omega_{bi}^{\mathrm{b}} = [\omega_1 \quad \omega_2 \quad \omega_3]^{\mathrm{T}} \tag{12}$$

The kinematic equation for the quaternion Q representing the rotation from inertial frame to body frame is given by the following Eqs. (13)–(16).

$$\dot{Q}(t) = \frac{1}{2} Q(t) * W(t) \tag{13}$$

$$Q(t + \Delta T) = Q(t) + \dot{Q}(t) * \Delta T \tag{14}$$

$$W(t) = [0 \quad p \quad q \quad r] \tag{15}$$

$$Q(t) = \begin{pmatrix} \frac{q1}{q_1^2+q_2^2+q_3^2+q_4^2} \\ \frac{q2}{q_1^2+q_2^2+q_3^2+q_4^2} \\ \frac{q3}{q_1^2+q_2^2+q_3^2+q_4^2} \\ \frac{q4}{q_1^2+q_2^2+q_3^2+q_4^2} \end{pmatrix} \tag{16}$$

3.2 Sensor Models

As stated before, the attitude is estimated by using three sensors:, three-axis gyroscope, three-axis magnetometer, and a Sun sensor. Gyroscope is used as the primary sensor, whereas the magnetometer and Sun sensor are used to aid the navigation.

If '*u*' is the gyroscope measurement, then the gyroscope sensor is modeled as shown in Eq. (17) [2].

$$u = \omega_{ib}^{\mathrm{b}} + b_{\mathrm{g}} + v_{\mathrm{g}} \tag{17}$$

$$\dot{b}_{\mathrm{g}} = w_{\mathrm{g}} \tag{18}$$

Here, b_{g} is the gyro drift, modeled as a Gaussian random walk. v_{g} and w_{g} are zero mean Gaussian white noise with σ_{vb} and σ_{g} as the standard deviation, respectively.

The magnetometer is modeled as shown in Eq. (19). $m_{\mathrm{e}}^{\mathrm{b}}$ is the earth's magnetic field vector in body frame, whereas $m_{\mathrm{b}}^{\mathrm{b}}$ is the magnetic field vector generated by the vehicle. However, it is assumed that the magnetic field due to the vehicle is known a priori and hence compensated in the measurement. Thus, the term $m_{\mathrm{b}}^{\mathrm{b}}$ is approximated to 0. v_{m} is a zero mean Gaussian noise with variance σ_{m}^2.

$$y_{\mathrm{m}} = m_{\mathrm{e}}^{\mathrm{b}} + m_{\mathrm{b}}^{\mathrm{b}} + v_{\mathrm{m}} \tag{19}$$

Sun sensor is also modeled similar to the magnetometer as shown in Eq. (20). The output of the Sun sensor is expected to be a vector pointing to the Sun with respect to the vehicle in the body frame. A zero mean Gaussian noise (v_{s}) with standard deviation σ_{s} is also expected.

$$y_{\mathrm{s}} = \vec{S}^{\mathrm{b}} + v_{\mathrm{s}} \tag{20}$$

3.3 Mechanization Equations

The state vector used for the estimation is $\hat{x} = [\hat{Q}^T, \hat{b_{\mathrm{g}}}^{\mathrm{T}}]^{\mathrm{T}}$. Equations (13–16) are used to propagate the quaternion. Based on Eq. (17), angular body rate is computed using the gyro measurement 'u' as follows:

$$\hat{\omega}_{ib}^{\mathrm{b}} = u - \hat{b_{\mathrm{g}}} \tag{21}$$

Using the state $\hat{Q}$, the rotation matrix $\hat{R}_i^{\mathrm{b}}$ is computed. The magnetometer and Sun sensor measurements are transformed to the inertial frame using $\hat{R}_i^{\mathrm{b}}$, when required.

$$\hat{m}^i = \hat{R}_{\mathrm{b}}^i y_{\mathrm{m}} \tag{22}$$

$$\hat{\vec{S}}^i = \hat{R}_b^i y_{\mathrm{s}} \tag{23}$$

As the true values of magnetic field and Sun vector is known from the IGRF model and VSOP87 model, respectively, the residual between the true and measured value is used in the Kalman filter to estimate the error states.

3.4 *Error Models*

The extended Kalman filter estimates the error states. The error state vector is defined as

$$\delta x = [\rho^{\mathrm{T}}, \delta b_{\mathrm{g}}^{\mathrm{T}}]^{\mathrm{T}} \tag{24}$$

The ρ vector is the misalignment of the true rotation matrix and the computed rotation matrix. $\rho = [\epsilon_x, \epsilon_y, \epsilon_z]^{\mathrm{T}}$ is the small angle rotations required with respect to the inertial to align the inertial frame with the computed inertial frame.

For the analysis, the skew symmetric matrix, P, of the small angle error can be used to align the computed and true rotation matrices.

$$P = [\rho^{\times}] = \begin{bmatrix} 0 & -\epsilon_z & \epsilon_y \\ \epsilon_z & 0 & -\epsilon_x \\ -\epsilon_y & \epsilon_x & 0 \end{bmatrix} \tag{25}$$

With the above definitions, following is established easily with the approximations of $P^T = -P$ and $(I - P)^{-1} = (I + P)$.

$$\hat{R}_{\mathrm{b}}^{i} = (I - P)R_{\mathrm{b}}^{i} \tag{26}$$

$$\hat{R}_{i}^{\mathrm{b}} = R_{i}^{\mathrm{b}}(I + P) \tag{27}$$

$$R_{i}^{\mathrm{b}} = \hat{R}_{i}^{\mathrm{b}}(I - P) \tag{28}$$

$$R_{\mathrm{b}}^{i} = (I + P)\hat{R}_{\mathrm{b}}^{i} \tag{29}$$

The entities with form $\hat{(.)}$ represents the computed value, whereas the others represents the true values.

3.5 *Measurement Error Model*

The magnetometer and Sun sensor is modeled as per Eqs. (19) and (20), respectively. The measurements are obtained in body frame, for which a value is known in the inertial frame using the IGRF model for magnetic field and VSOP87 model for Sun vector. Thus, residual of both can be obtained in body frame by transforming the value in inertial frame to the body frame.

Let $\hat{m}_{\mathrm{b}}$ be the magnetometer measurement and v_{m} be the magnetometer noise. Let m_{b} be the true magnetic field in body frame. The residual of magnetometer measurement is obtained as follows:

$$\delta\hat{m} = \hat{m}_{\mathrm{b}} - v_{\mathrm{m}} - m_{\mathrm{b}} \tag{30}$$

The true magnetic field in body frame m_b is obtained by transforming the reference magnetic field in inertial frame using the true DCM matrix R_i^{b}.

$$m_{\mathrm{b}} = R_i^{\mathrm{b}} \mathrm{b}_i^{\mathrm{ref}} = \hat{R}_i^{\mathrm{b}}(I - P)m_i^{\mathrm{ref}} \tag{31}$$

$$= \hat{m}_{\mathrm{b}} - \hat{R}_i^{\mathrm{b}} P m_i^{\mathrm{ref}} \tag{32}$$

$$\delta\hat{m} = \hat{R}_i^{\mathrm{b}} P m_i^{\mathrm{ref}} - v_{\mathrm{m}} \tag{33}$$

$$= -\hat{R}_i^{\mathrm{b}}[m_i^{\mathrm{ref}\times}]\rho - v_{\mathrm{m}} \tag{34}$$

As the Sun sensor model is similar to that of magnetometer, derivation is identical to Eqs. (30)–(34). Hence, the residue of Sun sensor is given by Eq. (35).

$$\delta\hat{s} = -\hat{R}_i^{\mathrm{b}}[s_i^{\mathrm{ref}\times}]\rho - v_{\mathrm{s}} \tag{35}$$

Unlike magnetometer, noise in the Sun sensor is correlated. This is due to the fact that Sun sensor measures the Euler angles, θ and ϕ, which align its bore-sight axis with the Sun vector. The Sun vector is computed from these angles. As a result, if $\Delta\theta$ and $\Delta\phi$ are the error on the measured angles, respectively, then the measured Sun vector is given by Eq. (36), where $[\hat{S}_x, \hat{S}_y, \hat{S}_z]^{\mathrm{T}}$ is the measured Sun vector and $[S_x, S_y, S_z]^{\mathrm{T}}$ is the true Sun vector.

$$\begin{bmatrix} \hat{S}_x \\ \hat{S}_y \\ \hat{S}_z \end{bmatrix} = \begin{bmatrix} 1 & 0 & -\Delta\theta \\ 0 & 1 & \Delta\phi \\ \Delta\theta & -\Delta\phi & 1 \end{bmatrix} \begin{bmatrix} S_x \\ S_y \\ S_z \end{bmatrix} \tag{36}$$

The measurement covariance matrix of the Sun sensor is derived using Eq. (36) with the approximation $\Delta\theta, \Delta\phi << 1$, as given in Eq. (37).

$$Cov(\hat{S}) = \begin{bmatrix} \hat{S}_z^2\sigma_\theta^2 & 0 & -\hat{S}_x\hat{S}_z\sigma_\theta^2 \\ 0 & \hat{S}_z^2\sigma_\phi^2 & -\hat{S}_y\hat{S}_z\sigma_\phi^2 \\ -\hat{S}_x\hat{S}_z\sigma_\theta^2 & -\hat{S}_y\hat{S}_z\sigma_\phi^2 & \hat{S}_x^2\sigma_\theta^2 + \hat{S}_y^2\sigma_\phi^2 \end{bmatrix} \tag{37}$$

Combining Eqs. (34) and (35), the observation matrix, H, for the EKF is shown in Eq. (38).

$$H = \begin{bmatrix} -\hat{R}_i^{\mathrm{b}}[m_i^{\mathrm{ref}\times}] & \mathbf{0} \\ -\hat{R}_i^{\mathrm{b}}[s_i^{\mathrm{ref}\times}] & \mathbf{0} \end{bmatrix} \tag{38}$$

3.6 Attitude Error Dynamics

In order to obtain the state space model of the system, it is essential to derive the attitude error dynamics. As stated before, ρ is defined as the delta rotations to align the computed and true body frame.

The differential equation for direction cosine matrix is

$$\dot{R}_{\mathrm{b}}^{i} = R_{\mathrm{b}}^{i}\Omega_{ib}^{\mathrm{b}} \tag{39}$$

Based on the kinematic equation, the estimate $\hat{R}_{\mathrm{b}}^{i}$ can be computed as

$$\dot{\hat{R}}_{\mathrm{b}}^{i} = \hat{R}_{\mathrm{b}}^{i}\hat{\Omega}_{ib}^{\mathrm{b}}, \quad \text{where, } \Omega = [\omega^{\times}] \tag{40}$$

$$R_{\mathrm{b}}^{i} = (I + P)\hat{R}_{\mathrm{b}}^{i} \tag{41}$$

Taking derivative,

$$\dot{R}_{\mathrm{b}}^{i} = (I + P)\dot{\hat{R}}_{\mathrm{b}}^{i} + \dot{P}\hat{R}_{\mathrm{b}}^{i} \tag{42}$$

By combining Eqs. (40) and (42), the dynamic model of ρ can be obtained as

$$\dot{P}\hat{R}_{\mathrm{b}}^{i} + (I + P)\dot{\hat{R}}_{\mathrm{b}}^{i} = (I + P)\hat{R}_{\mathrm{b}}^{i}(\Omega_{ib}^{\mathrm{b}}) \tag{43}$$

$$\delta\Omega_{ib}^{\mathrm{b}} = \Omega_{ib}^{\mathrm{b}} - \hat{\Omega}_{ib}^{\mathrm{b}} \tag{44}$$

$$\dot{P}\hat{R}_{\mathrm{b}}^{i} + (I + P)\dot{\hat{R}}_{\mathrm{b}}^{i} = (I + P)\hat{R}_{\mathrm{b}}^{i}(\Omega_{ib}^{\mathrm{b}}) \tag{45}$$

$$= (I + P)\hat{R}_{\mathrm{b}}^{i}(\delta\Omega_{ib}^{\mathrm{b}} + \hat{\Omega}_{ib}^{\mathrm{b}}) \tag{46}$$

$$= (I + P)\dot{\hat{R}}_{\mathrm{b}}^{i} + (I + P)\hat{R}_{\mathrm{b}}^{i}(\delta\Omega_{ib}^{\mathrm{b}}) \tag{47}$$

Thus,

$$\dot{P} = \hat{R}_{\mathrm{b}}^{i}(\delta\Omega_{ib}^{\mathrm{b}})\hat{R}_{i}^{\mathrm{b}} \tag{48}$$

$$\dot{\rho} = \hat{R}_{\mathrm{b}}^{i}(\delta\omega_{ib}^{\mathrm{b}}), \quad \text{as } \dot{P} = [\dot{\rho}^{\times}] \tag{49}$$

But,

$$\delta\omega_{ib}^{\mathrm{b}} = \omega_{ib}^{\mathrm{b}} - \hat{\omega}_{ib}^{\mathrm{b}} = \delta b_{\mathrm{g}} - v_{\mathrm{g}} \tag{50}$$

Finally,

$$\dot{\rho} = \hat{R}_{\mathrm{b}}^{i}\delta b_{\mathrm{g}} - \hat{R}_{\mathrm{b}}^{i}v_{\mathrm{g}} \tag{51}$$

3.7 State Space Error Model

Combining all the previous derivations, the state space error model is given below:

$$\begin{bmatrix} \dot{\rho} \\ \dot{b}_{\mathrm{g}} \end{bmatrix} = F \begin{bmatrix} \rho \\ \delta b_{\mathrm{g}} \end{bmatrix} + G \begin{bmatrix} v_{\mathrm{g}} \\ w_{\mathrm{g}} \end{bmatrix} \tag{52}$$

$$Y = \begin{bmatrix} \delta \hat{m} \\ \delta \hat{s} \end{bmatrix} = H \begin{bmatrix} \rho \\ \delta b_{\mathrm{g}} \end{bmatrix} + I \begin{bmatrix} v_{\mathrm{m}} \\ v_{\mathrm{s}} \end{bmatrix} \tag{53}$$

where,

$$F = \begin{bmatrix} 0 & \hat{R}^i_{\mathrm{b}} \\ 0 & 0 \end{bmatrix} \quad G = \begin{bmatrix} -\hat{R}^i_{\mathrm{b}} & 0 \\ 0 & I \end{bmatrix} \tag{54}$$

$$H = \begin{bmatrix} -\hat{R}^{\mathrm{b}}_i [m_i^{\mathrm{ref}\times}] & 0 \\ -\hat{R}^{\mathrm{b}}_i [s_i^{\mathrm{ref}\times}] & 0 \end{bmatrix} \tag{55}$$

3.8 Filter Equations

To estimate the state, a filter is implemented in discrete time where the subscript k represents the current time-step and $k-1$ the previous time-step. The states are predicted forward from the previous to current time-step using prior knowledge about the system dynamics and inputs [5]. States at the current time-step are predicted using the states at the previous time step and known system inputs.

$$x_k = \phi_k x_{k-1} \tag{56}$$

x_k, x_{k-1}: predicted states
$\phi_k = e^{\mathrm{FT}}$.
The covariance matrix is used to represent the confidence in the estimated states. Thus, the covariance matrix is updated at every time-step, when a new prediction is made. The uncertainty in the predicted states grows over time until a measurement fusion occurs. EKF computes the covariance matrix by linearising the system model at the current state estimate.

The covariance matrix at the current time is predicted using Eq. (57).

$$P_k = \phi_{k-1} P_{k-1} \phi_{k-1}^T + G_{k-1} Q_{k-1} G_{k-1}^T \tag{57}$$

ϕ_k, G_k: State and control matrices
Q_k: Process noise variance
P_k, P_{k-1}: Covariance matrix

The predicted state and covariance matrix are updated during the sensor fusion. Algorithm used for the same is shown in Eqs. (58)–(62).

$$K = P_k H_k^{\mathrm{T}} (R_k + H_k P_K H_k^{\mathrm{T}})^{-1} \tag{58}$$

$$Y_{\mathrm{cpt}} = H_k * X_k \tag{59}$$

$$Z = Y_{\mathrm{meas}} - Y_{\mathrm{cpt}} \tag{60}$$

$$X_k^{\mathrm{corrected}} = X_k + KZ \tag{61}$$

$$P_K^{\mathrm{corrected}} = (I - KH_k)P_k \tag{62}$$

where R_k is covariance of the measurement noise

4 Estimation Work-Flow

The navigation algorithm for the attitude estimation of the spacecraft is divided into multiple tasks. Few tasks are executed at a period of 500 ms, termed secondary tasks, and other tasks are executed at a period of 20 ms, termed primary tasks.

Attitude propagation is performed using the measured gyro rates at every $20ms$. The gyro drift and the error in the computed orientation is estimated using the Kalman filter at a frequency of 2 Hz (500 ms). The inertial reference measurement of the magnetometer and the Sun vector required for the Kalman filter is computed using the IGRF model and VSOP87 model, respectively, as a secondary task, using the GNSS position and time as input. The measurements from the respective sensors are also expected at the same frequency. A block diagram of the navigation system based on this scheme is shown in Fig. 1.

5 Performance Evaluation

In order to validate the performance of the algorithm, a spacecraft is simulated in an orbit of period 5000 s. The simulation is performed for a flight duration of 5000 s. Multiple reorientations and thruster firing are also simulated during the estimation.

The system is initialized with the following parameters. The IGRF model is initialized with the start time as September 26, 2016, and the order is chosen to be 13. The standard deviation of the gyro drift simulated is 10°/h, and the standard deviation of the zero mean gyro noise is taken as 0.0333°/h. The standard deviation of the noise in magnetometer and Sun sensor are initialized as $40nT$ and 0.1667^o, respectively. The run time is from 2500 to 7500 s. Time before 2500 s is used for

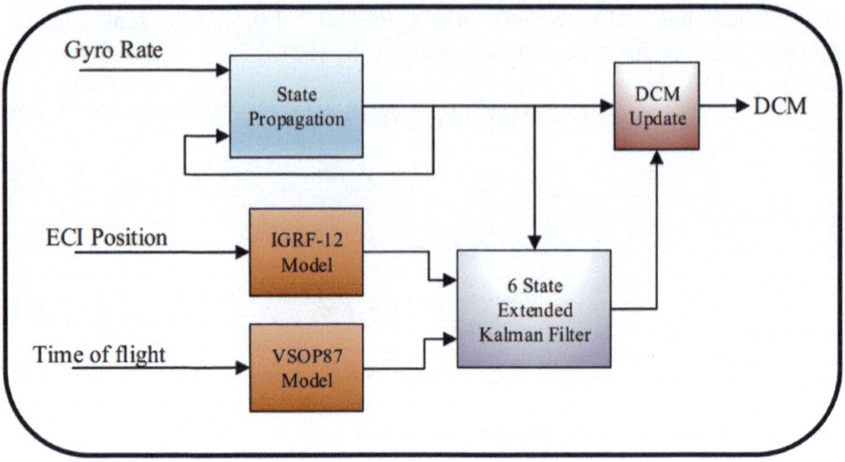

Fig. 1 Block diagram of attitude estimation using extended Kalman filter

spacecraft stabilization where the algorithm discussed is of limited use due to the high accelerations from the thrusters.

The performance of the algorithm is evaluated for multiple conditions. The details of the cases studied are as follows:

- Case 1: Both magnetometer and Sun sensor measurement are used for the measurement update.
- Case 2: The Sun sensor measurement is marked invalid between time of flight 4000 and 5000 s. Hence, only magnetometer measurement is available during this period.
- Case 3: The magnetometer measurement is marked invalid between time of flight 4000 and 5000 s. Hence, only Sun sensor measurement is available during this period.
- Case 4: Both the sensors are marked invalid between time of flight 4000 and 5000 s. Hence, there is no aiding during this period.

Table 1 Statistics of estimated states in various cases of simulation

Parameter	Case 1		Case 2		Case 3		Case 4	
	Mean	Std. dev.	Mean	Std. dev.	Mean	Std. dev.	Mean	Std. dev.
Yaw error (°)	−0.0001	0.0081	−0.0079	0.0486	−0.0003	0.0097	0.0074	0.0318
Pitch error (°)	−0.0054	0.0541	−0.0042	0.0778	−0.0045	0.0847	0.0113	0.0901
Roll error (°)	0.0123	0.0894	0.0384	0.1168	0.0184	0.1086	0.0518	0.1641
X Drift error (°/h)	14.38	5.48	14.00	5.84	14.28	5.61	14.06	5.91
Y Drift error (°/h)	−5.64	2.53	−5.54	3.15	−5.50	3.09	−5.53	3.15
Z Drift error (°/h)	11.73	1.43	11.98	1.24	11.73	1.43	12.45	1.16

The statistics of error in state estimation is tabulated in Table 1. In case 1, the filter converges with an error of less than 1° in attitude. The error in estimate is plotted with respect to time in Fig. 2a, b. Gyro drift is also estimated accurately, with a low noise variance on the estimate. The steady-state error in drift reduces after 5000 s. This is due to an increase in the body rates for short duration, which improves the convergence.

Case 2 simulates the effect of solar eclipse. Error in yaw has increased due to jerk. Error estimated about pitch and roll axes also have similar trend. The estimation error is within the stated bounds. Hence, the scheme can be used even in cases of eclipse. The estimate error is shown in Fig. 2c, d.

Results of case 3 show that short duration of magnetometer failure is of no concern. The error in estimate is plotted in Fig. 2e, f. The error between 4000*s* and 5000 s is comparable with other instances. Once the filter converges, data miss of one sensor

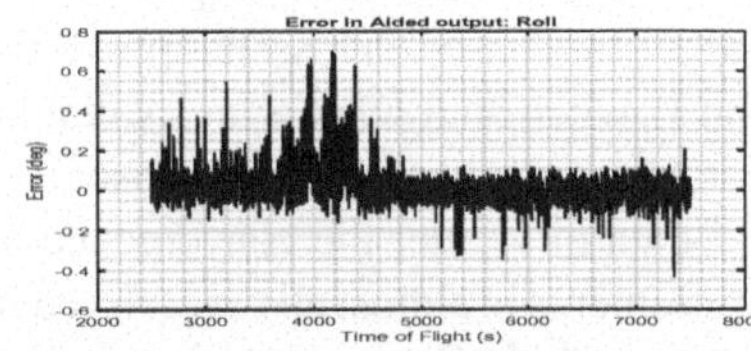

(a) Case 1: Error of EKF aided roll estimate

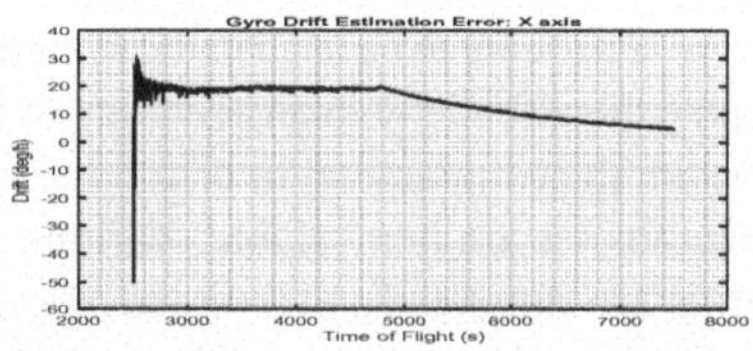

(b) Case 1: Error in estimation of gyro drift in X axis

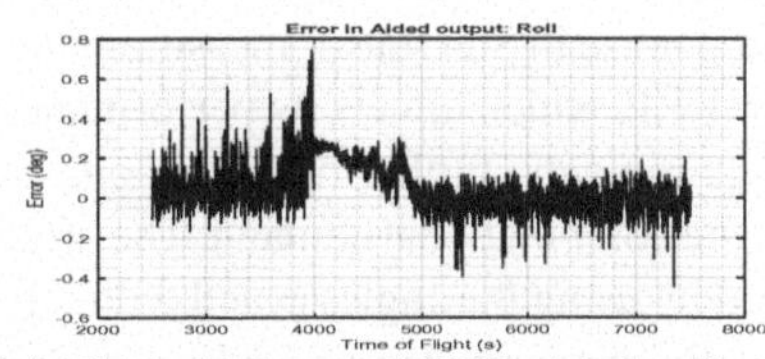

(c) Case 2: Error of EKF aided roll estimate

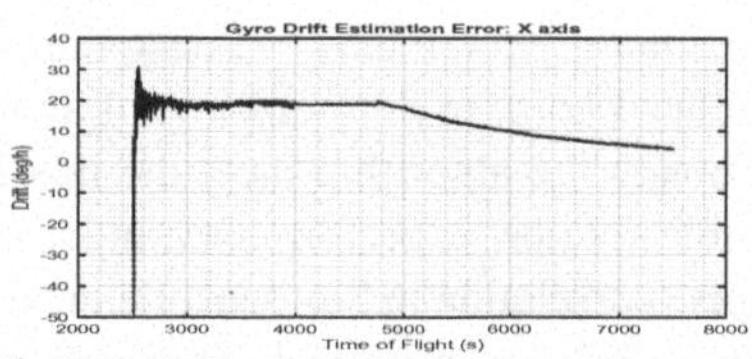

(d) Case 2: Error in estimation of gyro drift in X axis

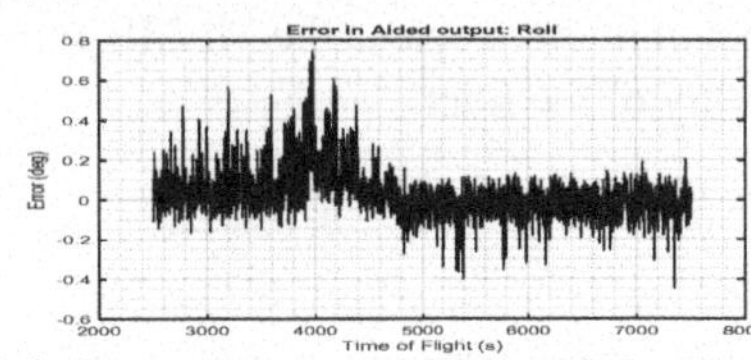

(e) Case 3: Error of EKF aided roll estimate

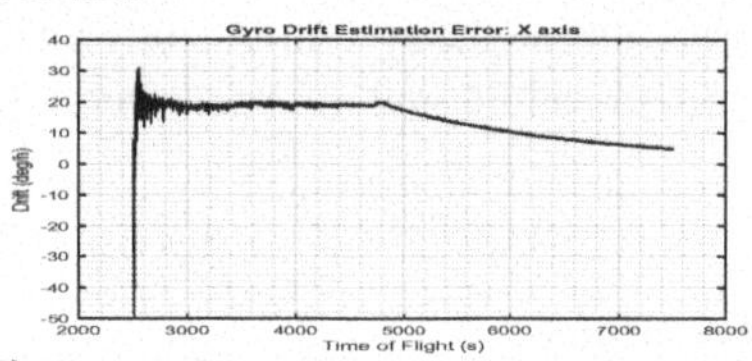

(f) Case 3: Error in estimation of gyro drift in X axis

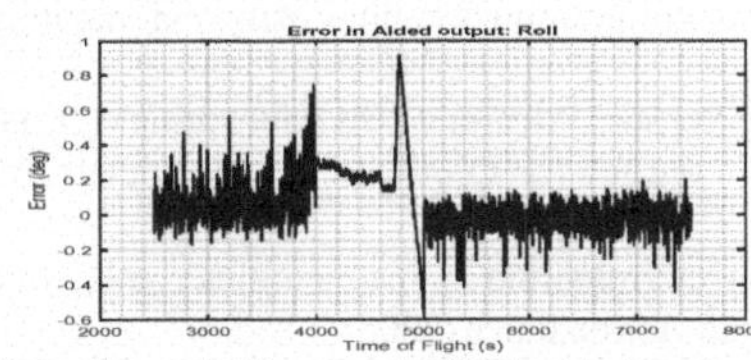

(g) Case 4: Error of EKF aided roll estimate

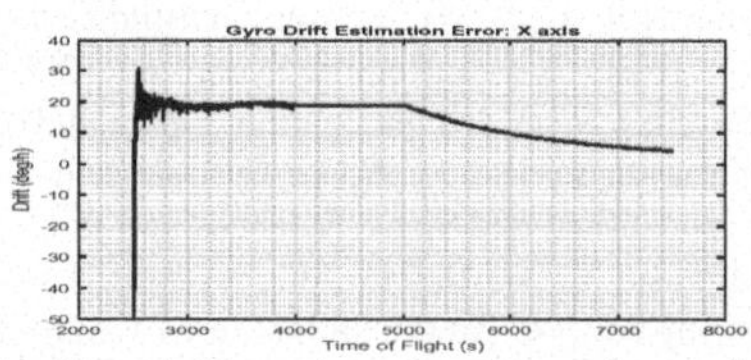

(h) Case 4: Error in estimation of gyro drift in X axis

Fig. 2 Error in EKF estimate in one channel

for short duration is tolerable. Even though the data miss happened at the period of high rates, the performance is nominal.

In case 4, between 4000s and 5000 s, the attitude is propagated using the gyro measurements with the last estimated gyro bias. The gyro drift estimate is held constant during this period, as there are no measurements to correct the drift. The error in estimate with respect to time is shown in Fig. 2g, h. Even in this case, the error in attitude estimate is less than 1°, as the filter had converged before the data miss.

6 Conclusion

The objective of the project is to design and develop an algorithm for the attitude estimation of a small spacecraft. Rate class MEMS gyros, three-axis magnetometer and Sun sensor are to be used in the system. The algorithm based on EKF has proved to estimate the attitude with an accuracy of less than 1° when both sensors are available. However, the estimated gyro drift has a steady-state error, when both Sun sensor and magnetometer are unavailable. Various cases of data miss after the convergence of the filter are simulated and found to meet the requirements in almost all cases. Even in cases where only magnetometer and gyroscope are available, the accuracy is less than 5°. However, when only Sun sensor and gyroscopes are available, the performance of the filter is poor. It may be due to the correlated noise in Sun sensor measurements. Correlated noise in Sun sensor is of an issue to EKF, which affects its performance.

Though the algorithm meet the mission requirements, further optimizations are possible to improve performance and to reduce the computational demand. The effect of noise correlation in Sun sensor can be tackled by using the angles ϕ and θ directly in the measurement model. This will improve the performance of EKF to a good extent. Further studies have to be performed to characterize the effect of higher angular rates on the system. A study on the effect of higher noise in the MEMS gyros can provide additional information on the stability of the filter.

References

1. Ahn, H.S., Lee, S.H.: Gyroless attitude estimation of sun-pointing satellites using magnetometers. IEEE Geosci. Remote Sens. Lett. (2005)
2. Farrell, J.A.: Aided Navigation: Gps with High Rate Sensors. McGraw-Hill Publications (2008)
3. Mimasu, Y., Ha, J.C.V.D.: Attitude determination concept for QSAT. In: Japan Society for Aeronautical and Space Sciences and ISTS (2009)
4. Mimasu, Y., VanDerHa, J.C., Narumi, T.: Attitude Determination by magnetometer and gyros during eclipse. American Institute of Aeronautics and Astronautics (2008)
5. Simon, D.: Optimal State Estimation: Kalman, h_∞, and Nonlinear Approaches. Wiley (2006)
6. Thébault, E., Finlay, C.C., Beggan, C.D., Alken, P., Aubert, J., Barrois, O., Bertrand, F., Bondar, T., Boness, A., Brocco, L., et al.: International geomagnetic reference field: the 12th generation. Earth Planets Space **67**(1), 79 (2015)

7. Thienel, J., Harman, R., Bar-Itzhack, I., Lambertson, M.: Results of the magnetometer navigation (MAGNAV) inflight experiment. In: AIAA/AAS Astrodynamics Specialist Conference and Exhibit, p. 4749 (2004)
8. Vallado, D.A.: Fundamentals of Astrodynamics and Applications, 2nd edn. Dordrecht Kluwer Academic Publishers (2001)

Characterisation of Multi-sensor 6D Pose Determination System for Underslung Winged Body

K. L. N. Sai Nitish, Jiljo K. Moncy, M. Dinesh Kumar, B. Karthik, V. T. Basker, and E. S. Padma Kumar

Abstract Micro electro-mechanical system (MEMS) based inertial sensors (accelerometers and gyros) are being widely used in recent times due to low cost and compact size. Navigation unit with MEMS based sensors are employed to measure the body rates and accelerations of a winged body, while being airlifted using an aircraft. The MEMS sensors are aided with global navigation satellite system (GNSS), barometer and magnetometer measurements, while using an extended kalman filter to attain good precision. Prior to the trial, thorough testing and characterisation of the navigation units are done to determine the performance of the navigation unit and repeatability of the results. Electromagnetic interference test ensured the functioning and acquisition of data from the navigation unit in the presence of various frequencies from the aircraft. The body rates and accelerations are measured by mounting the navigation unit on an Angular Motion Simulator (AMS). Deviation of the pose estimation by the MEMS navigation unit with respect to the reference inertial navigation system (INS) is found to be within bounds, justifying the selection of the navigation unit for instrumentation of the winged body. Post test results of the trial validates the pre-flight simulations with an accuracy as predicted from the pre-flight tests.

1 Introduction

A winged airframe is suspended from an aircraft and lifted to a predefined altitude. The aircraft is linked to the winged body through an interface system. Any body (in this case, a winged airframe and the interface system) suspended from an aircraft can be referred to as the underslung or the slung body. Transportation of jet trainers and gliders is some common examples of a slung body configuration. Winged bodies are also suspended from an aircraft prior to demonstration of autonomous landing. Trials

K. L. N. Sai Nitish and Jiljo K. Moncy—contributed equally to the work

K. L. N. Sai Nitish (✉) · J. K. Moncy · M. Dinesh Kumar · B. Karthik · V. T. Basker · E. S. Padma Kumar
Vikram Sarabhai Space Centre, I.S.R.O., Thiruvananthapuram, India

J. Gu et al. (eds.), *Communication and Control for Robotic Systems*, Smart Innovation, Systems and Technologies 229, https://doi.org/10.1007/978-981-16-1777-5_15

with the winged body are carried out to assess the underslung dynamics, directional stability of the underslung and to measure the body rates and accelerations of the underslung. Study of underslung dynamics is required to assess the combined stability of the aircraft-underslung system. In order to validate the predicted flight dynamics, the 6D pose of the rigid bodies is to be measured in-flight. A navigation unit is identified and tested to meet the requirements.

For the experiment, a configuration involving three navigation units is used. The units are placed close to the center of gravity of aircraft, winged body and interface system. The navigation unit contains multiple accelerometers, gyroscopes and magnetometers to measure body accelerations, rates and attitudes, respectively, in all three axes. As the noise characteristics of the MEMS sensors drifts with time, estimation of the gyro drift and accelerometer bias is of prime importance to achieve high precision. The estimation of the noise parameters is performed by sensor fusion. Sensor measurements are fused using an extended kalman filter (EKF). EKF is an extension of the Kalman filter for non-linear systems which uses the Jacobian for system linearization at the point of concern. GNSS, barometer and magnetometer are used as secondary sensors to estimate the noise in accelerometer and gyroscope measurements.

The flight software is coded in the open source robotic software framework PX4 [3]. It is run on a 32-bit STM32F427 cortex M4 core processor. Three redundant inertial measurement units (IMUs), MPU9250, ICM20948, and L3GD20 with LSM303D, are used for acceleration and rate measurements. The processor performs a best data voting to select the primary sensor. The unit also has two redundant barometers, MS5611. A GNSS module is also connected to the processor over UART. The GNSS module also houses an IMU (ICM20948) and a barometer (MS5611) for efficient localization utilizing an independent estimation algorithm. The module is capable of tracking and processing GLONASS, GPS, BeiDou, and Galileo constellations. The sampling rate of accelerometer and gyroscope is 1000 Hz, and the barometer is sampled at 130 Hz. GNSS module has an output frequency of 5 Hz. The state estimates from the extended kalman filter is available at 40 Hz frequency.

2 Preliminary Processing

IMU is the primary sensor for state estimation. However, as MEMS sensors are prone to varying bias and scale factor, dead reckoning using the IMU measurement will be erroneous. Hence, additional sensors are used as state observers. A sequential EKF is used to fuse all the sensor measurements to obtain an estimation with higher confidence. The bias and scale factor of IMU, barometer and magnetometer are estimated through calibration. Noise filtering is also performed on the measured data.

2.1 Calibration

Calibration of accelerometer, gyroscope and magnetometer is performed before flight. The sensor bias and scale factor are estimated during the calibration.

In order to calibrate the accelerometer, the gravity vector is aligned along all the axis vectors sequentially. Assuming that the measured values for each orientation is $\mathrm{accel}_{\mathrm{raw\_ref}}$, the offset in each axis can be computed using Eq. (1), where ($^{+}$) and ($^{-}$) denote positive and negative axis direction, respectively. The scale factor is calculated along with the rotation matrix for the sensor. The corrected sensor reading is as stated in Eq. (2), where $accel_T$ is the transformation matrix. The transformation matrix is solved from the same equation.

$$\mathrm{accel}_{\mathrm{off}} = \frac{\left(\mathrm{accel}^{+}_{\mathrm{raw\_ref}} + \mathrm{accel}^{-}_{\mathrm{raw\_ref}}\right)}{2} \tag{1}$$

$$\mathrm{accel}_{\mathrm{corr\_ref}} = \mathrm{accel}_{\mathrm{T}}\left(\mathrm{accel}_{raw\_\mathrm{ref}} - \mathrm{accel}_{\mathrm{off}}\right) \tag{2}$$

Gyroscope is calibrated using a number of reading when the system is at rest [4]. The offset is computed after correcting the measured data with default scale and bias values. Sample size is more than 5000 for each sensor. The algorithm used for calibrating the gyro bias is stated in Eq. (3).

$$\mathrm{gyro}_{\mathrm{off}} = \frac{\sum_i (\mathrm{sensor}_{\mathrm{meas}} - \mathrm{gyro}_{\mathrm{off\_def}}) * \mathrm{gyro}_{\mathrm{scale\_def}}}{\mathrm{number\_of\_measurements}} \tag{3}$$

Here, $\mathrm{sensor}_{\mathrm{meas}}$ is the measured sensor value, $\mathrm{gyro}_{\mathrm{off\_def}}$ is the default value of the gyro bias saved after the last usage, and $\mathrm{gyro}_{\mathrm{scale\_def}}$ is the default scale factor of the gyro. Detailed derivation of the calibration equations are given in [4].

The magnetometer is calibrated by generating a data cloud of the magnetic data read by the sensor in the 3D space. In order to generate the data, the sensor is rotated through various attitudes. The data collected is fit to a ellipsoidal least squares fit. The center of ellipse is the offset of sensor, whereas the radius in each axis is the estimated scale factor of the sensor. The relationship between the calibrated output (B^{r}) and measured magnetic field (B^{m}) is as stated in Eq. (4). The scale factor C and the bias b are calculated using the ellipsoidal fit algorithm derived in [1].

$$B^{\mathrm{r}} = C^{-1}(B^{\mathrm{m}} - b) \tag{4}$$

2.2 Noise Filtering

Before pushing the sensor data to the state estimation algorithm, preliminary data processing chain is executed to filter out the noise. The sequence of operation performed on IMU data is stated below:

- Sensor measurement is rotated to body frame (winged body).
- Fixed scale factor and offset, estimated by calibration is applied.
- Low-pass filtering is applied.
- The angular rates and accelerations are integrated using trapezoidal integration to calculate the $\Delta\alpha$ (angle change) and ΔV (velocity change). Compensation for coning errors is also applied. Coning error is induced by the angular oscillations with different phases on two axes of measurement.
- ΔV measurements are corrected for rotation and sculling errors. Sculling error is the result of sculling motion, which is defined as a sinusoidal angular motion on an axis with sinusoidal or linear acceleration motion on another axis. Equation (5) is used for the correction, where ($_\mathrm{p}$) represents data from previous instant.
- $\Delta\alpha$ is corrected for earth spin rate and coning errors. Earth spin rate in inertial frame (ω_{earth}) is calculated using Eq. (6). The latitude is known from GNSS data. Earth spin rate is transformed to the body frame from the inertial frame using the transformation matrix, R_n^{b}. The corrected $\Delta\alpha$ is shown in Eq. (7).
- At every 4 ms (250 Hz) $\Delta\alpha$ and ΔV are published.

$$\begin{aligned}\Delta V_{\text{cor}} &= \frac{1}{2}\Big[(\Delta\alpha_p + \Delta\alpha)(\Delta V_{\mathrm{p}} + \Delta V)\Big] \\ &+ \frac{1}{6}\Big[(\Delta\alpha_p + \Delta\alpha)\Big((\Delta\alpha_{\mathrm{p}} + \Delta\alpha)(\Delta V_{\mathrm{p}} + \Delta V)\Big)\Big] \\ &+ \frac{1}{12}\Big[\Delta\alpha_{\mathrm{p}}\Delta V + \Delta V_{\mathrm{p}} \times \Delta\alpha\Big]\end{aligned} \tag{5}$$

$$\Delta\omega_{\text{earth}} = \begin{pmatrix} 0.000072921\cos(\text{latitude}) \\ 0 \\ -0.000072921\sin(\text{latitude}) \end{pmatrix} \tag{6}$$

$$\begin{aligned}\Delta\alpha_{\text{cor}} &= \Delta\alpha - \frac{1}{12}(\Delta\alpha_{\mathrm{p}} \times \Delta\alpha) \\ &- R_n^{\mathrm{b}}\omega_{\text{earth}}\Delta t\end{aligned} \tag{7}$$

3 State Estimation

An extended kalman filter (EKF) is used to estimate the internal states from noisy measurements as the system is non-linear [7]. The measurement noise and process noise are assumed to be zero mean Gaussian processes for the estimation. The measurements from accelerometer and gyroscope are used for the state prediction. As MEMS based IMU is used for the same, the measured values drift considerably over time. Thus, secondary sensors-like GNSS, barometer and magnetometer are used to improve the confidence by fusing these secondary measurements to the predicted states. The estimation algorithm implemented by the authors of [5] is used in the navigation unit.

A 22 state EKF with multi sensor fusion is used for improved estimation [6]. Position and velocity information from GNSS, pressure and temperature data from the barometer, magnetic field data from the magnetometer are all used to produce the estimation and related covariance matrix. Few states are estimated in body frame (similar to aircraft body frame) (*XYZ*), while others are estimated in inertial (NED) frame. NED is North-East-Down frame, the origin fixed at the point of initialization. The body frame is defined as '*X*' in the forward direction (toward the direction of motion) , '*Y*' toward the right facing the *X* direction, and '*Z*' to complete the right handed system. The states estimated by the EKF are as follows.

- Quaternion (q_0, q_1, q_2, q_3)
- Velocity (V_N, V_E, V_D)
- Position (P_N, P_E, P_D)
- Gyro delta angle bias vector (Body frame: *XYZ*)
- Accelerometer bias (Body *XYZ*)
- Earth magnetic field vector (NED)
- Magnetometer bias errors (Body *XYZ*).

The pose (6D states) information is captured in the first ten states which uses a dynamic process model that defines the movement of the body in an inertial frame. Delta angle motion is sensed using the measurement (Δ_{α_meas}) from gyro measured angular velocity (ω). The measured value is subjected to a bias error (Δ_{α_bias}), which is estimated in the filter. Thus, the true delta angle movement (Δ_{α_truth}) is obtained by removing the bias. The related equations are shown in Eqs. (8)–(10).

$$\Delta_{\alpha\_meas} = \begin{pmatrix} \Delta_{\alpha\_x} \\ \Delta_{\alpha\_y} \\ \Delta_{\alpha\_z} \end{pmatrix} = \int_{t_{k-1}}^{t_k} \omega.\mathrm{d}t \tag{8}$$

$$\Delta_{\alpha\_bias} = \begin{pmatrix} \Delta_{\alpha\_bias\_x} \\ \Delta_{\alpha\_bias\_y} \\ \Delta_{\alpha\_bias\_z} \end{pmatrix} \tag{9}$$

$$\Delta_{\alpha\_truth} = \Delta_{\alpha\_meas} - \Delta_{\alpha\_bias} \tag{10}$$

The quaternion Δ_{quat} that defines the rotation from the quaternion at frame $k-1$ to k is calculated using small angle approximation using the Δ_{α_truth} as shown in Eq. (11).

$$\Delta_{quat} = \begin{pmatrix} \Delta q_0 \\ \Delta q_1 \\ \Delta q_2 \\ \Delta q_3 \end{pmatrix} = \begin{pmatrix} 1 \\ \frac{\Delta_{\alpha\_truth\_x}}{2} \\ \frac{\Delta_{\alpha\_truth\_y}}{2} \\ \frac{\Delta_{\alpha\_truth\_z}}{2} \end{pmatrix} \tag{11}$$

Quaternion product rule is used to update the quaternion, as shown in Eq. (12)

$$\begin{pmatrix} q_0 \\ q_1 \\ q_2 \\ q_3 \end{pmatrix} = \begin{pmatrix} q_0\Delta q_0 - q_1\Delta q_1 - q_2\Delta q_2 - q_3\Delta q_3 \\ q_0\Delta q_1 + q_1\Delta q_0 + q_2\Delta q_3 - q_3\Delta q_2 \\ q_0\Delta q_2 + q_2\Delta q_0 - q_1\Delta q_3 + q_3\Delta q_1 \\ q_0\Delta q_3 + q_3\Delta q_0 + q_1\Delta q_2 - q_2\Delta q_1 \end{pmatrix} \tag{12}$$

The next 6 states are the position and velocity in NED frame. Similar to the above case, the measurement from the accelerometer is used to propagate the position and velocity. The related equations are shown in Eqs. (13)–(17), where R_b^n is the transformation matrix from body to inertial frame. Measured acceleration is corrected for gravity as shown in Eq. (17), with the assumption that gravity acts along the 'Down' axis with a constant magnitude 'g' ($g = 9.8056\,\mathrm{m/s^2}$).

$$\Delta V_{\text{meas}} = \begin{pmatrix} \Delta V_x \\ \Delta V_y \\ \Delta V_z \end{pmatrix} = \int_{t_{k-1}}^{t_k} a.dt \tag{13}$$

$$\Delta V_{\text{bias}} = \begin{pmatrix} \Delta V_{\text{bias\_}x} \\ \Delta V_{\text{bias\_}y} \\ \Delta V_{\text{bias\_}z} \end{pmatrix} \tag{14}$$

$$\Delta V_{\text{truth}} = \Delta V_{\text{meas}} - \Delta V_{\text{bias}} \tag{15}$$

$$\begin{pmatrix} V_{\text{N}} \\ V_{\text{E}} \\ V_{\text{D}} \end{pmatrix}_k = \begin{pmatrix} V_{\text{N}} \\ V_{\text{E}} \\ V_{\text{D}} \end{pmatrix}_{k-1} + R_b^n \Delta_{V\text{\_truth}} + \begin{pmatrix} 0 \\ 0 \\ g \end{pmatrix} \Delta t \tag{16}$$

$$\begin{pmatrix} P_{\text{N}} \\ P_{\text{E}} \\ P_{\text{D}} \end{pmatrix}_k = \begin{pmatrix} P_{\text{N}} \\ P_{\text{E}} \\ P_{\text{D}} \end{pmatrix}_{k-1} + \begin{pmatrix} V_{\text{N}} \\ V_{\text{E}} \\ V_{\text{D}} \end{pmatrix}_{k-1} \Delta t \tag{17}$$

The IMU sensor bias, magnetic field in NED frame ($M_{\text{N}}, M_{\text{E}}, M_{\text{D}}$) and magnetometer bias in body frame (M_X, M_Y, M_Z) are modelled using a static process. Hence the state equations are as shown in Eq. (18).

$$\begin{pmatrix} \alpha_{\text{bias\_}x} \\ \alpha_{\text{bias\_}y} \\ \alpha_{\text{bias\_}z} \\ \Delta V_{\text{bias\_}x} \\ \Delta V_{\text{bias\_}y} \\ \Delta V_{\text{bias\_}z} \\ M_{\text{N}} \\ M_{\text{E}} \\ M_{\text{D}} \\ M_X \\ M_Y \\ M_Z \end{pmatrix}_k = \begin{pmatrix} \alpha_{\text{bias\_}x} \\ \alpha_{\text{bias\_}y} \\ \alpha_{\text{bias\_}z} \\ \Delta V_{\text{bias\_}x} \\ \Delta V_{\text{bias\_}y} \\ \Delta V_{\text{bias\_}z} \\ M_{\text{N}} \\ M_{\text{E}} \\ M_D \\ M_X \\ M_Y \\ M_Z \end{pmatrix}_{k-1} \tag{18}$$

To estimate the state, a filter is implemented in discrete time where the subscript k represents the current time-step and $k-1$ the previous time-step. The states are predicted forward from the previous to current time-step using prior knowledge about the system dynamics and inputs. The states at the current time-step are expressed as a function of the states at the previous time step and known system inputs.

$$x_k = f(x_{k-1}, u_k) \tag{19}$$

x_k: predicted states at the current time-step
x_{k-1}: states from previous time-step
u_k: known system inputs at current time-step.
The covariance matrix is used to represent the confidence of the estimated states. Thus, the covariance matrix is updated at every time-step, when a new prediction is made. The uncertainty in the predicted states grows over time until a measurement fusion occurs. The EKF computes the covariance matrix by linearising the system model at the current state estimate using Jacobian.

The covariance matrix is updated using the Eq. (20), where f is the system model, obtained using Eqs. (12), (16)–(18).

$$P_k = F_{k-1} P_{k-1} F_{k-1}^T + G_{k-1} Q_{k-1} G_{k-1}^T + Q_s \tag{20}$$

$$F_k = \left(\frac{\partial f}{\partial x}\right)_k \tag{21}$$

$$G_k = \left(\frac{\partial f}{\partial u}\right)_k \tag{22}$$

F_k, G_k: State and control Jacobians
Q_k: Process noise variance
Q_s: State noise variance
P_k, P_{k-1}: Covariance matrix

3.1 Measurement Fusion

The confidence in the estimated state decreases during the propagation, till a measurement fusion occurs. Multiple sensor data are fused to improve the certainty of the prediction. Both the state estimates and the covariance matrix are updated using the measurements. Covariance value decreases after the fusion, provided new measurement is used. GNSS and magnetometer are couple of examples of sensors used for the fusion.

The predicted state and covariance matrix are updated during the fusion, each sensor data weighted against its Jacobian [2]. Following procedure is used for the same:

- Computation of observation Jacobian:

$$H_k = \left(\frac{\partial z_p}{\partial x}\right)_k \tag{23}$$

- Computation of innovation variance:

$$S_k = H_k P_k^- H_k^T + R_k \tag{24}$$

  P_k^-: Predicted state covariance matrix
  R_k: Observation covariance
- Compute Kalman Gain:

$$K = P_k^- H_k' S_k^{-1} \tag{25}$$

- Compute innovation:

$$\mu = z - z_p \tag{26}$$

  z: Actual measurement
  z_p: Predicted measurement
- Update covariance matrix

$$P_k^+ = [I - K H_k] P_k^- \tag{27}$$

- Update state:

$$x_k^+ = x_k^- + K\mu \tag{28}$$

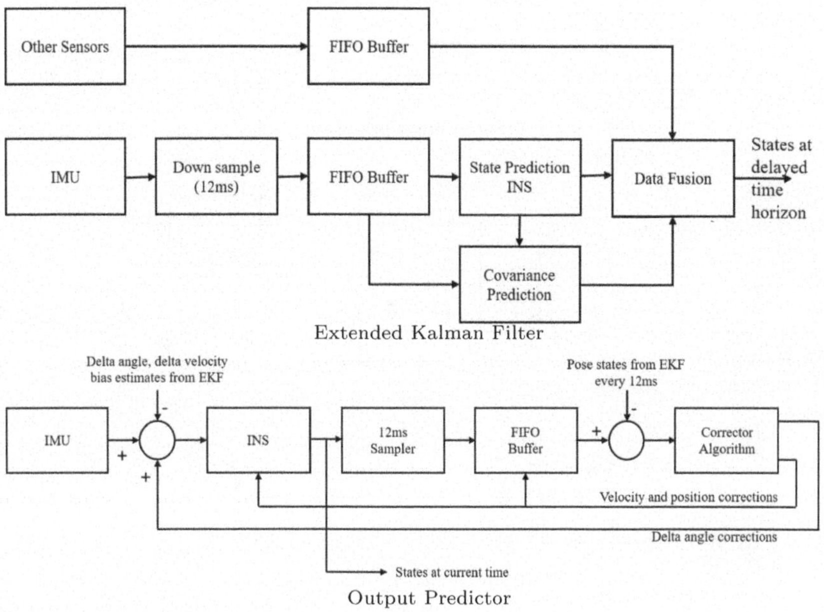

Fig. 1 Data fusion architecture [6]

3.2 *Output Predictor*

A major downfall of using measurements-like GNSS for state estimation is the lower operational frequency. For instance, in the navigation unit, the IMU measurements are sampled at 1000 Hz, while the GNSS measurements are obtained only at 5 Hz. Thus, GNSS observations are delayed relative to IMU data and are asynchronous. However, fusing observations that are delayed with respect to the predicted current state and covariance will increase the errors and reduces filter stability.

In order to tackle this issue, an output predictor as shown in Fig. 1 is used to obtain the data real-time [6]. The buffered IMU data is used for the fusion at the fusion time horizon, the output of which is used to correct the estimates at current time horizon. The strapdown inertial navigation algorithm is performed at the IMU sampling rate, as the output predictor. The states are buffered using the same length buffer as the EKF uses for it's IMU data. The delayed states are compared to the EKF states and the error is used to calculate corrections that cause the output predictor states to track the EKF states at the EKF time horizon.

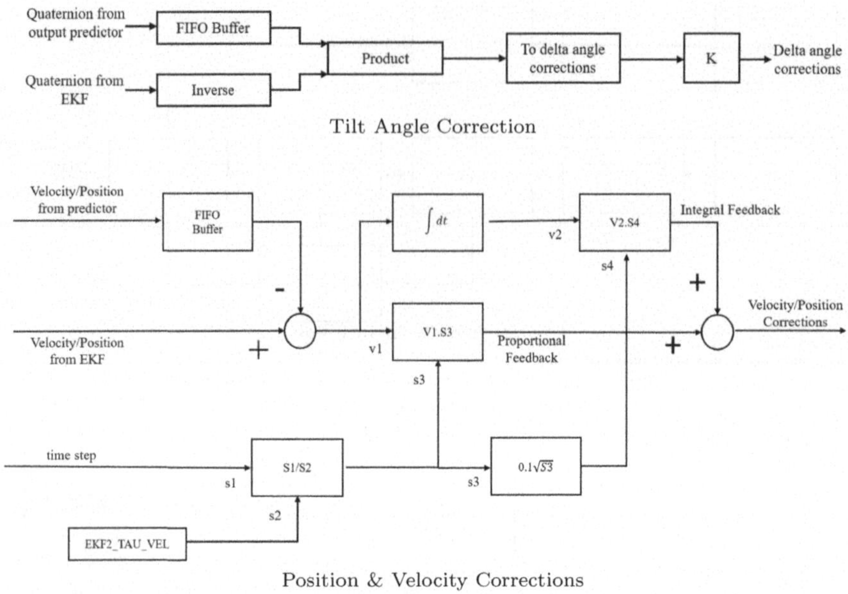

Fig. 2 Output predictor–corrector algorithm [6]

3.3 Corrector Algorithm

Specific algorithms are used to estimate the error in position, velocity and attitude due the output predictor. The EKF estimate is assumed as the true value, and the states from output predictor is updated to follow the same.

To correct the delta angle errors, the error quaternion between the output predictor estimated quaternion and EKF estimate is used. A PI feedback algorithm is used to correct the output state history so that the output velocity and position tracks the EKF velocity with 5% overshoot as shown in Fig. 2.

3.4 Initialization of States and Covariance

Once the system is powered on, the program initializes states and covariance. A feature is available to delay the initialization till a lock is attained for GNSS.

In order to initialize the attitude, both the acceleration vector and the magnetometer reading are used. The acceleration vector is used to determine the alignment relative to the gravity field. Following algorithm is used for computation which does not suffer from gimbal lock:

- Calculate the magnitude of the vehicle tilt using the acceleration measured in the body frame.
- The cross product of acceleration vector and negative of body Z-axis gives the axis of tilt in body frame.
- The magnitude of tilt and the tilt axis is used to compute the rotation matrix and the quaternion.

The magnetometer data is used to determine the initial heading direction. Following algorithm is used to align the heading:

- Initial attitude is obtained assuming yaw to be zero.
- The magnetometer data measured in body frame is transformed to world frame using the before quaternion.
- The projection of the field in NE plane is used to find the yaw relative to the magnetic declination. Thus, magnetic declination should be known beforehand.

The NED magnetic field states are initiated by transforming the magnetometer reading with the initial attitude states. The average of a few readings are used to avoid the effect of vibrations. Similarly, average of a few samples from barometer is also used to initialize the vertical position. Assuming that the GNSS data is provided in local frame, position and velocity are initialized to this value. GNSS data conversion happens in the driver code. All other states are initialized to zero.

The covariance matrix is initialized using the square of 1-sigma uncertainties specified as parameters for each sensor. It will be a diagonal matrix as shown in Eq. (29).

$$P_{k=0} = \begin{pmatrix} \sigma_1^2 & 0 & 0 & 0 \\ 0 & \sigma_2^2 & 0 & 0 \\ 0 & 0 & \dots & 0 \\ 0 & 0 & 0 & \sigma_n^2 \end{pmatrix} \tag{29}$$

4 Testing and Characterisation

In order to characterise the navigation unit , a series of rate tests are performed using the angular motion simulator (AMS). Also, the navigation unit is mounted alongside reference INS on the AMS for comparison and validation. The navigation unit is tested for interference due to aircraft frequencies through a electromagnetic interference test. Multiple navigation units are mounted onto a plate placed inside a motor vehicle and driven through prescribed circuits in order to cross verify the performance of the navigation units. Finally, the navigation units are used in the trial in order to measure the body rates and accelerations of the under-slung to validate the simulations.

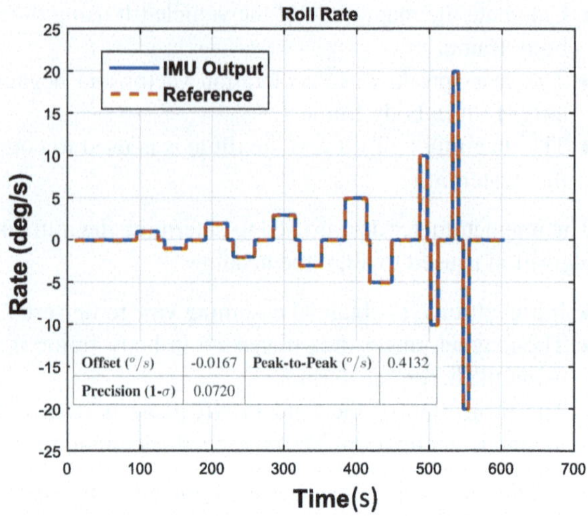

Fig. 3 Constant roll rate test

Table 1 RMS noise: data-sheet and experimental

Sensor	Datasheet	Experimental
Gyroscope	0.1 °/s	0.0934 °/s
Accelerometer	8 mg	2.2434 mg

4.1 Constant Rate Test

The AMS gimbals are commanded such that all the three axes of the navigation unit experience a constant rate, one axis at a time. The AMS input rates are 1, 2, 3, 5, 10, 15 and 20 °/s.

Constant roll rates are given to the navigation unit by rotating the middle gimbal of AMS at constant angular velocity. The output from navigation unit is plotted along with the AMS input as shown in Fig. 3. Navigation unit responds to different yaw rates as per the given input by AMS.

The RMS noise of sensors, stated in the datasheet and that observed during the tests is stated in Table 1.

4.2 Sine Tests

Sinusoidal body rates are given along the Yaw and Roll axes of the navigation unit by operating AMS gimbals. The navigation unit is excited with two sinusoidal signals as stated in Table 2.

Table 2 Sinusoidal excitation frequencies

Amplitude (θ)	Frequency (Hz)
1 °	1
0.25 °	2

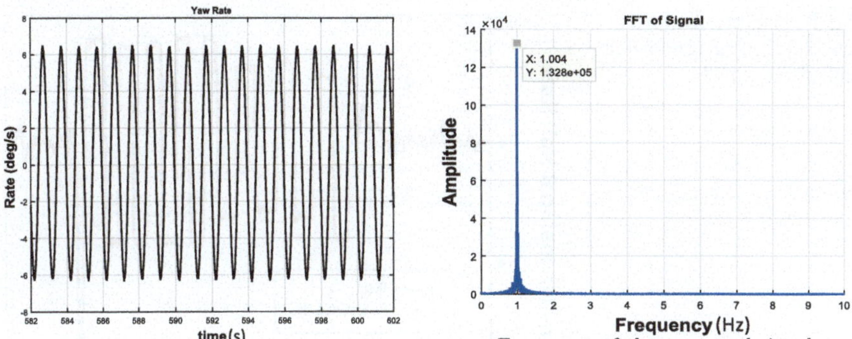

Yaw rate as measured by navigation unit

Frequency of the measured signal

Fig. 4 Result of sine tests

The input signal can be represented by Eqs. (30) and (31)

$$\theta = A\,\mathrm{Sin}(\omega t) \tag{30}$$

$$\dot{\theta} = A\omega\,\mathrm{Cos}(\omega t) \tag{31}$$

where, A is the amplitude of the signal and ω is the angular frequency.

Figure 4a, b shows the yaw rate and the FFT, respectively, of the output signal acquired from the navigation unit for the excitation with 1 Hz sinusoidal signal. The expected amplitude of the yaw rate is $A\omega = 2\,\pi$ and the same can be seen from Fig. 4a. The frequency also corresponds to the input AMS frequency.

4.3 Validation with Reference INS

Reference INS package is mounted on the AMS for the Hardware in loop simulations (HLS). Alongside INS, the navigation unit is mounted in order to compare the body rates. The acquired navigation unit data is filtered using a low-pass median filter with a moving window of size 150. Cross correlation of the filtered data and the INS data is done to remove the time mismatch between rates acquired from INS and navigation unit. Figure 5 compares the roll rates from the two units. Performance of the navigation unit against the reference INS is quantified in Table 3.

Table 3 Statistics of body rate measured by navigation unit in comparison with reference INS

Parameter channel	Roll	Pitch	Yaw
Gyro bias (°/s)	0.1275	−0.1485	−0.0144
Noise dev. (°/s)	0.2997	0.4622	0.2241

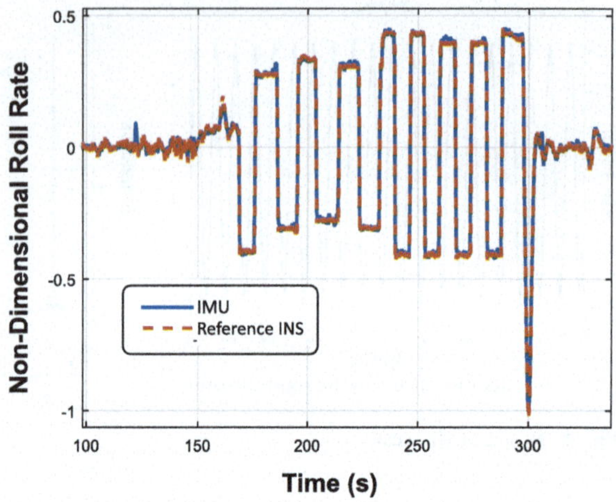

Fig. 5 Comparison of Roll rates between navigation unit and reference INS

4.4 Electromagnetic Interference (EMI) Test

Performance of the navigation unit in presence of aircraft transmitter frequencies are assessed. The test carried out are according to EMI Standard MIL 461 C and 462. Tests are carried out for various frequency ranges from 30 MHz to 30 GHz. It is observed that the performance of the navigation unit is normal without any loss of data for all the frequencies.

4.5 Drive Test

In order to compare the performance of the navigation units and the repeatability of the results, a set of tests are performed by mounting 5 navigation units (P1, P2, P3, P4 and P5) onto a plate inside a motor vehicle which is driven through specified circuits. Good match in body rates and accelerations are obtained from the test. Yaw rates for one of the motor vehicle tests is shown in Fig. 6. The x-axis is in hr:min which denotes the time of the day when the experiment was conducted in the 24 h format

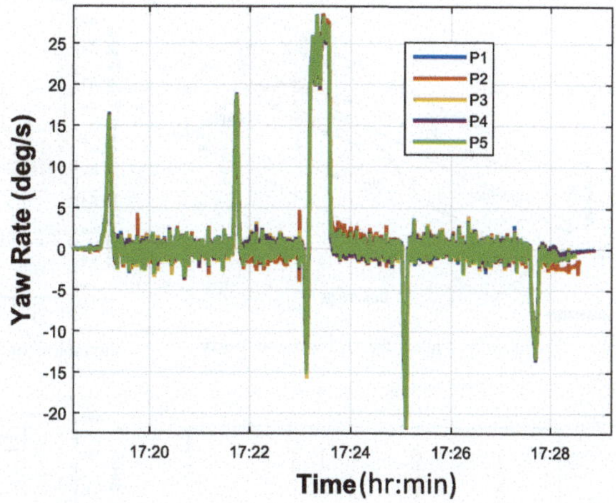

Fig. 6 Comparison of Yaw Rates among the navigation units during the motor vehicle test

5 Results and Discussion

For the lifting trial, the under-slung (an interface system and the winged body) is flown to a particular altitude using a aircraft in order to measure the body rates, accelerations and body attitudes. The aircraft carrying the underslung, the interface system and the winged body are equipped with one navigation unit each.

Post-processing algorithm is developed to read the navigation unit data, correlate the data obtained from different navigation units and understand the slung body dynamics. Body rates and Body accelerations processed using extended kalman filter are available at 40 and 10 Hz respectively while velocity, position and body attitude of the under-slung are available at 5 Hz. The sensor temperature for the entire duration of the sortie is within the operational range.

Figure 7 shows the trajectory traced during the trial. The blue star in the figure corresponds to the take-off point of the aircraft. The navigation unit captures all the phases of the flight namely, climb, turn and level flight with less than 5 m accuracy in the horizontal plane and 10 m accuracy in the vertical plane. Yaw rates and yaw angle of the under-slung are as shown in Fig. 7a, b respectively. From Fig. 7c, the body attitude of the under-slung tracks that of the aircraft during all phases of the flight. The difference in the azimuth between the aircraft and the under-slung is due to the natural trim of the under-slung in the yaw plane as predicted prior to the trial.

During the turn, centripetal acceleration acts on the under-slung. Hence, for an anticlockwise turn (7:22–7:25 in Fig. 7b) , the yaw rate is negative and for a clockwise turn (7:35–7:38 in Fig. 7c), and the yaw rate is positive. Figure 7d shows the velocity azimuth of the under-slung and the aircraft and the close match between them means that the velocity vectors of the aircraft and the underslung are always aligned during level flight and turning flight as per pre-flight predictions. The winged body and the interface system are connected to an intermediate joint using a bunch cable and the aircraft is attached to this intermediate joint through a single cable. This is a double

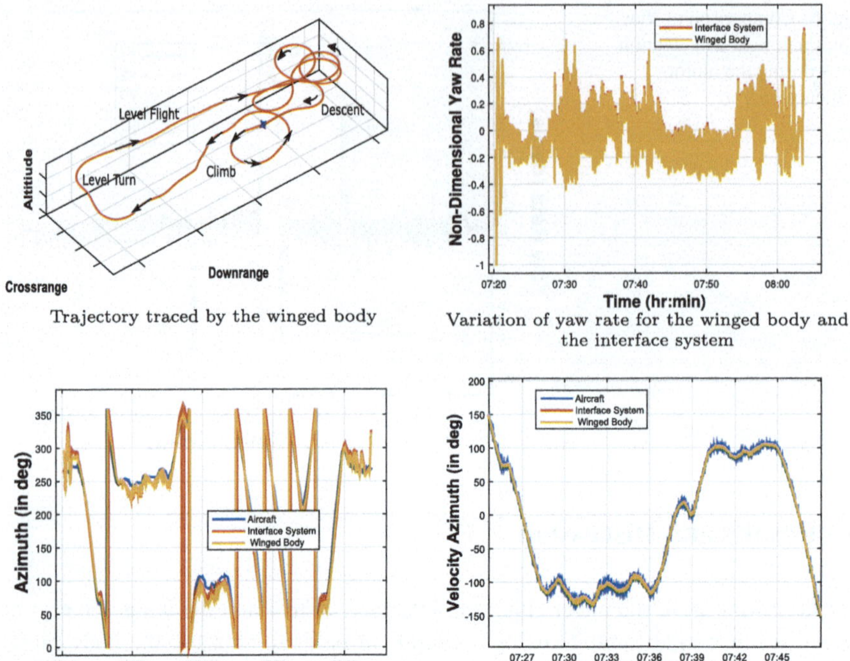

Trajectory traced by the winged body

Variation of yaw rate for the winged body and the interface system

Variation of azimuth with time for the winged body and the interface system w.r.t aircraft

Variation of velocity azimuth with time for the winged body and the interface system w.r.t aircraft

Fig. 7 Results of lifting trial experiment

pendulum configuration. The same is seen from the frequency modes obtained from the navigation unit data. The most dominant frequency is that of the first pendulum. Second pendulum mode exists, but its amplitude is one order less than the first pendulum mode.

6 Conclusion

Characterization of the navigation unit is successfully carried out through a series of tests. The various tests performed on the navigation unit concludes the following:

1. From the constant rate tests, it can be inferred that navigation unit can measure rates up to 20 °/s along all the three axes.
2. For a sinusoidal input rate, it is able to measure both the amplitude and frequency of the signal accurately.
3. Drive tests confirmed that the navigation unit can be subjected to higher rates than the AMS tested rates without sensor saturation.

It is to be noted that these measurements are taken from raw sensor data corrected for offset. Even though there is no other sensor measuring acceleration and angular rates, other than the IMU, an acceptable quality in measurement is observed. The attitude, position and velocity recorded by the unit has a higher confidence as other sensors-like GNSS and magnetometer is used in fusion.

The data procured by the navigation unit in the lifting trial can be used to validate the pre-flight simulations with good confidence. The noise observed during the test is less than that specified in the datasheet. The frequencies of the first and second pendulum corresponding to the motion of the under-slung matches with the simulations with an accuracy of 0.01 Hz. Thus, to conclude, the pre-flight and post-flight data show a good match for various phases of flight which demonstrates the robustness and stability of the navigation unit.

Acknowledgements The authors would like to express their gratitude to Shri. Sudar I., Smt. Sheelu Jose, Dr. M.Jayakumar, Shri. Navin M.S. and Dr. Rajeev U.P. for their guidance and motivation through the entirety of the work. The authors sincerely thank Shri. Sajichandrachood O.M., Smt. Suchitra Shenoy, Shri. Vijith Mukundan, Shri. Vishnu R. and Smt. Pallabi Sinha for their relentless support during sensor testing and characterisation.

References

1. Chi, C., Jun-Wei Lv, D.W.: Calibration of triaxial magnetometer with ellipsoid fitting method. IOP Conf. Seri.: Earth Environ. Sci. (ICAESEE) (2019)
2. Farrell, J.A.: Aided navigation: Gps with High Rate Sensors. McGraw-Hill Publications (2008)
3. Meier, L., Honegger, D., Pollefeys, M.: Px4: a node-based multithreaded open source robotics framework for deeply embedded platforms. In: 2015 IEEE International Conference on Robotics and Automation (ICRA), pp. 6235–6240. IEEE (2015)
4. Quan, Q.: Sensor calibration and measurement model. In: Introduction to Multicopter Design and Control, pp. 147–172. Springer (2017)
5. Riseborough, P.: Estimation and Control Library. GitHub Repository (2018)
6. Riseborough, P.: PX4 State Estimation. PX4 Developer Summit, Zurich (2019)
7. Simon, D.: Optimal State Estimation: Kalman, h_∞, and Nonlinear Approaches. Wiley (2006)

Tracking and Interception of a Ballistic Target on Reentry Using Adaptive Gaussian Sum Quadrature Filters

Rahul Radhakrishnan, Manika Saha, Shovan Bhaumik, and Nutan Kumar Tomar

Abstract In this work, new nonlinear filtering algorithms based on Gaussian sum framework is formulated and then applied to track a ballistic target on reentry. Other than representing the predicted and updated probability density function of states given measurements as a sum of weighted Gaussian densities, a weight update scheme in time update step is also considered. This weight update scheme is based on a quadratic optimization problem that minimizes the integral square error between the true and the approximated prior density. The Gaussian densities in the weighted sum is realized using various quadrature filters. These filters are applied to track a ballistic target on reentry using measurements from an inbuilt seeker of an interceptor missile, where a 6 degrees of freedom state-dependent coefficient model (SDC) of target–interceptor dynamics is considered. Further, the estimated states are provided to a guidance law for generating interceptor missile accelerations. For performance analysis, root mean square error (RMSE) of relative states and also the range between target and interceptor when they cross each other (miss-distance) are considered. It was found that the miss-distances obtained from the Gaussian filters are smaller than that of the conventional EKF. Further, the Gaussian sum and its weight adaptation scheme with quadrature filter proposals provided far better results, with the weight adaptation filters giving more accurate miss-distances.

R. Radhakrishnan (✉)
Department of Electrical Engineering, SVNIT, Surat, India
e-mail: r.rahul@eed.svnit.ac.in

M. Saha
Department of Electrical Engineering, Meghnad Saha Institute of Technology, Kolkata, India

S. Bhaumik
Department of Electrical Engineering, Indian Institute of Technology Patna, Patna, India

N. K. Tomar
Department of Mathematics, Indian Institute of Technology Patna, Patna, India

J. Gu et al. (eds.), *Communication and Control for Robotic Systems*, Smart Innovation, Systems and Technologies 229, https://doi.org/10.1007/978-981-16-1777-5_16

1 Introduction

Estimating the state of a nonlinear stochastic dynamic system using noisy measurements is a challenging problem because there is no optimal solution for it. The first suboptimal filter formulated was the extended Kalman filter (EKF) [1], but it may result in less accuracy and track divergence. To improve the accuracy of estimation, one approach is to assume the prior and posterior pdf as Gaussian and realize it with the help of a set of deterministic support points and weights. The first work in this direction uses unscented transformation to generate deterministic support points and weights, known as the unscented Kalman filter (UKF) [2, 3]. After UKF, many nonlinear filters which use deterministic sample points to approximate the pdfs were introduced. Some of these filters are cubature Kalman filter (CKF) [4], cubature quadrature Kalman filter (CQKF) [5], Gauss–Hermite filter (GHF) [6], etc.

Another approach to realize these non-Gaussian pdfs is to approximate it with the summation of many weighted Gaussian pdfs [7]. For more accurate approximation of prior densities, an optimization problem is formulated where weights of the individual Gaussian components are updated in the time prediction step [8]. The resulting filters were termed as adaptive Gaussian sum filters (AGSF). AGSF is reported to be more accurate than ordinary Gaussian sum filter (GSF). In this work, we use CQKF and GHF as a proposal of AGSF and expect that it will perform better than the existing filters.

Tracking a ballistic target and its interception preferably by a guided missile at a desired height is a highly challenging problem [9–11]. An interceptor missile is fired when an enemy object is detected by the ground RADAR. Along its flight, target information is obtained by the RADAR and is transmitted to the interceptor through a communication link. These values are processed by the on-board computer which steer the interceptor accordingly. The seeker is also constantly trying to obtain target information and when it is above a particular signal-to-noise ratio (SNR) threshold, target lock is achieved and the ground link is disabled. After target lock, seeker measurements are used by the processing unit to generate guidance commands. This phase is usually termed as terminal guidance phase or homing guidance.

In this work, the proposed AGSF with CQKF and GHF proposal are used to estimate the states of the target. The output of the estimator is used by the proportional navigation guidance (PNG) law. Here we assume an ideal autopilot which can achieve the guidance's demand. Figure 1 shows the interconnection between different blocks during homing guidance. Thus, we simulate a closed-loop target localization and interception scenario for a 6DOF state-dependent coefficient (SDC) model. By the term 'closed loop,' we mean to say that guidance law is incorporated with the estimation algorithm and the output of the guidance command is realized with an ideal autopilot, and the states of the interceptor are given back to the estimators. The performance of the interceptor is measured in terms of miss-distance and RMSE in states.

The structure of this paper is organized as follows: Sect. 2 discusses the Gaussian sum approach, whereas adaptive Gaussian sum filtering approach is mentioned in

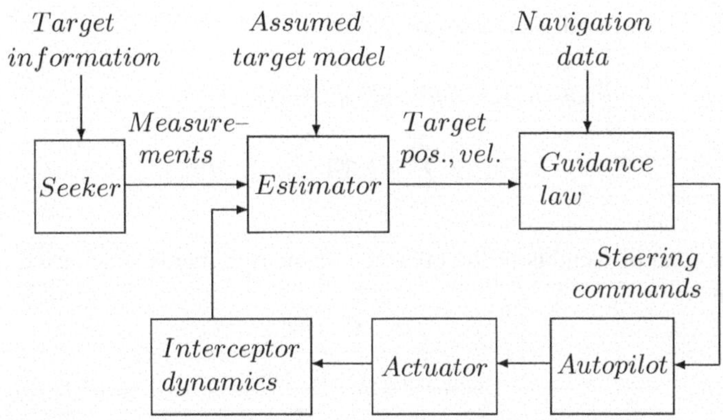

Fig. 1 Target–interceptor engagement simulation in terminal guidance phase

Sect. 3. Section 4 describes the quadrature filter proposals, and Sect. 5 explains the target and interceptor model. Finally, simulation results are given in Sect. 6 followed by conclusions in Sect. 7.

2 Gaussian Sum Filtering

Let us consider a stochastic dynamic system where the process and measurement models are given by

$$\mathcal{X}_k = f(\mathcal{X}_{k-1}) + \eta_k \tag{1}$$

$$y_k = h(\mathcal{X}_k) + v_k. \tag{2}$$

Here $\mathcal{X}_k \in \mathcal{R}^n$ represents the unknown state of the system to be estimated, and $y_k \in \mathcal{R}^q$ denotes the measurement at any time instant k. $f(\cdot)$ and $h(\cdot)$ are known non-linear functions of $\mathcal{X}_k$, and η_k, v_k are process and measurement noises respectively. These noises are assumed to be uncorrelated, Gaussian densities with zero mean and covariance Q_k and R_k, respectively.

The prior and posterior pdf in Gaussian sum approach are represented as weighted sum of individual Gaussian pdf [12–14]. That is, the prior pdf could be expressed as

$$p(\mathcal{X}_k|y_{k-1}) \approx \sum_{i=1}^{N_c} w^i_{k|k-1} \aleph(\mathcal{X}_k; \mu^i_{k|k-1}, P^i_{k|k-1}), \tag{3}$$

where N_c is the number of Gaussian components and w^i are their corresponding weights. The individual Gaussian densities $\aleph(\mathcal{X}_k; \mu^i_{k|k-1}, P^i_{k|k-1})$ can be generated with the help of quadrature filters. Here they are considered as proposal of GSF.

Now the posterior pdf can be expressed as

$$
\begin{aligned}
p(\mathcal{X}_k|y_k) &\propto p(y_k|\mathcal{X}_k)p(\mathcal{X}_k|y_{k-1}) \\
&\approx \sum_{i=1}^{N_c} w^i_{k|k} \aleph(\mathcal{X}_k; \mu^i_{k|k}, P^i_{k|k}).
\end{aligned}
\tag{4}
$$

Here the update of weights in the presence of measurements is achieved using the expression

$$
w^i_{k|k} = \frac{p(y_k|\mathcal{X}_k, i)w^i_{k|k-1}}{\sum_{i=1}^{N_c} p(y_k|\mathcal{X}_k, i)w^i_{k|k-1}}. \tag{5}
$$

The posterior means $\mu^i_{k|k}$ and covariances $P^i_{k|k}$ of ith Gaussian component can be determined with the help of various suboptimal filters. It can be noted from the weight equations mentioned above, the weights for various Gaussian components are evaluated only during the correction step. During time prediction, the weights calculated in the previous step are used.

3 Adaptive Gaussian Sum Quadrature Filtering

3.1 Adaptive Filtering

In this approach, a suitable weight update scheme is formulated by minimizing the integral square error in the true probability density function and its approximation as a mixture of Gaussian densities. So a cost function is formulated as

$$
\begin{aligned}
J &= \min_{w^i_{k|k-1}} \frac{1}{2} \int |p(\mathcal{X}_k|y_{k-1}) - \hat{p}(\mathcal{X}_k|y_{k-1})|^2 \, d\mathcal{X}_k, \\
&= \frac{1}{2} \int |p(\mathcal{X}_k|y_{k-1}) - \sum_{i=1}^{N_c} w^i_{k|k-1} \aleph(\mathcal{X}_k; \mu^i_{k|k-1}, P^i_{k|k-1})|^2 \, d\mathcal{X}_k.
\end{aligned}
$$

The above cost function can be written in the following form [15]

$$
J = \frac{1}{2} W^{\mathrm{T}}_{k|k-1} \mathbb{M} W_{k|k-1} - W^{\mathrm{T}}_{k|k-1} \mathbb{N} W_{k|k}, \tag{6}
$$

where $W_{k|k-1} = [w^1_{k|k-1}\ w^2_{k|k-1} \cdots w^{N_c}_{k|k-1}]^T$ is the prior weight matrix. The symmetric matrix $\mathbb{M} \in \mathcal{R}^{N_c \times N_c}$ can be expressed as

$$\begin{aligned}\mathbb{M}_{i,j} &= \int \aleph(\mathcal{X}_k; \mu^i_{k|k-1}, P^i_{k|k-1})\aleph(\mathcal{X}_k; \mu^j_{k|k-1}, P^j_{k|k-1})d\mathcal{X}_k \\ &= |2\pi(P^i_{k|k-1} + P^j_{k|k-1})|^{-1/2}\exp[-\frac{1}{2}(\mu^i_{k|k-1} - \mu^j_{k|k-1})^T \\ &(P^i_{k|k-1} + P^j_{k|k-1})^{-1}(\mu^i_{k|k-1} - \mu^j_{k|k-1})] \quad \text{for } i \neq j, \\ &= |4\pi(P^{i\ -1}_{k|k-1} + P^{j\ -1}_{k|k-1})^{-1}|^{-1/2} \quad \text{for } i = j.\end{aligned}$$

$\mathbb{N} \in \Re^{N_c \times N_c}$ is also a matrix which can be expressed as

$$\begin{aligned}\mathbb{N}_{ij} = \int & \aleph(f(\mathcal{X}_{k-1}); \mu^j_{k|k-1}, P^j_{k|k-1}) \\ & \times \aleph(\mathcal{X}_{k-1}; \mu^i_{k-1|k-1}, P^i_{k-1|k-1})\, \mathrm{d}\mathcal{X}_{k-1}.\end{aligned}$$

The entries in matrix $\mathbb{N}$ can be found using numerical methods.

Now, as we know that the weights must be positive and to represent a valid probability density function, the sum of all weights must be unity. Then the constrained optimization problem can be written as

$$\begin{aligned}&\min_{w^i_{k|k-1}} J = \frac{1}{2}W^T_{k|k-1}\mathbb{M}W_{k|k-1} - W^T_{k|k-1}\mathbb{N}W_{k|k} \\ &\text{s.t.} \sum_{i=1}^{N_c} w^i_{k|k-1} = 1 \\ &w^i_{k|k-1} \geq 0, \qquad i = 1, \cdots, N_c.\end{aligned} \tag{7}$$

4 Quadrature Filter Proposals

In this section, we briefly discuss two nonlinear filters which we intend to use as proposals of AGSF. Both the filters assume Gaussian distribution for the posterior and prior pdfs and represent them with points and weights. Both these filters were developed under the Bayesian framework and give suboptimal solution because of the intractable integrals in Bayesian recursion. Moreover, these filters rely on a unique numerical integration rule that approximate the intractable integral by using various quadrature points and their corresponding weights.

Consider an integral of any univariate function, $I = \int_{-\infty}^{\infty} f(\mathcal{X})e^{-\mathcal{X}^2}\mathrm{d}\mathcal{X}$. When the function is nonlinear, this integral is not tractable. Then a suitable approach to tackle this problem is to apply a numerical approximation technique. Suppose the integral can be evaluated numerically with N quadrature points $I \approx \sum_{j=1}^{N} f(q_j)\mathbf{w}'_j$, where q_j and $\mathbf{w}'_j$ are the jth quadrature point and its corresponding weight, respectively. Below we mention how CQKF and GHF provide a way for finding points and weights for multidimensional systems.

4.1 Cubature Quadrature Kalman Filter (CQKF)

In CQKF formulation, $2nn'$ quadrature points can be generated using the relation $\xi_j = \sqrt{2\epsilon_{i'}}[u]_i$, where $[u]_i$ are the cubature points located at the intersection of the unit hyper-sphere and its axes, and the quadrature points $\epsilon_{i'}$ are obtained from the roots of the following Chebyshev–Laguerre polynomial equation of order n'

$$L_{n'}^{\alpha}(\epsilon) = (-1)^{n'}\epsilon^{-\alpha}e^{\epsilon}\frac{\mathrm{d}^{n'}}{\mathrm{d}\epsilon^{n'}}\epsilon^{\alpha+n'}e^{-\epsilon} = 0,$$

where $\alpha = (n/2) - 1$, see [5] for more details. Moreover, their corresponding weights can be obtained from the relation

$$\mathbf{w}_j = \frac{n'!\Gamma(\alpha + n' + 1)}{2n\Gamma(n/2)\epsilon_{i'}[\dot{L}_{n'}^{\alpha}(\epsilon_{i'})]^2}.$$

4.2 Gauss–Hermite Filter

In GHF, the intractable integrals are numerically evaluated using the Gauss–Hermite quadrature rule. This approximation was first introduced in [6] for estimation problems. The quadrature points and their corresponding weights are calculated using a symmetric tridiagonal matrix J with $J_{j,j+1} = \sqrt{j/2}$; $1 \le j \le (N-1)$. The quadrature points are obtained as $q_j = \sqrt{2}\Psi_j$, where Ψ_j is the jth eigenvalue of matrix J. The jth weight $\mathbf{w}'_j$ can be defined as $\mathbf{w}'_j = k_{j1}^2$, where k_{j1} is the first element of the jth normalized eigenvector of J.

4.3 Proposed Algorithm

In this section, the AGSF algorithm with quadrature filter proposal is given.

Algorithm

- Initialize pdf as a sum of N_c Gaussian components with weight w^i: $p(\mathcal{X}_{k-1}|y_{k-1}) \approx \sum_{i=1}^{N_c} w^i_{k-1|k-1} \aleph(\mu^i_{k-1|k-1}, P^i_{k-1|k-1})$.
- Generate support points ξ_j and weights $\mathbf{w}_j$, $j = 1, \cdots, N_p$

Prediction step

- Evaluate quadrature points $\chi^i_{j,k-1|k-1} = \zeta^i_{k-1|k-1}\xi_j + \mu^i_{k-1|k-1}$, where $P^i_{k-1|k-1} = \zeta^i_{k-1|k-1}\zeta^{iT}_{k-1|k-1}$.
- Propagate quadrature points $\hat{\mathcal{X}}^i_{j,k|k-1} = f(\chi^i_{j,k-1|k-1})$.
- Find $\mu^i_{k|k-1}$ and $P^i_{k|k-1}$ using quadrature filters.
- Solve optimization problem (7) to find $w^i_{k|k-1}$.

Measurement update step

- Generate new quadrature points from $\mu^i_{k|k-1}$ and $P^i_{k|k-1}$.
- Measurement prediction $Y^i_{j,k|k-1} = h(\chi^i_{j,k|k-1})$.
- Predicted measurement $\hat{Y}^i_{k|k-1} = \sum_{j=1}^{N_p} \mathbf{w}_j Y^i_{j,k|k-1}$.
- Innovation covariance $P^i_{y,k|k-1} = \sum_{j=1}^{N_p} \mathbf{w}_j (Y^i_{j,k|k-1} - \hat{Y}^i_{k|k-1})(Y^i_{j,k|k-1} - \hat{Y}^i_{k|k-1})^T + R_k$.
- Cross-covariance $P^i_{\mathcal{X}y,k|k-1} = \sum_{j=1}^{N_p} \mathbf{w}_j (\chi^i_{j,k|k-1} - \mu^i_{k|k-1})(Y^i_{j,k|k-1} - \hat{Y}^i_{k|k-1})^T$.
- Kalman gain $K^i_k = P^i_{\mathcal{X}y,k|k-1} P^{i\,-1}_{y,k|k-1}$.
- Find $\mu^i_{k|k}$ and $P^i_{k|k}$.
- Determine weights $w^i_{k|k}$ using (5).
- Posterior point estimate is calculated using (4).

5 Ballistic Target Tracking Problem Formulation

Tracking of a ballistic target and its interception in three-dimensional space is discussed in this section. The measurements are assumed to be generated by an active seeker present in the interceptor missile. A six degrees of freedom state-dependent coefficient model [16–18] is considered. The advantage of this model is that the target accelerations are replaced in the state vector by a parameter, called the ballistic coefficient (β). After estimating β, target acceleration components are calculated.

5.1 Assumed Target Model

Let x^t, y^t, z^t be the positions v^{tx}, v^{ty}, v^{tz} be the velocities, and a^{tx}, a^{ty}, a^{tz} the accelerations of the target in three dimensions. Similarly, for the interceptor, x^m, y^m, z^m are the positions, v^{mx}, v^{my}, v^{mz} the velocities, and a^{mx}, a^{my}, a^{mz}, are the accelerations. Then the relative states (only position and velocity) can be defined as

$$\Delta x = x^t - x^m, \quad \Delta y = y^t - y^m, \quad \Delta z = z^t - z^m,$$
$$\Delta v^x = v^{tx} - v^{mx}, \ \Delta v^y = v^{ty} - v^{my}, \ \Delta v^z = v^{tz} - v^{mz}.$$

Then the rate of change of these states can be mentioned as [16]

$$\begin{aligned} &\dot{\Delta x} = \Delta v^x, \ \dot{\Delta y} = \Delta v^y, \ \dot{\Delta z} = \Delta v^z, \ \dot{\Delta v^x} = a^{tx} - a^{mx}, \\ &\dot{\Delta v^y} = a^{ty} - a^{my}, \ \dot{\Delta v^z} = a^{tz} - a^{mz}, \ \dot{1/\beta} = 0, \end{aligned} \tag{8}$$

where g is the acceleration due to gravity. Here it is considered to be a constant irrespective of the altitude.

By discretising Eq. (8) and incorporating various modeling errors, we get

$$\mathcal{X}_{k+1} = F_k \mathcal{X}_k - G B_k + \eta_k, \tag{9}$$

where $\mathcal{X}_k = [\Delta x_k \ \ \Delta y_k \ \ \Delta z_k \ \ \Delta v_k^x \ \ \Delta v_k^y \ \ \Delta v_k^z \ \ 1/\beta_k]^T$, $\eta_k \sim \aleph(0, Q_{k-1})$,

$$F_k = \begin{bmatrix} I_{3\times3} & TI_{3\times3} & 0_{3\times1} \\ 0_{3\times3} & I_{3\times3} & [f_1\, f_2\, f_3]^T \\ 0_{1\times3} & 0_{1\times3} & 1 \end{bmatrix},$$

$G = [T^2/2I_{3\times3}\ TI_{3\times3}\ 0_{1\times3}]^T$ and $B = [a_k^{mx} + g\ a_k^{my}\ a_k^{mz}]^T$, with $f_1 = -0.5\rho_k v_k^t v_k^{tx} T$, $f_2 = -0.5\rho_k v_k^t v_k^{ty} T$ and $f_3 = -0.5\rho_k v_k^t v_k^{tz} T$. Here the resultant or overall target velocity is v^t and T is the sampling time. ρ is the air density which is expressed as

$$\begin{aligned} \rho &= 0.36392 \exp(-0.0001516584(x^t - 11000)) \ x^t > 11 \text{ km} \\ &= 1.225(1 - 0.000022557695x^t)^{4.25587} \qquad x^t < 11 \text{ km}. \end{aligned} \tag{10}$$

It can be inferred from the above relation that ρ depends on the altitude of the target.

5.2 Seeker Measurement Model

The measurements generated by the seeker are range rate, gimbal angles, and line of sight (LOS) rates along yaw and pitch plane. Here the range rate is calculated in line of sight frame, whereas the gimbal angles are measured along fin-dash frame.

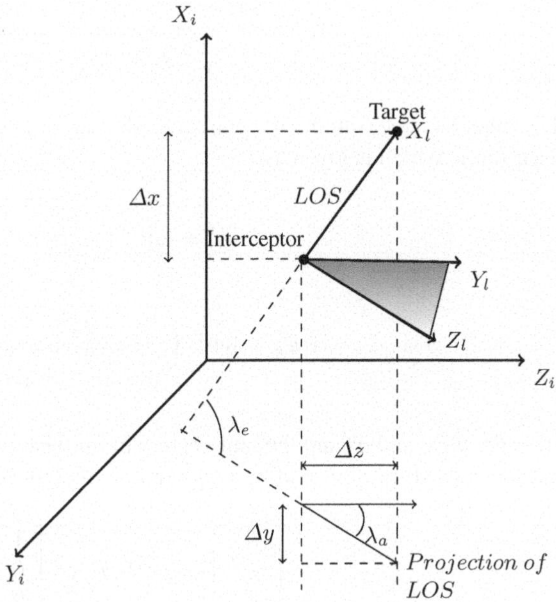

Fig. 2 Line of sight frame with respect to inertial frame

The remaining two measurements, the line of sight rates, are measured in the inner gimbal frame.

In Fig. 2, line of sight frame of reference is represented with X_l, Y_l, and Z_l-axes and X_i, Y_i, and Z_i-axes represent the inertial frame of reference. Here the azimuth angle is represented by λ_a and elevation by λ_e. They are expressed as [16, 17]

$$\lambda_{a,k} = \tan^{-1}\left(\frac{\Delta y_k}{\Delta z_k}\right) \tag{11}$$

$$\lambda_{e,k} = \tan^{-1}\left(\frac{\Delta x_k}{\sqrt{\Delta y_k^2 + \Delta z_k^2}}\right). \tag{12}$$

Let the rate of change of $\lambda_{a,k}$ and $\lambda_{e,k}$ with respect to time be

$$\dot{\lambda}_{a,k} = \frac{\Delta z_k \Delta v_k^y - \Delta y_k \Delta v_k^z}{\Delta y_k^2 + \Delta z_k^2} \quad \text{and} \tag{13}$$

$$\dot{\lambda}_{e,k} = \frac{\Delta v_k^x(\Delta y_k^2 + \Delta z_k^2) - \Delta x_k(\Delta y_k \Delta v_k^y + \Delta z_k \Delta v_k^z)}{(\Delta x_k^2 + \Delta y_k^2 + \Delta z_k^2)\sqrt{\Delta y_k^2 + \Delta z_k^2}}. \tag{14}$$

The rate of change of the range with respect to time, which is a measurement, is expressed as

$$\dot{r}_k = \frac{\Delta x_k \Delta v_k^x + \Delta y_k \Delta v_k^y + \Delta z_k \Delta v_k^z}{\sqrt{\Delta x_k^2 + \Delta y_k^2 + \Delta z_k^2}}. \tag{15}$$

The other two measurements, gimbal angles along pitch and yaw, calculated in fin-dash frame which is given as

$$\phi_k^y = \tan^{-1}\left(\frac{-n_k^{\mathrm{fd}}}{l_k^{\mathrm{fd}}}\right) \tag{16}$$

$$\phi_k^z = \tan^{-1}\left(\frac{m_k^{\mathrm{fd}}}{\sqrt{(l_k^{\mathrm{fd}})^2 + (n_k^{\mathrm{fd}})^2}}\right). \tag{17}$$

Here l_k^{fd}, m_k^{fd}, and n_k^{fd} are the unit vectors in fin-dash frame. The following transformation is used for representing range in line of sight frame to fin-dash frame

$$\begin{bmatrix} l_k^{\mathrm{fd}} \\ m_k^{\mathrm{fd}} \\ n_k^{\mathrm{fd}} \end{bmatrix} = D_{\mathrm{f}}^{\mathrm{fd}} D_{\mathrm{b}}^{\mathrm{f}} D_{\mathrm{i}}^{\mathrm{b}} D_{\mathrm{l}}^{\mathrm{i}} \begin{bmatrix} 1 \\ 0 \\ 0 \end{bmatrix}.$$

Here $D_{\mathrm{l}}^{\mathrm{i}}$, $D_{\mathrm{i}}^{\mathrm{b}}$, $D_{\mathrm{b}}^{\mathrm{f}}$, and $D_{\mathrm{f}}^{\mathrm{fd}}$ are the direction cosine matrices (DCM), which are defined as

$$D_{\mathrm{l}}^{\mathrm{i}} = \begin{bmatrix} \sin\lambda_{\mathrm{e}} & 0 & -\cos\lambda_{\mathrm{e}} \\ \cos\lambda_{\mathrm{e}}\sin\lambda_{\mathrm{a}} & \cos\lambda_{\mathrm{a}} & \sin\lambda_{\mathrm{e}}\sin\lambda_{\mathrm{a}} \\ \cos\lambda_{\mathrm{e}}\cos\lambda_{\mathrm{a}} & -\sin\lambda_{\mathrm{a}} & \sin\lambda_{\mathrm{e}}\cos\lambda_{\mathrm{a}} \end{bmatrix},$$

$$D_b^{\mathrm{f}} = \begin{bmatrix} 1 & 0 & 0 \\ 0 & \frac{1}{\sqrt{2}} & \frac{1}{\sqrt{2}} \\ 0 & -\frac{1}{\sqrt{2}} & \frac{1}{\sqrt{2}} \end{bmatrix}$$

and

$$D_{\mathrm{f}}^{\mathrm{fd}} = \begin{bmatrix} 1 & 0 & 0 \\ 0 & -1 & 0 \\ 0 & 0 & -1 \end{bmatrix}.$$

DCM $D_{\mathrm{i}}^{\mathrm{b}}$ is attained using the body frame quaternion with respect to the inertial frame. Defining the quaternion vector as $(\hat{q}, q_0) = q_1 + q_2 i + q_3 j + q_4 k$ gives

$$D_{\mathrm{i}}^{\mathrm{b}} = \begin{bmatrix} q_1^2 + q_2^2 - q_3^2 - q_4^2 & 2(q_2q_3 + q_4q_1) & 2(q_2q_4 - q_3q_1) \\ 2(q_2q_3 - q_4q_1) & q_1^2 - q_2^2 + q_3^2 - q_4^2 & 2(q_3q_4 + q_2q_1) \\ 2(q_2q_4 + q_3q_1) & 2(q_3q_4 - q_2q_1) & q_1^2 - q_2^2 - q_3^2 + q_4^2 \end{bmatrix}.$$

Now for the other two measurements, line of sight rates along roll, yaw, and pitch plane in inner gimbal frame is represented as

$$\begin{bmatrix} \Omega_{x,k} \\ \Omega_{y,k} \\ \Omega_{z,k} \end{bmatrix} = D^{\mathrm{g}}_{fd} D^{\mathrm{fd}}_{\mathrm{f}} D^{\mathrm{f}}_{\mathrm{b}} D^{\mathrm{b}}_{\mathrm{i}} D^{\mathrm{i}}_{\mathrm{l}} \begin{bmatrix} -\dot{\lambda}_{\mathrm{a},k} \sin \lambda_{\mathrm{e},k} \\ \dot{\lambda}_{\mathrm{e},k} \\ \dot{\lambda}_{\mathrm{a},k} \cos \lambda_{\mathrm{e},k} \end{bmatrix}.$$

Here $D^{\mathrm{g}}_{\mathrm{fd}}$ represents the direction cosine matrix to shift from fin-dash to inner gimbal frame where

$$D^{\mathrm{g}}_{\mathrm{fd}} = \begin{bmatrix} \cos\phi_y \cos\phi_y & \sin\phi_z & -\cos\phi_y \sin\phi_y \\ -\cos\phi_y \sin\phi_z & \cos\phi_z & \sin\phi_y \sin\phi_z \\ \sin\phi_y & 0 & \cos\phi_y \end{bmatrix}.$$

Hence, the complete set of noise corrupted seeker measurements can be represented as:

$$\dot{r}_k = \frac{\Delta x_k \Delta V_{x,k} + \Delta y_k \Delta V_{y,k} + \Delta z_k \Delta V_{z,k}}{\sqrt{\Delta x_k^2 + \Delta y_k^2 + \Delta z_k^2}} + \mathrm{v}_{1,k}\,,$$

$$\phi_{y,k} = \tan^{-1}\left(\frac{-n_{\mathrm{fd},k}}{l_{\mathrm{fd},k}}\right) + \mathrm{v}_{2,k}\,,$$

$$\phi_{z,k} = \tan^{-1}\left(\frac{m_{\mathrm{fd},k}}{\sqrt{l^2_{\mathrm{fd},k} + n^2_{\mathrm{fd},k}}}\right) + \mathrm{v}_{3,k}\,, \tag{18}$$

$$\begin{bmatrix} \Omega_{y,k} \\ \Omega_{z,k} \end{bmatrix} = \begin{bmatrix} 0 & 1 & 0 \\ 0 & 0 & 1 \end{bmatrix} \begin{bmatrix} \Omega_{x,k} \\ \Omega_{y,k} \\ \Omega_{z,k} \end{bmatrix} + \begin{bmatrix} \mathrm{v}_{4,k} \\ \mathrm{v}_{5,k} \end{bmatrix},$$

where $\mathrm{v}_k \sim \aleph(0, R)$.

5.3 Acceleration Model for the Target

Accurate model for the target acceleration is obtained by incorporating drag force and gravity [18]. Since the target is assumed to be freely falling, acceleration could be represented in terms of drag force a_{drag} as $a_{\mathrm{tx}} = a^{\mathrm{tx}}_{\mathrm{drag}} - g$, $a_{\mathrm{ty}} = a^{\mathrm{ty}}_{\mathrm{drag}}$, $a_{\mathrm{tz}} = a^{\mathrm{tz}}_{\mathrm{drag}}$.

The drag force magnitude is defined in terms of the dynamic pressure q, the cross-sectional area S^{t}, and the zero lift drag C^{d} as $D = qS^{\mathrm{t}}C^{\mathrm{d}}$, where $q = \frac{1}{2}\rho v^{t\,2}$. Then the ballistic coefficient $\beta = \frac{m^{\mathrm{t}}}{S^r C^{\mathrm{d}}}$, where m^{t} is the target's mass. Then drag force could be expressed as $D = \frac{\frac{1}{2}\rho v^{t\,2}}{\beta}$.

Now the target acceleration is defined as [16]

$$a_k^{\mathrm{tx}} = \frac{-0.5\rho_k v_k^{\mathrm{t}} v_k^{\mathrm{tx}}}{\beta_k} - g, \quad a_k^{\mathrm{ty}} = \frac{-0.5\rho_k v_k^{\mathrm{t}} v_k^{\mathrm{ty}}}{\beta_k}, \quad a_k^{\mathrm{tz}} = \frac{-0.5\rho_k v_k^{\mathrm{t}} v_k^{\mathrm{tz}}}{\beta_k}, \tag{19}$$

where $v_k^{\mathrm{t}} = \sqrt{(v_k^{\mathrm{tx}})^2 + (v_k^{\mathrm{ty}})^2 + (v_k^{\mathrm{tz}})^2}$ is the resultant target velocity. Here $v_k^{\mathrm{tx}} = \Delta v_k^x + v_k^{\mathrm{mx}}$, $v_k^{\mathrm{ty}} = \Delta v_k^y + v_k^{\mathrm{my}}$, and $v_k^{\mathrm{tz}} = \Delta v_k^z + v_k^{\mathrm{mz}}$, where Δv_k^x, Δv_k^y and Δv_k^z are the estimates and v_k^{mx}, v_k^{my} and v_k^{mz} are known, since they are velocities of the interceptor.

6 Simulation Results

The performance of the proposed filtering algorithms combined with PNG law is evaluated for the target–interceptor dynamics discussed in the above section. The performance comparison is done by calculating the RMSE of states and miss-distance, averaged over several Monte Carlo runs. All the simulation results were done on MATLAB software.

The initial state for the interceptor is $\mathcal{X}_{0|0}^{\mathrm{m}} = [8.69\,\mathrm{km}\ 9.46\,\mathrm{km}\ 6.34\,\mathrm{km}\ 800\,\mathrm{m/s}\ 363\,\mathrm{m/s}\ 418\,\mathrm{m/s}\ -7.84\,\mathrm{m/s^2}\ -60.31\,\mathrm{m/s^2}\ -23.27\,\mathrm{m/s^2}]^{\mathrm{T}}$. PNG law use these values in combination with the estimates attained from the filter for generating the interceptor states for the upcoming time instant. A discussion on PNG law is not included, and we suggest to refer [19], for its implementation in three-dimensional space. The initial estimate fed to the filter is assumed as $\aleph(\mathcal{X}_{0|0}, P_{0|0})$, whose values are mentioned in Table 1. From the estimated relative state, target state can be found out according to the relation $\mathcal{X}_k^{\mathrm{t}} = \mathcal{X}_k + \mathcal{X}_k^{\mathrm{m}}$. The measurement data is generated

Table 1 Values for $\mathcal{X}_{0|0}$ and $P_{0|0}$

States	$\mathcal{X}_{0\|0}$	$P_{0\|0}$
Δx	14, 296 m	100 m^2
Δy	846 m	100 m^2
Δz	4280 m	100 m^2
Δv_x	1946 m/s	10 m/s^2
Δv_y	262 m/s	10 m/s^2
Δv_z	662 m/s	10 m/s^2
$1/\beta$	1/3000	8/10^5

according to the measurement model defined. The covariance matrix for the process is $Q = \text{diag}(0.02\ 0.02\ 0.02\ 0.02\ 0.02\ 0.02\ 4 \times 10^{-11})$ and for the measurement is $R = \text{diag}((15\,\text{m})^2\ (0.3°)^2)\ (0.3°)^2(5°/\text{s})^2\ (5°/\text{s})^2)$. The relatively small covariance matrix Q shows the confidence in the assumed target model.

Here AGSF and GSF with EKF, UKF, CKF, CQKF, and GHF proposals have been implemented. To define the accuracy levels of CQKF and GHF, $n' = N = 3$ is considered. The total number of quadrature point requirement is also listed in Table 2. The number of quadrature points required for UKF, CKF, and CQKF are less when compared to GHF and comparable. To implement the Gaussian sum approach, the initial pdf is split into a total of 3 components, where each Gaussian density is realized by the filters mentioned. To define the weights for individual Gaussian densities, for realizing the AGS formulation, the quadratic optimization problem is solved. The measurements are attained at a constant time interval of $T = 0.025$ s. When the target and interceptor cross each other (this is inferred from the change of direction of the closing velocity $V_{c,k}$), the simulation is considered to be over. Here it lasted for about 7.25 s. Figure 6 illustrates the trajectory of target and interceptor in three dimensions. RMS range, the distance between the target and interceptor at each time step averaged over MC runs, is plotted in Fig. 7. This indicates that the interceptor missile was always approaching the target. However, the end error, which is termed as miss-distance, needs to be evaluated for all the nonlinear filters clubbed with the PNG law. RMSE in resultant target velocity v_k^t is plotted in Fig. 8. It can be inferred that it converges for all the filters implemented. As the interceptor's own velocity is known to the tracking computer, and relative velocities are computed by the estimator, the target velocity could be calculated. RMSE in $1/\beta$ is evaluated for 50 Monte Carlo runs and is shown in Fig. 9. From the figure, it could be noticed that $1/\beta$ converges to its true value before 20% of the total simulation time, or the homing guidance phase. This quick, and accurate estimation of $1/\beta$ help the guidance and the associated control to adjust with the initial uncertainties.

Target accelerations in all the three dimensions are calculated using estimates of $1/\beta$ and target velocity, according to Eq. (19). RMSEs of target accelerations in all the three dimensions are given in Figs. 3, 4, and 5. However, from these figures we shall not claim superiority of one filter over the other. Miss-distance incurred by all the proposed filters combined with PNG law, averaged over 50 Monte Carlo runs, is given Table 2. From this table, it can be inferred that the quadrature filters incurred miss-distance's which is less than that incurred by the EKF, around 20 m. Gaussian sum formulation with the quadrature filter proposals provided relatively higher accuracy with less miss-distance, around 16 m when compared to the ordinary Gaussian filters. Moreover, this is less than the miss-distance given by GS-EKF, which is around 22 m. Now coming to algorithms in AGS formulations, it can be understood that the miss-distance incurred is further less, when compared to the Gaussian sum and Gaussian filters. Also, it shall be noted that the proposed AGS formulations gave highly accurate estimates with miss-distance's around 13 m, when compared to the AGS-EKF.

Table 2 also indicates the relative computational time for different filters. It can be inferred that the slowest filter is AGS formulation with GHF proposal and the

Table 2 Miss-distance, number of points, and relative computational time for various filters

Filters	Miss-distance (m)	No. of points	Relative computational time
AGS-GHF	13.41	2187	1
GS-GHF	16.43		0.73
GHF	20.61		0.33
AGS-CQKF	13.52	42	0.10
GS-CQKF	16.13		0.017
CQKF	20.36		0.007
AGS-CKF	13.92	14	0.075
GS-CKF	16.35		0.008
CKF	20.52		0.003
AGS-UKF	13.85	15	0.082
GS-UKF	16.06		0.006
UKF	20.39		0.002
AGS-EKF	18.34		0.077
GS-EKF	21.54		0.004
EKF	22.26		0.001

fastest one is EKF. Execution time of all the algorithms based on AGS formulation, other than with the GHF proposal, is comparable. Since AGSF with UKF, CKF, and CQKF proposal performed with higher accuracy and moderate computational time, any of these filters in closed loop with PNG law could prove to be a better alternative for the problem considered.

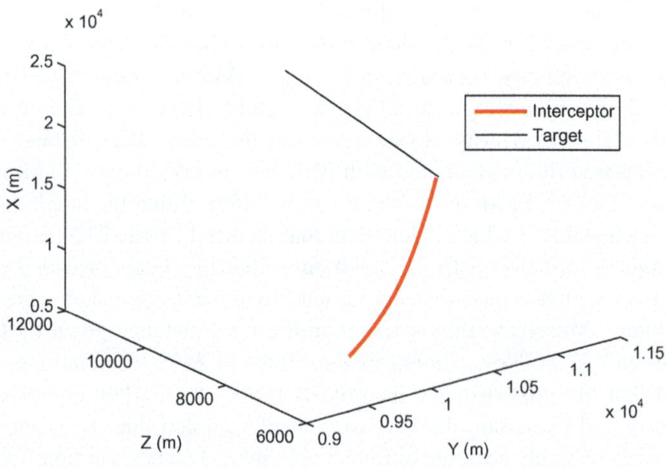

Fig. 3 3D Target-interceptor engagement scenario

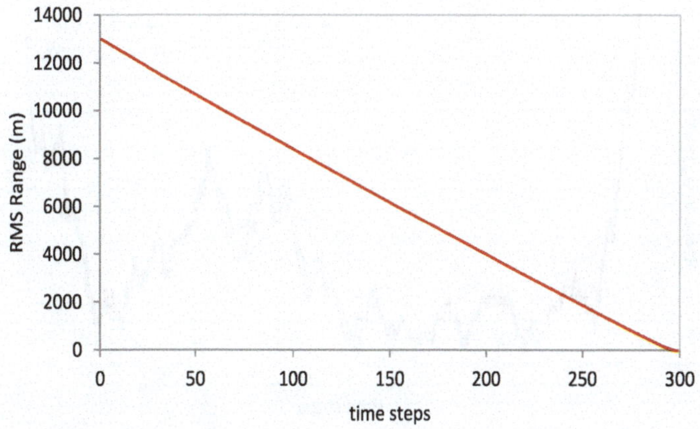

Fig. 4 RMS range between target and interceptor

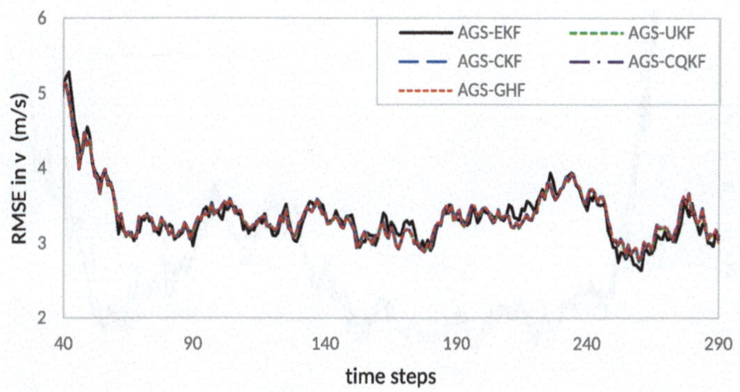

Fig. 5 RMSE in resultant target velocity

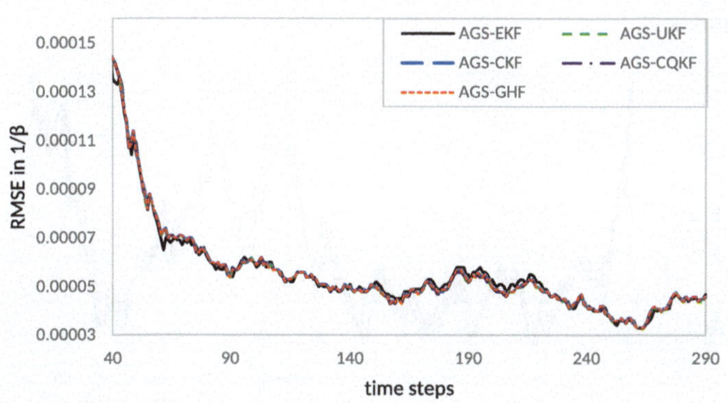

Fig. 6 RMSE in $1/\beta$

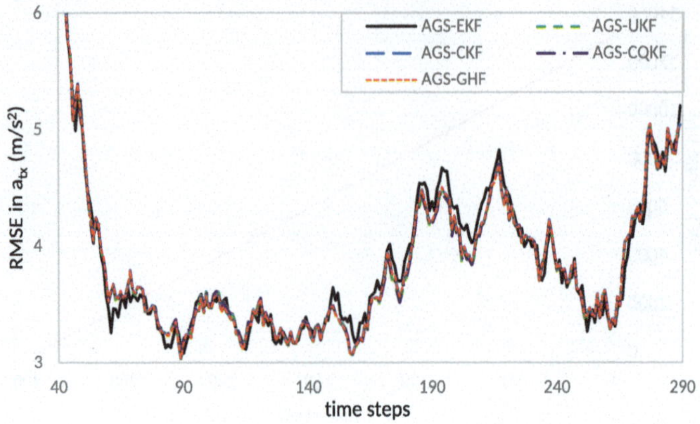

Fig. 7 RMSE in a_{tx}

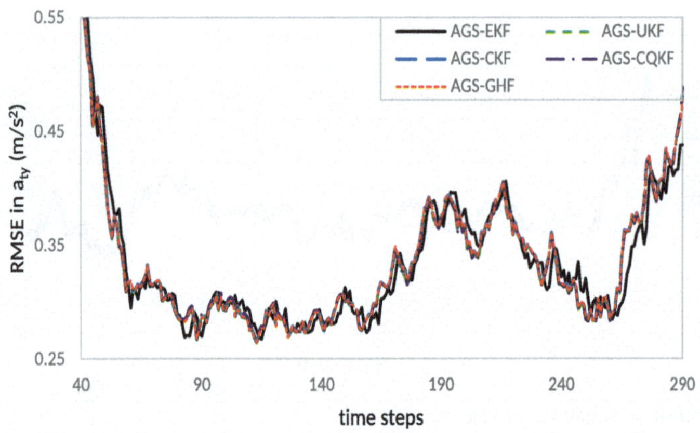

Fig. 8 RMSE in a_{ty}

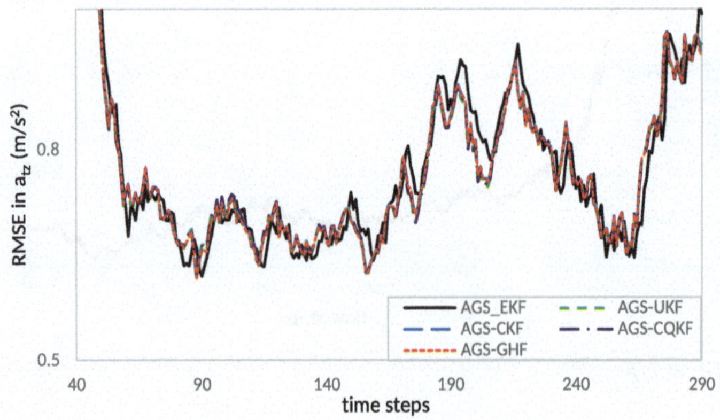

Fig. 9 RMSE in a_{tz}

7 Conclusion

In this work, AGSF with UKF, CQKF, and GHF proposal is formulated where the prior and posterior probability densities are estimated as a weighted sum of Gaussian densities. This framework incorporates a weight update scheme for the individual Gaussian densities by minimizing the square error between the true predicted pdf and its GS approximation. The proposed filters are termed as AGS-UKF, AGS-CQKF and AGS-GHF. These filters were applied to track a ballistic target and intercept it on reentry. A state-dependent coefficient (SDC) model was used for realizing the target–interceptor dynamics, and the performance of the estimators was studied in closed loop with the PNG law. From the results, it can be inferred that the AGSF-based quadrature filters performed with comparatively higher accuracy than the Gaussian sum filters. The relative computational time and the averaged miss-distance for all the filters have been calculated. It can be observed that AGSF-based quadrature filters are more accurate with all the filters giving acceptable miss-distance. Hence, a combination of AGS-UKF, AGS-CKF, and AGS-CQKF with PNG law for accurately tracking and intercepting a ballistic target on reentry may be considered over the existing techniques.

References

1. Bar-Shalom, Y., Li, X.R., Kirubarajan, T.: Estimation with Application to Tracking and Navigation: Theory Algorithms and Software. Wiley (2004)
2. Julier, S., Uhlmann, J., Durrant-Whyte, H.F.: A new method for the nonlinear transformation of means and covariances in filters and estimators. IEEE Transa. Autom. Control **45**(3), 477–482 (2000)
3. Radhakrishnan, R., Yadav, A., Date, P., Bhaumik, S.: A new method for generating sigma points and weights for nonlinear filtering. IEEE Control Syst. Lett. **2**(3), 519–524 (2018)
4. Arasaratnam, I., Haykin, S.: Cubature Kalman filters. IEEE Trans. Autom. Control **54**(6), 1254–1269 (2009)
5. Bhaumik, S., Swati.: Cubature quadrature Kalman filter. IET Signal Process. **7**(7), 533–541 (2013)
6. Ito, K., Xiong, K.: Gaussian filters for nonlinear filtering problems. IEEE Trans. Autom. Control **45**(5), 910–927 (2000)
7. Alspach, D., Sorenson, H.: Nonlinear Bayesian estimation using Gaussian sum approximations. IEEE Trans. Autom. Control **17**(4), 439–448 (1972)
8. Terejanu, G., Singla, P., Singh, T., Scott, P.D.: Adaptive Gaussian sum filter for nonlinear Bayesian estimation. IEEE Trans. Autom. Control **56**(9), 2151–2156 (2011)
9. Mehra, R.: A comparison of several nonlinear filters for reentry vehicle tracking. IEEE Trans. Autom. Control **16**(4), 307–319 (1971)
10. Siouris, G.M., Chen, G., Wang, J.: Tracking an incoming ballistic missile using an extended Kalman filter. IEEE Trans. Aerosp. Electron. Syst. **33**(1), 232–240 (1997)
11. Dwivedi, P.N., Bhale, P., Bhattacharyya, A., Padhi, R.: Generalized Estimation and Predictive Guidance for Evasive Targets. IEEE Trans. Aerosp. Electron. Syst. **52**(5), 2111–2122 (2016)
12. Leong, P.H., Arulampalam, S., Lamahewa, T.A.: A Gaussian sum based cubature Kalman filter for bearings-only tracking. IEEE Trans. Aerosp. Electron. Syst. **49**(2), 1161–1176 (2013)

13. Kotecha, J.H., Djuric, P.M.: Gaussian sum particle filtering. IEEE Trans. Signal Process. **51**(10), 2602–2612 (2003)
14. Radhakrishnan, R., Bhaumik, S., Tomar, N.K.: Gaussian sum shifted Rayleigh filter for underwater bearings-only target tracking problems. IEEE J. Ocean. Eng. **44**(2), 492–501 (2019)
15. Terejanu, G., Singla, P., Singh, T.: Uncertainty propagation for nonlinear dynamic systems using Gaussian mixture models. J. Guid. Control Dyn. **31**(6), 1623–1633 (2008)
16. Dwivedi, P., Bhale, P., Bhattacharyya, A.: Quick and accurate state estimation of RV from RF seeker measurements using EKF. In: Proceedings of AIAA Guidance Navigation and Control Conference, AIAA (2006)
17. Radhakrishnan, R., Bhaumik, S., Tomar, N.K.: Continuous-discrete quadrature filters for intercepting a ballistic target on reentry using seeker measurements. IFAC-PapersOnLine **51**(1), 383–388 (2018)
18. Li, X.R., Jilkov, V.P.: A survey of manoeuvring target tracking-Part II: Ballistic target models. In: Proceedings of SPIE Conference on Signal and Data Processing of Small Targets, SPIE (2001)
19. Zarchan, P.: Tactical and Strategic Missile Guidance, 6th edn. AIAA (2012)

Three-Dimensional Bearings-Only Target Tracking: Comparison of Few Sigma Point Kalman Filters

U. Asfia, R. Radhakrishnan, and S. N. Sharma

Abstract Three-dimensional bearings-only target tracking problem is considered in this work. To solve this problem, highly accurate nonlinear filtering algorithms are required. This is because the nonlinearities in the noise corrupted measurements, combined with the other uncertainties, make the estimation problem highly challenging. In this work, the extended Kalman filter (EKF), the cubature Kalman filter (CKF), the unscented Kalman filter (UKF), and the new sigma point Kalman filter (NSKF) are used to solve this state estimation problem. The performance criteria chosen are root mean square error (RMSE) in position and velocity. To have a more meaningful study, RMSE at the end of the observation period and time-averaged RMSE after the observer manoeuvre is calculated, for varying measurement noise and initial uncertainty.

1 Introduction

The bearings-only tracking (BOT) problem has been a popular area of research for a sufficient period of time. In particular, it is a nonlinear filtering problem with a goal to estimate the target states (positions and velocities) using only bearing measurements that are corrupted with noise [1]. These measurements are obtained by a moving platform, also called as observer, with the help of one or more sensors, whereas the target can be a ship, an aircraft, or any mobile platform. The problem of target tracking, which considers only bearing angles as measurements, is sometimes referred to as bearings-only target motion analysis (TMA). The nonlinearities in the measurement equation and the associated noises only make the problem even

U. Asfia (✉) · R. Radhakrishnan · S. N. Sharma
Department of Electrical Engineering, SV National Institute of Technology Surat, Surat, India
e-mail: ds19el006@eed.svnit.ac.in

R. Radhakrishnan
e-mail: r.rahul@eed.svnit.ac.in

S. N. Sharma
e-mail: sns@eed.svnit.ac.in

J. Gu et al. (eds.), *Communication and Control for Robotic Systems*, Smart Innovation, Systems and Technologies 229, https://doi.org/10.1007/978-981-16-1777-5_17

more challenging. Another significant factor is the observability. This problem is not observable, or the filtering solution does not converge until and unless the observer makes a manoeuvre [2].

Significant research has been conducted for BOT problem formulated in two-dimensional space [3, 4]. Since the measurement is nonlinear, the assumption that prior and posterior probability densities follow Gaussian distribution may not be true. Hence, for 2D BOT problems, nonlinear filters with both Gaussian as well as non-Gaussian assumption have been implemented for the solution. For nonlinear systems, the most widely used estimator is the extended Kalman filter (EKF) [5]. The EKF uses the traditional Kalman filter equations and applies it to the nonlinear system by linearising the nonlinear model [6]. In doing so, EKF proves to be computationally efficient over other advanced Kalman filters. Later, various deterministic sampling point filters such as the unscented Kalman filter (UKF) [7], cubature Kalman filter (CKF) [8], cubature quadrature Kalman filter (CQKF) [9], Gauss–Hermite filter (GHF) [10], etc., have been implemented. All these filters assume Gaussianity for the prior and posterior density functions. Filtering accuracy and their performance comparison are widely reported in the literature [11].

Now coming to the filters that follow non-Gaussian assumption, we have the particle filter (PF) [12] and algorithms that make use of Gaussian sum approach [13]. Most of the nonlinear filters mentioned above were implemented in the Gaussian sum approach and evaluated for solving BOT problems [11] In the majority of scenarios described for the study of 2D BOT problems, it was observed that there was significant track-loss associated with the estimator performance. Hence, a track-loss condition was formulated and implemented so that the most accurate estimated values were considered for performance analysis. It was noted that without this condition, certain filters such as EKF diverged or performed with very less accuracy [4]. All these works focused on tracking a target that is assumed to be following a constant velocity motion. There are also numerous works available for tracking a manoeuvring target with bearings-only measurements, in 2D [4].

A pseudolinear estimator for 2D case was introduced [14] to deal with the nonlinearity and convergence problem directly. This approach reformulates the nonlinear bearing measurement into a linear form. But the disadvantage was that the correlation between the new form and the bearing noise leads to bias problems [15, 16]. To overcome this bias problem, some attempts were made, which resulted in the proposal of iterative instrumental variable estimator (IV) [17], and also a bias compensated IV [18]. Now, these were all recursive algorithms. With respect to IV, a batch iterative algorithm was proposed, termed as modified IV estimator [19]. Hence, from this discussion, it is obvious that BOT problem in 2D scenarios is widely studied and addressed.

However, there are relatively less number of works which addresses the BOT problem in 3D geometry [20–23]. Recently, much focus is on this problem, as researchers have decided to not to ignore the depth, or the height between the target and observer. So, the basic difference here is the addition of one more measurement, the elevation angle. Some of the nonlinear filters which were used for solving BOT problem have already been applied for 3D geometry [24]. In [25], in addition to the problem formulation in cartesian coordinates, tracking algorithms were developed for modified spherical coordinates too. Robust tracking algorithms were also developed for 3D BOT problems [26]. Similar to the PLE and IV estimators developed for 2D geometry, the same concepts with certain modifications were applied to problem formulation in 3D geometry [27].

Here, the authors focus their work on solving 3D BOT problem formulated in Cartesian coordinate system. A comparative study on the performance evaluation of certain sigma point Kalman filters was done. NSKF [28] was a recent addition to the sigma point Kalman filter family, which claims more accuracy than the conventional UKF. Rather than assigning same weights to all the sigma points, NSKF assigns more weights to the sigma points that are nearer to the mean value. The inherent disadvantage in this formulation is the increase in the number of sigma points.

This paper deals with a 3D bearings-only target tracking scenario for which the performance of various filters like EKF, UKF, CKF, and NSKF is studied. The filtering accuracy is analysed using RMSE of resultant position and velocity. To further analyse the accuracy, the effect of initial condition and varying measurement noise is taken into account. The rest of the paper is organised as follows: Sect. 2 formulates the 3D BOT problem in Cartesian coordinates, and the filtering algorithms are mentioned in Sect. 3. The simulation results are discussed in Sect. 4 followed by concluding remarks in Sect. 5.

2 Problem Formulation

The purpose of the 3D bearings-only tracking problem studied in this paper is to accurately estimate the position and velocity of the target from bearing (β) and elevation (ϵ) measurements obtained by an observer at fixed-time intervals. The target motion is assumed to be a constant velocity motion during the observation period. We assume that the observer's motion is deterministic, with known position and velocity. Figure 1 is a Cartesian coordinate frame representing the position of the target and observer.

Let the target and observer state vector be defined as

$$X_k^{\mathrm{t}} = \left[x_k^{\mathrm{t}}\ y_k^{\mathrm{t}}\ z_k^{\mathrm{t}}\ \dot{x}_k^{\mathrm{t}}\ \dot{y}_k^{\mathrm{t}}\ \dot{z}_k^{\mathrm{t}}\right]^{\mathrm{T}}$$
$$X_k^{\mathrm{o}} = \left[x_k^{\mathrm{o}}\ y_k^{\mathrm{o}}\ z_k^{\mathrm{o}}\ \dot{x}_k^{\mathrm{o}}\ \dot{y}_k^{\mathrm{o}}\ \dot{z}_k^{\mathrm{o}}\right]^{\mathrm{T}}.$$

Now, a relative state vector is defined as

$$\begin{aligned} \mathcal{X}_k &= X_k^{\mathrm{t}} - X_k^{\mathrm{o}} \\ &= \left[x_k^{\mathrm{t}} - x_k^{\mathrm{o}}\ y_k^{\mathrm{t}} - y_k^{\mathrm{o}}\ z_k^{\mathrm{t}} - z_k^{\mathrm{o}}\ \dot{x}_k^{\mathrm{t}} - \dot{x}_k^{\mathrm{o}}\ \dot{y}_k^{\mathrm{t}} - \dot{y}_k^{\mathrm{o}}\ \dot{z}_k^{\mathrm{t}} - \dot{z}_k^{\mathrm{o}}\right]^{\mathrm{t}} \\ &= \left[x_k\ y_k\ z_k\ \dot{x}_k\ \dot{y}_k\ \dot{z}_k\right]^{\mathrm{T}}. \end{aligned} \tag{1}$$

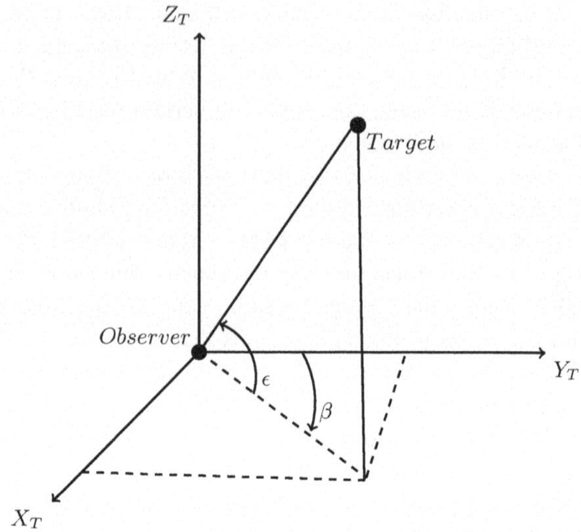

Fig. 1 Target–observer in Cartesian coordinate frame

Let the range vector be $\mathtt{r}$ and is defined as

$$\begin{aligned} \mathtt{r} &= \left[x_k^t - x_k^o \; y_k^t - y_k^o \; z_k^t - z_k^o\right]^T \\ &= \left[x_k \; y_k \; z_k\right]^T. \end{aligned} \tag{2}$$

Now, the actual range or the slant range at any instant is defined as $r_k = \sqrt{x_k^2 + y_k^2 + z_k^2}$. From the target–observer geometry shown in Fig. 1, $\mathtt{r}$ can be expressed in terms of β and ϵ as

$$\mathtt{r} = r \begin{bmatrix} \cos\epsilon \sin\beta \\ \cos\epsilon \cos\beta \\ \sin\epsilon \end{bmatrix}.$$

Here, $\beta \in [0, 2\pi]$ and $\epsilon \in [-\frac{\pi}{2}, \frac{\pi}{2}]$. Moreover, the ground range or the range in x-y plane is defined as $r_g = r\cos\epsilon$.

2.1 Process Model

The discrete time state equation decribing the constant velocity target dynamics can be expressed as

$$X^t{}_k = F_{k-1} X_{k-1}^t + \mathrm{w}_{k-1}^t, \tag{3}$$

where F_{k-1} is the state transition matrix defined as

$$F_{k-1} = \begin{bmatrix} 1 & 0 & 0 & T & 0 & 0 \\ 0 & 1 & 0 & 0 & T & 0 \\ 0 & 0 & 1 & 0 & 0 & T \\ 0 & 0 & 0 & 1 & 0 & 0 \\ 0 & 0 & 0 & 0 & 1 & 0 \\ 0 & 0 & 0 & 0 & 0 & 1 \end{bmatrix}.$$

T is the sampling interval, and $\mathrm{w}^{\mathrm{t}}_{k-1}$ is a zero mean Gaussian noise, uncorrelated with X.

Now, the relative state vector dynamics using (3) and (1) is defined as [24]

$$\mathcal{X}_k = F_{k-1}\mathcal{X}_{k-1} + \mathrm{w}^{\mathrm{t}}_{k-1} - X^{\mathrm{o}}_k + F_{k-1}X^{\mathrm{o}}_{k-1}, \tag{4}$$

where $\mathrm{w^t}_{k-1} \sim N(0, Q_{k-1})$ and

$$Q_{k-1} = \begin{bmatrix} \frac{T^3}{3}q_x & 0 & 0 & \frac{T^2}{2}q_x & 0 & 0 \\ 0 & \frac{T^3}{3}q_y & 0 & 0 & \frac{T^2}{2}q_y & 0 \\ 0 & 0 & \frac{T^3}{3}q_z & 0 & 0 & \frac{T^2}{2}q_z \\ \frac{T^2}{2}q_x & 0 & 0 & Tq_x & 0 & 0 \\ 0 & \frac{T^2}{2}q_y & 0 & 0 & Tq_y & 0 \\ 0 & 0 & \frac{T^2}{2}q_z & 0 & 0 & Tq_z \end{bmatrix}.$$

Here, q_x, q_y, q_z are the power spectral densities of the process noise along the X, Y, and Z axes, respectively. Figure 2 shows the target–observer dynamics considered in this study. It can be seen that the target follows a constant velocity model and the observer manoeuvre is modelled in different constant velocity phases.

2.2 Measurement Model

The measurement model involving the bearing and elevation angle is defined as

$$\mathrm{z}_k = h(\mathcal{X}_k) + \mathrm{v}_k, \tag{5}$$

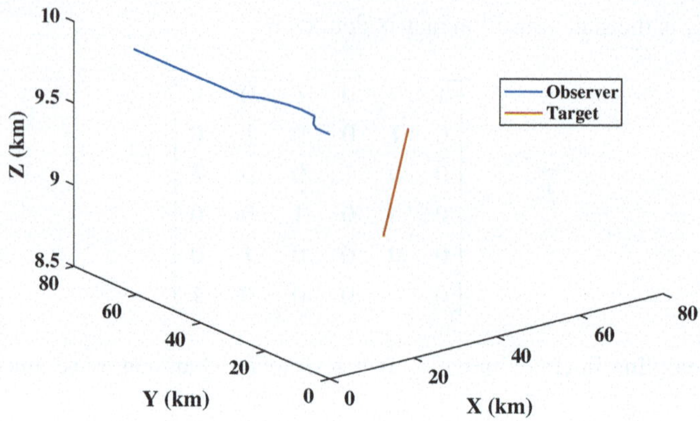

Fig. 2 3D BOT scenario

where

$$h(\mathcal{X}_k) = \begin{bmatrix} \beta_k \\ \epsilon_k \end{bmatrix} = \begin{bmatrix} \tan^{-1}(\frac{x_k}{y_k}) \\ \tan^{-1}\left(\frac{z_k}{\sqrt{x_k{}^2 + y_k{}^2}}\right) \end{bmatrix}.$$

Here, $v_k = [v_{1,k}\ v_{2,k}]^T$ is a zero mean white Gaussian noise with covariance matrix R_k, that is $v_k \sim N(0, R_k)$, $R_k = \text{diag}(\sigma_\beta{}^2, \sigma_\epsilon{}^2)$. Here, σ_β and σ_ϵ are the standard deviations of error in bearing and elevation angles, respectively.

3 Recursive Bayesian Algorithms

Even though certain batch estimation algorithms have been proposed recently for 3D BOT problem [27], we resort to the implementation of three recursive deterministic sampling point filters. They are the cubature Kalman filter (CKF), unscented Kalman filter (UKF), and the recently proposed new sigma point Kalman filter (NSKF). In addition to this, the extended Kalman filter is also implemented. In this section, Bayesian filtering, the formulation of these filtering algorithms, and their implementation with respect to the problem considered are discussed.

3.1 Bayesian Approach to Filtering

Suppose the discrete state-space process and measurement models are represented as in (4) and (5). In Bayesian filtering, the unknown states and parameters to be estimated are assumed as random variables that follow a particular probability distribution. This probability distribution is termed as prior probability density. The prior information regarding the states to be estimated is combined with the measurements received, via a conditional density function of the states, given the measurements. This probability density function (pdf) is usually termed as posterior probability density function.

In this framework, the two steps involved in calculating the prior and posterior pdfs are the time update step and the measurement update step.

Time update step: The prior pdf, $p(\mathcal{X}_k|z_{1:k-1})$, is obtained using the Chapman–Kolmogorov equation

$$p(\mathcal{X}_k|z_{1:k-1}) = \int p(\mathcal{X}_k|\mathcal{X}_{k-1})p(\mathcal{X}_{k-1}|z_{1:k-1})d\mathcal{X}_{k-1}. \tag{6}$$

Measurement update step: The measurement at time instant k, z_k and the prior pdf is used to obtain the posterior pdf using Bayes' rule

$$p(\mathcal{X}_k|z_{1:k}) = \frac{p(z_k|\mathcal{X}_k)p(\mathcal{X}_k|z_{1:k-1})}{\int p(z_k|\mathcal{X}_k)p(\mathcal{X}_k|z_{1:k-1})d\mathcal{X}_k}. \tag{7}$$

When the process or measurement model becomes nonlinear, the integrals become intractable, and we need to look for suboptimal solutions. This is due to the fact that prior and posterior pdfs may not be Gaussian anymore. To overcome this, one approach is to approximate them with Gaussian distribution and take the mean as a point estimate. Under this assumption, the prior and posterior pdfs can be expressed as $p(\mathcal{X}_k|z_{1:k-1}) = N(\hat{\mathcal{X}}_{k|k-1}, P_{k|k-1})$ and $p(\mathcal{X}_k|z_{1:k}) = N(\hat{\mathcal{X}}_{k|k}, P_{k|k})$. The posterior mean $\hat{\mathcal{X}}_{k|k}$ is the estimated value of state $\mathcal{X}$.

3.2 Extended Kalman Filter

The extended Kalman filter is an immediate extension of the Kalman filter to nonlinear state estimation problems. The nonlinear process and measurement models are linearised, and then the Kalman filter equations are used. In this problem, since the process model is linear, the same Kalman filter equations for the time update step can be implemented. But for the nonlinear measurement model, a corresponding linearised model has to be computed in the measurement update step. With a suitable assumption of the posterior estimate $\hat{\mathcal{X}}_{k-1|k-1}$ and covariance matrix $P_{k-1|k-1}$ at time $k-1$, the recursive algorithm can be given as

$$\begin{aligned}\hat{\mathcal{X}}_{k|k-1} &= F_{k-1}\hat{\mathcal{X}}_{k-1|k-1} - X_k^{\mathrm{o}} + F_{k-1}X_{k-1}^{\mathrm{o}}\\ \bar{P}_{k|k-1} &= F_{k-1}P_{k-1|k-1}F_{k-1}^{\mathrm{T}} + Q\end{aligned}$$

Now, the measurement update step involves the formation of Jacobian matrix followed by innovation covariance and Kalman gain, given as

$$\begin{aligned} H_k &= \frac{\partial h(\hat{\mathcal{X}}_{k|k-1})}{\partial \hat{\mathcal{X}}_{k|k-1}} \\ S_k &= H_k \bar{P}_{k|k-1} H_k^{\mathrm{T}} + R_k \\ K_k &= \bar{P}_{k|k-1} H_k^{\mathrm{T}} S_k^{-1} \end{aligned}$$

Then, the posterior mean and covariance are found out as

$$\begin{aligned} \hat{\mathcal{X}}_{k|k} &= \hat{\mathcal{X}}_{k|k-1} + K_k[\mathrm{z}_k - h(\hat{\mathcal{X}}_{k|k-1})] \\ P_{k|k} &= \bar{P}_{k|k-1} - K_k S_k K_k^{\mathrm{T}} \end{aligned}$$

3.3 *Unscented Kalman filter*

Unscented Kalman filter (UKF)[7] was the first suboptimal filter that introduced the concept of numerical integration of the intractable integrals using a set of deterministic points and weights. The major advantage over the already existing EKF was that the algorithm was derivative free. It proved to be more accurate and less sensitive to initial condition mismatch when compared with the EKF. The main concept was to implement the numerical approximation using the unscented transformation.

Suppose that the integral to be approximated is of the form

$$\mathtt{I}(\mathcal{X}) = \int h(\mathcal{X}) p_{\mathcal{X}}(\mathcal{X})\, \mathrm{d}\mathcal{X}$$

where $\mathcal{X}$ is a random variable that is assumed to follow a Gaussian density $p_{\mathcal{X}}(\mathcal{X}) \sim N(\hat{\mathrm{x}}, P)$. Now, since $\mathtt{I}(\mathcal{X})$ is not a tractable one, a numerical approximation with the help of unscented transformation is defined as

$$\mathtt{I}(\mathcal{X}) \approxeq \sum_{i=1}^{2n+1} w_i h(\hat{\mathcal{X}}_i), \quad \text{where } n \text{ is the order of the system.}$$

Here

$$\begin{aligned} &\hat{\mathcal{X}}_1 = \hat{\mathrm{x}} \\ &\hat{\mathcal{X}}_i = \hat{\mathrm{x}} + \left(\sqrt{(n+\kappa)P}\right)_i, \quad i = 1, \ldots, n \\ &\hat{\mathcal{X}}_i = \hat{\mathrm{x}} - \left(\sqrt{(n+\kappa)P}\right)_i, \quad i = 1, \ldots, n \quad \text{and} \\ &w_1 = \frac{\kappa}{n+\kappa}, \quad w_i = \frac{1}{2(n+\kappa)}, \quad \text{for } i = 1, \ldots, 2n. \end{aligned} \tag{8}$$

3.4 New Sigma Point Kalman Filter

In UKF, the sigma point equal to mean is given the maximum weight, whereas all other sigma points are assigned equal weights. In [28], this weight assignment was revisited because of the fact that the sigma points which are nearer to the mean value should have higher probability of occurrence. Accordingly, a new sigma point transformation was formulated where sigma points closer to the mean were given more weights when compared to the sigma points that are further away from mean. This new transformation also took care of the criteria that first two moments are matched exactly. Taking the same assumption above, $\mathcal{X} \sim N(\hat{\mathrm{x}}, P)$, the new transformation is defined as

$$
\begin{aligned}
\hat{\mathcal{X}}_1 &= \hat{\mathrm{x}}, \quad w_1 = 1 - \frac{\sum_{i=1}^{n} \alpha_i}{2(\sum_{i=1}^{n} \alpha_i + \mathrm{b})} \\
\hat{\mathcal{X}}_{i+1} &= \hat{\mathrm{x}} + \sqrt{\frac{\sum_{i=1}^{n} \alpha_i + \mathrm{b}}{\mathrm{m}\alpha_i}} S_i, \quad w_{i+1} = \frac{\mathrm{m}\alpha_i}{4(\sum_{i=1}^{n} \alpha_i + \mathrm{b})}, \; i = 1, \ldots, n \\
\hat{\mathcal{X}}_{i+1} &= \hat{\mathrm{x}} - \sqrt{\frac{\sum_{i=1}^{n} \alpha_i + \mathrm{b}}{\mathrm{m}\alpha_{i-n}}} S_{i-n}, \quad w_{i+1} = \frac{\mathrm{m}\alpha_{i-n}}{4(\sum_{i=1}^{n} \alpha_i + \mathrm{b})}, \; i = n+1, \ldots, 2n \\
\hat{\mathcal{X}}_{i+1} &= \hat{\mathrm{x}} + \sqrt{\frac{\sum_{i=1}^{n} \alpha_i + \mathrm{b}}{(1-\mathrm{m})\alpha_{i-2n}}} S_{i-2n}, \; w_{i+1} = \frac{(1-\mathrm{m})\alpha_{i-2n}}{4(\sum_{i=1}^{n} \alpha_i + \mathrm{b})}, \; i = 2n+1, \ldots, 3n \\
\hat{\mathcal{X}}_{i+1} &= \hat{\mathrm{x}} - \sqrt{\frac{\sum_{i=1}^{n} \alpha_i + \mathrm{b}}{(1-\mathrm{m})\alpha_{i-3n}}} S_{i-3n}, \; w_{i+1} = \frac{(1-\mathrm{m})\alpha_{i-3n}}{4(\sum_{i=1}^{n} \alpha_i + \mathrm{b})}, \; i = 3n+1, \ldots, 4n.
\end{aligned}
\tag{9}
$$

Here, S is defined such that $SS^{\mathrm{T}} = P$, and S_i and P_i denote the ith column of S and P, respectively. The variable α is defined as $\alpha_i = \frac{|<\mathrm{x}_i, P_i>|}{\|\mathrm{x}_i\|_2 \|P_i\|_2}$. b is a real constant such that $\mathrm{b} > \{\frac{1}{4} \max(\mathrm{m}\alpha_i) - \frac{1}{2}\sum_{i=1}^{n} \alpha_i\}$, and m is chosen as $\mathrm{m} \in (0.5, 1)$.

If we represent the total number of sigma points with $\mathcal{N}$, then from this transformation, the value of $\mathcal{N}$ becomes $4n + 1$. But in case of the UKF, value of $\mathcal{N} = 2n + 1$. Compared to the UKF, NSKF has an advantage of having two tuning parameters, b and m. The algorithm for implementing UKF and NSKF for the 3D BOT problem is given below.

4 Simulation Results

The simulation and results comprise a performance comparison of EKF, UKF, CKF, and NSKF for 1000 Monte Carlo runs. The entire tracking scenario is implemented and simulated in MATLAB software. The initial parameter values required for gen-

Algorithm 1: For UKF and NSKF

`Initialise` $\hat{\mathcal{X}}_{k-1|k-1}$ and $P_{k-1|k-1}$

$\hat{\mathcal{X}}_{k|k-1} = F_{k-1}\hat{\mathcal{X}}_{k-1|k-1} - X_k^o + F_{k-1}X_{k-1}^o$

$\bar{P}_{k|k-1} = F_{k-1}P_{k-1|k-1}F_{k-1}^T + Q$

`Calculate` $\hat{\mathcal{X}}_i$ `using` (8) `or` (9), $i = 1, \cdots, \mathcal{N}$

$\mathrm{z}_{i,k|k-1} = h(\hat{\mathcal{X}}_i)$

$\hat{\mathrm{z}}_k = \sum_{i=1}^{\mathcal{N}} w_i \mathrm{z}_{i,k|k-1}$

$P_{\mathrm{z}_k} = \sum_{i=1}^{\mathcal{N}} w_i [\mathrm{z}_{i,k|k-1} - \hat{\mathrm{z}}_k][\mathrm{z}_{i,k|k-1} - \hat{\mathrm{z}}_k]^T + R_k$

$P_{\mathcal{X}_k \mathrm{z}_k} = \sum_{i=1}^{\mathcal{N}} w_i [\hat{\mathcal{X}}_i - \hat{\mathcal{X}}_{k|k-1}][\mathrm{z}_{i,k|k-1} - \hat{\mathrm{z}}_k]^T$

$K_k = P_{\mathcal{X}_k \mathrm{z}_k} P_{\mathrm{z}_k}^{-1}$

$\hat{\mathcal{X}}_{k|k} = \hat{\mathcal{X}}_{k|k-1} + K_k[\mathrm{z}_k - \hat{\mathrm{z}}_k]$

$P_{k|k} = \bar{P}_{k|k-1} - K_k P_{\mathrm{z}_k} K_k^T$

Table 1 Target and observer initial parameters

Parameters	Value
Initial target position	$\begin{bmatrix} 138/\sqrt{2} & 138/\sqrt{2} & 9 \end{bmatrix}$ (km)
Initial target speed (s)	0.297 (km/s)
Target course	−135 (°)
Elevation angle	0.415 (°)
Initial observer position	$\begin{bmatrix} 0 & 0 & 10 \end{bmatrix}$ (km)
Initial observer speed (s)	0.297 (km/s)

erating the scenario shown in Fig. 2 are given in Table 1. Here, target and observer initial position in the three coordinates are given. The resultant velocity of the target and observer is given, which is used for finding the velocities in the three coordinate axes. From Fig. 2, it should be noted that target position in Z position is decreasing. The bearing angle β is calculated with reference to the Y-axis.

For each Monte Carlo run, the target is assumed to start at a position defined by the initial measurement received. According to this measurement, a suitable assumption for range r, target speed s, bearing, and elevation angle is considered. Here, these values are mentioned in Table 2. The variables in the process noise, defined for calculating process noise covariance matrix Q, are taken as $q_x = q_y = 0.01\,\mathrm{m}^2/\mathrm{s}^3$ and $q_z = 10^{-4}\mathrm{m}^2/\mathrm{s}^3$. The motion of the observer is deterministic and moves in a plane parallel to the X-Y plane at a fixed height of 10 km.

Table 2 Parameters for defining $\hat{\mathcal{X}}_{0|0}$ & $P_{0|0}$

Parameter	Mean	Standard deviation
Range (km)	150	13.6
Speed (m/s)	258	41.6
Bearing heading (rad/s)	$\beta + \pi$	$\pi/\sqrt{12}$
Elevation heading (rad/s)	0	$\pi/60$

4.1 *Filter Initialisation*

All the filters in this paper are initialised using the method given in [24] with the parameters shown in Table 2. The relative state is initialised depending on the first measurement of bearing and elevation angle and the prior range estimate of the target. The $\bar{r} \sim N(r, \sigma_r{}^2)$ is the prior range estimate of the target, with r as the assumed true initial range of the target from the observer. The true initial bearing and elevation measurement estimate are $\hat{\beta}_1$ and $\hat{\epsilon}_1$ with headings $\overline{\alpha}_1$ and $\overline{\gamma}_1$, respectively. Also, $\bar{s} \sim N(s, \sigma_s{}^2)$ is the initial speed estimate for the target.

The initial estimate ($\hat{\mathcal{X}}_{0|0}$) for relative state vector is given as [24]

$$\hat{\mathcal{X}}_{0|0} = \begin{bmatrix} \bar{r}\, \eta_{1,0}(\hat{\epsilon}_1, \sigma_\epsilon{}^2)\, \eta_{0,1}(\hat{\beta}_1, \sigma_\beta{}^2) \\ \bar{r}\, \eta_{1,0}(\hat{\epsilon}_1, \sigma_\epsilon{}^2)\, \eta_{1,0}(\hat{\beta}_1, \sigma_\beta{}^2) \\ \bar{r}\, \eta_{0,1}(\hat{\epsilon}_1, \sigma_\epsilon{}^2) \\ \bar{s}\, \eta_{1,0}(\overline{\gamma}_1, \sigma_\gamma{}^2)\, \eta_{0,1}(\overline{\alpha}_1, \sigma_\alpha{}^2) - \dot{x}_1^{\mathrm{o}} \\ \bar{s}\, \eta_{1,0}(\overline{\gamma}_1, \sigma_\gamma{}^2)\, \eta_{1,0}(\overline{\alpha}_1, \sigma_\alpha{}^2) - \dot{y}_1^{\mathrm{o}} \\ \bar{s}\, \eta_{0,1}(\overline{\gamma}_1, \sigma_\gamma{}^2) - \dot{z}_1^{\mathrm{o}} \end{bmatrix},$$

where $\eta_{1,0}(\mu, \sigma^2) = \cos\mu\ exp(-\sigma^2/2)$ and $\eta_{0,1}(\mu, \sigma^2) = \sin\mu\ exp(-\sigma^2/2)$. Also, $(\dot{x}_1^{\mathrm{o}}, \dot{y}_1^{\mathrm{o}}, \dot{z}_1^{\mathrm{o}})$ is the initial velocity components of the observer state vector.

Let P be the covariance matrix, whose (i, j)th element is represented as $P_{i,j}$. The upper triangular nonzero elements of the covariance matrix P are [24]

$$\begin{aligned}
P_{11} &= (\sigma_r{}^2 + \bar{r}^2)\, \eta_{2,0}(\hat{\epsilon}_1, \sigma_\epsilon{}^2)\, \eta_{0,2}(\hat{\beta}_1, \sigma_\beta{}^2) - \bar{r}^2\, \eta_{1,0}^2(\hat{\epsilon}_1, \sigma_\epsilon{}^2)\, \eta_{0,1}^2(\hat{\beta}_1, \sigma_\beta{}^2) \\
P_{12} &= (\sigma_r{}^2 + \bar{r}^2)\, \eta_{2,0}(\hat{\epsilon}_1, \sigma_\epsilon{}^2)\, \eta_{1,1}(\hat{\beta}_1, \sigma_\beta{}^2) - \bar{r}^2\, \eta_{1,0}^2(\hat{\epsilon}_1, \sigma_\epsilon{}^2)\, \eta_{1,0}(\hat{\beta}_1, \sigma_\beta{}^2)\, \eta_{0,1}(\hat{\beta}_1, \sigma_\beta{}^2) \\
P_{13} &= [(\sigma_r{}^2 + \bar{r}^2)\, \eta_{1,1}(\hat{\epsilon}_1, \sigma_\epsilon{}^2) - \bar{r}^2\, \eta_{1,0}(\hat{\epsilon}_1, \sigma_\epsilon{}^2)\, \eta_{0,1}(\hat{\epsilon}_1, \sigma_\epsilon{}^2)]\, \eta_{0,1}(\hat{\beta}_1, \sigma_\beta{}^2) \\
P_{22} &= (\sigma_r{}^2 + \bar{r}^2)\, \eta_{2,0}(\hat{\epsilon}_1, \sigma_\epsilon{}^2)\, \eta_{2,0}(\hat{\beta}_1, \sigma_\beta{}^2) - \bar{r}^2\, \eta_{1,0}^2(\hat{\epsilon}_1, \sigma_\epsilon{}^2)\, \eta_{1,0}^2(\hat{\beta}_1, \sigma_\beta{}^2) \\
P_{23} &= [(\sigma_r{}^2 + \bar{r}^2)\, \eta_{1,1}(\hat{\epsilon}_1, \sigma_\epsilon{}^2) - \bar{r}^2\, \eta_{1,0}(\hat{\epsilon}_1, \sigma_\epsilon{}^2)\, \eta_{0,1}(\hat{\epsilon}_1, \sigma_\epsilon{}^2)]\, \eta_{1,0}(\hat{\beta}_1, \sigma_\beta{}^2) \\
P_{33} &= (\sigma_r{}^2 + \bar{r}^2)\, \eta_{0,2}(\hat{\epsilon}_1, \sigma_\epsilon{}^2) - \bar{r}^2\, \eta_{0,1}(\hat{\epsilon}_1, \sigma_\epsilon{}^2) \\
P_{44} &= (\sigma_s{}^2 + \bar{s}^2)\, \eta_{2,0}(\overline{\gamma}_1, \sigma_\gamma{}^2)\, \eta_{0,2}(\overline{\alpha}_1, \sigma_\alpha{}^2) - \bar{s}^2\, \eta_{1,0}^2(\overline{\gamma}_1, \sigma_\gamma{}^2)\, \eta_{0,1}^2(\overline{\alpha}_1, \sigma_\alpha{}^2)
\end{aligned}$$

$$P_{45} = (\sigma_s{}^2 + \bar{s}^2)\ \eta_{2,0}(\overline{\gamma}_1, \sigma_\gamma{}^2)\ \eta_{1,1}(\overline{\alpha}_1, \sigma_\alpha{}^2) - \bar{s}^2\ \eta_{1,0}^2(\overline{\gamma}_1, \sigma_\gamma{}^2)\ \eta_{1,0}(\overline{\alpha}_1, \sigma_\alpha{}^2)\ \eta_{0,1}(\overline{\alpha}_1, \sigma_\alpha{}^2)$$
$$P_{46} = [(\sigma_s{}^2 + \bar{s}^2)\ \eta_{1,1}(\overline{\gamma}_1, \sigma_\gamma{}^2) - \bar{s}^2\ \eta_{1,0}(\overline{\gamma}_1, \sigma_\gamma{}^2)\ \eta_{0,1}(\overline{\gamma}_1, \sigma_\gamma{}^2)]\ \eta_{0,1}(\overline{\alpha}_1, \sigma_\alpha{}^2)$$
$$P_{55} = (\sigma_s{}^2 + \bar{s}^2)\ \eta_{2,0}(\overline{\gamma}_1, \sigma_\gamma{}^2)\ \eta_{2,0}(\overline{\alpha}_1, \sigma_\alpha{}^2) - \bar{s}^2\ \eta_{1,0}^2(\overline{\gamma}_1, \sigma_\gamma{}^2)\ \eta_{1,0}^2(\overline{\alpha}_1, \sigma_\alpha{}^2)$$
$$P_{56} = [(\sigma_s{}^2 + \bar{s}^2)\ \eta_{1,1}(\overline{\gamma}_1, \sigma_\gamma{}^2) - \bar{s}^2\ \eta_{1,0}(\overline{\gamma}_1, \sigma_\gamma{}^2)\ \eta_{0,1}(\overline{\gamma}_1, \sigma_\gamma{}^2)]\ \eta_{1,0}(\overline{\alpha}_1, \sigma_\alpha{}^2)$$
$$P_{66} = (\sigma_s{}^2 + \bar{s}^2)\ \eta_{0,2}(\overline{\gamma}_1, \sigma_\gamma{}^2) - \bar{s}^2\ \eta_{0,1}^2(\overline{\gamma}_1, \sigma_\gamma{}^2),$$

where

$$\eta_{2,0}(\mu, \sigma^2) = [\sinh \sigma^2 + \cos^2 \mu\ exp(-\sigma^2)]\ exp(-\sigma^2)$$
$$\eta_{0,2}(\mu, \sigma^2) = [\sinh \sigma^2 + \sin^2 \mu\ exp(-\sigma^2)]\ exp(-\sigma^2)$$
$$\eta_{1,1}(\mu, \sigma^2) = \cos \mu\ \ \sin \mu\ exp(-2\sigma^2).$$

Since P is a symmetric matrix, the property that $P_{ij} = P_{ji}$ gives the lower triangular elements of the covariance matrix. Hence, $P_{0|0}$ is defined as

$$P_{0|0} = \begin{bmatrix} P_{11} & P_{12} & P_{13} & 0 & 0 & 0 \\ P_{12} & P_{22} & P_{23} & 0 & 0 & 0 \\ P_{13} & P_{32} & P_{33} & 0 & 0 & 0 \\ 0 & 0 & 0 & P_{44} & P_{45} & P_{46} \\ 0 & 0 & 0 & P_{45} & P_{55} & P_{56} \\ 0 & 0 & 0 & P_{64} & P_{56} & P_{66} \end{bmatrix}.$$

4.2 Performance Criteria

Two measures of performance using root mean square error (RMSE) are used to compare the various filters. RMSE was calculated for the target states, which were found out from the estimated relative states and from the deterministic observer values. We found out the RMSE for both resultant position as well as for the resultant velocity. The first measure defined was the RMSE for position and velocity averaged from the end of the last observer manoeuvre to the end of observation or simulation period. The final RMSE for position and velocity at the end of the simulation period was considered to be the second measure of performance. These statistics are computed by averaging over 1000 realisations, for bearing and elevation measurement noise standard deviations of 0.057°, 0.65°, 0.8°, and 1°. The effect of increase in initial uncertainty on filter accuracy is also studied. The expression for RMSE in position is given as

$$\text{RMSE}_k = \sqrt{\frac{1}{M}\sum_{j=1}^{M}[(x^{t}{}_{j,k} - \hat{x}^{t}_{j,k})^2 + (y^{t}{}_{j,k} - \hat{y}^{t}_{j,k})^2 + (z^{t}{}_{j,k} - \hat{z}^{t}_{j,k})^2]}$$

where k and M denote the total number of time steps and Monte Carlo runs, respectively. A similar expression for RMSE in velocity was implemented for the study. As compared to the 2D BOT scenarios widely available in the literature, the 3D scenario considered in this study did not incur any track-loss. Hence, the need for implementing any track-loss condition for performance analysis is ruled out.

4.3 Performance Comparison

For simulation, the sampling time is considered as $T = 1$ s, and the total observation period lasted for $T = 210$ s. The tracking performance of NSKF is shown in Fig. 3 for a single run. The estimated target path is plotted along with the truth target path and the observer path in this figure. It is observed that the NSKF accurately tracks the target.

RMSE in position and velocity is plotted in Figs. 4 and 5 to compare the estimation accuracy of various filters. From these figures, it can be observed that all filters performed with comparable accuracy. The results are shown in Tables 3 and 4. For the particular target–observer scenario discussed in the paper, it can be seen that all the filters give comparable results. The effect of increase in the measurement noise can be seen in the increase of the RMSE value at the final time for both position and velocity resulting in lower filter accuracy. From Table 3, when the measurement noise is low (i.e. standard deviation is 0.057°), EKF gives comparable response. For higher measurement noise standard deviations (i.e. 1°), it can be seen that NSKF

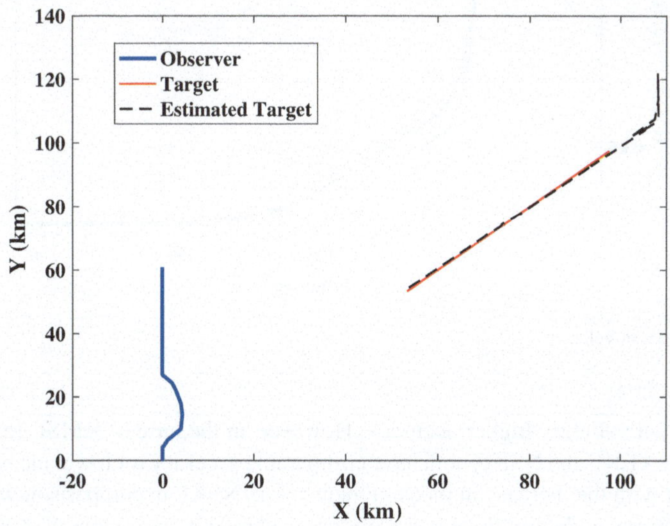

Fig. 3 Estimated and truth target path in X-Y plane

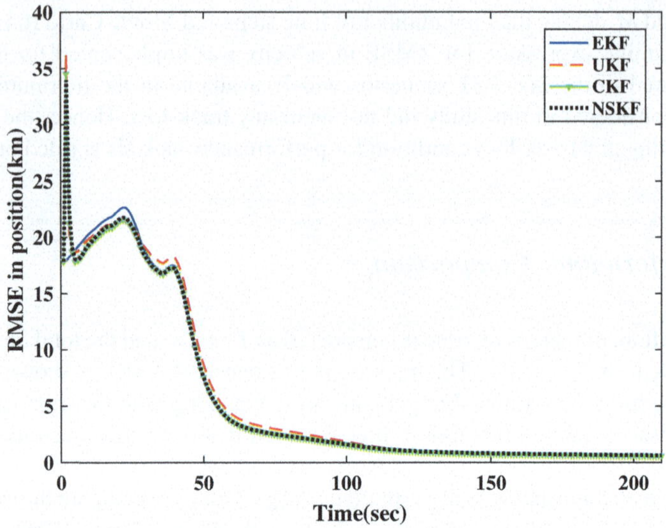

Fig. 4 RMSE in position

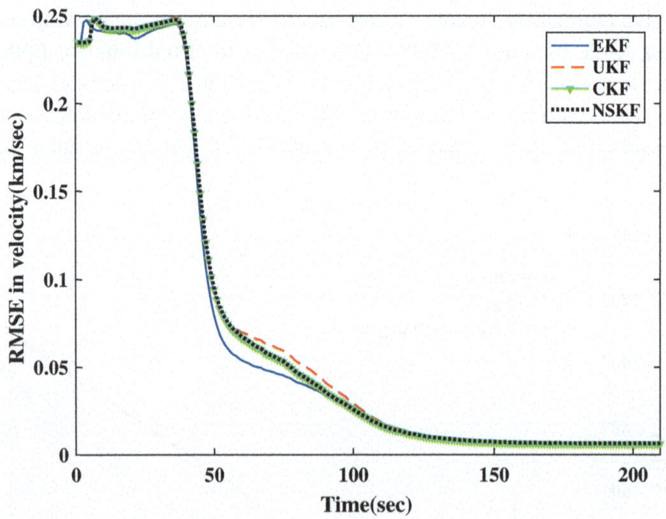

Fig. 5 RMSE in velocity

performs with slightly higher accuracy. However, in the case of RMSE in velocity from Table 4 EKF and NSKF, both have comparable accuracy for lower measurement noise. But with the increase in measurement noise, NSKF in comparison with other filters performs with slightly better accuracy. It should be also noted that, when the noise in bearing and elevation angles is increased from 0.057° to 1°, the error in

Table 3 RMSE in position

Final RMS position error (km)

Filter	Measurement noise Standard deviation (°) 0.057	0.65	0.8	1
EKF	0.626	4.932	5.467	5.979
UKF	0.671	4.968	5.483	5.955
CKF	0.634	4.951	5.482	5.946
NSKF	0.631	4.948	5.477	5.944

Time-averaged RMS position error (km)

Filter	Measurement noise Standard deviation (°) 0.057	0.65	0.8	1
EKF	0.900	6.710	7.859	9.293
UKF	0.983	6.821	7.995	9.450
CKF	0.906	6.802	7.993	9.441
NSKF	0.903	6.787	7.979	9.226

Table 4 RMSE in velocity

Final RMS velocity error (km/s)

Filter	Measurement noise Standard deviation (°) 0.057	0.65	0.8	1
EKF	0.0063	0.0491	0.0539	0.0578
UKF	0.0067	0.0494	0.0538	0.0575
CKF	0.0063	0.0492	0.0538	0.0575
NSKF	0.0063	0.0491	0.0537	0.0575

Time-averaged RMSE in velocity (km/s)

Filter	Measurement noise Standard eviation (°) 0.057	0.65	0.8	1
EKF	0.0099	0.0566	0.0618	0.0662
UKF	0.108	0.0572	0.0622	0.0665
CKF	0.010	0.0570	0.0622	0.0667
NSKF	0.0099	0.0570	0.0622	0.0660

Table 5 RMSE in position with varying $P_{0|0}$

Final RMS position error (km)

Filter	$P = 1.5 * P$	$P = 2 * P$	$P = 3 * P$
EKF	6.247	7.012	8.217
UKF	6.171	6.775	7.731
CKF	6.169	6.785	7.858
NSKF	6.204	6.835	7.970

Time-averaged RMS position error (km)

Filter	$P = 1.5 * P$	$P = 2 * P$	$P = 3 * P$
EKF	9.072	9.799	10.912
UKF	9.176	9.727	10.569
CKF	9.177	9.710	10.681
NSKF	9.212	9.760	10.815

RMS position shifted from 631 m to 5.9 km. This indicates a significant decrease in the accuracy of the nonlinear filters implemented.

The initial covariance matrix was computed as $P = \nu \times P_{0|0}$ where the value of $\nu = 1.5, 2, 3$ to analyse the effect of initial uncertainty. Here, the measurement standard deviation for bearing and elevation angle was set to be 1°. From Tables 5 and 6, it can be seen that EKF in comparison with NSKF shows more error with the increase in the uncertainty. But still, it cannot be considered as a significant increase in estimation accuracy, as it is comparable.

Table 6 RMSE in velocity with varying $P_{0|0}$

Final RMSE in velocity (km/s)				Time-averaged RMSE in velocity (km/s)			
Filter	$P = 1.5 * P$	$P = 2 * P$	$P = 3 * P$	Filter	$P = 1.5 * P$	$P = 2 * P$	$P = 3 * P$
EKF	0.0618	0.0690	0.0820	EKF	0.0678	0.0740	0.0839
UKF	0.0609	0.0663	0.0765	UKF	0.0679	0.0734	0.0807
CKF	0.0608	0.0664	0.0779	CKF	0.0680	0.0732	0.0815
NSKF	0.0612	0.0669	0.0790	NSKF	0.0684	0.0737	0.0827

5 Conclusion

This paper presents a comparative study of certain nonlinear sigma point Kalman filters for solving the problem of 3D bearings-only target tracking. The filters used were EKF, UKF, CKF, and NSKF. The filtering accuracy is analysed using the RMSE in both position and velocity. From the RMSE, we focused on two criteria: (1) RMS error in position and velocity at the final simulation time step (2) the time-averaged RMS position and velocity error, after the end of observer manoeuvre. To make it more meaningful, the two criteria were evaluated by increasing the measurement noise standard deviation in bearing as well as elevation angles. It is observed that the accuracy of all filters decreases with the increase in measurement noise, but at a similar rate without any significant difference. A very small increase in accuracy was observed for the sigma point filters when compared to the EKF. We would also like to emphasise the fact that these results are valid only for the particular tracking scenario considered, and not a generalised one for 3D BOT problems. Except for highly nonlinear scenarios (where the rate at which the bearing angle rate is very high)[11], we expect the filters to perform in a similar fashion as described in the paper. An attempt was made to evaluate the two criteria with varying initial uncertainty. Here also, comparable results were obtained for sigma point filters with respect to the EKF.

References

1. Nardone, S., Lindgren, A.G., Gong, K.: Fundamental properties and performance of conventional bearings-only target motion analysis. IEEE Trans. Autom. Control **29**(9), 775–787 (1984)
2. Radhakrishnan, R., Singh, A.K., Bhaumik, S., Tomar, N.K.: Quadrature filters for underwater passive bearings-only target tracking. In: 2015 Sensor Signal Processing for Defence (SSPD), pp. 1–5. IEEE, Edinburgh (2015). https://doi.org/10.1109/SSPD.2015.7288519
3. Blackman, S., Popoli, R.: Design and Analysis of Modern Tracking Systems. Artech House, Norwood, MA (1999)
4. Ristic, B., Arulampalam, S., Gordon, N.: Beyond the Kalman Filter: Particle Filters for Tracking Applications. Artech House, Boston (2003)

5. Bar-Shalom, Y., Li, X.R., Kirubarajan, T.: Estimation, Tracking and Navigation: Theory, Algorithms and Software. Wiley, New York (2001)
6. Peach, N.: Bearings-only tracking using a set of range-parameterised extended Kalman filters. In: IEE Proceedings-Control Theory and Applications, vol. 142, issue 1, pp. 73–80. IET (1995)
7. Simon, J., Uhlmann, J., Hugh, F., Durrant, W.: A new method for the nonlinear transformation of means and covariances in filters and estimators. IEEE Trans. Autom. Control **45**(3), 477–482 (2000)
8. Arasaratnam, I., Haykin, S.: Cubature Kalman filter. IEEE Trans. Autom. Control **54**(6), 1254–1269 (2009)
9. Bhaumik, S., Swati.: Cubature quadrature Kalman filter. IET Signal Process. **7**(7), 533–541 (2013). https://doi.org/10.1049/iet-spr.2012.0085
10. Ito, K., Xiong, K.: Gaussian filters for nonlinear filtering problems. IEEE Trans. Autom. Control **45**(5), 910–927 (2000)
11. Radhakrishnan, R., Bhaumik, S., Tomar, N.K.: Gaussian sum shifted Rayleigh filter for underwater bearings-only target tracking problems. IEEE J. Ocean. Eng. **44**(2), 492–501 (2018)
12. Kotecha, J.H., Djuric, P.M.: Gaussian sum particle filtering. IEEE Trans. Signal Process. **51**(10), 2602–2612 (2003)
13. Alspach, D., Sorenson, H.: Nonlinear Bayesian estimation using Gaussian sum approximations. IEEE Trans. Autom. Control **17**(4), 439–448 (1972)
14. Lingren, A.G., Gong, K F.: Position and velocity estimation via bearing observations. IEEE Trans. Aerosp. Electron. Syst. **AES-14**(4), 564–577 (1978)
15. Doğançay, K.: On the bias of linear least squares algorithms for passive target localization. Signal Process. **84**(3), 475–486 (2004) (Elsevier)
16. Aidala, V.J., Nardone, S.C.: Biased estimation properties of the pseudolinear tracking filter. IEEE Transa. Aerosp. Electron. Syst. **AES-18**(4), 432–441 (1978)
17. Chan, Y.T., Rudnicki, S.W.: Bearings-only and Doppler-bearing tracking using instrumental variables. IEEE Trans. Aerosp. Electron. Syst. **28**(4), 1076–1083 (1992)
18. Doğançay, K.: Passive emitter localization using weighted instrumental variables. Signal Processi. **84**(4), 487–497 (2004) (Elsiever)
19. Lindgren, A.G., Gong, K.F.: Properties of a bearings-only motion analysis estimator: an interesting case study in system observability. In: Proceedings of the 12th Asilomar Conference on Circuits Systems, vol. 78, issue 1, pp. 61–78. Monterey, CA (1978)
20. Badriasl, L., Arulampalam, S., Finn, A.: A novel batch Bayesian WIV estimator for three-dimensional TMA using bearing and elevation measurements. IEEE Trans. Signal Process. **66**(4), 1023–1036 (2017)
21. Mallick, M., Morelande, M., Mihaylova, L., Arulampalam, S., Yan, Y.: Comparison of angle-only filtering algorithms in 3D using Cartesian and modified spherical coordinates. In: 2012 15th International Conference on Information Fusion, pp. 1392–1399. IEEE (2011)
22. Badriasl, L., Doğançay, K.: Three-dimensional target motion analysis using azimuth/elevation angles. IEEE Trans. Aerosp. Electron. Syst. **50**(4), 3178–3194 (2014)
23. Badriasl, L., Sathyan, T., Arulampalam, S., Finn, A.: A novel closed-form estimator for 3D TMA using heterogeneous sensors. IEEE Trans. Signal Process. **63**(8), 1895–1910 (2015)
24. Mallick, M., Morelande, M., Mihaylova, L., Arulampalam, S., Yan, Y.: Angle-only filtering in three dimensions. In: Integrated Tracking, Classification, and Sensor Management. Wiley Online Library (2013)
25. Gupta, S.D., Yu, J.Y., Mallick, M., Coates, M., Morelande, M.: Comparison of angle-only filtering algorithms in 3D using EKF, UKF, PF, PFF, and ensemble KF. In: 18th International Conference on Information Fusion (Fusion), pp. 1649–1656. IEEE, Washington (2015)
26. Mehrjouyan, A., Alfi, A.: Robust adaptive unscented Kalman filter for bearings-only tracking in three dimensional case. Appl. Ocean Res. **87**, 223–232 (2019)
27. Badriasl, L., Arulampalam, S., Hoek, J., Finn, A.: Bayesian WIV estimators for 3-D bearings-only TMA with speed constraints. IEEE Trans. Signal Process. **67**(13), 3576–3591 (2019)
28. Radhakrishnan, R., Yadav, A., Date, P., Bhaumik, S.: A new method for generating sigma points and weights for nonlinear filtering. IEEE Control Syst. Lett. **2**(3), 519–524 (2018)

Performance Comparison of EKF and UKF for Offshore Boom Crane System

Manash Jyoti Deori, Nabanita Adhikary, and Krishna Jyothi Pallacherla

Abstract In this paper, a performance comparison between the extended Kalman filter (EKF) and the unscented Kalman filter (UKF) is shown for nonlinear underactuated systems. The filters are designed and tested for the benchmark cart pendulum system and the underactuated offshore boom crane system. The EKF and UKF are designed for both state estimation and noise removal of measured output in the presence of disturbance in the system. Simulation results from the MATLAB environment have shown that both estimators can produce the required result, but UKF gives higher accuracy than that of EKF when system nonlinearity is higher.

1 Introduction

Nonlinearity is present in all the systems in the real world. Estimation and filtering are two of the most prevalent tools for the application control theory in such complex nonlinear dynamics. When the sensor information becomes noisy in the reading of the state variable, then estimators are employed to assemble a more precise result of the noise-free system state. One of the most commonly and widely used nonlinear estimators is the extended Kalman filter (EKF) that gives required results when there is low nonlinearity and noise level in the system [1, 2]. But when the nonlinearity and the noise level in the system become more, then EKF does not hold well. This is because EKF uses the Kalman filter equations in the applied nonlinear systems by linearising the process and measurement equations around the state which is being currently estimated. Problems regarding linearisation are the derivation of the Jacobian matrix which can be an erroneous and difficult process [3, 4]. At the end of the last century, Julier and Uhlmann introduced a paper [5] introducing a new filter termed as unscented Kalman filter (UKF). The methodology behind UKF is that it is not based on the linearisation around the states, but it is based on the unscented transformation of the states.

M. J. Deori · N. Adhikary (✉) · K. J. Pallacherla
Department of Electrical Engineering, National Institute of Technology Silchar, Silchar, India
e-mail: nabanita@ee.nits.ac.in

J. Gu et al. (eds.), *Communication and Control for Robotic Systems*, Smart Innovation, Systems and Technologies 229, https://doi.org/10.1007/978-981-16-1777-5_18

This paper aims to examine the effects of EKF and UKF applied to offshore boom crane systems [6]. These cranes are used for the transfer of cargo from one ship to another. Due to its motion and the unsteadiness of the sea, ship motion has six degrees of freedom (DOF) in space, which are roll, pitch, yaw, surge, sway and heave [7]. In most of the practical mechanical systems, only position sensors are available, since velocity sensors increase the overall cost and weight of the system. One of the popular ways to obtain the velocity signal is the numerical differentiation of the signal provided by the position sensors. But that differentiation may lead to the addition of more noise and distorted signals in practical applications. In such a case, the Kalman filter and its variants are expected to produce the estimates of all the states (velocity information) from the output measurement. The study presented in this paper shows the performance of both EKF and UKF for the underactuated nonlinear cart pendulum model and the offshore boom crane model.

The paper is organised as follows. The two nonlinear filters EKF and UKF are introduced in Sect. 2. The problem of state estimation which is based on underactuated nonlinear dynamics is formulated in Sect. 3. Numerical simulations of the two filters are done in Sect 4. Finally, conclusions based on the simulated results are drawn in Sect 5.

2 Nonlinear Filters

Kalman filter is a recursive filter, and it can estimate the states of a dynamic system from the noisy measurements [8]. With only a few derivative steps, it is almost a straightforward filter. It is assumed that the input data is in the form of a Gaussian probability distribution function (pdf). It can be stated that it is a predictive tool for both states and parameters with consecutive cycles of prediction and correction derived in the framework of a Gaussian PDF.

The following linear non-autonomous dynamic system is considered:

$$x(k+1) = f(x(k|k), u(k), w(k), k) \tag{1}$$

$$z(k) = h(x(k|k), v(k), k) \tag{2}$$

where it is assumed that $x(k) \in \mathbb{R}^n$ is the state vector at every time step k, $u(k)$ is the input vector, $v(k) \in \mathbb{R}^q$ is the process noise vector, $z(k)$ is the measurement vector, and $w(k)$ is the measurement noise. It is also assumed that both the process and measurement noise vectors are zero mean white Gaussian noise, and hence it can be written that

$$E(v(p)v^T(q)) = \delta_{pq}Q(p) \tag{3}$$

$$E[w(p)w^T(q)] = \delta_{ij}R(p) \tag{4}$$

$$E[v(p)w^T(q)] = 0, \quad \forall p, q \tag{5}$$

where $R()$ and $Q()$ are the process and measurement noise covariances, respectively.

2.1 Extended Kalman Filter (EKF)

The EKF works on the principle of linearising the system around the desired point and then performing the estimation and filtering operation. It is considered that $\hat{x}(k|k)$ and $p(k|k)$ are the state estimate and error covariance at every time step k. If $F(k)$ is the Jacobian of the system and $H(k)$ is the Jacobians of the measurement equations, respectively, as they are linearised, then the prediction and the estimation equations can be written as follows [9]:
Prediction:

$$\hat{x}(k+1|k) = f(\hat{x}(k|k), u(k), w(k), k) \tag{6}$$

$$p(k+1|k) = F(k)p(k)F^T(k) + G(k)QG^T(k) \tag{7}$$

$$F(k) = \frac{\partial f(x(k|k), u(k), w(k), k)}{\partial x}\Big|_{x=\hat{x}(k|k)} \tag{8}$$

$$G(k) = \frac{\partial f(x(k|k), u(k), w(k), k)}{\partial w}\Big|_{x=\hat{x}(k|k)} \tag{9}$$

Correction:

$$\hat{x}(k|k) = \hat{x}(k|k-1) + K(k)[z(k) - \hat{x}(k|k-1)] \tag{10}$$

$$K(k) = p(k|k-1)H^T(k)[H(k)p(k|k-1) + R]^(- 1) \tag{11}$$

$$p(k|k) = [I - K(k)H(k)]p(k|k-1) \tag{12}$$

$$H(k) = \frac{\partial h(x(k|k), v(k), k)}{\partial x}\Big|_{x=\hat{x}(k|k)} \tag{13}$$

The scientific diagram of the EKF is shown in Fig. 1

2.2 Unscented Kalman Filter

According to Julier and Uhlmann [5], UKF is another recursive estimator that is based on unscented transformation which is a method to approximate the required mean and covariance of the posterior distribution [10]. One of the main advantages of UKF over EKF is that this method need not linearise the system and measurement equations. UKF algorithm is a deterministic sampling technique. Initially, a set of sample points called the sigma points are selected in distribution with pre-selected

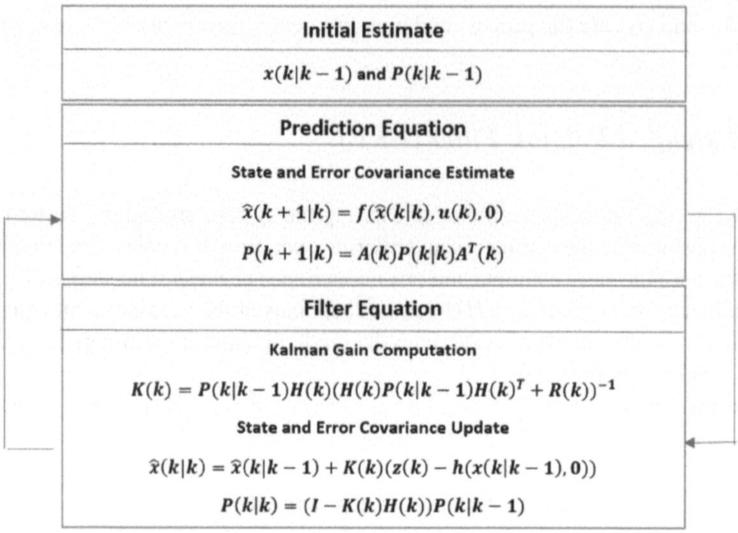

Fig. 1 Scientific diagram of EKF

mean and covariance. Then at every step, these sigma points are used to predict and correct, and based on the nonlinear function a new mean and covariance are estimated. Thus, the sigma points get propagated via the nonlinear function.

Let $x(k)$ be the n-dimensional state vector with mean $\hat{x}(k|k)$ and error covariance $p(k|k)$ at every time step k. Then, the UKF algorithm may be defined as follows:
Sigma points generation: $2n + 1$ weighed sigma points are defined which are as follows:

$$x_0(k|k) = \hat{x}(k|k) \tag{14}$$

$$x_i(k|k) = \hat{x}(k|k) + (\sqrt{(n+\lambda)P(k|k)})_i \tag{15}$$

$$x_{i+n}(k|k) = \hat{x}(k|k) + (\sqrt{(n+\lambda)P(k|k)})_i, \quad i = 1, 2, 3, \cdots, n \tag{16}$$

where $\lambda = \alpha^2(n + \kappa) - n, 0 < \alpha < 1$ denotes the spread of the sigma points around $\hat{x}(k|k)$, and κ is a secondary scaling parameter. $(\sqrt{(n+\lambda)P(k|k)})_i$ is the ith row or column of the matrix $(\sqrt{(n+\lambda)P(k|k)})$, and the square root is calculated using Cholesky decomposition. After the creation of the sigma or sample point, the general KF equations with weights added to them are applied as shown below.
Prediction:

$$x_i(k+1|k) = f(x_i(k|k), u(k), w(k), k) \tag{17}$$

$$\hat{x}(k+1|k) = \sum_{i=0}^{2n} W_i^c(x_i(k+1|k)) \tag{18}$$

$$P(k+1|k) = \sum_{i=0}^{2n} W_i^{c}(x_i(k+1|k) - \hat{x}(k+1|k))(x_i(k+1|k) - \hat{x}(k+1|k))^T + Q_k \tag{19}$$

where W_i^{m} and W_i^{c} represent the weights of the mean and error covariance and are defined as follows:

$$W_0^{m} = \frac{\lambda}{n+\lambda}$$

$$W_0^{c} = \frac{\lambda}{n+\lambda}(1-\alpha+\beta^2)$$

where $\beta > 0$ depends upon the prior knowledge of the distribution function.

$$W_i^{c} = W_i^{m} = \frac{1}{2(n+\lambda)}; \quad i = 1, 2, 3, \cdots, 2n$$

Correction:

$$y_i(k+1|k) = h(x_i(k|k), v(k), k) \tag{20}$$

$$\hat{z}(k+1|k) = \sum_{i=0}^{2n} W_i^{m} y_i(k+1|k) \tag{21}$$

$$\hat{x}(k+1|k+1) = \hat{x}(k+1|k) + K(k)(z(k) - \hat{z}(k+1|k)) \tag{22}$$

$$P(k+1|k+1) = P(k+1|k) - K(k+1)p_{zz}(k+1|k+1)K^T(k+1) \tag{23}$$

$$K(k+1) = P_{xz}(k+1|k+1)P_{zz}(k+1|k+1)^{-1} \tag{24}$$

$$P_{zz}(k+1|k+1) = \sum_{i=0}^{2n} W_i^{c}(y_i(k|k) - \hat{z}(k|k))(y_i(k|k) - \hat{z}(k|k))^T + R_k \tag{25}$$

$$P(k+1|k+1) = \sum_{i=0}^{2n} W_i^{c}(x_i(k+1|k) - \hat{x}_i(k+1|k))(y_i(k+1|k) - \hat{z}(k+1|k))^T \tag{26}$$

For all the above equations, it is assumed that process and measurement equations are linear with respect to noise. But in most of the real-life nonlinear systems, the noise may enter the process and measurement equations nonlinearly. In those cases, UKF demonstrated above will not be rigorous as it will treat the noise as additive. To handle those kinds of situations, noise vector is augmented with the state vector as in [11], i.e.

$$x_a(k|k) = \begin{bmatrix} \hat{x}(k|k) \\ w_k \\ v_k \end{bmatrix} \tag{27}$$

Then, the initialisation to the UKF will become

$$x_a(0|0) = \begin{bmatrix} \hat{x}(0|0) \\ 0 \\ 0 \end{bmatrix} \tag{28}$$

$$P_a(0|0) = \begin{bmatrix} P(0|0) & 0 & 0 \\ 0 & Q_0 & 0 \\ 0 & 0 & R_0 \end{bmatrix} \tag{29}$$

The above presented UKF algorithm can be used to estimate the mean and covariance of the process and measurement equations. These two algorithms, i.e. EKF and UKF, will be applied in the next section to the cart pendulum system and the boom crane system to test their efficacy in estimation and noise reduction.

3 Problem Formulation

For an underactuated and highly nonlinear system such as the boom crane, the controller performance and stability are of utmost importance. Generally for such systems, feedback controller is necessary so that the unactuated degrees of freedom can also be controlled. However, most of the feedback taken via sensors is often noisy, and the use of the noisy feedback in the controller can impact the system adversely. Moreover, often if a full state feedback controller is used, often all the state information may not be available. Given these two factors, EKF and UKF can be used so that noise-free state estimations can be obtained [1, 2, 5] as shown in Fig. 2.

As seen from the descriptions of both the EKF and UKF algorithms, it is evident that EKF relies on the linearisation about one point and UKF linearises the system about the sigma points. Hence, it is expected that with the increase in nonlinearity, the performance of EKF will deteriorate, whereas UKF should perform better.

This study aims to observe the efficacy of EKF and UKF in state estimation for highly nonlinear and underactuated systems so that the estimated noise-free states can be used by the controller. The two nonlinear estimators, EKF and UKF, described in the above sections are applied on the cart pendulum system and offshore boom crane system with some noise level in the MATLAB environment. The dynamics of both systems are discussed in this section.

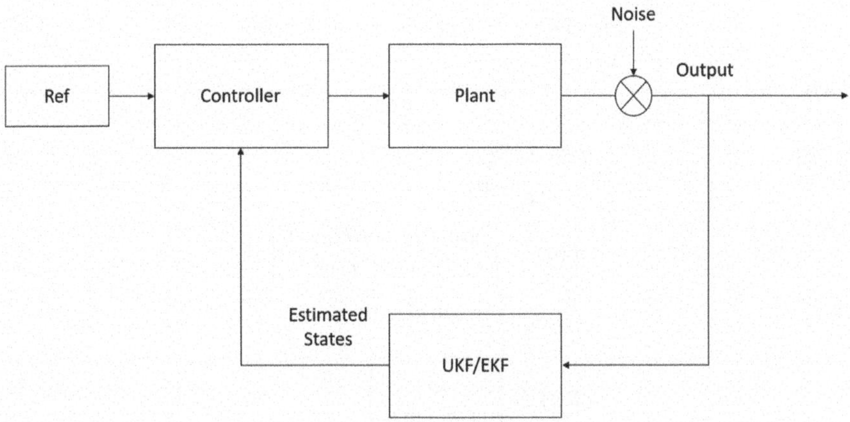

Fig. 2 Block diagram of filter positioning with controller and plant

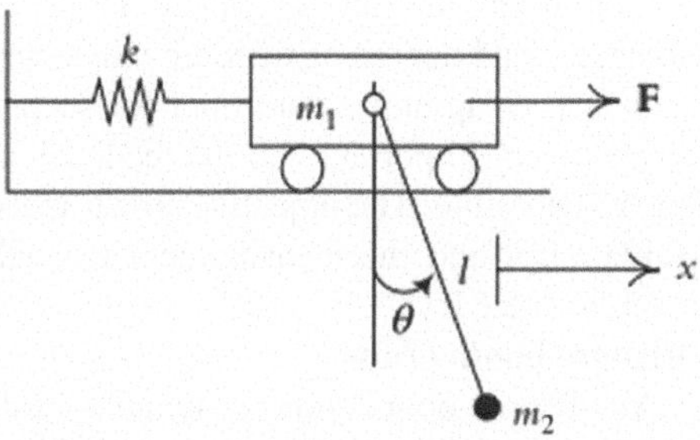

Fig. 3 Cart pendulum system

3.1 Dynamic Development

Cart pendulum system: Cart pendulum system is a nonlinear underactuated system which can be extensively used for the different control experiment. It has been used as a benchmark system for testing the algorithms related to the estimation of the states and removal of the measurement noise [12–14].

The crane system also has a similar behaviour as the cart pendulum system, and hence in this paper, the cart pendulum dynamics having cart mass m_1, the mass of pendulum m_2 with a rope of length l as shown in Fig. 3 is used to represent the crane movement. Using Lagrange's method, the dynamics of the cart pendulum system can be derived as follows [15]:

$$F = (m_1 + m_2)\dot{x} + m_2\dot{\theta}\cos\theta - m_2 l\dot{\theta}^2 \sin\theta + kx \tag{30}$$

$$0 = m_2 l^2\dot{\theta} + m_2 l\dot{x}\cos\theta + m_2 g l \sin\theta \tag{31}$$

Above dynamics can be made more comprehensible as follows:

$$M(q)\ddot{q} + c(q,\dot{q})\dot{q} + G(q) = u + f_{\text{dist}} \quad (32)$$

where

$$q = [x \ \ \theta]^T$$
$$u = [F \ \ 0]^T$$
$$M(q) = \begin{bmatrix} m_1 + m_2 & m_2 l \cos\theta \\ m_2 l \cos\theta & m_2 l^2 \end{bmatrix}$$
$$C(q) = \begin{bmatrix} 0 & -m_2\dot{\theta}\sin\theta \\ 0 & 0 \end{bmatrix}$$
$$G(q) = \begin{bmatrix} kx \\ m_2 g l \sin\theta \end{bmatrix}$$

As evident from the above dynamics, the position of the cart is the actuated degree of freedom where the force is applied. However, the angular motion of the pendulum is the unactuated degree of freedom whose motion is by the linear motion of the cart. Evidently, designing a controller for such a system is a challenging task, and this makes it very important that the feedback controller uses noise-free information as much as possible. The objective now is to estimate the cart position x, the pendulum oscillation θ and the measurement through the state having cart position x with different noise levels.

Offshore boom crane

An offshore boom crane system has 12 degrees of freedoms (DOFs) in total. The crane has six DOFs itself, the boom's luff and slew angles, the 2D swing of payload, cable length and luff angle of the jib. Again, the attached ship has six DOFs, i.e. roll, pitch, yaw, surge, sway and heave [6].

The seaboard crane system considered throughout this paper comprises a boom crane installed in a seaboard as shown in Fig. 4a. The coordinates of the offshore boom crane are shown in Fig. 4b, where (o_N, x_N, y_N, z_N) and (o_B, x_B, y_B, z_B) represent the coordinates of the earth and the system itself, respectively. The seaboard boom crane motion is represented by the following generalised coordinates:

α = slew angle,
β = luff angle,
l = length of the rope,
θ_1 = tangential pendulation due to motion of the tower,
θ_2 = radial sway due to motion of the boom.

The values of the tower height h and the boom length L_b are considered to be constant.

Applying Lagrange's analysis, the dynamics of the seaboard crane system are developed in [7] which will be used for simulation in this paper. The simplified dynamics are as follows:

$$M(x)\ddot{x} + c(x,\dot{x})\dot{x} + B\dot{x} + G(x) = u + f_{dist} \quad (33)$$

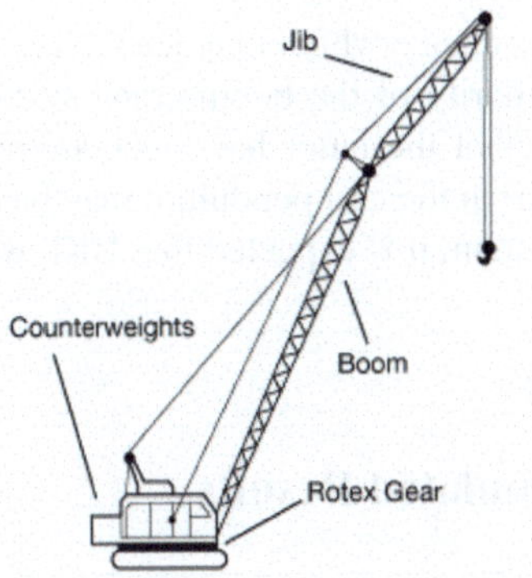

(a) Land fixed crane structure

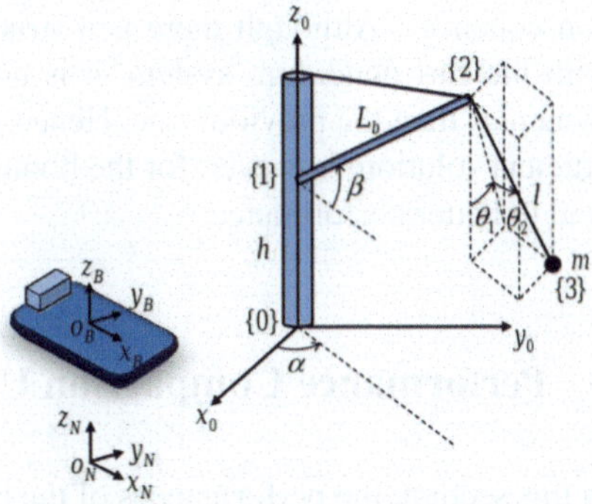

(b) Offshore boom crane motion

Fig. 4 Two different crane structures [6]

where

$$x = [\alpha(t)\ \ \beta(t)\ \ l(t)\ \ \theta_1(t)\ \ \theta_2(t)]^T$$
$$u = [u_\alpha\ \ u_\beta\ \ u_l\ \ 0\ \ 0]^T$$
$$M(x) = \begin{bmatrix} m_{11} & m_{12} & m_{13} & m_{14} & m_{15} \\ m_{21} & m_{22} & m_{23} & m_{24} & m_{25} \\ m_{31} & m_{32} & m_{33} & m_{34} & m_{35} \\ m_{41} & m_{42} & m_{43} & m_{44} & m_{45} \\ m_{51} & m_{52} & m_{53} & m_{54} & m_{55} \end{bmatrix}$$
$$C(x, \dot{x}) = \begin{bmatrix} c_{11} & c_{12} & c_{13} & c_{14} & c_{15} \\ c_{21} & c_{22} & c_{23} & c_{24} & c_{25} \\ c_{31} & c_{32} & c_{33} & c_{34} & c_{35} \\ c_{41} & c_{42} & c_{43} & c_{44} & c_{45} \\ c_{51} & c_{52} & c_{53} & c_{54} & c_{55} \end{bmatrix}$$
$$G(x) = \begin{bmatrix} g_{11} \\ g_{12} \\ g_{13} \\ g_{14} \\ g_{15} \end{bmatrix}$$

where f_{dist} is the external lumped disturbances. The details of all the nonzero entities of the boom crane matrices are defined in the Appendix. In the case of the boom crane system dynamics, the tangential sway of the tower (θ_1) and the radial sway due to the motion of the boom (θ_2) are the unactuated states. For the crane operation, keeping these two motions in check is imperative for the performance as well as safety of operation. However, a controller designed based on noisy measurement data in such cases may lead to a reduction in performance which may lead to costly consequences. Although there is a structural similarity of the boom crane system with the cart pendulum system, it is noteworthy that the latter has much simpler dynamics than the previous one. Hence, using EKF in the cart pendulum may be an efficient solution; however, for the boom crane system, it is expected that UKF will lead to better performance.

4 Performance Comparison Using Simulated Results

In this section, the performances of the EKF and UKF algorithms are verified using MATLAB simulations. The dynamics are first applied with both the EKF and the UKF, and then their results are compared.

Cart pendulum system

In this section, the performances of the filters are evaluated by simulating the EKF and UKF equations using dynamics in Eqs. (30) and (31). The parameters of the cart pendulum system used for the MATLAB simulation are as follows:

Mass of cart, $m_1 = 2\ \text{kg}$

Mass of pendulum, $m_1 = 0.1\ \text{kg}$

Length of the rope, $l = .25$ m

Spring constant, $k = 1$

Gravitational constant, $g = 9.8\ \text{kg}\,\text{ms}^{-2}$

The initial conditions of the system and both the estimators are given as follows:

$$q_0 = \begin{bmatrix} 0 \\ 0 \end{bmatrix} \qquad p_0 = \begin{bmatrix} 0.1 & 0 \\ 0 & 0.01 \end{bmatrix}$$

The sampling time to discretise the model dynamics and the EKF and UKF equations are 0.01 s. Simulation results are taken after 1500 iterations, and each simulation is carried out for 10 secs. A step input is applied to the system. Process noise of variance 0.1 and measurement noise of variance 0.8 are applied to the state crane position. It is to be noted that measurement of the output is supposed to be taken from the first state, i.e. the crane position. For UKF, the assumptions for the tuning parameters are as follows:

$$\alpha = 0.01 \qquad \beta = 2 \qquad \kappa = 0 \qquad n = 5$$

The cart position and pendulum oscillation results for both EKF and UKF are shown in Fig. 5. Figures 6 and 7 show the simulated result for error in estimated results both for crane position and pendulum oscillation.

From Fig. 6, it is very much clear that the UKF has performed better than EKF in terms of the linear position of the cart; however, for the angular motion (i.e. the underactuated degree of freedom) the error of the UKF increases with time. This can be attributed to the fact that there is no feedback controller on the system. A

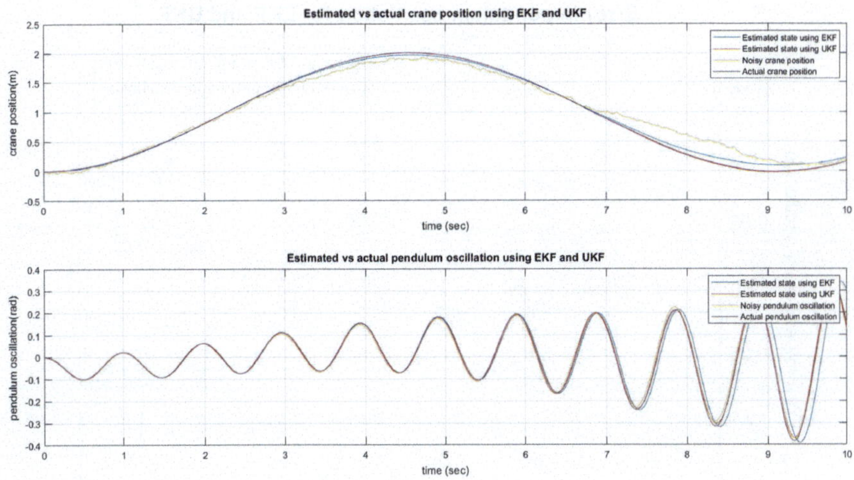

Fig. 5 Performance Comparison of EKF and UKF linear position and oscillation of pendulum

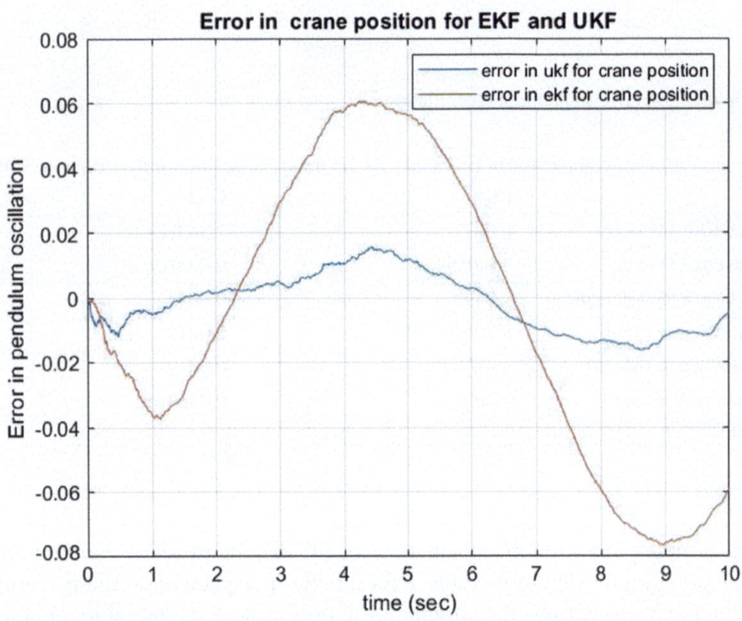

Fig. 6 Estimation error in cart position

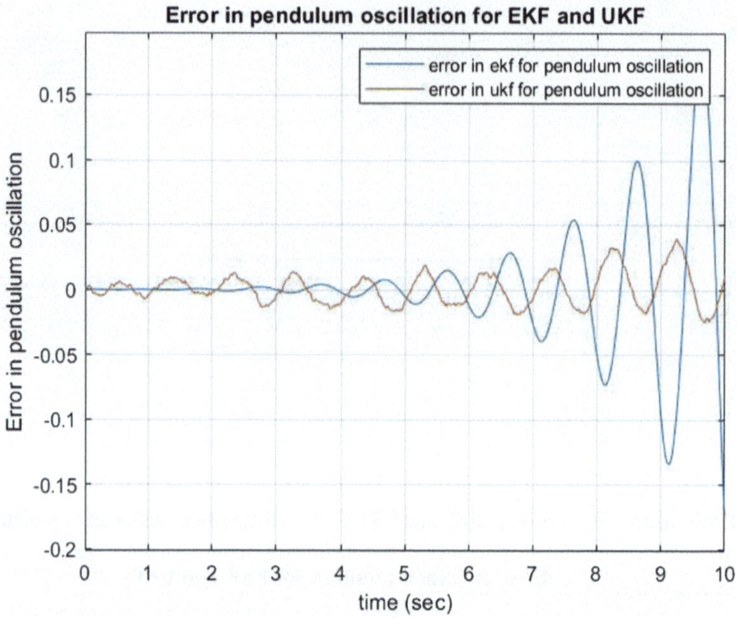

Fig. 7 Estimation error in pendulum oscillation

Table 1 Performance comparison of EKF and UKF for the crane position and oscillation estimation

	EKF	UKF
Position error (MSE)	0.0027	2.3438e−05
Position error (Mean)	−0.0154	−0.0019
Position error (Std deviation)	0.0499	0.0045
Oscillation error (MSE)	8.2063e−05	0.0019
Oscillation error (Mean)	−4.7469e−04	0.0019
Oscillation error (Std Deviation)	0.0091	0.0439

detailed comparison of the mean square error (MSE), the mean of the error and the standard deviation is shown in Table 1 for a better perspective on the performance of the filters. As seen from the simulation figures as well as the table, although in the underactuated degree, the UKF is not showing much improvement; however, in the actuated degree of freedom, i.e. the cart position, the estimation by UKF is much better with a mean of the error nearer to zero and a low standard deviation.

Offshore boom crane:

In the second experiment, the above procedure is extended to deal with higher-order dynamics, i.e. for an offshore boom crane whose dynamics are mentioned in Eq. (33). There are five states where two of them are underactuated, viz. tangential sway and

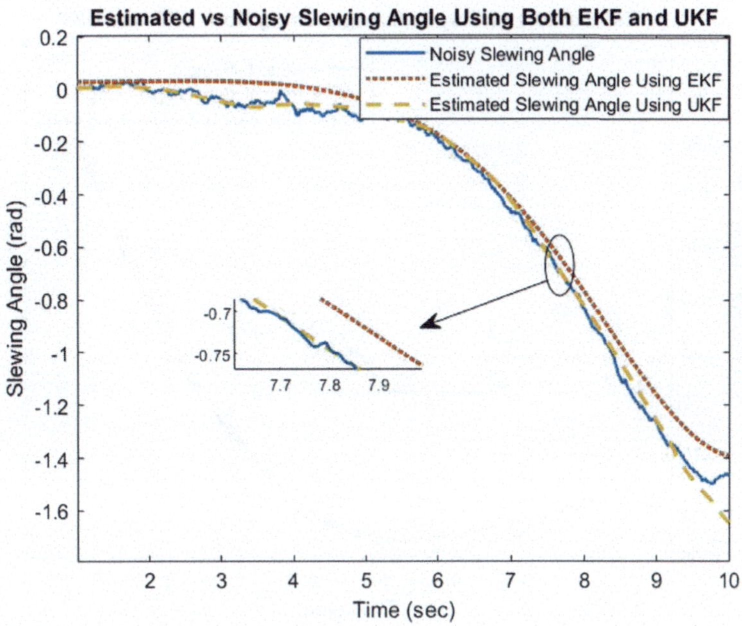

Fig. 8 Estimated versus noisy slewing angle

radial sway angles. Unit step signals are applied to the three input channels slewing angle, luffing angle and rope length. Here also to avoid the direct differentiation of the states, they are transformed into ten states while simulation. The initial positions and velocities of the states are $x = [0, 0, 1.57, 0, 0, 0, 0, .09, 0, 0]$. The initial error covariance for both the estimators is set to $P = \text{diag}(0.001, 0.0001, 3.5, 0.001, 0.001)$.

The estimation results for all the five states are shown in Figs. 8, 9, 10, 11 and 12, with the zoomed figure shown in inset of each plot. These plots clearly show that UKF performs better than EKF when the nonlinearity in the system increases. Figures 13, 14, 15, 16 and 17 show the estimated error for all the states using both the EKF and UKF. From Figs. 8, 10, 11 and 12, it can be observed that the EKF deviates from the actual states and is noisier compared to the UKF output. This shows the better filtering performance of UKF for a highly nonlinear system. As opposed to the cart pendulum system which had only two states and lower nonlinearities as compared to the boom crane system, the UKF can give better tracking even for the unactuated degrees of freedoms as can be observed in Figs. 11 and 12.

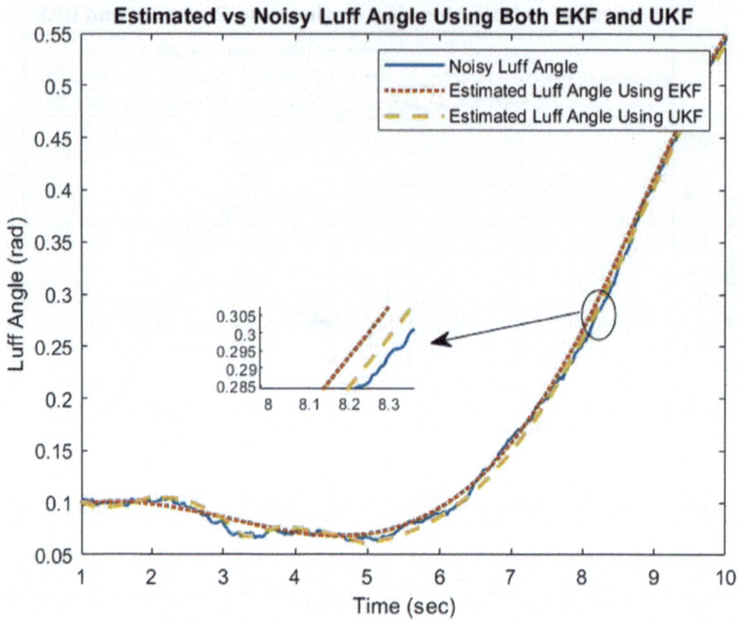

Fig. 9 Estimated versus noisy luffing angle

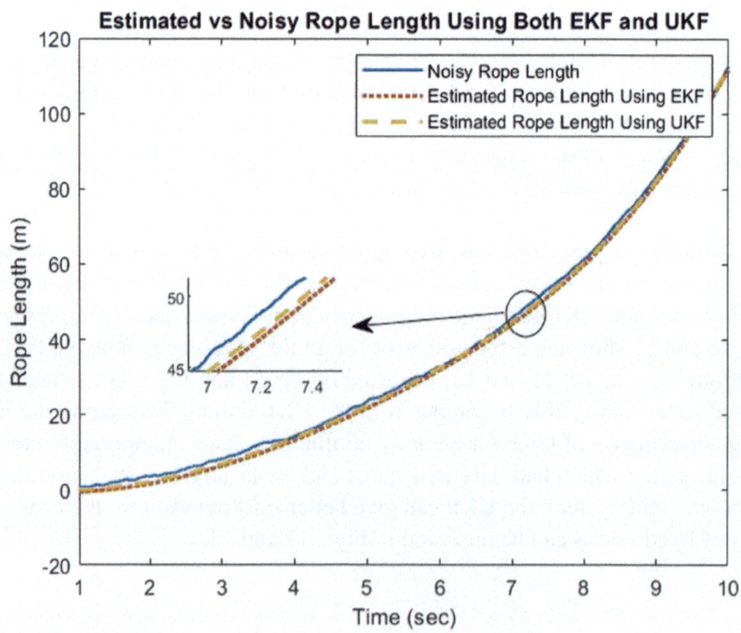

Fig. 10 Estimated versus noisy rope length

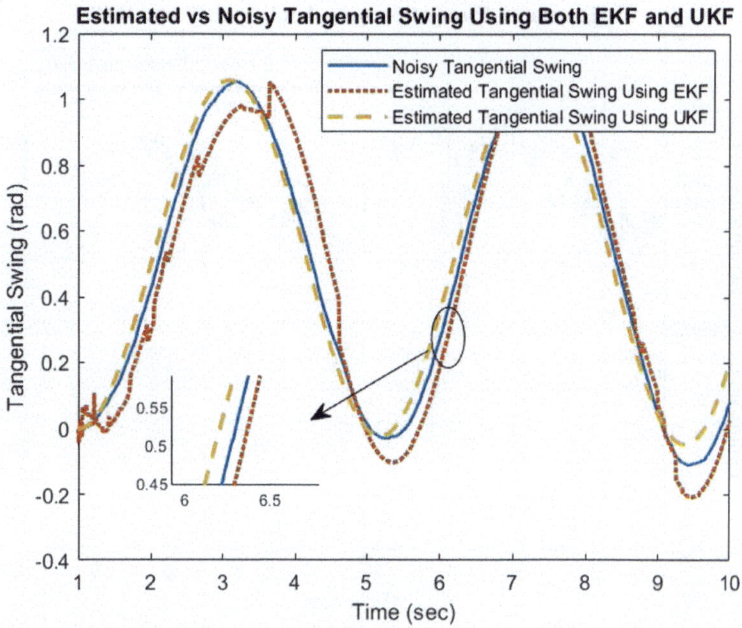

Fig. 11 Estimated versus noisy tangential sway

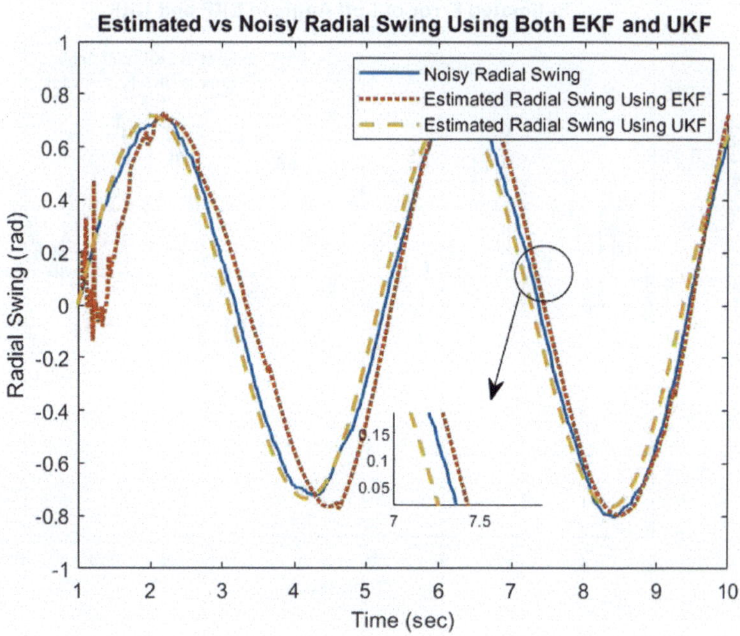

Fig. 12 Estimated versus noisy radial sway

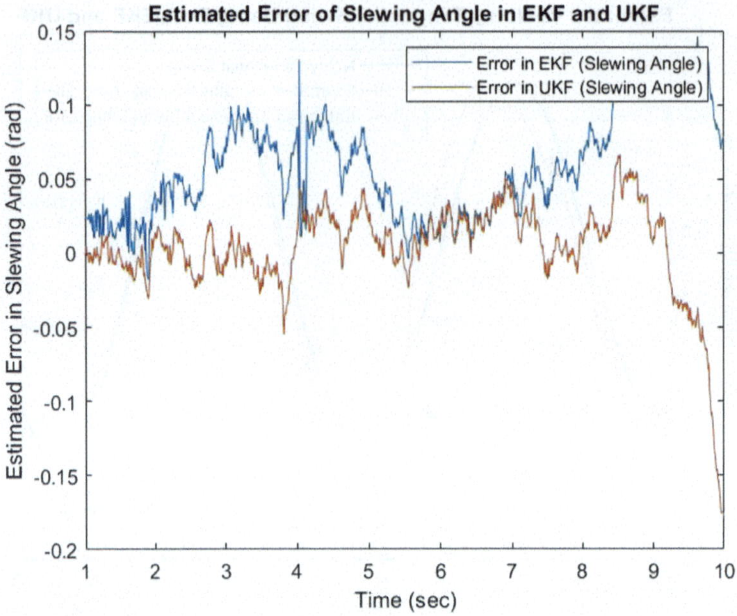

Fig. 13 Estimated error in slew angle

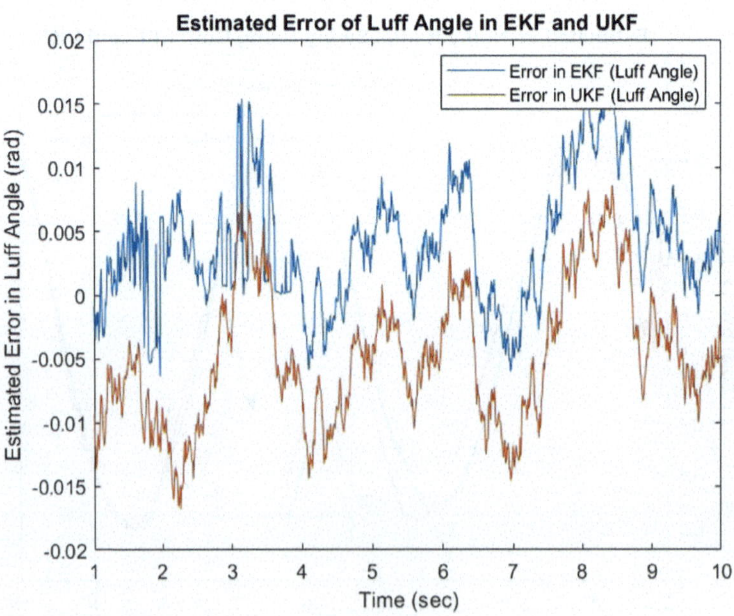

Fig. 14 Estimated error in luff angle

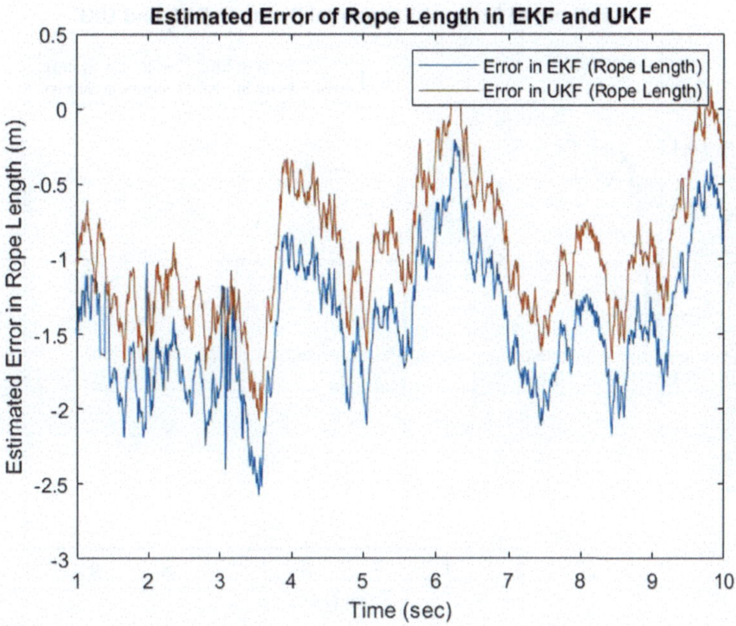

Fig. 15 Estimated error in rope length

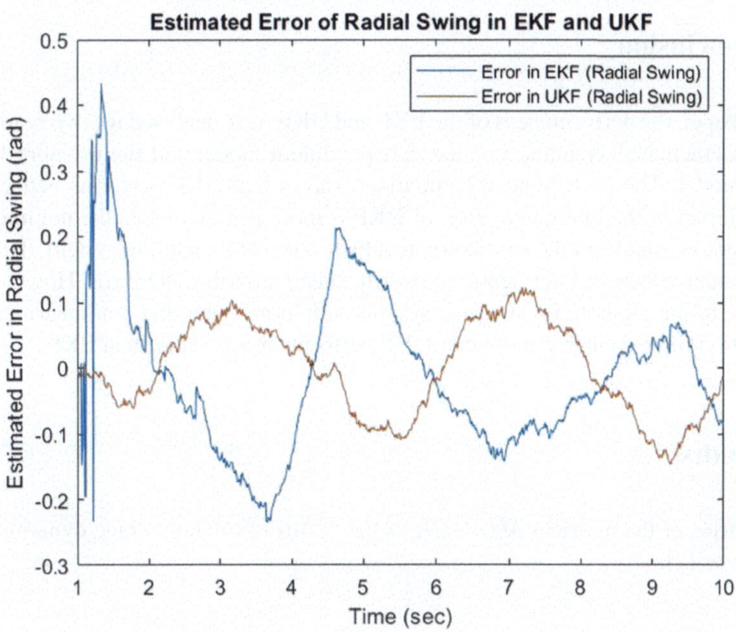

Fig. 16 Estimated error in radial angle

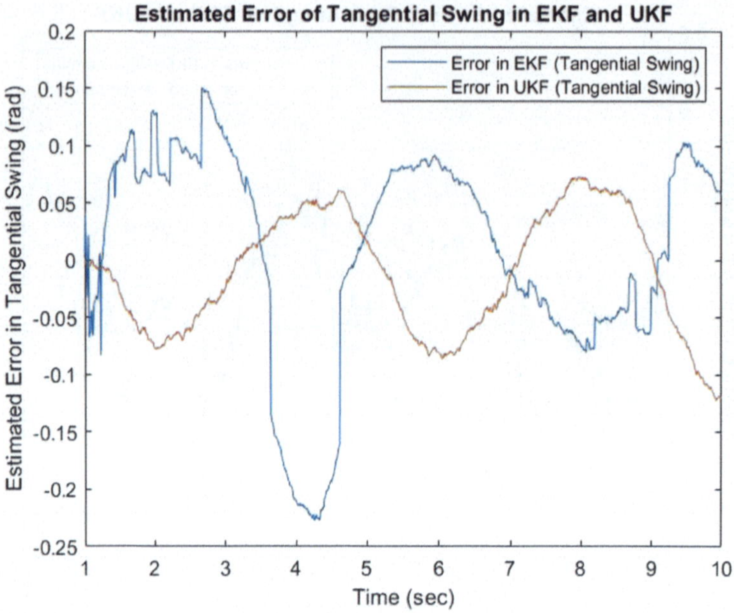

Fig. 17 Estimated error in tangential angle

5 Conclusion

In this paper, the performances of the EKF and UKF were analysed for two nonlinear and underactuated systems, viz. the cart pendulum model and the offshore boom crane model. The performance comparison shows that UKF performs better than EKF. However, the implementation of UKF is more justified when the nonlinearity is higher. As seen from the simulation results for the cart pendulum system, the EKF is sufficient, whereas UKF tends to overfit, thus increasing the error. However, in the case of the offshore boom crane system with more states and nonlinearities, the simulation figures clearly show that UKF performs much better than EKF.

Appendix

The entries of the matrices $M(q)$, $C(q, \dot{q})$,B, $G(q)$ of offshore crane dynamics are as follows [7]:

$$m_{11} = J_\alpha + mx_3^2 + mL^2\cos^2 x_2 + 2mLx_3\cos x_2 sinx_5 - mx_3^2\cos^2 x_4\cos^2 x_5$$
$$m_{22} = J_\beta + mL^2$$

$$m_{33} = m + M$$

$$m_{44} = mx_3^2 \cos^2 x_5$$

$$m_{55} = mx_3^2$$

$$m_{12} = m_{21} = mLx_3 \sin x_2 \sin x_4 \cos x_5$$

$$m_{13} = m_{31} = mL \cos x_2 \sin x_4 \cos x_5$$

$$m_{14} = m_{41} = m(Lx_3 \cos x_2 + x_3^2 \sin x_5) \cos x_4 \cos x_5$$

$$m_{15} = m_{51} = -m(Lx_3 \cos x_2 \sin x_5 + x_3^2) \sin x_4$$

$$m_{23} = m_{32} = -mL(\sin x_2 \sin x_5 + \cos x_2 \cos x_4 \cos x_5)$$

$$m_{24} = m_{42} = mLx_3 \cos x_2 \sin x_4 \cos x_5$$

$$m_{25} = m_{52} = mLx_3(-\sin x_2 \cos x_5 + \cos x_2 \cos x_4 \cos x_5)$$

$$\begin{aligned} c_{11} = {} & mx_3x_8(1 - \cos^2 x_4 \cos^2 x_5) + mLx_8 \cos x_2 \sin x_5 + m(x_3^2 \cos^2 x_4 \sin x_5 \\ & + Lx_3 \cos x_2) \cos x_5 x_{10} - m(L^2 \cos x_2 + Lx_3 \sin x_5) \sin x_2 x_7 \\ & + mx_3^2 \sin x_4 \cos x_4 \cos^2 x_5 x_9 \end{aligned}$$

$$c_{12} = mLx_3 \cos x_2 \sin x_4 \cos x_5 x_7 - m \sin x_2 x_6 (L^2 \cos x_2 + Lx_3 \sin x_5)$$

$$\begin{aligned} c_{13} = {} & mx_3 \cos x_4 \cos x_5 \sin x_5 x_9 + mL \cos x_2 (\sin x_5 x_6 + \cos x_4 \cos x_5 x_9 \\ & - \sin x_4 \sin x_5 x_1 0) + mx_3 x_6 (1 - \cos^2 x_4 \cos^2 x_5) - mx_3 \sin x_4 x_{10} \end{aligned}$$

$$\begin{aligned} c_{14} = {} & mx_3^2 \sin x_4 \cos x_4 \cos^2 x_5 x_6 + mx_8 \cos x_4 \cos x_5 (L \cos x_2 + x_3 \sin x_5) \\ & - m(Lx_3 \cos x_2 + x_3^2 \sin x_5) \sin x_4 \cos x_5 x_9 - m(Lx_3 \cos x_2 \sin x_5 \\ & + x_3^2 \sin^2 x_5) \cos x_4 x_{10} \end{aligned}$$

$$\begin{aligned} c_{15} = {} & m(Lx_3 \cos x_2 + x_3^2 \cos^2 x_4 \sin x_5) \cos x_5 x_6 - mx_8 \sin x_4 (x_3 + L \cos x_2 \sin x_5) \\ & - mLx_3 \cos x_2 \sin x_4 \cos x_5 x_{10} - m(Lx_3 \cos x_2 \sin x_5 + x_3^2 \sin^2 x_5) \cos x_4 x_9 \end{aligned}$$

$$\begin{aligned} c_{21} = {} & m \sin x_2 (L^2 \cos x_2 x_6 + Lx_3 \sin x_5 x_6 + Lx_8 \sin x_4 \cos x_5 + Lx_3 \cos x_4 \cos x_5 x_9 \\ & - Lx_3 \sin x_4 \sin x_5 x_{10}) \end{aligned}$$

$$\begin{aligned} c_{23} = {} & mL \sin x_4 \cos x_5 (\sin x_2 x_6 + \cos x_2 x_9) + mL(\cos x_2 \sin x_4 \cos x_5 x_9 \\ & - \sin x_2 \cos x_5 x_{10} + \cos x_2 \cos x_4 \sin x_5 x_{10}) \end{aligned}$$

$$\begin{aligned} c_{24} = {} & mLx_3 \cos x_4 \cos x_5 (\sin x_2 x_6 + \cos x_2 x_9) + mL \cos x_2 x_4 (x_8 \cos x_5 \\ & - x_3 \sin x_5 x_{10}) \end{aligned}$$

$$\begin{aligned} c_{25} = {} & -mLx_3 \sin x_2 \sin x_4 \sin x_5 x_6 - mLx_8 (\sin x_2 \cos x_5 + \cos x_2 \cos x_4 \sin x_5) \\ & - mLx_3 \cos x_2 \sin x_4 \sin x_5 x_9 + mLx_3 x_{10} (\sin x_2 \sin x_5 + \cos x_2 \cos x_4 \sin x_5) \end{aligned}$$

$$\begin{aligned} c_{31} = {} & mx_6(-x_3 + x_3 \cos^2 x_4 \cos^2 x_5 - L \cos x_2 \sin x_5) \\ & - mL \sin x_2 \sin x_4 \cos x_5 x_7 - mx_3 \cos x_4 \sin x_5 \cos x_5 x_9 + mx_3 \sin x_4 x_{10} \end{aligned}$$

$$c_{32} = mL(\sin x_2 \cos x_4 \cos x_5 x_7 - \sin x_2 \sin x_4 \cos x_5 x_6 - \cos x_2 \sin x_5 x_7$$

$$c_{34} = -mx_3(\cos^2 x_5 x_9 + \cos x_4 \sin x_5 \cos x_5 x_6)$$

$$c_{35} = mx_3(\sin x_4 x_6 - x_{10})$$

$$c_{41} = m\cos x_4 \cos x_5(-x_3^2 \sin x_4 \cos x_5 x_6 - Lx_3 \sin x_2 x_7 + x_3 x_8 \sin x_5 + x_3^2 \cos x_5 x_{10})$$

$$c_{42} = -mLx_3 \sin x_2 \cos x_5(\cos x_4 x_6 + \sin x_4 x_7)$$

$$c_{43} = mx_3(\cos x_4 \sin x_5 \cos x_5 x_6 + \cos^2 x_5 x_9)$$

$$c_{44} = mx_3 x_8 \cos^2 x_5 - mx_3^2 \sin x_5 \cos x_5 x_{10}$$

$$c_{45} = mx_3^2(\cos x_4 \cos^2 x_5 x_6 - \sin x_5 \cos x_5 x_9)$$

$$c_{51} = -m\cos x_5 x_6(Lx_3 \cos x_2 + x_3^2 \cos^2 x_4 \sin x_5) + m\sin x_4(Lx_3 \sin x_2 \sin x_5 x_9 - x_3 x_8) - mx_3^2 \cos x_4 \cos^2 x_5 x_9$$

$$c_{52} = mLx_3(\sin x_2 \sin x_4 \sin x_5 x_6 - \cos x_2 \cos x_5 x_7 - \sin x_2 \cos x_4 \sin x_5 x_7)$$

$$c_{53} = mx_3(-\sin x_4 x_6 + x_{10})$$

$$c_{54} = mx_3^2(-\cos x_4 \cos^2 x_5 x_6 + \sin x_5 \cos x_5 x_9)$$

$$c_{55} = mx_3 x_8$$

$$g_1 = 0 \qquad g_2 = mgl\cos x_2$$

$$g_3 = -mgL\cos x_4 \cos x_5 \qquad g_4 = mgx_3 \sin x_4 \cos x_5$$

$$g_5 = mgx_3 \cos x_4 \sin x_5$$

$$b_1 = 0.1\,\mathrm{Nm/rads^{-1}} \qquad b_2 = 0.1\,\mathrm{Nm/rads^{-1}} \qquad b_3 = 0.2\,\mathrm{Nm/rads^{-1}}$$

$$b_4 = 0.1\,\mathrm{Nm/rads^{-1}} \qquad b_5 = 0.1\,\mathrm{Nm/rads^{-1}}$$

where,

M = Mass of the boom = m = mass of the pendulum
L = Length of the boom g = gravitational constant
J_α = Moment of inertia of α J_β = Moment of inertia of β

References

1. Kim, Y., Bang, H.: Introduction to Kalman filter and its applications. In: Introduction and Implementations of the Kalman Filter 5 Nov 2018. IntechOpen
2. Yang, S., Li, H.: Application of EKF and UKF in target tracking problem. In: 2016 8th International Conference on Intelligent Human-Machine Systems and Cybernetics (IHMSC) 27 Aug 2016, vol. 1, pp. 116–120. IEEE

3. Niazi, S.: Estimation of los rates for target tracking problems using ekf and ukf algorithms-a comparative study. Int. J. Eng. **28**(2), 172–9 (2015)
4. Toloei, A., Niazi, S.: State estimation for target tracking problems with nonlinear Kalman filter algorithms. Int. J. Comput. Appl. **98**(17) (2014)
5. Julier, S.J., Uhlmann, J.K.: Unscented filtering and nonlinear estimation. Proc. IEEE **92**(3), 401–22 (2004)
6. Lu, B., Fang, Y., Sun, N., Wang, X.: Antiswing control of offshore boom cranes with ship roll disturbances. IEEE Trans. Control Syst. Technol. **26**(2), 740–7 (2017)
7. Ismail, R.R., Ha, Q.P.: Trajectory tracking and anti-sway control of three-dimensional offshore boom cranes using second-order sliding modes. In: 2013 IEEE International Conference on Automation Science and Engineering (CASE), 17 Aug 2013, pp. 996–1001. IEEE
8. Faragher, R.: Understanding the basis of the Kalman filter via a simple and intuitive derivation [lecture notes]. IEEE Signal Processi. Mag. **29**(5), 128–32 (2012)
9. Ribeiro, M.I.: Kalman and Extended Kalman Filters: Concept, Derivation and Properties, vol 43. Institute for Systems and Robotics (2004)
10. Julier, S., Uhlmann, J., Durrant-Whyte, H.F.: A new method for the nonlinear transformation of means and covariances in filters and estimators. IEEE Trans. Autom. Control **45**(3), 477–82 (2000)
11. Sarkka, S.: On unscented Kalman filtering for state estimation of continuous-time nonlinear systems. IEEE Trans. Autom. Control **52**(9), 1631–41 (2007)
12. Chatterjee, D., Patra, A., Joglekar, H.K.: Swing-up and stabilization of a cart-pendulum system under restricted cart track length. Syst. Control Lett. **47**(4), 355–64 (2002)
13. Adhikary, N., Mahanta, C.: Integral backstepping sliding mode control for underactuated systems: swing-up and stabilization of the Cart-Pendulum System. ISA Trans. **52**(6), 870–80 (2013)
14. Prasad, L.B., Tyagi, B., Gupta, H.O.: Optimal control of nonlinear inverted pendulum system using PID controller and LQR: performance analysis without and with disturbance input. Int. J. Autom. Comput. **11**(6), 661–70 (2014)
15. Niku, S.B.: Introduction to Robotics: Analysis, Systems, Applications. Prentice Hall, New Jersey (2001)

Performance Analysis of Speech Command Recognition Using Support Vector Machine Classifiers

M. Venkata Subbarao, Akhendra Kumar Padavala, and Kudupudi Durga Harika

Abstract Recently attention of researchers towards Automatic Speech Recognition (ASR) is tremendously increased than ever because of its applications in security applications, home applications, military applications, health care applications, language learning applications, and marketing applications. Initially, most of the researches in respect of ASR is confined to robotics and security applications. Further, attention is also motivated towards home applications. Now industrial automation is at the tipping point with ASR. Motivated by this, the proposed paper deals with the Speech Command Recognition (SCR) for next-generation automated vehicles. For SCR, a set of features such as Gammatone Cepstrum Coefficient (GTCC), Mel-Frequency Cepstrum Coefficients (MFCC), and Pitch, Spectral Flux, and Spectral Entropy are considered for training the proposed Support Vector Machine (SVM) classifier. Performance analysis is carried with different training rates and with different noisy conditions. Simulation results depict the superiority of the proposed classifier than the existing methods.

1 Introduction

Speech is the easiest way and natural means of communication used by humans to communicate with each other. SCR allows us to communicate with machines such as laptops, computers, mobile phones, etc. SCR is essential because there is a whole raft of applications where speech is the most efficient and convenient way of communicating. Speech recognition has become essential in our day-to-day life without even knowing that it is invading our lives. This section presents a detailed survey of different existing ASR approaches starting from very early approaches to the most accurate recent methods. The pros and cons of various ASR techniques are also presented in this section to identify the further direction of potential research. It includes various techniques like Machine Learning, Neural Networks, Deep Learning, and different feature extraction techniques like MFCC, GTCC, LPC,

M. V. Subbarao (✉) · A. K. Padavala · K. D. Harika
Department of Electronics and Communication Engineering, Shri Vishnu Engineering College for Women (Autonomous), Bhimavaram, AP, India

J. Gu et al. (eds.), *Communication and Control for Robotic Systems*, Smart Innovation, Systems and Technologies 229, https://doi.org/10.1007/978-981-16-1777-5_19

etc. proposed by various authors. Guiming et al. [1] have proposed that speech recognition plays a vital role in establishing a man–machine interface, and it can be achieved using artificial intelligence. Neural networks are the best approach to realize speech recognition as it eliminates the limitations of traditional techniques. This paper uses the five-layer Convolutional Neural Network (CNN) principles for speech recognition and the backpropagation method to train the network. When compared with Deep Neural Networks (DNN), CNN reduces the model complexity to a greater extent. Zhao et al. [2] have proposed symmetric orthogonalization in Independent Component Analysis (ICA) for extracting new speech features to strengthen the speech recognition systems. These features were examined in eight various noise environments. The results demonstrated that the recognition rate of these features is higher when compared to that of MFCC features by 6.17%, especially in low Signal to Noise Ratios. This method has better performance when compared to other conventional means. Solera-Urena et al. [3] have proposed a hybrid WLS-SVC/HMM system. The system uses a weighted least squares (WLS) procedure for training, which reduces the complexity and other practical issues related to the SVM. Performance accuracy with this procedure is improved in a clean environment as well as in a noisy environment. Ganapathiraju et al. [4] have demonstrated the usage of SVM for continuous speech. In this paper, SVMs are also applied to continuous speech recognition and yields a reduction in error rate for a continuous speech. The obtained results in the experiment indicate the high classification rate for SVM. Al-Alaoui et. al. [5] have developed a speech recognition component that would assist adult illiterates for better reading capabilities. This paper uses Neural Networks (NN) for classification purposes, and they are trained using the Al-Alaoui Algorithm. This method was proven to be the best in sentence recognition when compared with the HMM method.

Kumar et al. [7] have proposed that Machine Learning (ML) algorithms like Tree and KNN are used in speech recognition. Accuracy of identification will be less for more extensive database and training dataset. The combined application of MLSVM, a Machine Learning Algorithm, and feature vector MFCC improves the accuracy for a vast database, which are approximately twenty words that can generate 40 commands. MFCC shown 7% more efficiency for words with the same phonetics proves that MLSVM and feature vector MFCC can be a better choice for big and similar phonetics datasets. Sonkamble et al. [8], have proposed the Linear-Polynomial kernel (LP), which is the new form of kernel function used for designing the SVMs for SCR. There is a performance improvement in this speech recognition system when compared to linear kernel function and other polynomial kernel functions. Gupta et al. [9] have proposed two approaches for speech recognition. In one method, for Supervised Learning, a Bi-directional Recurrent Neural Network (RNN) with Long Short-Term Memory model (LSTM) can be done efficiently. In another method, a Restricted Boltzmann Machine (RBM) based model is designed for Unsupervised Learning. Dhankar [10] has given an overview of machine learning, deep learning, CMU sphinx for speech recognition. GMM-HMM model is used in CMU Sphinx to detect a specific word or a group of words that were spoken. DL approach is more accurate, and it can detect the words with 100% accuracy, whereas CMU Sphinx has an accuracy of 54.54% with a limited vocabulary.

In ML approaches accuracies are getting decreased when size of data set or commands increases. Feature extraction techniques can be improved. Most of the papers considered less no. of features for training and testing. SVM algorithm is widely used in more papers but, the basic SVM formalism suffers from three fundamental problems: scalability, sparsity, and Bayesianness. Error rates are high. They have not considered many noise conditions. Accuracies of systems are less. Training and testing rates of datasets effecting accuracy of the system, when ML techniques are applied. In order to overcome these limitations a wide variety of SVM classifiers are implemented in this work with different kernel functions.

The organization of the rest of the paper is as follows. Section 2 deals with the proposed framework for SCR is elaborated. Section 3 deals with the different SVM classifiers that are used for SCR. Section 4 deals with the performance analysis of proposed SVM classifiers under different noise conditions, training, and testing rates. Finally, Sect. 5 depicts the important conjectures of the research work that are derived through simulation.

2 Frame Work

The framework of the proposed classifier approach is shown in Fig. 1. Training phase includes generation of multiple copies of different voice commands under various noisy conditions, feature extraction, feature selection, and training mechanism of the proposed SVM classifier. In feature extraction stage, initially a set of 82 statistical

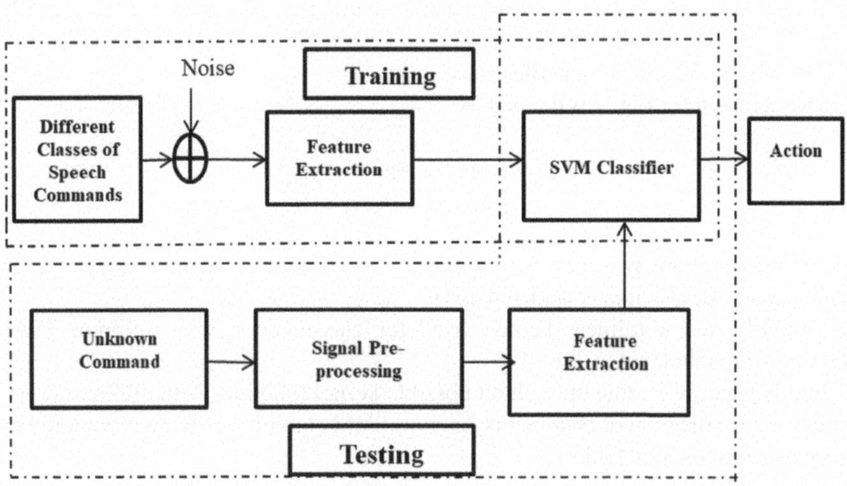

Fig. 1 Framework of the proposed approach

features considering MFCC, GTCC, Pitch, Spectral Flux and Spectral Entropy are extracted for each class of voice 9 command under varying noise conditions. In the last stage of training phase, proposed SVM classifiers are trained with the features of all voice command classes. In the testing phase, the selected features of the signal are extracted and passed through a trained SVM classifier to recognize the command of an unknown voice signal.

3 Support Vector Machine Classifiers

The SVM having the ability to classify data even it noisy. It is a supervised ML algorithm (SMLA) that able to classify the data using a subset of training data called support vectors. With the help of known set of data SMLA builds a model. Thereafter, if a new set of data is given to model it predicts an accurate output. The basic impression behind SVM classifier is that it makes a feature space via the characteristics in the training data. Thereafter, it tries to identify a hyper-plane or a decision boundary that splits the feature space into two halves. Here each part contains data points belonging to a single category in the data [10].

To distinguish between the classes, a linear kernel is considered in linear SVM and it is given as,

$$F(a, w) = a^T w \tag{1}$$

where a is the input feature vector and w is the weight vector.

The hyper plane with linear kernel and a constant w_0 is given by:

$$h(a) = a^T w + w_0 \tag{2}$$

The binary classification is illustrated in Fig. 2.

The decision for Fig. 2 is defined as

$$M = \begin{cases} M(P), \text{if } h(a) \geq 0 \\ M(Q), \text{if } h(a) < 0 \end{cases} \tag{3}$$

where P and Q are two classes, the class is recognized as member (M) of P if $a^T w + w_0 \geq 0$, else it is considered as Q.

Similarly, the non-linear kernels used for classification with different SVM classifiers are tabulated in Table1.

In this paper, we came up with all possible kernel functions with different properties for classification of commands. Various SVM classifiers that are considered in this paper is shown in Table 1.

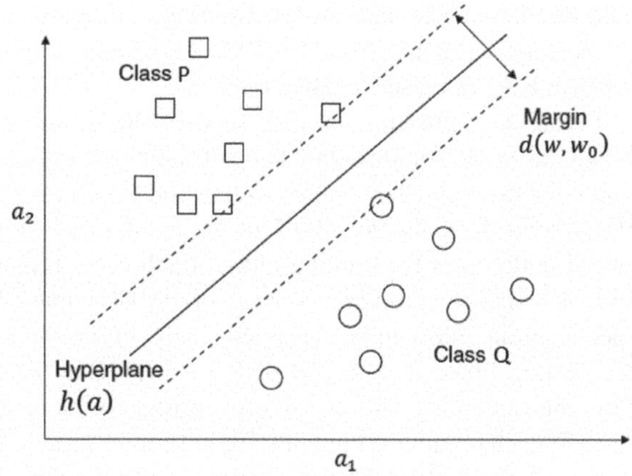

Fig. 2 Classification using linear SVM

Table 1 Characteristics of SVM classifiers

SVM classifier	Kernel	Kernel function	γ/σ
Linear (LSVM)	Linear	$F(a, w) = a^T w$	–
Quadratic (QSVM)	Polynomial	$K(X,Y) = (\gamma.X^TY + r)^d$	$\gamma > 0, d = 4$
Cubic (CSVM)		$K(X,Y) = (\gamma.X^TY + r)^d$	$\gamma > 0, d = 3$
Fine Gaussian (FGSVM)	Gaussian	$K(X,Y) = \exp(\|X–Y\|^2/2\sigma^2)$	$\sigma = \sqrt{p}/4$
Medium Gaussian (MGSVM)			$\sigma = \sqrt{p}$
Coarse Gaussian (CGSVM)			$\sigma = 4^*\sqrt{p}$

4 Simulation Results

Table 2 represents the simulation parameters for the simulation environment. It consists of 9 Speech commands which are down, happy, house, left, off, on, right, stop, up [11]. The performance is measured with additive white aussian noise as disturbance ranging from 0 to 40 dB. The size of data set is 90,000 × 82 which consists of 9 Speech commands each with 10,000 copies and 81 statistical features

Table 2 Simulation parameters

Parameter	Description
Speech commands	down, happy, house, left, off, on, right, stop, up
Features	MFCC, GTCC, Pitch, Spectral Flux, Spectral Entropy
Size of dataset	90,000*82
SNR	0–40 dB
Data set for training	90–50%
Data set for testing	10–50%

with one label. The data set for Training is ranging from 90–50% and the data set for Testing is ranging from 10–50%. Performance indices contain number of True Positive Rate & False Negative rate.

Figure 3a represents the data sets for the Training Phase at 0–40 dB SNR for MFCC at 90% Training. For better training, every command is generated 10,000 times for random SNR value, so that the total feature set size becomes 90,000 * 39. Among these features, 90% of the features are considered for training so there are 81,000 copies for training phase. Each color represents each speech command and each speech command consists of 10,000 copies. Since it is 0–40 dB SNR the speech commands can be classified easily. Figure 3b represents the Scatter Plot for the Testing Phase at 0–40 dB SNR for MFCC at 90% Training. Since it has 90% Training remaining 10% is used for testing so it has 9000 copies for testing phase. Each Speech command contains 10,000 copies represented in different colors. Since it is 0–40 dB SNR all the copies are correctly identified.

Figure 3a represents the dataset for training and testing purposes. Figure 3b represents the scatter plot after testing with LSVM. Here dot indicates the copies of each modulation class that are correctly identified and cross indicates the number of copies of each modulation class that are incorrectly identified. Figure 3c indicates that each command is tested with how many copies and out of which how many copies are classified correctly and incorrectly. Figure 3d represents the percentage of Classification accuracy for 0–40 dB SNR at 90% training it also indicates the True Positive Rate and False Negative Rate.

Table 3 represents the performance of all SVM classifiers with different sets of features and at different training rates.

Figure 4 depicts the performance of LSVM classifier with three different feature sets at different training rates. Also, LSVM classifier works better with the GTCC features when compared to MFCC and Spectral Features. Though MFCC has obtained less accuracy, it is not far away from GTCC but the spectral features are greatly affected by noise.

Figures 5, 6, 7 and 8, represents the performance accuracy of QSVM, CSVM, FGSVM, and MGSVM. The accuracy of these models remained 100% throughout all training rates for both MFCC and GTCC features, whereas the accuracy of these models is greatly reduced with spectral features.

Figure 9 depicts the performance of COSVM classifier. This model also works better with GTCC features when compared with MFCC features while the accuracy with the spectral features is greatly reduced to 20%.

From the Table 3, MFCC features are better features for classification when compared with GTCC and spectral features such as pitch, spectral flux, and spectral entropy. From Table 3 it is cleared that spectral features are useful for classification of commands and they are useful only along with MFCC and GTCC features.

Table 4 represents the classification accuracy of all classifiers with MFCC features and at the training rate of 90%. QSVM, CSVM, and FGSVM have best classification accuracy even at 0 dB noise case.

From the Table 5, it is evident that the performance of proposed classifiers is better than the superior approaches in the existing methods.

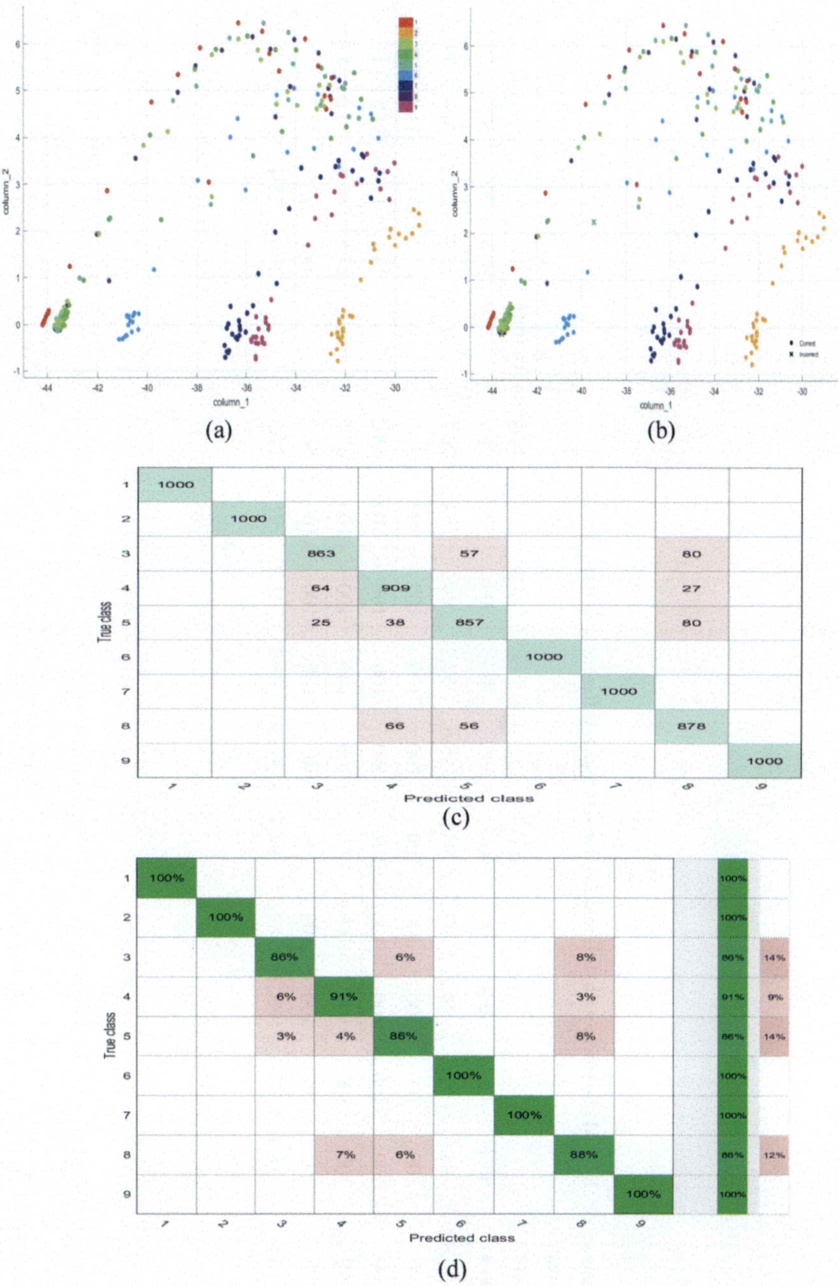

Fig. 3 Simulation results for linear SVM (LSVM) for MFCC at 0–40 dB SNR, **a** data set, **b** scatter plot, **c** confusion matrix-number of observations, **d** Confusion matrix—classification percentage

Table 3 Performance of all SVM classifiers with different feature sets and at different tartaning rates

Classifier	Features														
	MFCC					GTCC					Spectral features				
	90%	80%	70%	60%	50%	90%	80%	70%	60%	50%	90%	80%	70%	60%	50%
LSVM	94.8	94.3	95.1	94.2	92.5	95.8	95.6	96.2	94.7	94.6	76.1	75.4	75.2	74.4	74.1
QSVM	100	100	100	100	100	100	100	100	100	100	23.2	11.1	11.1	11.1	11.1
CSVM	100	100	100	100	100	100	100	100	100	100	18.2	11.1	11.1	11.1	11.1
FGSVM	100	100	100	100	100	100	100	100	100	100	87.5	87.8	86.3	85.2	84.3
MGSVM	100	100	100	100	100	100	100	100	100	100	50.2	49.6	48.5	47.1	47.3
CGSVMM	88.4	88.4	85.5	83.7	83.7	94.4	94.3	93.2	91.2	91.1	27.3	26.4	25.4	24.6	23.2

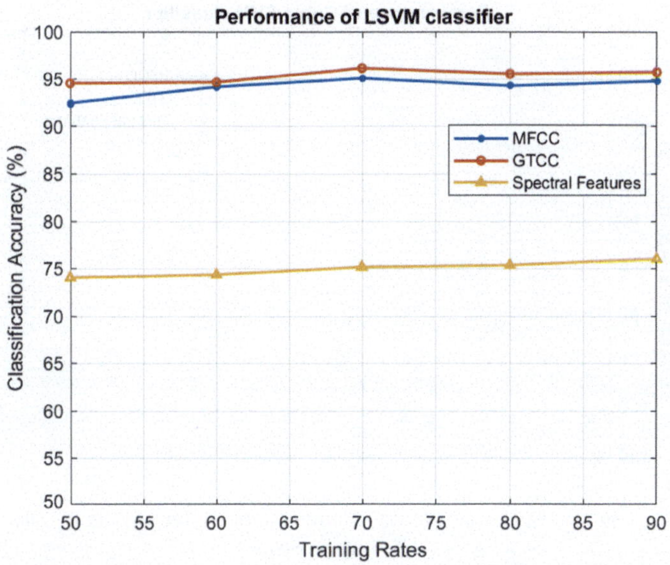

Fig. 4 Performance of LSVM

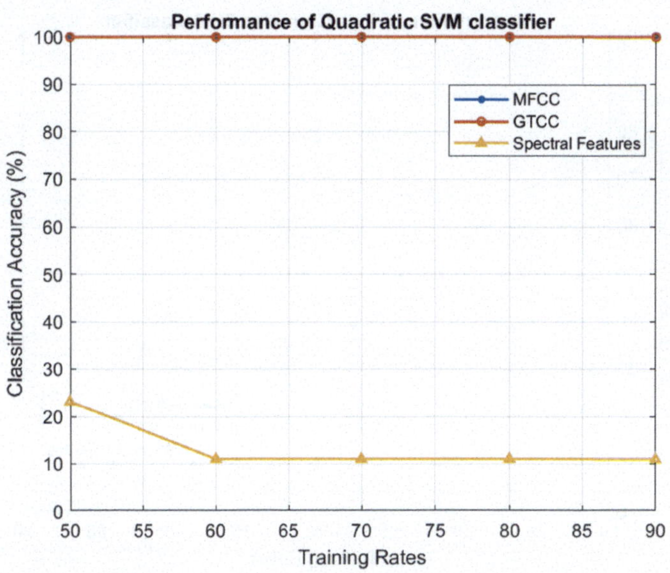

Fig. 5 Performance of QSVM

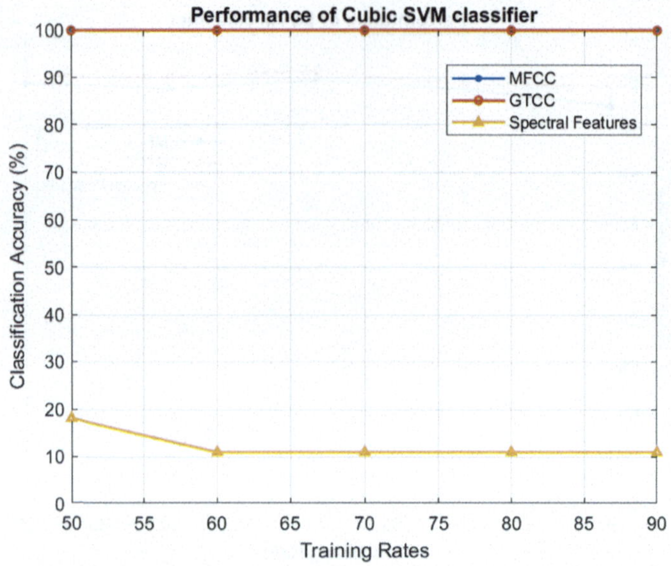

Fig. 6 Performance of CSVM

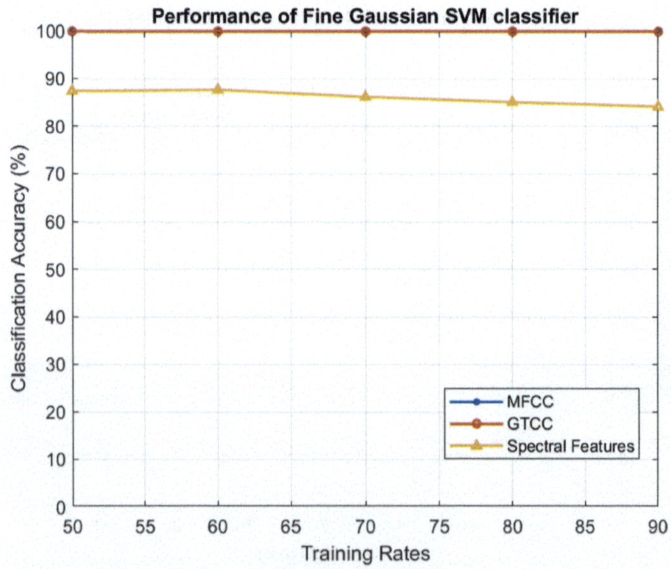

Fig. 7 Performance of FGSVM

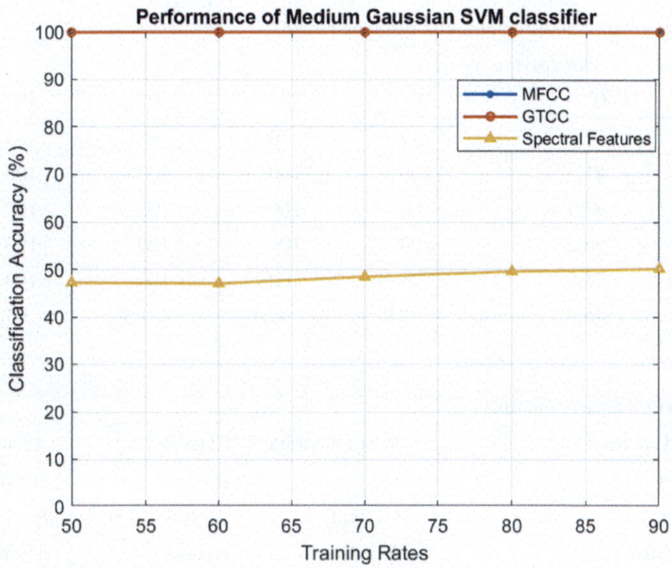

Fig. 8 Performance of MGSVM

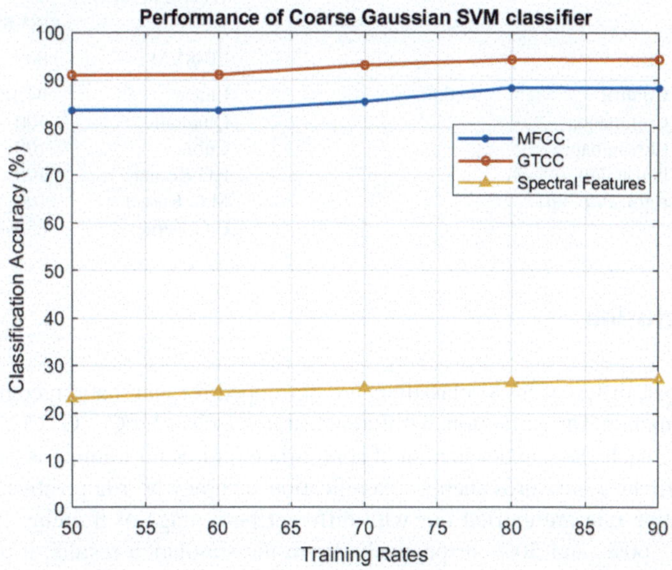

Fig. 9 Performance of CGSVM

Table 4 Performance at different SNRs (training rate 90%)

Classifier	SNR (dB)				
	0	10	20	30	40
LSVM	62.1	77.1	70.3	86	88.5
QSVM	91.2	100	100	100	100
CSVM	97.9	100	100	100	100
FGSVM	96.2	100	100	100	100
MGSVM	73.5	98.3	100	100	100
CGSVM	46.9	77.9	65.4	91.1	95.8

Table 5 Performance comparison

Ref. No.	Classes	No. of recordings	Classifier	Accuracy (%)
[2]	8	4680	HMM	60–70
[3]	12	160,000	WLS-SVM/HMM	82–94
[4]	Vowels, Alpha digits and switchboard	90,000	Hybrid SVM/HMM	45–50
[6]	180*129		WLS-SVC/HMM	92.5 (88.8% T.R)
[7]	TI-20		LP-SVM	80
Proposed	Control Commands (down, happy, house, left, off, on, right, stop, up)	90,000	Linear Quadratic Cubic F.G Kernel M.G Kernel C.G Kernel	94.18 100 100 100 100 85.94

5 Conclusion

In this paper, different SVM classifiers are developed for automatic recognition of voice commands. The extraction of different features such as MFCC, GTCC, Spectral features for each class, and selection of appropriate futures for training is presented. The performance measures such as classification accuracy or true positive rate and false-negative rate are carried out with different percentage of training, i.e. 90%, 80%, 70%, 60%, and 50%, respectively. From the simulation results, it is proved that all nonlinear SVM classifiers other than CGSVM achieve more classification accuracy even at lower SNRs than that of existing approaches. From simulations, it is also proved that the classification performance of proposed classifiers SVM are better for different feature sets, even with less training.

References

1. Guiming, D., Xia, W., Guangyan, W., Yan, Z., Dan, L.: Speech recognition based on convolutional neural networks. In: 2016 IEEE International Conference on Signal and Image Processing (ICSIP), pp. 708–711. Beijing (2016). https://doi.org/10.1109/SIPROCESS.2016.7888355.
2. Zhao, H., Zhao, K., Liu, H., Yu, F.: Improved MFCC feature extraction combining symmetric ICA algorithm for robust speech recognition. J. Multimedia **7**, (2012). https://doi.org/10.4304/jmm.7.1.74-81
3. Solera-Ureña, R., Garcia-Moral, A., Peláez-Moreno, C., Martinez-Ramon, M., Díaz-de-María, F.: Real-time robust automatic speech recognition using compact support vector machines. IEEE Trans. Audio Speech Lang. Process. **20**, 1347–1361 (2012). https://doi.org/10.1109/TASL.2011.2178597
4. Ganapathiraju, A., Hamaker, J., Picone, J.: Application of support vector machines to speech recognition. IEEE Trans. Signal Process. **52**, 2348–2355 (2004). https://doi.org/10.1109/TSP.2004.831018
5. Al-Alaoui, M., Mouci, R., Mansour, M., Ferzli, R.: A cloning approach to classifier training. Syst. Man Cybernet. Part A: Syst. Humans IEEE Trans. **32**, 746–752 (2002). https://doi.org/10.1109/TSMCA.2002.807035
6. Kumar, A., Rout, S.S., Goel, V.: Speech Mel frequency Cepstral coefficient feature classification using multi-level support vector machine. In: 2017 4th IEEE Uttar Pradesh Section International Conference on Electrical, Computer and Electronics (UPCON), pp. 134–138. Mathura (2017). https://doi.org/10.1109/UPCON.2017.8251036
7. Sonkamble, B., Doye, D.: Use of support vector machines through linear-polynomial (LP) kernel for speech recognition, pp. 46–49 (2012). https://doi.org/10.1109/MNCApps.2012.14
8. Gupta, A., Joshi, A.: Speech recognition using artificial neural network. In: 2018 International Conference on Communication and Signal Processing (ICCSP), pp. 0068–0071. Chennai (2018). https://doi.org/10.1109/ICCSP.2018.8524333
9. Dhankar, A.: Study of deep learning and CMU sphinx in automatic speech recognition, pp. 2296–2301 (2017). https://doi.org/10.1109/ICACCI.2017.8126189
10. Venkata Subbarao, M., Samundiswary, P.: Performance analysis of modulation recognition in multipath fading channels using pattern recognition classifiers. Wirel. Pers. Commun. (2020). https://doi.org/10.1007/s11277-020-07564-z
11. Warden, P.: Speech commands: a public dataset for single-word speech recognition, (2017). Available from http://download.tensorflow.org/data/speech_commands_v0.01.tar.gz

Parameter Identification of Coulomb Oscillator from Noisy Sensor Data

Guddu Kumar, Vikash Kumar Mishra, R. Swaminathan, and Abhinoy Kumar Singh

Abstract Coulomb oscillator is used for analyzing several real-life systems, which demand a precise modeling of the oscillation. The modeling is based on the stochastic estimation of unknown parameters of the model representing the oscillation. This paper introduces a Bayesian approach for the estimation of unknown parameters from sensor-generated noisy data. Among several Bayesian approaches, Gaussian filtering approach is most popular. A major challenge that appeared with the Gaussian filtering is intractable integral, which is approximated numerically. Several Gaussian filters have been reported by using different numerical approximation methods. This paper implements a popular Gaussian filter, named as cubature Kalman filter (CKF), for the estimation of unknown parameters. The CKF uses a third-degree spherical radial rule for the numerical approximation of the intractable integrals. Simulation results conclude a high accuracy of the CKF-based estimate of unknown parameters.

Keywords Bayesian filtering · Cubature Kalman filter · Coulomb oscillator · Parameter estimation · Nonlinear filtering

1 Introduction

Parameter estimation is a stochastic method of determining the unknown parameters of a system or a process model based on noisy measurements [1]. It is useful

G. Kumar · R. Swaminathan · A. K. Singh (✉)
Discipline of Electrical Engineering, Indian Institute of Technology Indore, Indore 453552, India
e-mail: abhinoy.singh@iiti.ac.in

G. Kumar
e-mail: phd1901102010@iiti.ac.in

R. Swaminathan
e-mail: swamiramabadran@iiti.ac.in

V. K. Mishra
Electrical and Electronics Engineering, Government Engineering College Raipur, Raipur 492015, India

J. Gu et al. (eds.), *Communication and Control for Robotic Systems*, Smart Innovation, Systems and Technologies 229, https://doi.org/10.1007/978-981-16-1777-5_20

in analyzing the behavior of a system in several aspects, like the damage and fault occurrence [2]. Its frequent applications are observed in the analysis of Lorenz attractor [3], Euler-Bernoulli beam [1], Coulomb and Duffing oscillators [4–6], etc. The focus of this paper is limited to Coulomb oscillator.

The Coulomb oscillator is a chaotic nonlinear dynamic oscillator with a single degree of freedom [4, 7]. Its applications appear in several domains of science and technology, such as vibration analysis [8] and orbital momentum analysis [9]. Various applications of the oscillator are based on accurate knowledge of the oscillator model. As discussed above, the stochastic parameter estimation can be an efficient solution to this problem.

In the literature, the solution to parameter estimation problems is mostly discussed as filtering under Bayesian framework [10, 11]. The filtering is a recursive process of estimation. The Bayesian estimation is the characterization of prior and posterior probability density functions (PDF). The popularly known Kalman filter is an optimal Bayesian estimator for linear dynamical systems with white Gaussian noises [12]. However, the oscillatory behavior of Coulomb oscillator exhibits nonlinear dynamics and the Kalman filter fails. An early nonlinear extension of the Kalman filter, named as extended Kalman filter (EKF) [10, 11], was developed in the nineties. The EKF is based on derivative-based linearization of system models at the current estimate. Due to derivative-based implementation, it suffers from several limitations [10, 11], like smoothness requirement of system models and poor stability. Moreover, the linear approximation of nonlinear system models further leads to poor estimation accuracy and frequent track failures.

A derivative-free era of nonlinear filtering begins in the nineties with the development of unscented Kalman filter (UKF) [13]. It is followed by several derivative-free developments in later years, like cubature Kalman filter (CKF) [14], Gauss-Hermite filter (GHF) [15, 16] and cubature quadrature Kalman filter (CQKF) [17]. The derivative-free filters approximate the prior and posterior PDFs as Gaussian. Subsequently, the PDFs can be characterized by mean and covariance, such as the first two moments. The moment computation encounters Gaussian weighted intractable integrals, which are numerically approximated using deterministically chosen sample points and weights. The Gaussian approximation-based filters are popularly known as Gaussian filters.

Among the several derivative-free Bayesian filters available in the literature, the CKF draws huge attention due to its high estimation accuracy at a significantly low computational cost. Therefore, we choose to implement the CKF for parameter estimation of the Coulomb oscillator. We also compare the performance of the CKF with UKF in terms of root mean square error (RMSE). The performance analysis shows a reduced RMSE, i.e., an enhanced estimation accuracy, for the CKF.

2 Problem Formulation

In this section, we formulate the discrete-time process dynamics and measurement model for the Coulomb oscillator. The continuous-time process dynamics follows [4]

$$\ddot{x} + \mu \,\text{sgn}(\dot{x}) + \omega^2 x = g(t), \tag{1}$$

where x, $\dot{x}$, and $\ddot{x}$ are position, velocity, and acceleration, respectively, μ is friction damping parameter, sgn$(\cdot)$ is signum function, ω is frequency of oscillation, and $g(t) = A\cos\lambda t$ is external input with A and λ being the amplitude and frequency, respectively. The friction damping coefficient μ regulates the nonlinearity of dynamic process model. It also regulates the span of transient time. The problem is to estimate the parameter μ based on some noisy measurements. The dynamics modeling of μ requires the information on $\dot{x}$ and $\ddot{x}$ also. Therefore, we define the state variable as $\mathbf{x} = [x\ \dot{x}\ \mu]^{\text{T}}$, which we denote as $\mathbf{x} = [\mathbf{x}_1\ \mathbf{x}_2\ \mathbf{x}_3]^{\text{T}}$ in the remaining part of this paper. Subsequently, the simplified dynamics can be represented as [4]

$$\begin{aligned}
\text{d}\mathbf{x}_1 =& \mathbf{x}_2 \text{d}t, \\
\text{d}\mathbf{x}_2 =& (-\mathbf{x}_3\,\text{sgn}(\dot{x}) - \omega^2 \mathbf{x}_1 + A\cos\lambda t)\text{d}t + \sigma \text{d}W_1(t), \\
\text{d}\mathbf{x}_3 =& \text{d}W_2(t),
\end{aligned} \tag{2}$$

where W_1 and W_2 are independent Brownian motions.

The measurements model is given as [4],

$$\mathbf{y}_k = \mathbf{x}_k + \sigma_m I_m; \qquad I_m \sim N(0, 1) \tag{3}$$

where y_k represents measurement variable, σ_m is constant, and N denotes normal Gaussian distribution.

The Bayesian filters are designed for discrete-time state-space model; however, the process dynamics shown in Eq. (2) is in continuous-time domain. A discretized process model at an instant t_k is given in [4], as

$$\begin{aligned}
\mathbf{x}_{1,k+1} =& \mathbf{x}_{1,k} + \mathbf{x}_{2,k}\delta + \mathbf{a}_{2,k}\frac{\delta^2}{2} - \omega^2 \mathbf{x}_{2,k}\frac{\delta^3}{6} \mathbf{x}_{3,k} b_1(k)\frac{\delta^3}{6} \\
&+ A\cos\lambda t_k \frac{\delta^3}{6} + \sigma I_{10} + \sigma b_2(k) I_{100}.
\end{aligned} \tag{4}$$

$$\begin{aligned}
\mathbf{x}_{2,k+1} =& \mathbf{x}_{2,k} - \mathbf{x}_{3,k}\,\text{sgn}(\frac{1}{2}(\mathbf{x}_{2,k} + \mathbf{x}^e_{2,k+1}))\delta - \omega^2(\,\mathbf{x}_{1,k}\delta \\
&+ \mathbf{x}_{2,k}\frac{\delta^2}{2}) + \frac{A}{\lambda}(\sin\lambda t_{k+1} - \sin\lambda t_k) + \sigma I_1.
\end{aligned} \tag{5}$$

$$\mathbf{x}_{3,k+1} = \mathbf{x}_{3,k} + \sigma_\mu I_2. \tag{6}$$

where

$$\mathbf{a}_{2,k} = -\mathbf{x}_{3,k}\,\mathrm{sgn}(x_{2,k}) - \omega^2 \mathbf{x}_{1,k} + A\cos\lambda t_k$$

$$b_1(k) = \frac{\mathrm{sgn}(\mathbf{x}_{2,k+1}) - \mathrm{sgn}(\mathbf{x}_{2,k})}{\mathbf{x}_{2,k+1} - \mathbf{x}_{2,k}}$$

$$b_2(k) = \frac{\mathbf{a}^e_{2,k+1} - \mathbf{a}_{2,k}}{\mathbf{x}_{2,k+1} - \mathbf{x}_{2,k}}$$

$$\mathbf{x}^e_{2,k+1} = \mathbf{x}_{2,k} + \mathbf{a}_{2,k}\delta + \sigma I_1,$$

$$\mathbf{a}^e_{2,k+1} = -\mathbf{x}_{3,k}\,\mathrm{sgn}(\mathbf{x}_{2,k+1}) - \omega^2 \mathbf{x}^e_{1,k+1} + A\cos\lambda t_{k+1},$$

$$\mathbf{x}^e_{1,k+1} = \mathbf{x}_{1,k} + \frac{1}{2}(\mathbf{x}_{2,k} + \mathbf{x}_{2,k+1})\delta,$$

with $\delta = t_{k+1} - t_k$ be the discretization interval, $I_1 = \int_{t_k}^{t_{k+1}} dW_1(t)$ and $I_2 = \int_{t_k}^{t_{k+1}} dW_2(t)$. The terms I_1, I_{10}, I_{100} and I_2 are normally distributed random variables, with zero mean and covariance $\mathbf{q}$, i.e., [4]

$$\mathbf{q} = \begin{bmatrix} \delta & \frac{\delta^2}{2} & \frac{\delta^3}{6} & 0 \\ \frac{\delta^2}{2} & \frac{\delta^3}{3} & \frac{\delta^4}{8} & 0 \\ \frac{\delta^3}{6} & \frac{\delta^4}{8} & \frac{\delta^5}{20} & 0 \\ 0 & 0 & 0 & \delta \end{bmatrix}.$$

The problem, at this end, is to recursively estimate the state $\mathbf{x}_k$ from a sequence of measurements $\mathbf{y}_1, \mathbf{y}_2, \ldots, \mathbf{y}_k$. We use the Gaussian filtering, under the Bayesian framework for this purpose.

3 Bayesian Filtering Framework

As discussed earlier, the Bayesian filtering is implemented over the dynamic state-space model of a system, represented by process and measurement models. Let us consider the dynamic state-space model shown below:

$$\mathbf{x}_k = \Phi(\mathbf{x}_{k-1}) + \eta_{k-1}, \tag{7}$$

$$\mathbf{y}_k = \varphi(\mathbf{x}_k) + \vartheta_k, \tag{8}$$

where $\mathbf{x}_k \in \mathbb{R}^n$ and $\mathbf{y}_k \in \mathbb{R}^m$ are process and measurement vectors, respectively; $\Phi(\cdot)$ and $\varphi(\cdot)$ are standard representation of nonlinear functions, respectively. $k \in \{0, 1, ..., N\}$, η_k, and ϑ_k represent process and measurement noises. The noises are assumed to be Gaussian with zero mean and covariance Q_k and R_k, respectively.

The Bayesian framework is performed in two steps: prediction and update. The prediction step makes a prior believe about the state before the measurement is received. In this regard, it computes the prior pdf using the Chapman-Kolmogorov equation, as [18]

$$p(\mathbf{x}_k|\mathbf{y}_{1:k-1}) = \int p(\mathbf{x}_k|\mathbf{x}_{k-1})p(\mathbf{x}_{k-1}|\mathbf{y}_{1:k-1})\mathrm{d}\mathbf{x}_{k-1}. \tag{9}$$

The update step, on the other hand, corrects the prior believe based on the measurement $\mathbf{y}_k$. For this purpose, it computes posterior PDF using Baye's rule, as

$$p(\mathbf{x}_k|\mathbf{y}_{1:k}) = \frac{p(\mathbf{y}_k|\mathbf{x}_k)p(\mathbf{x}_k|\mathbf{y}_{1:k-1})}{p(\mathbf{y}_k|\mathbf{y}_{1:k-1})}, \tag{10}$$

where the normalized constant

$$p(\mathbf{y}_k|\mathbf{y}_{1:k-1}) = \int p(\mathbf{y}_k|\mathbf{x}_k)p(\mathbf{x}_k|\mathbf{y}_{1:k-1})\mathrm{d}\mathbf{x}_k. \tag{11}$$

Hereafter, we use the notations $p(\mathbf{x}_{k|k-1})$ and $p(\mathbf{x}_{k|k})$ for $p(\mathbf{x}_k|\mathbf{y}_{1:k-1})$ and $p(\mathbf{x}_k|\mathbf{y}_{1:k})$, respectively.

As discussed earlier, our interest is to apply Gaussian filtering under the Bayesian framework. The Gaussian filters approximate the prior and posterior pdfs as Gaussian [14], i.e., $p(\mathbf{x}_k|\mathbf{y}_{k-1}) \approx \aleph(\mathbf{x}_k; \hat{\mathbf{x}}_{k|k-1}, P_{k|k-1})$ and $p(\mathbf{x}_k|\mathbf{y}_k) \approx \aleph(\mathbf{x}_k; \hat{\mathbf{x}}_{k|k}, P_{k|k})$. Therefore, $p(\mathbf{x}_k|\mathbf{y}_{k-1})$ can be characterized by $\hat{\mathbf{x}}_{k|k-1}$ and $P_{k|k-1}$, while $p(\mathbf{x}_k|\mathbf{y}_k)$ can be characterized by $\hat{\mathbf{x}}_{k|k}$ and $P_{k|k}$. Consequently, the prediction step is about computing $\hat{\mathbf{x}}_{k|k-1}$ and $P_{k|k-1}$, and the update step aims to compute $\hat{\mathbf{x}}_{k|k}$ and $P_{k|k}$. The computational aspects of Gaussian filtering are provided herewith.

3.1 Prediction

The prediction step parameters, $\hat{\mathbf{x}}_{k|k-1}$ and $P_{k|k-1}$, can be computed as [14, 17]

$$\begin{aligned}\hat{\mathbf{x}}_{k|k-1} &= \int \Phi(\mathbf{x}_{k-1})p(\mathbf{x}_{k-1}|\mathbf{y}_{1:k-1})d\mathbf{x}_{k-1}\\ &= \int \Phi(\mathbf{x}_{k-1})\aleph(\mathbf{x}_{k-1}; \hat{\mathbf{x}}_{k-1|k-1}, P_{k-1|k-1})\mathrm{d}\mathbf{x}_{k-1}.\end{aligned} \tag{12}$$

$$P_{k|k-1} = E[(\mathbf{x}_k - \hat{\mathbf{x}}_{k|k-1})(\mathbf{x}_k - \hat{\mathbf{x}}_{k|k-1})^{\mathrm{T}}|\mathbf{y}_{1:k-1}],$$
$$= \int [\Phi(\mathbf{x}_{k-1})\Phi^{\mathrm{T}}(\mathbf{x}_{k-1})\aleph(\mathbf{x}_{k-1}; \hat{\mathbf{x}}_{k-1|k-1}, P_{k-1|k-1}) \tag{13}$$
$$\mathrm{d}\mathbf{x}_{k-1}] - \hat{\mathbf{x}}_{k|k-1}\hat{\mathbf{x}}^{\mathrm{T}}_{k|k-1} + Q_k.$$

3.2 *Update*

The update step parameters, $\hat{\mathbf{x}}_{k|k}$ and $P_{k|k}$, are computed as [14, 17]

$$\hat{\mathbf{x}}_{k|k} = \hat{\mathbf{x}}_{k|k-1} + K_k(\mathbf{y}_k - \hat{\mathbf{y}}_{k|k-1}),$$

$$P_{k|k} = P_{k|k-1} - K_k P_{\mathbf{yy},k|k-1} K_k^{\mathrm{T}},$$

where

$$\hat{\mathbf{y}}_{k|k-1} = \int \varphi(\mathbf{x}_k)\aleph(\mathbf{x}_k; \hat{\mathbf{x}}_{k|k-1}, P_{k|k-1})\mathrm{d}\mathbf{x}_k \tag{14}$$

$$P_{\mathbf{yy},k|k-1} = \int \varphi(\mathbf{x}_k)\varphi^{\mathrm{T}}(\mathbf{x}_k)\aleph(\mathbf{x}_k; \hat{\mathbf{x}}_{k|k-1}, P_{k|k-1})\mathrm{d}\mathbf{x}_k - \hat{\mathbf{y}}_{k|k-1}\hat{\mathbf{y}}^{\mathrm{T}}_{k|k-1} + R_k, \tag{15}$$

$$P_{\mathbf{xy},k|k-1} = \int \mathbf{x}_k\varphi^{\mathrm{T}}(\mathbf{x}_k)\aleph(\mathbf{x}_k; \hat{\mathbf{x}}_{k|k-1}, P_{k|k-1})\mathrm{d}\mathbf{x}_k - \hat{\mathbf{x}}_{k|k-1}\hat{\mathbf{y}}^{\mathrm{T}}_{k|k-1}. \tag{16}$$

$$K_k = P_{\mathbf{xy},k|k-1}P^{-1}_{\mathbf{yy},k|k-1}.$$

The integrals appeared in Eqs. (12)–(16) are intractable for most of the nonlinear systems. As discussed earlier, the integrals are approximated numerically. Therefore, the estimation accuracy of a Gaussian filter depends on the numerical approximation accuracy of the intractable integrals. The literature witnesses several Gaussian filters designed with the same filtering structure presented in Eqs. (12)–(16), but using different numerical approximation approach.

4 CKF: Multidimensional Integral Approximation

In this section, we present the numerical approximation method used by the CKF for approximating the desired intractable integrals that appeared in Eqs. (12)–(16). The CKF decomposes the intractable integrals into a spherical and a radial integral. The spherical integral is approximated using third-degree spherical cubature rule; however, the radial integral is approximated using first-order Gauss-Laguerre quadrature rule. The combined rule is named as a third-degree spherical radial rule.

4.1 Decomposition on Intractable Integral

The intractable integrals, appeared in Eqs. (12)–(16), are of a general form [14]

$$
\begin{aligned}
I(g) &= \int_{-\infty}^{\infty} g(X)\aleph(X,\mu,\Sigma)\mathrm{d}X \\
&= \frac{1}{\sqrt{|\Sigma|(2\pi)^n}} \int_{\Re^n} \left(g(X)\mathrm{e}^{-(1/2)(X-\mu)^{\mathrm{T}}\Sigma^{-1}(X-\mu)} dX \right),
\end{aligned}
\tag{17}
$$

where $g : \mathbb{R}^n \rightarrow \mathbb{R}^n$ represents a nonlinear function, and X is an n-dimensional random variable with mean μ and covariance Σ. The above integral can be transformed in to spherical coordinate system, as presented in [14]. The transformed integral is appeared in the form of [14]

$$
I(g) = \frac{1}{\sqrt{(2\pi)^n}} \int_{r=0}^{\infty} \int_{U_n} [g(CrY+\mu)d\sigma(Y)]\, r^{n-1}\mathrm{e}^{-r^2/2}\mathrm{d}r, \tag{18}
$$

where $X = CrY + \mu$, C is Cholesky decomposition of Σ, $r \in [0, \infty)$ is radius, and $\| Y \| = 1$ and U_n are surface of an unit hypersphere. Without losing the generality, we simplify the above integral with respect to a zero mean and unity covariance system. The simplified integral is of the form

$$
I_o(g) = \frac{1}{\sqrt{(2\pi)^n}} \int_{r=0}^{\infty} \int_{U_n} [g(rY)\mathrm{d}\sigma(Y)]\, r^{n-1}\mathrm{e}^{-r^2/2}\mathrm{d}r. \tag{19}
$$

Subsequently, the spherical and radial integrals can be represented as

$$
S_I = \int_{U_n} g(rY)\mathrm{d}\sigma(Y) \tag{20}
$$

and

$$
R_I = \frac{1}{\sqrt{(2\pi)^n}} \int_{r=0}^{\infty} S_I r^{n-1}\mathrm{e}^{-r^2/2}\mathrm{d}r, \tag{21}
$$

respectively.

4.2 Third-Degree Spherical Cubature Rule

Using third-degree spherical cubature rule, the spherical integral could be approximated as [14]

$$S_I \approx \frac{2\sqrt{\pi^n}}{2n\Gamma(n/2)} \sum_{i=1}^{2n} g(r\zeta_i), \tag{22}$$

where $\zeta_i\ \forall\ i \in \{1, 2...., 2n\}$ are known as cubature points, which are located at the intersection points of unit hypersphere and coordinate axes. For a two-dimensional system, $\zeta = \{(1, 0), (0, 1), (-1, 0), (0, -1)\}$.

4.3 First-Degree Gauss-Laguerre Quadrature Rule:

The integrals of the form $g(\cdot)$ can be approximated as [14]

$$\int_{\lambda=0}^{\infty} g(\lambda)\lambda^{n-1}\mathrm{e}^{-\lambda^2}\mathrm{d}\lambda \approx \omega_1 g(\lambda_1), \tag{23}$$

where $\lambda_1 = \sqrt{n/2}$ is sample point, and $\omega_1 = \Gamma(n/2)/2$ is the associated weight.

4.4 Third-Degree Spherical Radial Rule

Substituting Eq. (22) into (19) and replacing $t = r/\sqrt{2}$, we get

$$I_o(g) = \frac{2}{2n\Gamma(n/2)} \int_{t=0}^{\infty} \left(\sum_{i=1}^{2n} g[\sqrt{2}tu]_i \right) t^{n-1}\mathrm{e}^{-t^2}\mathrm{d}t. \tag{24}$$

This is a radial integral, which can be approximated using the first-order Gauss-laguerre quadrature rule, as [14]

$$I_o(g) \approx \frac{1}{2n} \left[\sum_{i=1}^{2n} g(\sqrt{n}[\zeta_i]) \right]. \tag{25}$$

The above rule is derived for approximating $I_o(g)$, which is the simplified intractable integral for a system with zero mean and unity covariance. However, the same rule can be extended for a system with arbitrary mean μ and covariance Σ, by transforming the cubature quadrature points with the same mean and covariance.

Therefore, the desired intractable integral $I(g)$ could be approximated as

$$I_o(g) \approx \frac{1}{2n} \sum_{j=1}^{2n} g(\mu + C\sqrt{n}[\zeta_j]). \tag{26}$$

The CKF, like the other Gaussian filters, uses the same filtering structure presented in Eqs. (12)–(16). However, it implements the third-degree spherical radial rule for approximating the intractable integrals.

5 Simulation and Results

In this section, we present the simulation results obtained for the CKF in estimation of the unknown parameters of the Coulomb oscillator. The simulation is performed in MATLAB 2019 over a personal computer with i5 core, 3.40 GHz processor, and 4 GB RAM. For the implementation purpose, the oscillator and noise parameters were considered as: $\delta = (0.8\pi/25\lambda)$s, $\omega = 4\pi$ rad/s, $A = 10$, $\sigma = A/10$, $\lambda = 3/4\omega$, $\sigma_m = 0.0023$, $\sigma_\mu = 0.5$, $\sigma_{10} = 0.02$, and $\sigma_{20} = \lambda\sigma_{10}$. Simulated datasets for the true state and measurement were generated by considering the initial state as $\mathbf{x}_0 = [2.5, -3, 5]$. A poorly approximated initial state was obtained as two times of the multivariate Gaussian random number generated from $\aleph(\mathbf{x}_0, P_{0|0})$, where $P_{0|0} = \text{diag}([0.4, 0.2\ 3, 1])$ is initial covariance. The filtering is performed over 200 time steps. The estimated states using the CKF are plotted in Fig. 1 and compared with the true states. The figure shows a close match between the true and estimated states, which concludes a high estimation accuracy. The further analysis is based on RMSE obtained over 500 Monte Carlo simulations. The RMSEs obtained from the CKF and UKF for various states are plotted in Fig. 2, The plots start from 20th time-step considering the initial convergence time of filters. The RMSE analysis shows a high accuracy for the parameters estimated using the CKF as compared to UKF.

6 Discussions and Conclusions

The Coulomb oscillator is useful in analyzing several real-life systems and processes. Modelling of this oscillator is based on friction damping parameter, and the dynamics of the parameter is defined with displacement and velocity. Subsequently, the parameter estimation problem of Coulomb oscillator could be formulated as an estimation problem with friction damping parameter, displacement, and velocity being the unknown states. This paper implements CKF for parameter estimation of the Coulomb oscillator. The performance of the CKF for this parameter estimation problem is compared with another popular Gaussian filter, UKF. The performance analysis shows a reduced RMSE for the CKF. In conclusion, we can say that the

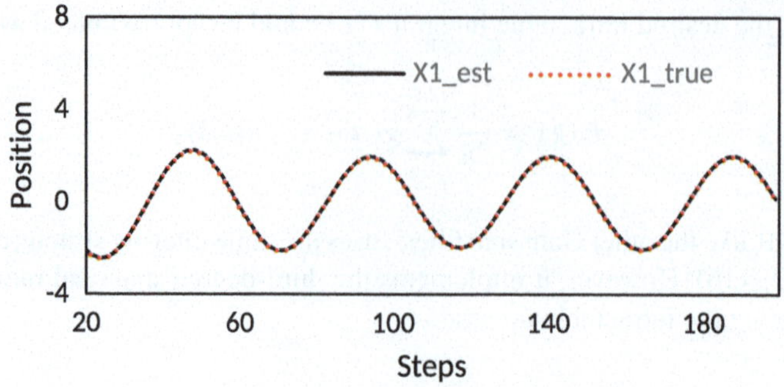

(a) Position

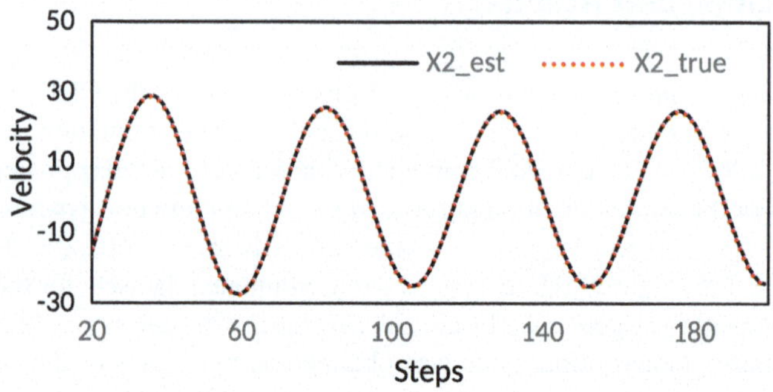

(b) Velocity

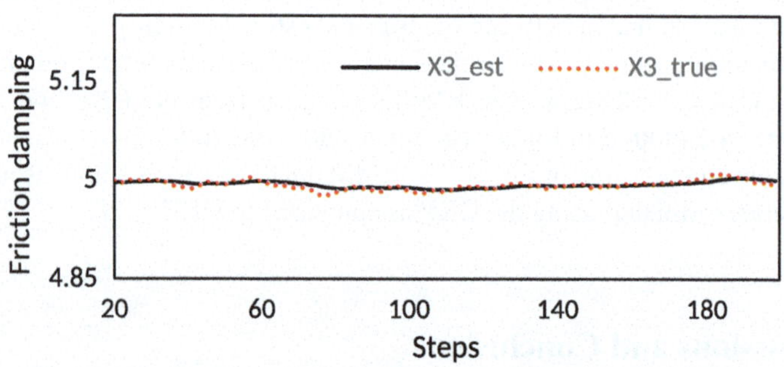

(c) Friction damping

Fig. 1 True and estimate plots of different states

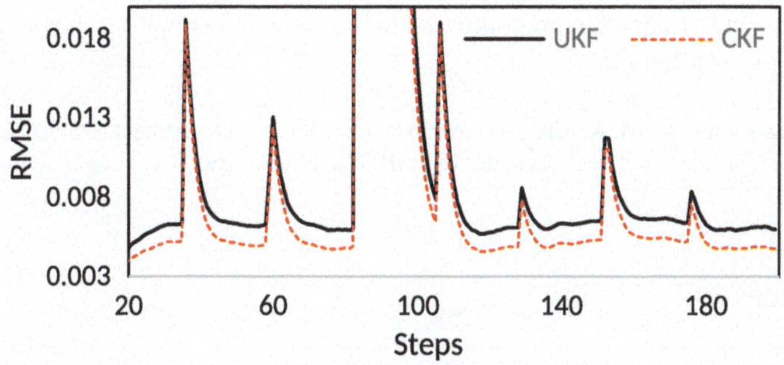

(a) Position

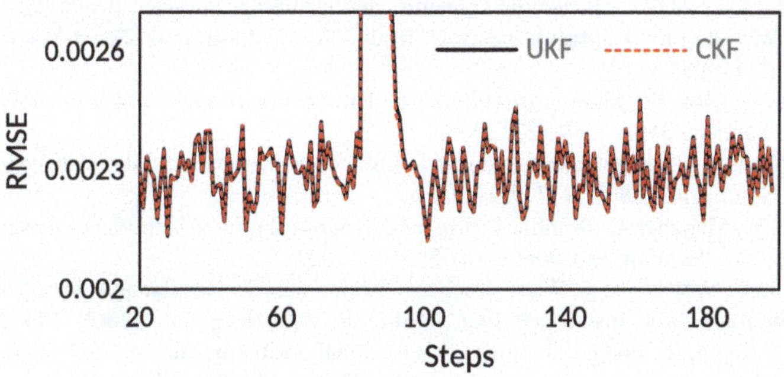

(b) Velocity

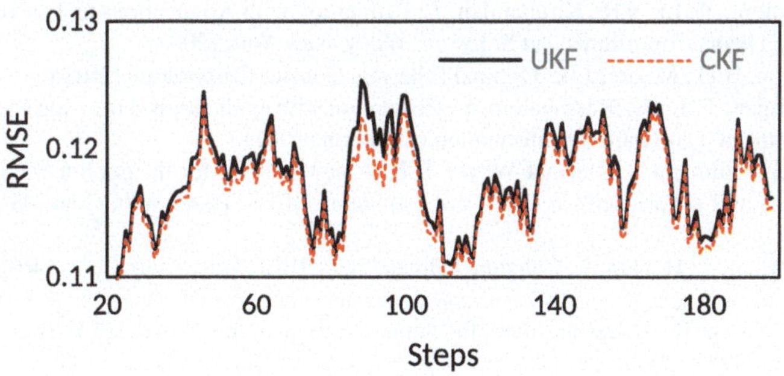

(c) Friction damping

Fig. 2 RMSE plots for obtained for three different states using the CKF and UKF

CKF gives a better accuracy as compared to the UKF in estimating the parameter of the Coulomb oscillator.

Acknowledgements Dr. A. K. Singh's research is funded by the Department of Science and Technology (DST), Government of India under INSPIRE Faculty Award.

References

1. Ghosh, S.J., Manohar, C., Roy, D.: A sequential importance sampling filter with a new proposal distribution for state and parameter estimation of nonlinear dynamical systems. Proc. Roy. Soc. A: Math. , Phys. Eng. Sci. **464**(2089), 25–47 (2007)
2. Jauregui, J.C.: Nonlinear Structural Dynamics and Damping. Springer, Berlin (2019)
3. Tucker, W.: The lorenz attractor exists. C. R. de l'Académie des Sci.-Ser. I-Mathe. **328**(12), 1197–1202 (1999)
4. Manohar, C., Roy, D.: Monte carlo filters for identification of nonlinear structural dynamical systems. Sadhana **31**(4), 399–427 (2006)
5. Hu, N., Wen, X.: The application of duffing oscillator in characteristic signal detection of early fault. Jr. Soun. Vib. **268**(5), 917–931 (2003)
6. Namdeo, V., Manohar, C.: Nonlinear structural dynamical system identification using adaptive particle filters. Jr. Soun. Vib. **306**(3–5), 524–563 (2007)
7. Stefański, A., Wojewoda, J., Wiercigroch, M., Kapitaniak, T.: Regular and chaotic oscillations of friction force. Proc. Inst. Mech. Eng. , Part C: Jr. Mech. Eng. Sci., **220**(3), 273–284 (2006)
8. Duan, C., Singh, R.: Forced vibrations of a torsional oscillator with coulomb friction under a periodically varying normal load. Jr. Soun. Vib. **325**(3), 499–506 (2009)
9. Bucher, M.: Coulomb oscillation in the hydrogen atom and molecule ion. arXiv preprint physics/0605258 (2006)
10. Bar-Shalom, Y., Li, X.R., Kirubarajan, T.: Estimation with Applications to Tracking and Navigation: Theory Algorithms and Software. Wiley, New York (2004)
11. Anderson, B.D., Moore, J.B.: Optimal Filtering. Courier Corporation (2012)
12. Kirubarajan, T., Li, X., Bar-Shalom, Y.: Estimation with application to tracking and navigation: algorithms and software for information extraction (2001)
13. Julier, S., Uhlmann, J., Durrant-Whyte, H.F.: A new method for the nonlinear transformation of means and covariances in filters and estimators. IEEE Tran. Auto. Cont. **45**(3), 477–482 (2000)
14. Arasaratnam, I., Haykin, S.: Cubature kalman filters. IEEE Tran. Auto. Cont. **54**(6), 1254–1269 (2009)
15. Ito, K., Xiong, K.: Gaussian filters for nonlinear filtering problems. IEEE Tran. Auto. Cont. **45**(5), 910–927 (2000)
16. Arasaratnam, I., Haykin, S., Elliott, R.J.: Discrete-time nonlinear filtering algorithms using gauss-hermite quadrature. Proc. IEEE **95**(5), 953–977 (2007)
17. Bhaumik, S., et al.: Cubature quadrature kalman filter. IET Sig. Proc. **7**(7), 533–541 (2013)
18. Julier, S.J., Uhlmann, J.K.: New extension of the Kalman filter to nonlinear systems. In: Signal Processing, Sensing Fusion, Target Recording VI, vol. 3068. International Society for Optics and Photonics, pp. 182–193 (1997)

Power Electronics Application for Transportation

Parameter Extraction of Dye-Sensitized Solar Cell: Improved Method and Comparative Analysis

Soumik Chakraborty, Ranjith G. Nair, and Lalu Seban

Abstract In this article, with the aim of improving the efficiency of dye-sensitized solar cells (DSSCs), three different algorithms are used for extraction of the parameters. An improved electrical equivalent model of DSSC is used for parameter extraction and hence to achieve an optimal solution. Algorithms used in this article are particle swarm optimization (PSO), genetic algorithm (GA) and artificial bee colony (ABC), and a comparative study is done among all three algorithms regarding higher parameter precision, which will eventually give better energy conversion efficiency and better computational speed. In the considered electrical equivalent model, there are some fixed parameters with variable constraints, and the values of the variables will be extracted by the algorithms used. The upper and lower bounds are defined for each of the variables according to the standard fabrication process. The final results of each of the algorithm will be compared with a recently published practical results, which is a standard TiO_2-based DSSC with N719 as dye.

1 Introduction

Due to the depletion of nonrenewable energy sources and with a rise in environmental issues, renewable energy sources are best suited as an alternative. Among all the energy sources, solar energy is considered to be the most convenient for implementation as it readily converts solar radiation into a usable form of energy, without any residual. Because of higher manufacturing as well as installation cost, silicon-based photovoltaic cells are one of the costly options, as an alternative dye-sensitized solar cells (DSSCs) are considered as a competitive alternative to the silicon-based photovoltaic cells [9]. The various parameters of DSSCs are modeled using an improved electrical equivalent model, and the parameters of this electrical equivalent model are

S. Chakraborty · L. Seban
Department of Electronics and Instrumentation Engineering, National Institute of Technology,Silchar, Cachar, Assam 788010, India

S. Chakraborty (✉) · R. G. Nair
Solar Energy Materials Research and Testing Laboratory, (Smart Lab) Department of Physics, National Institute of Technology, Silchar, Cachar, Assam 788010, India

J. Gu et al. (eds.), *Communication and Control for Robotic Systems*, Smart Innovation, Systems and Technologies 229, https://doi.org/10.1007/978-981-16-1777-5_21

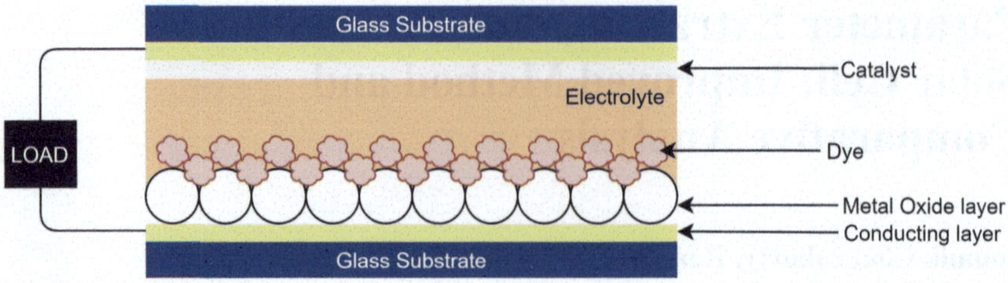

Fig. 1 Schematic diagram of a dye-sensitized solar cell [18]

then optimized by various algorithms. By using the optimized result, the fabrication process will be less time consuming and further reduces time and cost. The most commonly used electrical equivalent models are one-diode model and two-diode model. In this article, an improved three-diode models have been used for extraction of the parameters using particle swarm optimization (PSO) genetic algorithm (GA) and artificial bee colony (ABC). A DSSC consists of photo-anode, the sensitizer, the electrolyte, the catalyst and finally the cathode. The photo-anode consists of the transparent conducting glass along with a metal oxide layer [3]. A layer of sensitizer is adsorbed on the surface of the metal oxide. The electrons of the dye get excited from HOMO to LUMO on incident radiation. The excited electron gets transferred to the metal oxide layer. The electrons are collected by the cathode having a layer of catalyst, via an electrolyte the electron is back to the sensitizer. TiO_2 is widely used as the metal oxide in the photo-anode. FTO/ITO is generally used as the conducting glass, and generally, iodide (I-/I3-) electrolyte is used to transfer the electrons [18] (Fig. 1).

2 Electrical Equivalent Model

Generally, the standard one-diode model is used to study the I-V characteristics of a standard silicon-based photovoltaic cell. In DSSC, one-diode model can also be used, but the results will be more precise if we use the two-diode model [16]. The recombination of excited electrons in DSSC is modeled using the second diode, and its ideality factor is generally considered in such a way that it should mimic the trap assisted recombination occurring in a standard DSSC. In this article, improved three-diode model is used for parameter extraction using the three mentioned algorithms. J_{ph} denotes the photocurrent due to incident radiation. Since calculation is done in terms of current density, the output current equation will be in terms of current density. The rectifying diode is having an ideality factor of 'n'. The output current equation for a standard one-diode model will be [11]

$$J = J_{\text{ph}} - J_i \left(e^{\left(\frac{q(v+JR_s)}{nK_B T_C}\right)} - 1 \right) - \left(\frac{v + JR_s}{R_{\text{sh}}} \right) \tag{1}$$

Similarly, the output current equation for two-diode model will be [22]

$$J = J_{\mathrm{ph}} - J_i\left(e^{\left(\frac{q(v+JR_S)}{nK_BT_C}\right)} - 1\right) - J_r\left(e^{\left(\frac{q(v+JR_S)}{mK_BT_C}\right)} - 1\right) - \left(\frac{v + JR_s}{R_{\mathrm{sh}}}\right) \tag{2}$$

where J is the final output current, J_{ph} is the photocurrent per unit area, J_i and J_r is the temperature-dependent saturation current, n and m are the ideality factors of the two diodes, K_B is the Boltzmann's constant, R_s denotes the series resistive losses, R_{sh} denotes the parallel resistive losses, q is the elementary charge, and T_c is the cell temperature.
In DSSC, in the charge transport mechanism, the photo-generation and all possible recombination pathways are shown in Fig. 2.

At first, the dye is excited due to incident irradiance from HOMO to LUMO denoted by '1', then the excited electron is transferred to TiO_2 conduction band (C.B) denoted by '2', and the electron is passed through the external circuit and then collected back by the counter electrode, passed to the redox electrolyte and subsequently used for regeneration of the dye, denoted by '3'. A, B and C are the possible electron recombination or loss paths.

The standard two-diode model is known to be more accurate, but to model all the recombination pathways, a third diode can be included, and hence in this article, the proposed model is used for parameter extraction using the algorithms mentioned (Fig. 3).

For all the tree diode, the values of E_g will be different value as each of them will represent the possible recombination pathways. During the charge transfer at the TiO2/dye/electrolyte, a diode D1 denotes the recombination loss LUMO to HOMO

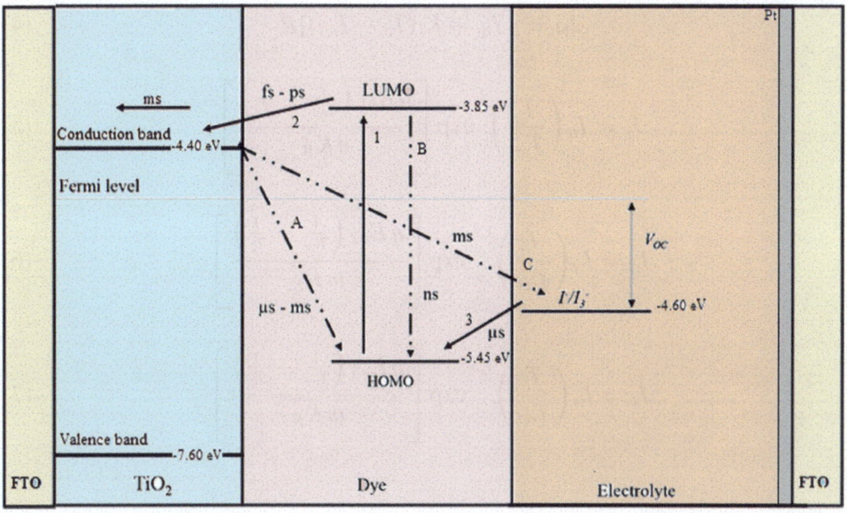

Fig. 2 Schematic diagram showing the kinetic processes at the TiO_2/dye/electrolyte interface [13] [21]

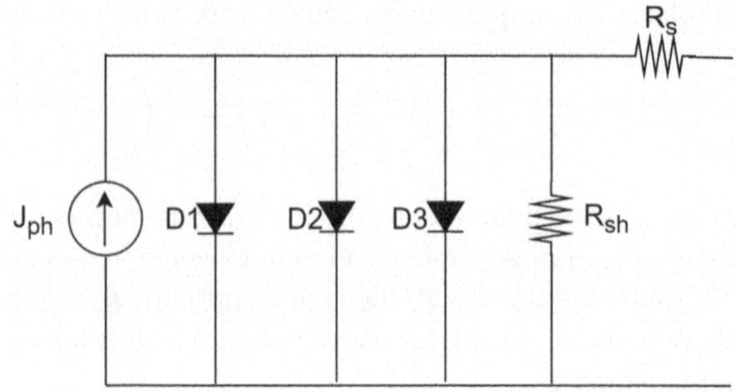

Fig. 3 Proposed three-diode electrical equivalent model

of the dye, the second diode D2 with an ideality factor m is employed to denote the interfacial charge recombination losses from TiO2 to redox electrolyte, and final diode D3 with an ideality factor m is employed to denote charge recombination losses from TiO2 to HOMO of the dye. The short circuit current is given by [3]

$$J_{\text{sh}} = \frac{[1 - r]q\phi L_n\alpha[-L_n\alpha cosh\frac{d}{L_n} + sinh\frac{d}{L_n} + L_n\alpha e^{(-d\alpha)}]}{(1 - L_n{}^2\alpha^2)cosh\frac{d}{L_n}} \tag{3}$$

where J_{sh} is short circuit current, r represents reflection losses, q is elementary charge, ϕ is flux intensity, $L_n$is electron diffusion length, α is light absorption coefficient, and d is TiO_2 layer thickness. The resistor R_s in the model represents the contact resistances present in a DSSC, such as the resistance of the conducting layer of the FTO/ITO glass. The resistor R_{sh} in the model represents all the parallel resistive losses across the DSSC [19].

$$J_{\text{ph}} = [J_{\text{sh}} + K_i(T_c - T_{\text{ref}})]R \tag{4}$$

$$J_i = I_{\text{rr}}\left(\frac{T}{T_{\text{ref}}}\right)^3 \exp\left[\frac{qE_{g1}\left(\frac{1}{T_{\text{ref}}} - \frac{1}{T_c}\right)}{nK_B}\right] \tag{5}$$

$$J_r = I_{\text{rr}}\left(\frac{T}{T_{\text{ref}}}\right)^{1.5} \exp\left[\frac{qE_{g2}\left(\frac{1}{T_{\text{ref}}} - \frac{1}{T_c}\right)}{mK_B}\right] \tag{6}$$

$$J_s = I_{\text{rr}}\left(\frac{T}{T_{\text{ref}}}\right)^{1.5} \exp\left[\frac{qE_{g3}\left(\frac{1}{T_{\text{ref}}} - \frac{1}{T_c}\right)}{mK_B}\right] \tag{7}$$

Equation obtained by solving above electrical equivalent model:

$$J = J_{\text{ph}} - J_i \left(e^{\left(\frac{q(v+JR_S)}{nK_BT_C}\right)} - 1 \right) - J_r \left(e^{\left(\frac{q(v+JR_S)}{mK_BT_C}\right)} - 1 \right) - J_s \left(e^{\left(\frac{q(v+JR_S)}{mK_BT_C}\right)} - 1 \right) - \left(\frac{v + JR_s}{R_{\text{sh}}} \right) \quad (8)$$

where J_i, J_r and J_s are the temperature-dependent saturation current. Some of the parameters are kept variable, and others are kept constants. However, the standard parameters and their values are chosen such that it mimics a practical DSSC device. All the constants are listed in Table 1

where $R(\text{mW/cm}^2)$ is irradiance considered 1000 W/m^2 also considered as A.M. 1.5, $\tau(\text{ms})$ is life time of electron in conduction band of the semiconductor used, K_i is short circuit current temperature coefficient, and $D(\text{cm}^{-1}s^{-1})$ is diffusion coefficient. The parameters which will be extracted by the algorithms to achieve the optimal solution are defined within a range of values. The range of electrode thickness(d) is considered from 10×10^{-4} cm to 15×10^{-4} cm, and at this range, the short circuit current is maximum within this range [14]. Ideality factor(n) is considered from 2 to 4.5 of diode D1, and for high-efficiency DSSC, the n should be within the given range [5] [14]. The range of ideality factor (m) of diode D2 and D3 is considered from 1 to 2 for no trapping states to recombination of charge carriers through trap states [1]. R_{sh} range is considered from 1000Ω to 3000Ω such that the value of V_{oc} is 0.7 [19, 22]. The range of R_s considered from 10Ω to 50Ω as the fill factor of a standard DSSC is in between 0.6 to 0.8 [5], and the cell temperature T_c is considered from 298.15K to 300.15K. The band gap range is considered from 1.4 to 1.6 as photosensitizers which react with visible and near infrared spectrum are included [17].

Table 1 Constant parameters used with respective values

Parameters Used	Values
$\alpha(\text{cm}^{-1})$	5000 [15]
$\tau(\text{ms})$	10 [2]
$T_{\text{ref}}(K)$	298.15
$D(\text{cm}^{-1}s^{-1})$	5.0×10^{-4} [15]
$R(\text{mW/cm}^2)$	100
Refractive index of FTO	1.5
K_i	0.0023 [12]
$L_n(\text{cm}^{-1}s^{-1})$	2.2361×10^{-3} [6]
$\phi(\text{cm}^{-2}s^{-1})$	1.0×10^{17} [10, 19]
Porosity, P	0.41 [15]

3 Description of Algorithms

3.1 Particle Swarm Optimization

The particle swarm optimization (PSO) algorithm is based on observations of the behaviors of certain animals whose are driven by their instincts such as bird flocking, fish schooling, etc. Algorithm is initialized by defining the population size, the fitness function, the fixed parameters and the variables along with their upper and lower bounds. The population size will also denote the number of possible solutions. Every individual entity is called a particle, and every particle of the population tries to achieve the best position. Every particle changes its position in order to get the personal best. Finally, the global best is considered to be the best possible solution. The position and velocity of each and every particle are updated according to the following equations:

$$\begin{aligned} v(i, j) = w * v(i, j) + c1 * \text{rand} * \{P.\text{best}(i, j) - x(i, j)\} \\ + c2 * \text{rand} * \{G.\text{best}(1, j) - x(i, j)\}\ [4] \end{aligned} \tag{9}$$

$$x(i, j) = x(i, j) + v(i, j)\ [4] \tag{10}$$

where $x(i, j)$ is the position of each particle, rand() is a random function, $c1$ and $c2$ are acceleration factors usually $c1 = c2 = 2$, and w is an inertia weight (Fig. 4).

3.2 Artificial Bee Colony

In this algorithm, there are three groups of bees, named employed bees, onlooker bees and scout bees. The colony is first segregated in two halves, first halves include the employed bees, and second half comprises the onlooker bees. Every employed bee is aware of one food sources each, and therefore, the number of employed bees also signifies the number of food sources. The employed bee having information of an abandoned source becomes a scout bee. At first, the employed bee finds a food source, shares the information about the food source with the onlooker bees, and onlooker bees select one of the food sources. The employed bees of whose food sources are abandoned start to find new sources and become scouts. The onlooker bees will give preference to a food source having a greater amount of nectar, amount of nectar is denoted by $X(L)$ where L denotes the location of the food source, and the probability(P_i) of a food source located at a particular location L_i by an onlooker bee can be expressed as

$$P_i = \frac{X(L_i)}{\sum_{k=1}^{N} X(L_k)}\ [8] \tag{11}$$

Fig. 4 Flowchart of PSO algorithm [4]

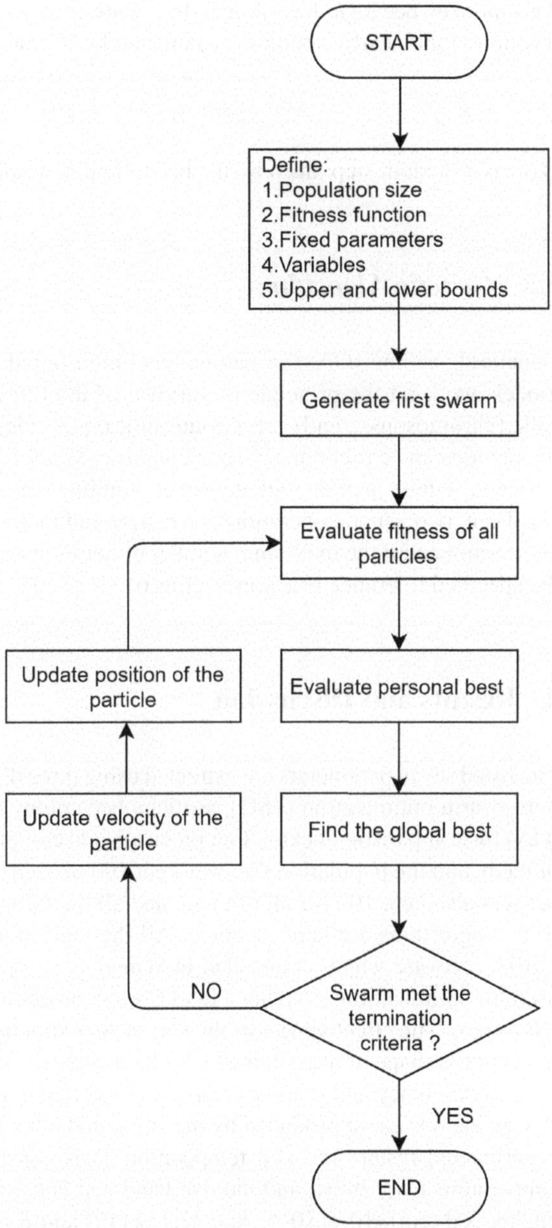

An onlooker bee searches the nearby location of L_i for better food sources. The position of the nearby sources is calculated by following:

$$L_{(i+1)} = L_i \pm Q(c) \tag{12}$$

$Q(c)$ is a random step taken by the bee to find new source around L_i (Fig. 5).

3.3 Genetic Algorithm

Genetic algorithms (GAs) are nature's evolution-based, computational models. These models work on the principle of survival of the fittest. First the initial population called chromosomes, and each chromosome is possible solution. Generation on initial population can be random or fixed in number. Methods of selecting the solutions are numerous which include roulette wheel sampling, sigma scaling, rank selection, etc. [7]. Then, two genetic operations, crossover and mutation take place. The algorithm gets terminated if the maximum number of iterations is performed or solution within the specified tolerance is achieved (Fig. 6).

4 Results and Discussion

The listed seven parameters are extracted using three different algorithms named particle swarm optimization (PSO), artificial bee colony (ABC) and genetic algorithm (GA) for comparison of extraction precision and computational performance of each of them, and the population size was kept 100 in each algorithm. The iteration number was also kept 100 for all of them, and all the initial conditions for running each of the algorithms are kept identical. All the algorithms are coded with MATLAB 2019A software which is installed in Windows 10 operating system. The machine used for optimization is having a Intel Core i5 processor clocked at 2.50 GHz with 8 GB of ram. The effectiveness of the parameter extraction process will be assessed in terms of the shape of the obtained I-V characteristics, fitness function value, computational efficiency and relative error of extracted parameters. To verify the proposed model, the I-V curve produced by the simulated model is compared with a standard experimental result. The cell temperature T_c is considered equal to the reference temperature T_{ref}= 298K, and the dye band gap is taken 1.6 eV. The semiconductor thickness d (cm)=10×10^{-4}, R_s= 22.2 Ω [19] and R_{sh}= 3000 Ω [19] (Fig. 7).

The simulated results are in well agreement with the experimental results proving the correctness of the proposed model. The output parameters of each of the results are listed in Table 2.

To verify the correctness of the proposed model, no algorithm is used, and standard values of each variables are used. After verifying the proposed model, algorithms are used, their parameters are extracted from a given standard ranges, and the extracted

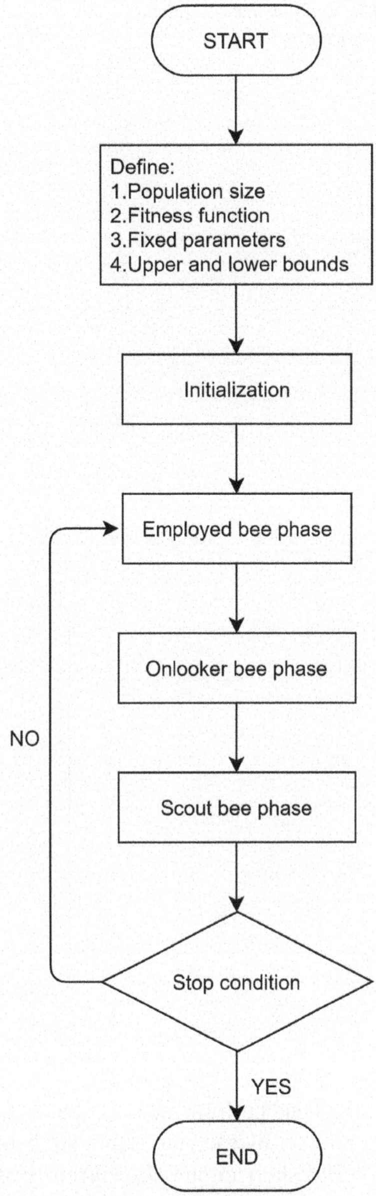

Fig. 5 Flowchart of ABC algorithm [8]

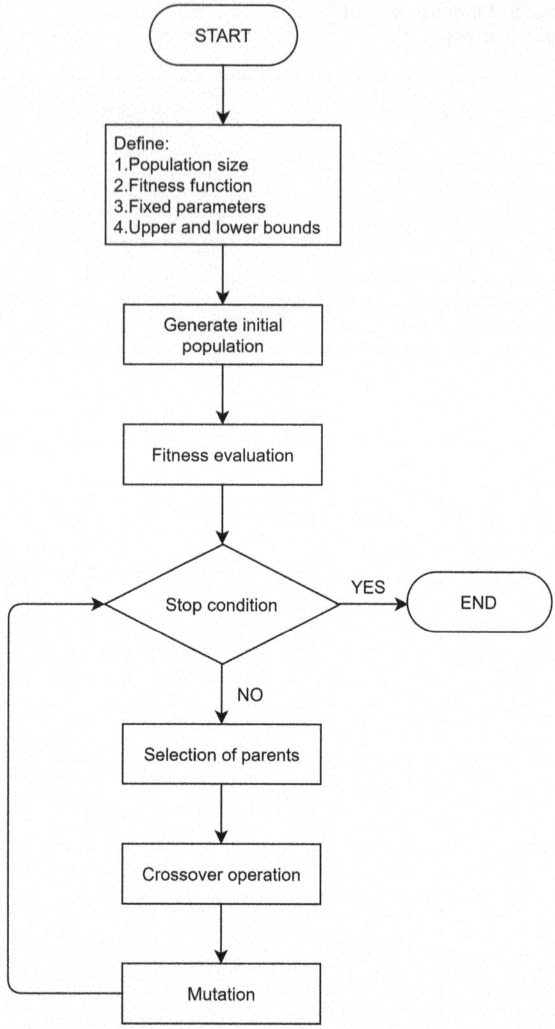

Fig. 6 Flowchart of genetic algorithm [7]

values and performance of each algorithm are discussed. The extracted values of the parameter by each algorithm are listed In Table 3:

The short circuit J_{sc} current was found to be 15.35 mA/cm^2 in PSO, 15.08 mA/cm^2 in ABC and 15.14 mA/cm^2 in GA. The efficiency of 9.0235 % was found highest in GA. The results achieved are Table 4.

The final fitness value of the PSO algorithm was achieved after 23rd iteration, making it the quickest to reach the final value. Fitness function value versus iteration graph for all three algorithms is shown in Fig. 8

Fig. 7 Comparison graph of simulated and experimental results [20]

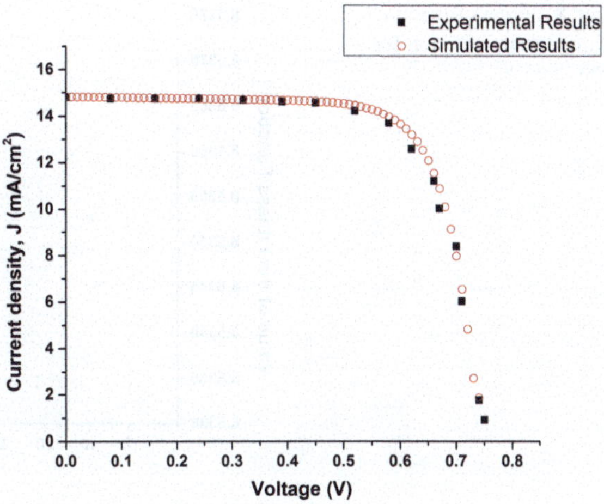

Table 2 Comparison table of theoretical and practical results

	V_{oc}(V)	J_{sc} (mA/cm^2)	FF(%)	η(%)
Theoretical	0.75	14.83	73	8.225
Practical	0.76 [20]	14.75 [20]	70 [20]	7.85 [20]

Table 3 Extracted parameters by different algorithms

Parameters	Extracted vales by PSO	Extracted vales by GA	Extracted vales by ABC
d(cm)	10×10^{-4}	10.15×10^{-4}	10.255×10^{-4}
n	4.5	4.495	4.5
m	2	1.9943	2
$R_{sh}(\Omega)$	2997	2979.1	2998
$R_s(\Omega)$	10	40.67	20.108
T_c(K)	300.15	300.0920	300.15
E_g(eV)	1.59	1.4017	1.4

Table 4 Comparison table of results achieved

Algorithm	V_{oc}(V)	J_{sc} (mA/cm^2)	$FF(\%)$	η(%)
PSO	0.75	15.35	74.09	8.53
ABC	0.74	15.08	76.59	8.53
GA	0.73	15.14	81.6	9.0235

The global best value of the ABC algorithm was achieved after 26th iteration, making it the second quickest to reach its global best. The final efficiency value of ABC and PSO was found same (Fig. 9).

Fig. 8 Fitness function value versus iteration for PSO

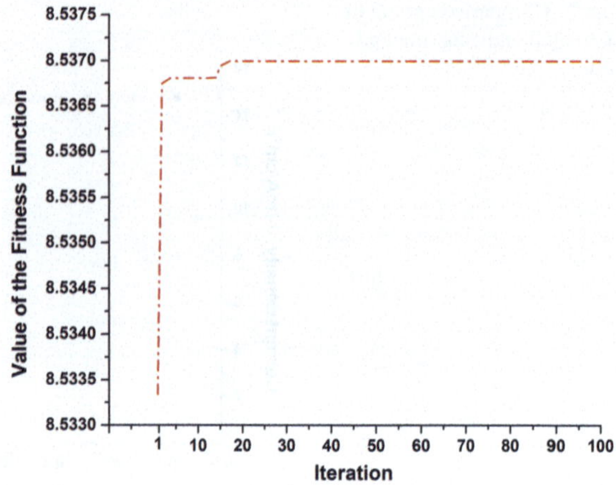

Fig. 9 Fitness function value versus iteration for ABC

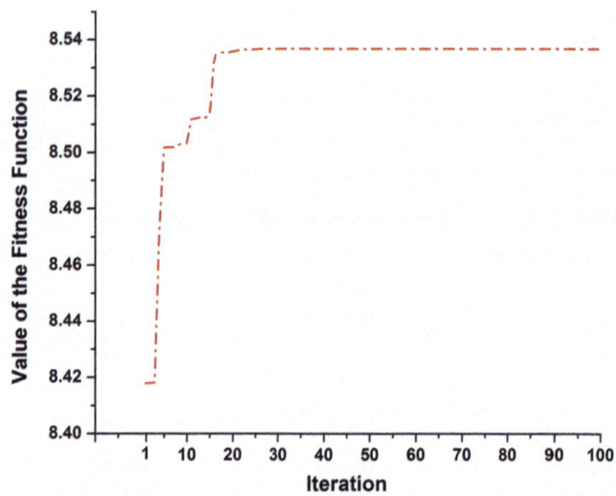

Table 5 Computational time for different algorithms

Algorithm	Computational time (s)
PSO	8.05
ABC	16.22
GA	35.36

In case of GA, the global best was achieved after 49th iteration making the algorithm slowest of all; however, the efficiency achieved is the highest among all three. The computational time for each of the algorithm is listed in Table 5 (Fig. 10).

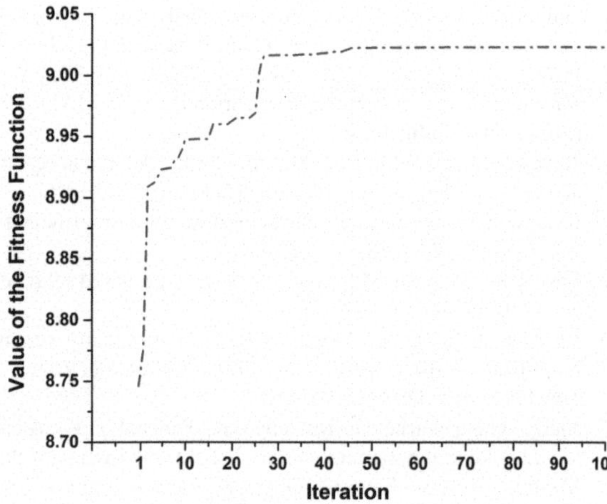

Fig. 10 Fitness function value versus iteration for GA

5 Conclusion

By using the proposed model, PSO algorithm has higher computational speed as compared to others in this article. GA is found to be more efficient in parameter extraction which leads to greater efficiency. ABC algorithm results gave similar efficiency with PSO but with a high computational time. Finally, GA shows the highest of all efficiency, with highest computational time, and the fill factor achieved by GA was the highest of all. The initial conditions were kept same in all the three algorithms, to make a fair comparison between them.

Acknowledgements The work in this paper was supported by Department of Electronics and Instrumentation Engineering, and research facilities were provided by Solar Energy Materials Research and Testing Laboratory (SMaRT Lab), Department of Physics, National Institute of Technology, Silchar

References

1. Duan, L., Yi, H., Xu, C., Upama, M.B., Mahmud, M.A., Wang, D., Shabab, F.H., Uddin, A.: Relationship between the diode ideality factor and the carrier recombination resistance in organic solar cells. IEEE J. Photovoltaics **8**(6), 1701–1709 (2018)
2. Gómez, R., Salvador, P.: Photovoltage dependence on film thickness and type of illumination in nanoporous thin film electrodes according to a simple diffusion model. Solar Energy Mater. Solar Cells **88**(4), 377–388 (2005)
3. Gong, J., Krishnan, S.: Mathematical modeling of dye-sensitized solar cells. In: Dye-Sensitized Solar Cells, pp. 51–81. Elsevier (2019)
4. Gong, L., Cao, W., Zhao, J.: An improved PSO algorithm for high accurate parameter identification of PV model. In: 2017 IEEE International Conference on Environment and Electrical Engineering and 2017 IEEE Industrial and Commercial Power Systems Europe (EEEIC/I&CPS Europe). pp. 1–5. IEEE (2017)

5. Guliani, R., Jain, A., Kapoor, A.: Exact analytical analysis of dye-sensitized solar cell: improved method and comparative study. Open Renew. Energy J. textbf5(1), (2012)
6. Habieb, A.M., Irwanto, M., Alkian, I., Sya'diyah, F.K., Widiyandari, H., Gunawan, V.: Dye-sensitized solar cell simulation performance using MATLAB. J. Phys. Conf. Seri. 1025, 012001 (2018). (IOP Publishing)
7. Jervase, J.A., Bourdoucen, H., Al-Lawati, A.: Solar cell parameter extraction using genetic algorithms. Measur. Sci. Technol. **12**(11), 1922 (2001)
8. Karaboga, D., Basturk, B.: On the performance of artificial bee colony (ABC) algorithm. Appl. Soft Comput. **8**(1), 687–697 (2008)
9. Lau, K.K., Soroush, M.: Overview of dye-sensitized solar cells. In: Dye-Sensitized Solar Cells, pp. 1–49. Elsevier (2019)
10. Lee, J.J., Coia, G.M., Lewis, N.S.: Current density versus potential characteristics of dye-sensitized nanostructured semiconductor photoelectrodes. 2. simulations. J. Physi. Chem. B **108**(17), 5282–5293 (2004)
11. Mehta, H.K., Warke, H., Kukadiya, K., Panchal, A.K.: Accurate expressions for single-diode-model solar cell parameterization. IEEE J. Photovoltaics **9**(3), 803–810 (2019)
12. Meshram, S., Agnihotri, G., Gupta, S.: Modeling of grid connected dc linked pv/h ydro hybrid system. Electri. Electron. Eng. Int. J. (ELELIJ) **2**(3), 13–27 (2013)
13. Mingsukang, M.A., Buraidah, M.H., Arof, A.K., Das, N.: Third-generation-sensitized solar cells. Nanostruct. Solar Cells (2017)
14. Ni, M., Leung, M.K., Leung, D.Y.: Theoretical modelling of the electrode thickness effect on maximum power point of dye-sensitized solar cell. Can. J. Chem. Eng. **86**(1), 35–42 (2008)
15. Ni, M., Leung, M.K., Leung, D.Y., Sumathy, K.: An analytical study of the porosity effect on dye-sensitized solar cell performance. Solar Energy Mater. Solar Cells **90**(9), 1331–1344 (2006)
16. Shannan, N.M.A.A., Yahaya, N.Z., Singh, B.: Single-diode model and two-diode model of pv modules: A comparison. In: 2013 IEEE International Conference on Control System, Computing and Engineering. pp. 210–214. IEEE (2013)
17. Shirkavand, M., Bavir, M., Fattah, A., Alaei, H.R., Tayarani Najaran, M.H.: The construction and comparison of dye-sensitized solar cells with blackberry and n719 dyes. J. Optoelectronical Nanostruct. **3**(1), 79–92 (2018)
18. Soroush, M., Lau, K.K.: Insights into dye-sensitized solar cells from macroscopic-scale first-principles mathematical modeling. In: Dye-Sensitized Solar Cells, pp. 83–119. Elsevier (2019)
19. Tripathi, B., Yadav, P., Kumar, M.: Charge transfer and recombination kinetics in dye-sensitized solar cell using static and dynamic electrical characterization techniques. Solar energy **108**, 107–116 (2014)
20. Umale, S.V., Tambat, S.N., Sudhakar, V., Sontakke, S.M., Krishnamoorthy, K.: Fabrication, characterization and comparison of DSSC using anatase tio2 synthesized by various methods. Adv. Powder Technol. **28**(11), 2859–2864 (2017)
21. Wei, L., Yang, Y., Fan, R., Na, Y., Wang, P., Dong, Y., Yang, B., Cao, W.: N, n-bis ((6-methoxylpyridin-2-yl) methylene)-p-phenylenediimine based d 10 transition metal complexes and their utilization in co-sensitized solar cells. Dalton Trans. **43**(29), 11361–11370 (2014)
22. Yong, V., Ho, S.T., Chang, R.P.: Modeling and simulation for dye-sensitized solar cells. Appl. Phys. Lett.**92**(14) (2008)

Role of Battery Management System (BMS) in Sustainable Transportation

B. S. Sagar, Rajashekar P. Mandi, B. P. Divakar, and Santoshkumar Hampannavar

Abstract Battery management system (BMS) manages and monitors the overall action of the battery pack. BMS has a vital role to play in sustainable transportation. The depleting fossil fuels and serious environmental concerns have opened the doors for development and promotion of Electric Vehicle (EV) at large. Sustainable transportation is possible using EVs and supports the clean air and reduced carbon emissions and footprint. State of charge (SOC) of EV battery is to be computed or estimated before EVs are integrated into the grid for charging/discharging. In this paper, different SOC estimation methods are discussed and experimental setup for battery equalization is proposed.

1 Introduction

Sustainable transportation is the way forward for cleaner and greener world. Transportation on offer should adhere to the basic accessibility and also meet the requirements of social development. The transportation system must be with a proper balance between the usage of renewable and/or non-renewable resources with its generation or availability respectively.

Sustainable transportation can thus be called for those systems that provides required transportation for the developing society which is affordable, operates fairly and efficiently. Also, the system should not be hazardous to the environment with its emissions and eventually avoiding impacts on the climate. A move towards reducing the accidents also makes transportation system sustainable. Figure 1 represents the framework of sustainable transport system.

To name a few, sustainable transport modes are public transport in motorized category and in non-motorized category bicycle and walking. Considering the practical aspects, these modes can be complemented by using cleaner and greener technology by switching over to electric mobility.

B. S. Sagar (✉) · R. P. Mandi · B. P. Divakar · S. Hampannavar
REVA Universtiy, Bengaluru, India

J. Gu et al. (eds.), *Communication and Control for Robotic Systems*, Smart Innovation, Systems and Technologies 229, https://doi.org/10.1007/978-981-16-1777-5_22

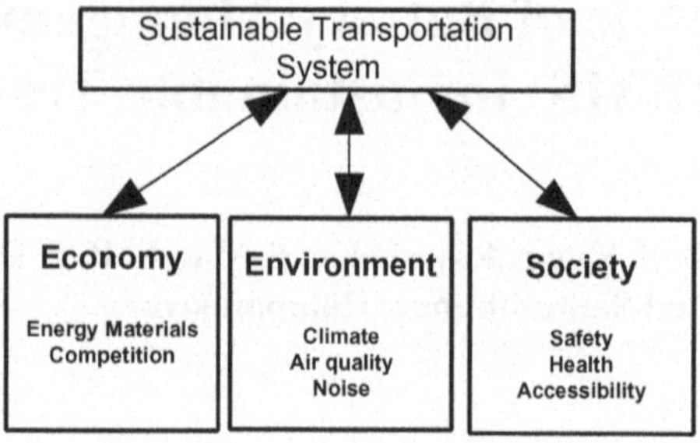

Fig. 1 Sustainable transportation socio-economic environmental framework

Recent trends show there is a steady increment in moving towards sustainable transportation system in India and across the world. The major challenge in electrification of transport sector is in managing of energy source which is the battery pack. Battery management system (BMS) is a critical system that address the issues with batteries which are powering the vehicles and ensures that the batteries are operated in a safe zone and its life is enhanced thus enhancing the overall efficiency of the system.

Penetration of EVs to the users depends on the technicality of EVs in terms of its characteristics as compared to IC engine vehicle like the range, battery capacity, and others. In order to get the information about entire status of the battery pack, Battery Management System (BMS) is essential.

BMS an electronic system that performs a set of continuous operation that monitors the overall action of battery pack including the identification of chemistry of the battery, logging various charging and discharging cycles information, estimating the state of charge (SOC), state of health (SOH), charge equalization, reporting the data and temperature monitoring. Block diagram of BMS is shown in Fig. 2. In [1] detailed information on the BMS components is provided. Insightful and in-depth guide to varied cases of application-oriented BMS is presented in [2]. BMS monitors the battery by various parameters of the battery such as voltage, current, temperature, SOC, and SOH.

The functions of a BMS are as follows:

- BMS protects the battery pack from overcharging, undercharging, deep discharge, temperature monitoring, and short circuit

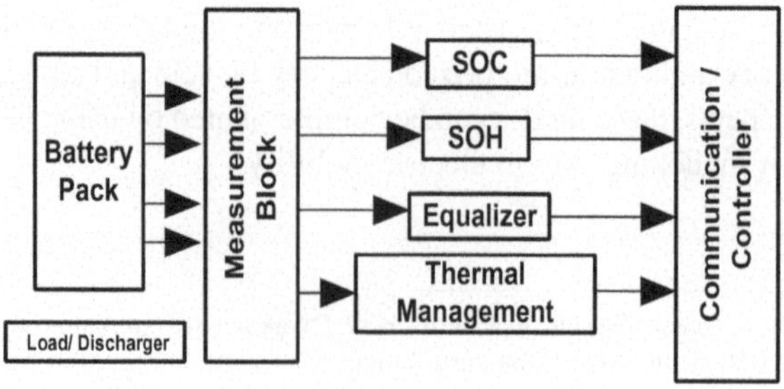

Fig. 2 Block diagram of BMS

- BMS ensures all the cells in battery are charged to its SOC level
- BMS communicates with various devices and collects data which will then be used for analysis and communicating to the user
- BMS will monitor the temperature of battery pack and maintains the safe operating limits
- BMS will monitor the overall activities of the battery pack for analysis.

Data acquisition unit captures electrical parameters like voltage and current. The system consists of sensors which will be connected to a computer. The software embedded with it picks up these data and logs into the device. Various analyses can be carried out on the data and required information will be communicated to the user. The most important blocks of data acquisition system are the sensors, ADC, signal conditioners, processor, and the communication unit.

Measurement of voltage consists of total voltage of the cell/battery pack, voltage of individual cell/battery pack, minimum and maximum voltage that each cell can hold. Measurement of current can be either the current flowing inside the battery during charging period or current flowing out of the battery during the discharge phase. SOC indicates the current capacity of the battery in holding the charge as opposed to the manufacturer details. Reaming capacity of the battery is estimated using the parameter SOH as opposed to the manufacturer details. The temperature of the cell/battery pack varies during the charging or the discharging phase due to flow of electrons. The temperature of the pack will play a major role in estimating the parameters and therefore monitoring of the temperature also plays a major role. The adverse effects caused by the temperature on battery like thermal runaway, performance degradation of the battery pack are discussed in depth [3]. Various cooling methods associated with battery pack to mitigate the effects of temperature on the battery pack are discussed in [4]. A complete technical overview of the BMS for a given system is presented in [5].

One of the requirements of BMS is in ensuring the safety of the battery pack. As mentioned earlier, the battery pack measurement parameters are to be monitored in maintaining the safety of the battery pack. Battery pack is to be protected from over charging under charging, over current, and voltage during the charging phase and under current and voltage during the discharging phase. The temperature of the battery pack essentially sets the performance of the battery pack. High-temperature results in higher discharge rate and low-temperature results in lower output current from the battery.

Depending on the arrangement of the controllers, BMS can be classified into following types:

i. Centralized
ii. Semi-centralized
iii. Localized.

Centralized type of BMS consists of a single controller will monitor the overall activities of the battery pack. This type offers a very simplistic design defect in a single module will eventually bring down the performance of the entire system. The semi-centralized BMS has there are two or more master controllers with multiple individual battery controllers. Each module communicates to other module and also to its individual controllers. In localized type, each battery has its own controller and will be given with complete provision to monitor and take necessary actions while maintaining the working of the battery pack. However there is no hard and fast rule in setting up of BMS. BMS always depends on the requirement of the user and on the applications. Requirement from stationary application like in uninterrupted power supplies (UPS) are different from that of high power application like EVs.

2 Sate of Charge Estimators

State of charge (SOC) estimation is most challenging for the battery usage. It is most predominant parameter for finding the capacity of the battery and its discharge.

Mathematically, SOC can be defined as

$$\mathrm{SOC}(t) = \left(\frac{Q(t)}{Q(n)}\right) \tag{1}$$

where, $Q(t)$ is current capacity and $Q(n)$ is nominal capacity.

There are various methods to estimate the SOC and some of the methods are

- Open circuit voltage method
- Terminal voltage method
- Impedance method
- Coulomb counting method
- Kalman filtering method
- Artificial Neural networks(ANN)
- Hybrid models.

Open circuit voltage method: This is one of the simple ways to estimate the battery of SOC. There exists almost a linear relationship between open-circuit voltage and SOC of the battery.

SOC of the battery at any given instant of time 't' is given by the equation

$$V_t = A\,\mathrm{SOC}(t) + B \tag{2}$$

In the equation, A is obtained from the value of B which itself is the battery terminal voltage a 0% SOC. The relationship can be experimentally tested and corresponding SOC can be estimated. Even though the method is simple, there are irregularities associated with this method. This method can be applicable only when the batteries are subjected to longer settling time. The method is more suitable for lead acid type batteries and has found errors for lithium based batteries.

Terminal voltage method: This method approximates the EMF of the battery closely to SOC of the battery as during the discharge internal impedance of the battery affects the current flow out of the battery. The major drawback of the method is that the error that accumulates over a period of time is very large.

In impedance method, impedance offered by the battery terminals provides information about several other parameters through which the SOC can be estimated. But it has to be noted that impedance parameters and their variations with SOC is not common for all chemistries of batteries.

Coulomb counting method measures the discharging current of the battery and integrates over the time period and thus estimates the battery SOC.

SOC is obtained by the equation

$$\mathrm{SOC}(t) = \mathrm{SOC}(t-1) + \frac{i(t)}{Qn}\Delta t \quad (3)$$

Experimental results are proven that this method provides values of SOC. But the fact that several other parameter that affect the accuracy of estimation like temperature, cycle life, and others are not considered.

Kalman filters and extended Kalman filters are also in place to estimate the SOC of the battery. With the advancement of algorithms and artificial intelligence, neural methods, fuzzy methods are being employed to best estimate the SOC by reducing error and including the dynamism that a battery pack can offer.

Hybrid models are being employed to further enhance the battery SOC estimation by including advantages of different techniques and reducing the time taken to estimate SOC while increasing the accuracy of the estimate. Future research orients towards hybrid models to estimate the SOC.

3 Battery Equalization

3.1 Charge Control of Batteries and Its Importance

Charging of batteries plays an important role in defining the performance of the battery over a period of time. The charging process differs for different chemistries of battery and manufacturing technologies. Charging methods needs to be optimized to avoid overcharging leading to damaging the batteries.

Multistage charging processes of lead acid batteries approximately take about 8–10 h to completely charge whereas Lithium-ion batteries take about 2–3 h. The charging process of lead acid battery and lithium-ion battery can be considered similar to a certain extent but for float charging stage which is different. Charging stages of lead acid battery is shown in Fig. 3, which constitutes of constant current stage, constant voltage stage, and floating stage.

Manufacturers of these cells/batteries recommend charging at C/10 or C/20 to prolong the battery life.

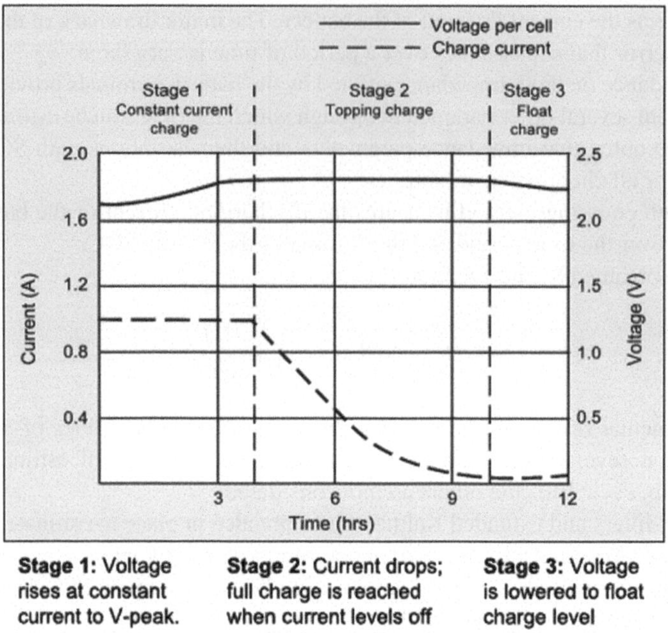

Fig. 3 Charging stages of a lead acid battery [6]

Similarly, to highlight for lithium-ion batteries, constant current stage and constant voltage stage are the two stages which is shown in Fig. 4. Manufacturers of these cells/batteries recommend charging at 0.8 C or less although these can be charged at 0.5 C or 1 C to prolong the battery life.

3.2 SOC-Based Charge Equalizer

As deeper we get into the batteries, the other important aspect to consider is the charge equalizing. As discussed earlier, SOC of the battery changes due to various parameters and it is found that the batteries must be charged to its 100% SOC than its rated voltage. In real-time applications, a pack of batteries either in series or in parallel or in combination of series parallel will be employed to meet the requirements of the load. When this pack is considered as a single unit and subjected to charge then there comes the problem of mismatching of voltage to which each battery/cell is charged. Charge equalizers should be employed to avoid this mismatching of voltage levels. In reference [7], the effects of over-charging on battery is discussed. This clearly paves way for the need of charge equalization scheme in a battery management system to avoid the adverse effects on the battery leading to malfunction of the system and degrading the life of the battery.

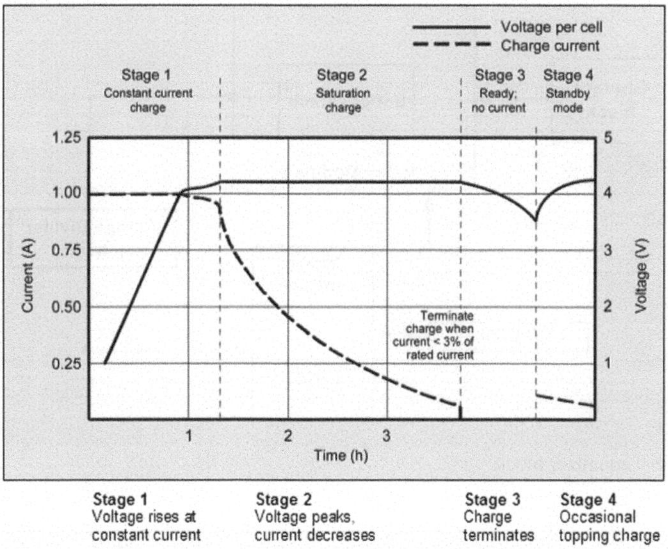

Fig. 4 Charging stages of a lithium ion battery (Courtesy Battery University)

Charge equalization is a technique in which all the batteries in a stack is brought to the same level. This protects the stack from being overcharged and/or undercharged and thus enhancing reliability and life cycle of the battery. Reference [8–10] provides effective charging schemes to over the coming the irregularity associated with the regular charging scheme. However, these schemes consider the battery pack as a two-terminal unit. But the requirement to meet the individual battery/cell inside the pack has to be met.

Reference [11] discuss in detail the importance of charge equalizers and also proposes new and simple technique to equalize the batteries in a pack. Charge equalization techniques can broadly be classified into dissipative and non-dissipative schemes. However, the power dissipation associated with dissipative schemes limits its applications. Non-dissipative scheme is a viable option as it either transfers the energy from strong batteries to weaker batteries or employing algorithms so as to charge the batteries without the criteria of overcharging or undercharging of the batteries in the pack. One such algorithm is series differential algorithm as proposed in [7]. Battery equalization block is shown in Fig. 5.

Three 12 V, 42 Ah valve regulated lead acid (VRLA) batteries are employed in the battery stack. The batteries are connected with relays which provide the path for the current between the batteries and the charger. Voltage divider helps in reducing the voltage appropriately so as to be measured by the controller and take the necessary actions. Simulation circuit consisting of three batteries, for equalizing the battery bank is shown as in Fig. 6.

Sequential difference algorithm provides the required voltage level of the battery inside the battery stack. The algorithm is one of the easiest and simple way of

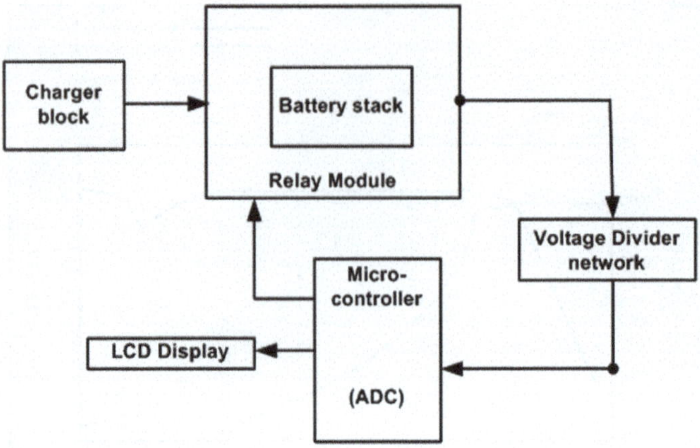

Fig. 5 Battery equalizer block

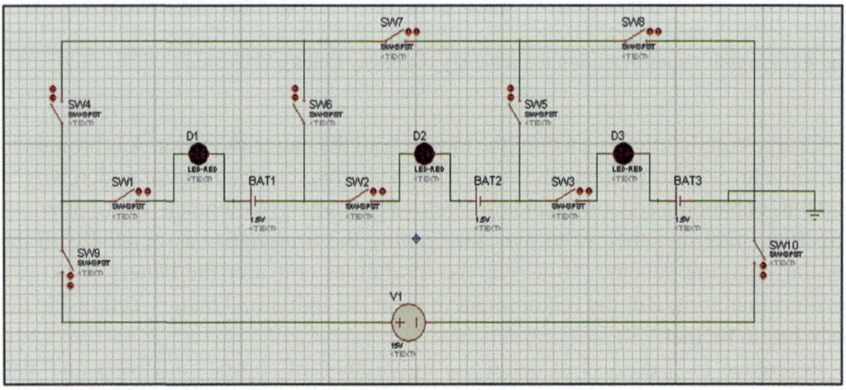

Fig. 6 Simulation circuit for battery equalizing

measuring the individual battery voltage which eliminates the problems associated with battery terminal voltage measurement when batteries are connected in series.

Assuming individual battery voltage to be V_1, V_2 and V_3 and measured voltage be V_A, V_B, V_C then the required individual battery pack voltage be obtained by employing the sequential difference algorithm as in the equations below:

$$V_{R1} = V_A = V_1 \tag{4}$$

$$V_{R2} = V_B - V_A = (V_1 + V_2) - V_1 \tag{5}$$

$$V_{R3} = V_C - V_B = (V_1 + V_2 + V_3) - (V_1 - V_2) \tag{6}$$

The controller will identify the weakest battery among the stack and thus routes charging current to that identified weak battery and also virtually isolates the identified weak battery from the stack.

3.3 Experimental Setup and Methodology

The experiment module was setup to gain firsthand experience of implementing sequential difference algorithm to equalize the battery pack which contained ten 12 V, 42 Ah VRLA batteries.

The methodology to implement the sequential difference algorithm to equalize the battery pack is given in the steps below:

Step 1: The entire charging period of the battery pack is split into two parts namely charging time and equalizing time
Step 2: After the completion of charging time, equalization has to be employed
Step 3: During the equalization time period, battery terminal voltages are logged by the controller
Step 4: Controller decides which battery is at the lease potential in the stack and connects the charger to that identified weak battery. Further charging of the battery is continued for a set time period
Step 5: After the weak battery is allowed to further charge for defined period of time, controller again checks all the battery voltages and repeats the process till all battery voltages are balanced.

4 Results and Discussion

The sequential difference algorithm with the equalizer circuit is employed for battery pack containing ten 12 V, 42 Ah valve regulated lead acid (VRLA) batteries in the laboratory. The experimental findings are given below. The switching action is mentioned in Table 1. The working setup is shown as mentioned in Fig. 7. The initial battery voltages of the batteries in the battery pack is tabulated in Table 2. Figure 8 shows the imbalance in the charge levels of the batteries in the stack. With the help of equalizer, the battery voltages are equalized and are found to be as tabulated in Table 3. Figure 8 shows the balanced charge levels of the batteries in the stack which is achieved due to charge equalization. Thus, the battery pack is equalized which ensures no damage to any of the battery inside the battery pack due to either overcharging or under-charging as shown in Figure 9.

Table 1 Switching action

Switches\battery	B1	B2	B3
S1	1	0	0
S2	0	1	0
S3	0	0	1
S4	0	1	1
S5	1	1	0
S6	0	1	1
S7	1	0	1
S8	1	1	0

Fig. 7 Hardware setup of the charge equalizer

Table 2 Battery charge level before equalization

Battery	Battery voltage without equalization in volts
B1	12.44
B2	11.82
B3	10.9
B4	12.12
B5	11.45
B6	11.05
B7	10.98
B8	12.01
B9	12.22
B10	12.82

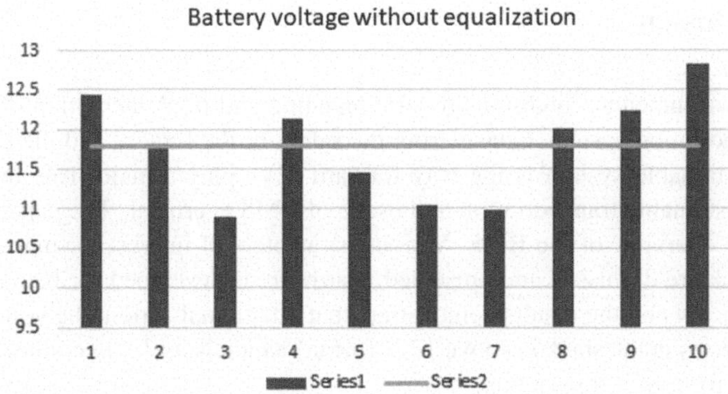

Fig. 8 Battery voltages without equalization

Table 3 Battery charge level after equalization

Battery	Battery voltage with equalization in volts
B1	12.58
B2	12.59
B3	12.60
B4	12.80
B5	12.75
B6	12.72
B7	12.68
B8	12.62
B9	12.32
B10	12.49

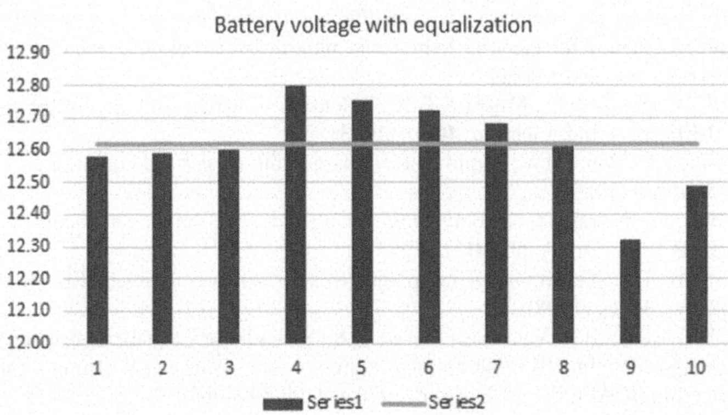

Fig. 9 Battery voltages with equalization

5 Conclusion

Present environmental conditions require planning and implementation of sustainable transportation system. Considering the safety of the society and the economic impact sustainable system is the way forward. EVs play a major role in moving towards sustainable transportation and usage of BMS is crucial. The paper deliberates on the overview of the BMS. Various parameters of interest for managing the battery pack are discussed and presented. Hardware is developed for battery equalization concept and the results demonstrate that after equalization, the voltage level of the batteries in the stack is above 12 V (mean) which is useful in connecting EVs to the grid for power transaction.

References

1. Chang, W.-Y.: The State of Charge Estimating Methods for Battery: A Review. Hindawi Publishing Corporation ISRN Applied Mathematics, vol. 2013, Article ID 953792, 7 p. https://doi.org/10.1155/2013/953792
2. Barsukov, Y., Qian, J.: Battery Power Management for Portable Devices (2013). ISBN 9781608074914
3. Liu, H., Wei, Z., He, W., Zhao, J.: Thermal issues about Li-ion batteries and recent progress in battery thermal management systems: a review. Energy Convers. Manag. **150**, 304–330. https://doi.org/10.1016/j.enconman.2017.08.016. ISSN 0196–8904
4. Chen, D., Jiang, J., Kim, G.-H., Yang, C., Pesaran, A.: Comparison of different cooling methods for lithium ion battery cells. Appl. Thermal Eng. **94**, 846–854 (2016). https://doi.org/10.1016/j.applthermaleng.2015.10.015. ISSN 1359–4311
5. Sagar, B.S., Mandi, R.P.: Estimation of battery management system, 978-1-5386-0569-1, IEEE Xplore, May 2018, https://doi.org/10.1109/SmartTechCon.2017.8358569
6. Naseri, F., Farjah, E., Ghanbari, T.: An efficient regenerative braking system based on battery/supercapacitor for electric, hybrid and plug-in hybrid electric vehicles with BLDC motor. 0018-9545 © 2016 IEEE
7. Hung, S.T., Hopkins, D.C., Mosling, C.R.: Extension of battery life via charge equalization control. IEEE Trans. Ind. Electron. **40**(1), (1993)
8. Podrazhansky, Y., Popp, P.W.: Rapid battery charger, discharger and conditioner. U.S. Patent 4,829,225, May 9 (1989)
9. Kutkut, N.H. Deepakraj, M, Novotny, D.W.: Charge equalization for series connected battery strings. IEEE Trans. Ind. Appl. **31**(3), (1995)
10. Ye, Z., Stuart, T.A.: Sensitivity of ramp equalizer for series batteries. IEEE Trans. Aerosp. Electron. Syst. **34**(4), (1998)
11. Sagar, B.S., Divakar, B.P., Venkatesh Prasad, K.S.: Series battery equalization using sequential difference algorithm. In: 2014 International conference on advance in electronic, computer and communication (ICAECC), IEEE Xplore, 978-1-4799-5496-4/14

A Fundamental Study on Electric Vehicle Model for Longitudinal Control

K. Indu and M. Aswatha Kumar

Abstract Stricter emission norms need to drift toward being environment friendly have shifted the concentration in the automobile sector toward electric vehicles. This research article highlights the fundamental modeling steps required for an electric vehicle control system design following a simulation approach using MATLAB/Simulink software. From an electric vehicle design perspective, this approach offers an excellent solution to give insights into EV research involving multidisciplinary engineering aspects. The study presents longitudinal control technique, relevant observations and results to bring out the differences in open-loop and closed-loop case studies. It also intends to provide better understanding toward the need for a feedback, realization of an expected path profile for students and researchers in this field of interest. The steps involved in transforming the mathematical model into a simulation model and analysis of the simulation results are detailed in this article.

1 Introduction

Ever-increasing environmental concerns and depletion of fossil fuels have led to serious focus toward greener and friendlier environment conditions. Moreover, research and development activities for quite some time now have started concentrating toward fewer emission or zero emission transport. In the automobile sector, one of the major contributors in terms of pollution, as studies suggest, is from vehicles (light or heavy). Among these, a significant amount still remains from emissions of light vehicles like the two/four wheelers. Slowly, people and governments are realizing the need for electric vehicles, policies are being drafted, and research and innovative product developments are taking place in this field. However, there exists

K. Indu (✉) · M. Aswatha Kumar
Christ (Deemed to be University), Bengaluru, Karnataka, India
e-mail: indu.k@res.christuniversity.in

M. Aswatha Kumar
e-mail: aswathakumara.m@christuniversity.in

J. Gu et al. (eds.), *Communication and Control for Robotic Systems*, Smart Innovation, Systems and Technologies 229, https://doi.org/10.1007/978-981-16-1777-5_23

a gap in terms of knowledge transfer between the academic and industrial research. This can be reduced by understanding the industrial need, frequent interactions with industrial professionals or experts. This article tries to focus on industry-based problems and tries to implement efficient electric vehicle (EV) design through continuous simulation approach with the help of software tools like MATLAB/Simulink.

Electric vehicle (EV) modeling and simulation involve several steps from drafting the requirements of the vehicle up to validation/testing. Simulation can be carried out using different software to analyze the behavior of the designed EV model. Modeling is a fundamental step in simulation to help determine a vehicle's reaction to certain situations. Mathematical modeling can be carried out using variety of techniques, for example, linear, nonlinear modeling, dynamic systems, multibody-based systems, lumped models, etc. Besides, any EV's behavior during its initial modeling stages can be explored in three different ways, namely

- Dynamics of vehicle longitudinally (straight line behavior)
- Dynamics of vehicle laterally (during a turn)
- Dynamics of vehicle in the vertical direction.

Reactions from the EV model are analyzed from a longitudinal control point of view in this work. This article tries to interpret the vehicle model from a control system perspective, EV behavior when subjected different external conditions. Next few sessions will be highlighting the significance of previous research, mathematical approach followed, simulation steps and finally the result interpretations.

1.1 Few Notable Contributions

As mentioned earlier, longitudinal behavior of a vehicle involves its response or reaction when the electric vehicle movement is along a straight line. Several researchers have worked on different control strategies.

For example, [1] focuses on designing control technique based on longitudinal forces, considering tire force parameters with a lower acceleration rate. For validation purpose, three different methods in terms of longitudinal force distribution have been compared in the study. Study carried out in [2] highlights the importance of a safety distance model in a vehicle since most accidents as mentioned in the article happen mainly due to serious error from driver's side. Such mistakes as stated can be avoided by actively ensuring the performance of an efficient control scheme or strategy. In another work [3], researchers have proposed an economical design for sliding mode controller incorporating Bayes theorem for traction allocation. Similarly, in [4], authors have used fusion approach by neglecting values for the pitch, other vertical and roll parameters. Furthermore, in [5], researchers have used the driver characteristics for controlling the EV using the parametric model. Recently, researchers in [6] have identified the importance of energy-saving (quadratic dependency)-based scheme. They have combined the energy-saving aspects with the vehicle control scheme.

2 Modeling Approach

Mathematical modeling is carried out in this research article as per the block diagram shown in Fig. 1. In general, EV model includes the EV dynamics, powertrain and battery specifications. Block diagram depicts four different sub-systems, namely driving pedal, EV dynamics sub-system, electrical sub-system and feedback controller sub-system. Acceleration or deceleration command from the driving pedal sub-system serves as the input to the EV dynamics sub-system. Forces and tire-related parameters are also set according to the requirement specification within the dynamics sub-system. Electrical sub-system consists of the motor and battery components of the EV.

Efficiency of the vehicle at desired velocity will be the output from the electrical sub-system. Battery state of charge (SoC), motor torque and motor speed will be the feedback to the controller module. During any unexpected situation, the controller will decide and take the necessary steps for the proper performance improvement of EV. Based on the decision taken, torque allocation happens within the electrical sub-system.

As depicted in the figure, vehicle control process is carried out as a coordinated control between feedforward and feedback control. Vehicle-related parameters are fed through the EV model. Environmental parameters can be set as per the requirements. Reference velocity required for tracking the vehicle is being fed through the reference model.

Torque allocation process will be the summation of output from the control designs. Furthermore, the imminent instances can be captured prior to the occurrence of any situation, thereby aiding the vehicle control process through allocation of torque. Proportional and integral gains also contribute in the performance of electric vehicle. Output magnitudes are measured in terms of saturated and nominal outputs. Few equations involved in the simulation are explained below:

$$[Y_n = \left(V_f/N_v\right)^* R_v + \left(K_p * V_e\right)/N_v + \int \left(\left(K_i^* V_e\right)/N_v + K_w^* D_o\right)\mathrm{d}t + G_f^* \beta \quad (1)$$

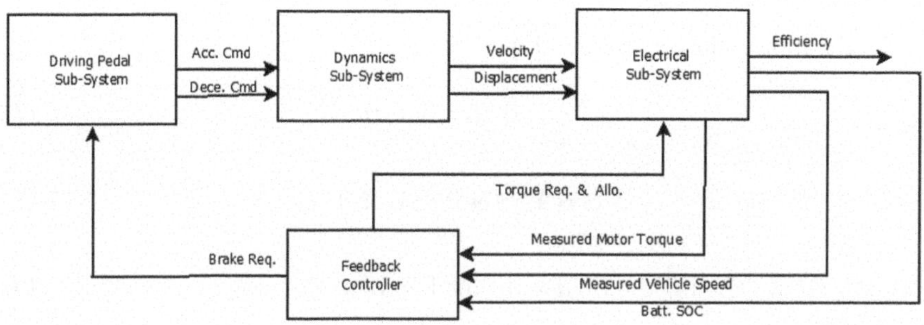

Fig. 1 Block diagram of EV design

Table 1 Nomenclature of variables

a	Acceleration	Y_s	Saturated output	R_v	Reference velocity
d	Deceleration	V_f	Velocity feedforward gain	V	Velocity feedback signal
Y_n	Nominal output	N_v	Speed nominal	D_o	Difference between nominal and saturated output
K_p	Proportional gain	V_e	Error in velocity	K_i	Integral gain
K_w	Gain against wind	G_f	Grade feedforward gain	β	Grade angle
F_t	Tractive force	a_r	Rolling resistance	b_r	Driving resistance with rolling effect
g	Gravitational force	m	Vehicle mass	t_r	Response time

Nomenclature details for the variables used through the article have been summarized in Table 1.

Nominal output Y_n and saturated output Y_s are related as stated below:

where

$$Y_s = -1 : \text{for } Y_n < -1 \tag{2}$$

$$Y_s = Y_n : \text{for } -1 \leqslant Y_n \leqslant 1 \tag{3}$$

$$Y_s = 1 : \text{for } 1 < Y \tag{4}$$

Error in velocity V_e and difference in nominal and saturated outputs (D_o) are estimated as below:

$$V_e = R_v - V \tag{5}$$

$$D_o = Y_s - Y_n \tag{6}$$

Acceleration command follows Eqs. 7–9.

$$a = 0 : \text{for } Y_s < 0 \tag{7}$$

$$a = Y_s : \text{for } 0 \leqslant Y_s \leqslant 1 \tag{8}$$

$$a = 1 : \text{for } 1 < Y_s \tag{9}$$

Deceleration command follows Eqs. 10–12.

$$d = 0 : \text{for } Y_s > 0 \tag{10}$$

$$d = -Y_s : \text{for } -1 \leqslant Y_s \leqslant 0 \tag{11}$$

$$d = 1 : \text{for } Y_s < -1 \tag{12}$$

Outputs and interpretations from these measurements are elaborated in the following session.

3 Simulation Outcome

Simulation of the longitudinal control was carried out using MATLAB/Simulink software. In order to perceive the performance of the EV, few parameters have been plotted. Figure 2 illustrates the EV response for open-loop condition considering the lowest accelerations. Responses obtained from the modeling can be divided into three sets of plots, namely for EV dynamics sub-system, electrical sub-system and the battery responses.

In general, EV has been designed to cover a distance of 19 km, and velocity is determined based on the road condition and drive cycle input requirements. Acceleration, deceleration, velocity and displacement of the EV are plotted under vehicle dynamics plot set. Considerations are over a given time period 't' with 80% state of charge (SOC) assumed for battery at time $t = 0$ s. Furthermore, motor torque,

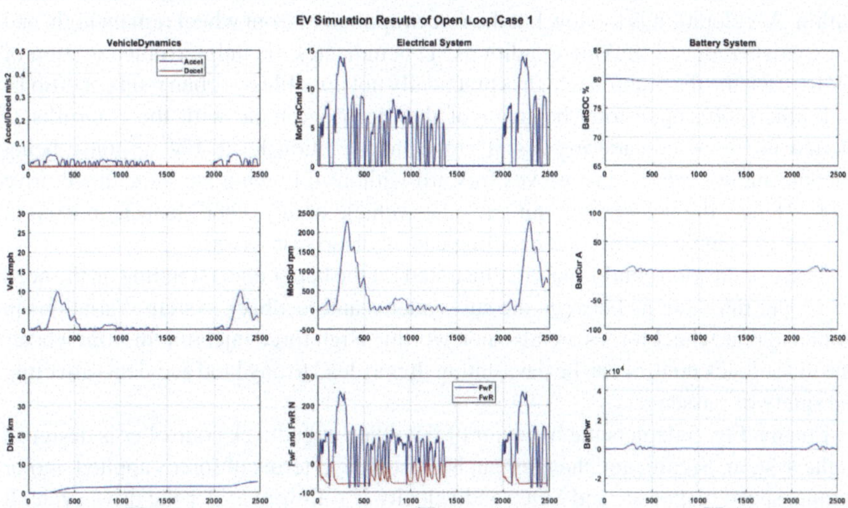

Fig. 2 Open-loop simulation output for Case 1

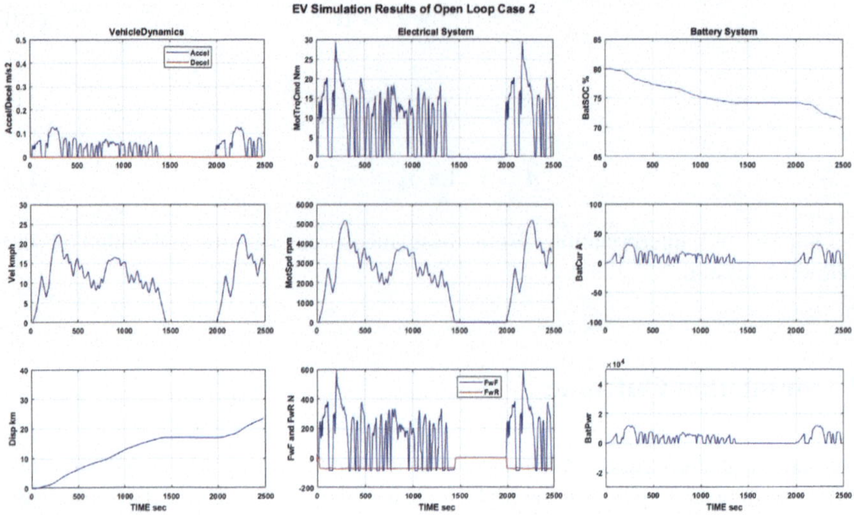

Fig. 3 Open loop for Case 2

speed and force on wheels in the front and rear tires are plotted. Finally, battery SOC during the travel, battery current and power-related details have also been recorded for the analysis purpose. The destination cannot be reached during the specified time period due to the lack of continuous feedback in the case 1 of open-loop condition.

Figure 3 depicts the open-loop response for case 2 at higher acceleration level than case 1. Response of the EV, in terms of distance covered, velocity, acceleration, motor speed and torque estimations followed by battery SOC and efficiency are plotted here within. Acceleration is too low but the forces applied to front wheel remain high, and there exists reduced braking conditions. This indicates the uninterrupted motion of EV to reach the destination which is practically not possible on urban road conditions.

Furthermore, open-loop behavior of the electric vehicle with the controller is plotted in Fig. 4 considering a still higher acceleration level. EV response being uncontrollable, vehicle just moves forward without following any predefined drive cycle. Hence, the vehicle may hit any other vehicle or any other obstacle in front of it. Lack of vehicle control and handling can be clearly observed.

These open-loop conditions are illustrated for the better understanding on the need for proper interactions between the sub-systems and feedback system. Variations in front and rear wheel forces are also noteworthy. High force injection in front wheels due to feedback control can be devastating. It may lead to any kind accidents affecting the safety of passengers.

Finally, Fig. 5 depicts the behavior of EV when a feedback controller is designed in the system. Significant changes can be observed in terms of forces applied, motor torque, speed responses and SOC, and velocity is followed as per the drive cycle. It is worth noting in the closed-loop case that the desired speed profile is achieved by appropriate acceleration and deceleration inputs.

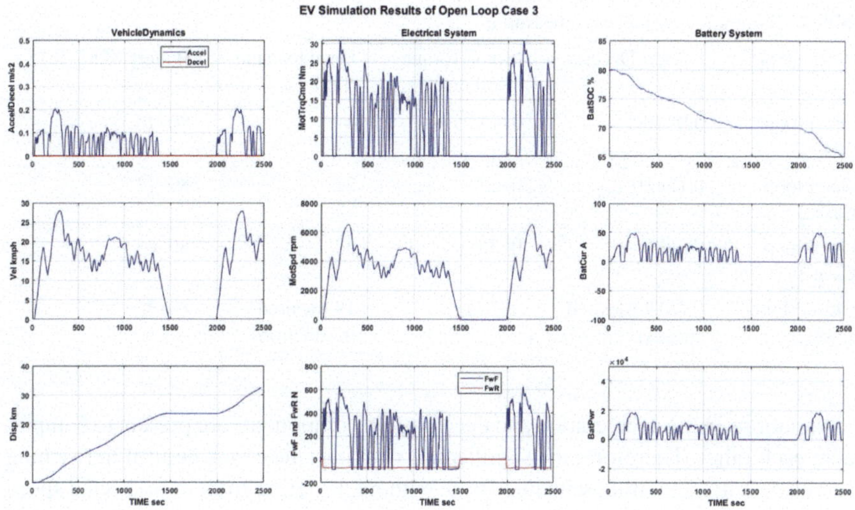

Fig. 4 Open-loop simulation output for Case 3

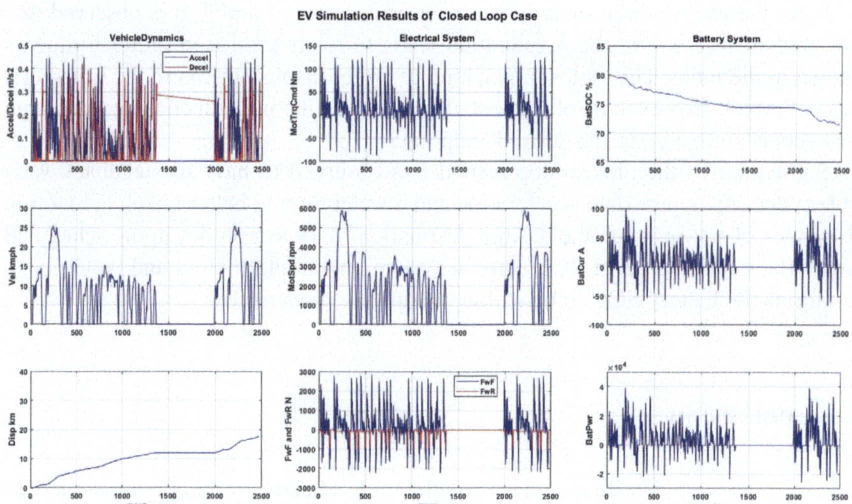

Fig. 5 Closed-loop case response

This case is more realistic wherein speeding up and braking of vehicle are considered. Clearly, from the closed-loop response figure, it can be perceived that EV reaches the desired destination of 19 km, with SOC value being maintained between 80 and 72%, with a motor torque ranging between the positive and negative peaks. It is only in the closed-loop case, and it is observed that necessary amount of force is applied on both front and rear wheels, to realize the exact speed profile.

Table 2 Simulation variables as measured

Parameters	Accel./ Dece. (m/s^2)	Motor Torque (Nm) range	Displacement (km)	Battery SOC (%)
Open-loop Case 1:	0.05 /0.0	0–14	4	80–78
Open-loop Case 2:	0.15/0.0	0–29	22	80–72
Open-loop Case 3:	0.2/0.0	0–30	32	80–65
Closed-loop Case:	0.4/0.4 (desired speeds)	± 100	19 (desired destination)	80–72

Variations in the acceleration and deceleration components are present to complement each other. Therefore, the significance of proper feedback control in reaching desired destination within scheduled time without deviating from the expected speed profile is justified in this article.

Table 2 provides details on the changes in simulation variables for both open-loop and closed-loop response in an EV.

From the responses of open-loop studies in Cases 1, 2 and 3, it is observed that the gradual increase in the acceleration leads to relative raise in motor torque as shown in the table. This helps in reaching to larger displacements in the scheduled time of travel. However, displacement affected is ending up with considerably large overshoots compared to the desired endpoint.

Subsequently, the closed-loop response is observed to have the feedback variables effecting appropriate acceleration and deceleration to better match the motion dynamics of EV to reach the desired destination of 19 km on dot in the scheduled time. The performance of the electrical system containing motors and battery also conforms the battery SOC to be within the range satisfactorily.

4 Conclusion

The work presented in this article is an attempt to effectively bring out the nuances in interpreting electric vehicle model following a simple to complex modeling approach. Unlike other research works, this paper intends to provide a thorough understanding of the background and fundamentals involved in modeling an electric vehicle with essential mathematical equations. The study concludes in a concise form along with the figures illustrating the variations of an expected speed profile, vehicle dynamics, control variables and the relevant data of electric motor and battery. Need for feedback-based control system is presented with appropriate graphs. This study is being continued further with increased levels of complexities in the model.

Acknowledgements Authors would like to sincerely thank Christ (Deemed to be University) for providing the necessary facilities for carrying out research.

References

1. Luo, Y., Cao, K., Li, K.: Coordinated control of longitudinal/lateral/vertical tire forces for distributed electric vehicles. In: 2014 American Control Conference (ACC). pp. 3905–3910. IEEE, Oregon, USA (2014).
2. Lian, Y., Zhao, Y., Hu, L., Tian, Y.: Longitudinal collision avoidance control of electric vehicles based on a new safety distance model and constrained-regenerative-braking-strength-continuity braking force distribution strategy. IEEE Trans. Veh. Technol. **65**, 4079–4094 (2016)
3. Zhang, X., Göhlich, D.: Integrated traction control strategy for distributed drive electric vehicles with improvement of economy and longitudinal driving stability. Energies **10**(1) (2017)
4. Chen, T., Xu, X., Chen, L., Jiang, H., Cai, I., Li, Y.: Estimation of longitudinal force, lateral vehicle speed and yaw rate for four-wheel independent driven electric vehicles. Mech. Syst. Signal Process. **101**, 377–388 (2018)
5. Sim, G., Ahn, S., Park, I., Youn, J., Yoo, S., Min, K.: Automatic longitudinal regenerative control of evs based on a driver characteristics-oriented deceleration model. World Electr. Veh. J. **10**, 58 (2019)
6. Han, J., Vahidi, A., Sciarretta, A.: Fundamentals of energy efficient driving for combustion engine and electric vehicles: an optimal control perspective. Automatica **103**, 558–572 (2019)

Supercapacitor-Based Automated Fast Charging System for Electric Vehicles

Saswati Pattnaik, Mano Ranjan Kumar, and Sunil Kumar Mishra

Abstract As the demand for electric vehicles (EVs) is increasing nowadays, the need for customer-friendly, effective, and innovative solutions for charging is rapidly enhancing. The recent advancements in this direction not only include the fast and efficient charging mechanism but also the automation in the process. Frequent plugging in the EV's charging cable to the socket maybe tiresome that can be overcome by robotic technology, where this repeated task can be done using autonomous mobile robots. This paper deals with a fully automated system of charging EVs via robots with V2X communication. Further, a supercapacitor-based fast and efficient charging method has been proposed, which can be adopted for the autonomous charging process to improve its effectiveness.

1 Introduction

Recently electric and hybrid vehicles have gained immense popularity in comparison to conventionally powered vehicles around the world due to increased focus on renewable energy. Electric vehicles (EVs) are expected to increase its global market share from 2% in 2016 to 22% in 2030. Few predictions reveal that if the worldwide sales of EVs keep on increasing as such then its price may fall making it more affordable. This has led to many developments related to the EV charging infrastructures [1], as it is one of the prime aspects of any EV to have an efficient charging system. In [2], a study of automated charging system has been done, where an initial concept for automated charging station for various types of vehicles has been discussed. Effective concepts of functionalities like parking lot navigation, EV plug guidance, and robot battery swapping used by mobile autonomous robots which charges parked EVs at any location has been reported in [3]. Also, a Markov chain model has been designed to describe the operation of the autonomous robot-like mobile chargers, which can move freely to charge vehicles [4]. Further, the overcharging/undercharging of battery is a common problem in EVs. This occurs due

S. Pattnaik · M. R. Kumar (✉) · S. K. Mishra
School of Electronics Engineering, Kalinga Institute of Industrial Technology, Bhubaneswar, India
e-mail: mano.kumarfet@kiit.ac.in

J. Gu et al. (eds.), *Communication and Control for Robotic Systems*, Smart Innovation, Systems and Technologies 229, https://doi.org/10.1007/978-981-16-1777-5_24

to defective voltage regulator or alternator, improper wire settings, extreme heat or incorrect labeling of charges, which can significantly decrease its lifespan, lower its efficiency and may lead to battery explosion. Hence this autonomous charging of EV using robots is an innovative and effective method where robots can control and prevent the battery from overcharging. Unlike the fixed charging, mobile charging could help the users save their time wasted in a charging station when their electric vehicles are being charged [4]. Automated charging systems can create a revolution when it comes to the smart and efficient charging of EVs.

Although the reported works, in this direction, have focused on the concept from several dimensions, less attention has been given to the time taken to charge the EVs effectively. This is because of the maximum throughput of a mobile charger is dependent only on the effective charging powers and efficient charging mechanism [5]. Also, majority of the reported works have employed Lithium-ion battery-based energy storage system, which suffers from the drawbacks such as low specific power density, less shelf and cycle life. On the other hand, supercapacitors are known for relatively high specific power density leading to charging and discharging at very high rate, making the process less time consuming [6]. This is because of the fact unlike batteries' chemical reaction, free charge storage mechanism takes place in supercapacitors. These are also capable of sustaining almost 10,000 times more charging and discharging cycles with respect to battery. However, inspite of having several advantages of supercapacitors, it is not in applications at par with that of battery. Instead these are mostly used for the applications where there is peak power requirement for short duration of time [7]. Also, supercapacitors have relatively very small equivalent series resistance (ESR), resulting in very small or negligible leakage current, therefore better retention of stored energy. All these benefits of supercapacitors are attributed to the unusual phenomena that occurs within due to the porous interface of electrode and electrolytes [8].

Therefore, motivated by the enormous performance especially the fast charging and discharging capabilities and keeping in view of providing fast and efficient charging mechanism, this paper mainly focuses on obtaining an effective robot-based automatic charging system for EVs.

2 A Brief on Automated Charging System

2.1 Automated Charging System by Volkswagen

Recently Volkswagen has claimed that electric car owners won't need to drive to charging stations in future because the charger will be delivered to them via robots [9]. These robots are aimed at providing charging solution in multistory and underground car parks where space is at minimum. The car owners just need to send an alert using an app that their car needs to charge. Self-driving robots will tow a mobile energy storage device known as battery wagon on a trailer to the car. Robots will be able

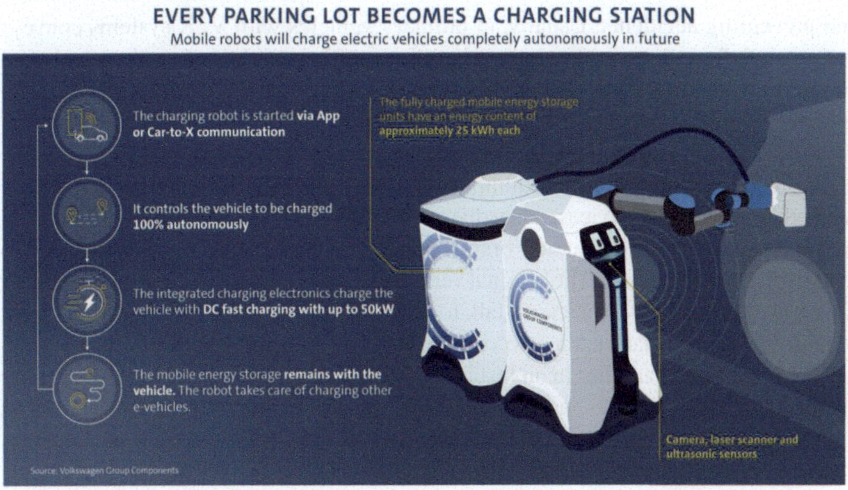

Fig. 1 Process of Volkswagen's autonomously charging mobile robots for EVs [9]

to open the vehicle charging flap and plug-in the port and decouple it once the batteries have been replenished. The battery wagon stays with the vehicle during the whole charging process takes place while the robot in the meantime transports more systems to other vehicles in that car park that need electricity. Once the charging service is completed, the robot collects the battery wagon and brings it back to the on-site charging stations where it can be plugged into the mains to boost it back to full capacity, ready to service other charge demanding cars in the vicinity. New Volkswagen models equipped with the brand's latest V2X communication system [10], that will be able to request the services of one of these robots. Figure 1 depicts the entire process carried out.

The robot has arms that can connect the battery wagon. The charging robots can move several battery wagons at the same time. The robots themselves are fitted with the latest self-driving systems powered by cameras, laser scanners, and ultrasonic sensors. With this, almost every car park can be made electric without any complex infrastructural measures. Nevertheless, two major aspects that can be focused on for effective functioning of this entire system is proper communication between the vehicles and the robots and suitable charging system.

2.2 V2X Communication

Vehicle to everything (V2X) is a new communication technique that allows vehicles to interact with any other moving or stationary systems around them using the latest generation of information technology [10]. This communication strategy is mainly used for the purpose of improving traffic efficiency, reducing pollution, and preventing accidents. Usually, in ordinary vehicles with V2X systems conveys important information to the driver regarding traffic, accidents, weather, and road condition, whereas, in autonomous vehicles, V2X plays a vital role in navigating the vehicle. In addition to that, the technology also serves other purposes like automatic payments for tolls and parking fees.

It is to note that the vehicles equipped with this technology can only communicate with other V2X based devices. Also, this technology has remained confined to limited car models, especially the luxury brands but it is expected to eventually be added to lower-priced vehicles too. The original version of Volkswagen's V2X system is called Car2x [11]. This allows the car to talk to other models with the system installed. In order to share information, the two vehicles must be within 800 m distance. Data can be sent from one to another in an instant and it can send and receive alerts regarding accidents, broken vehicles, traffic jams, or roadworks nearby. The flow diagram of the entire charging process carried out by robots using V2X technology has been depicted in Fig. 2.

Dc fast chargers are designed for EV to give an electric output ranging from 50 to 350 kW. With higher power operation the ac-dc converter, the dc-dc converter, and the power control circuits become larger and quite expensive. That is why the dc fast chargers are implemented as off-board chargers rather than on-board chargers so that it does not take space within the vehicles and the fast charger can be shared by many users. Figure 3 depicts the power flow of dc charging [12].

The ac power provided by the ac grid is converted to dc power using a rectifier in the dc charging station. Then the power control unit appropriately adjusts the voltage and current of a dc converter to control the variable dc power delivered to charge the

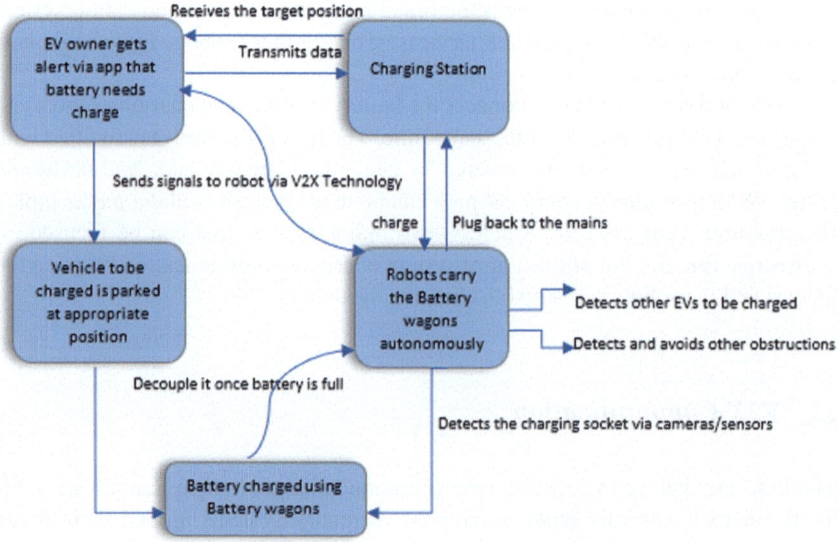

Fig. 2 Flow diagram of the complete process using V2X Communication

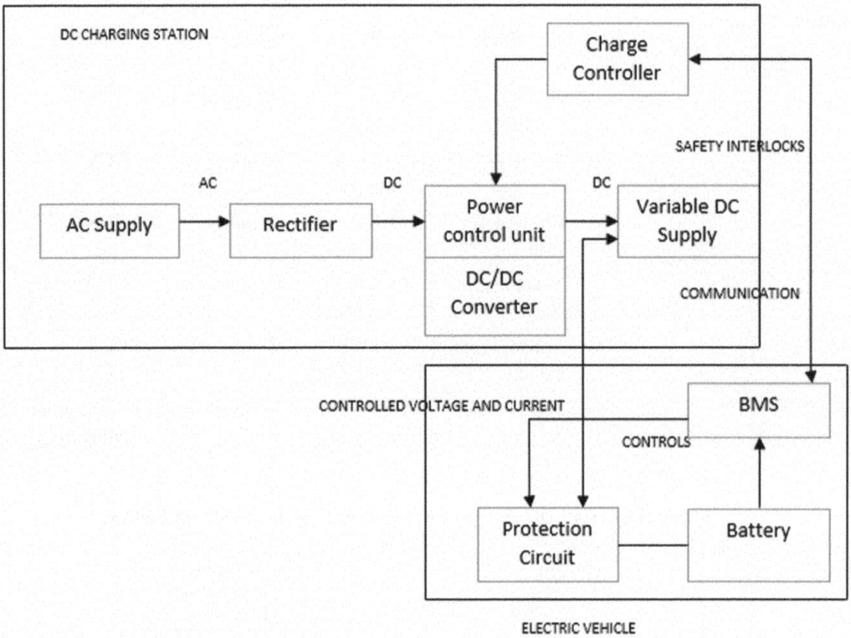

Fig. 3 Power flow of DC charging [12]

battery. There are safety interlocks and protection circuits used to de-energize the EV connector and to stop the charging process whenever there is any fault connection. The BMS plays a key role between communicating with the charging station and controlling the voltage and current delivered to the battery. Almost all popular brands of EV using DC fast charging standards use lithium-ion batteries only for charging purposes. Table 1 shows some popular brands of EV and their specifications [13].

Lithium-ion batteries have higher energy density, longer life span, and higher power density than most other practical batteries. In the past, nickel metal hydride batteries were used in EV but they have been considered outdated due to their tendencies to self-discharge in the heat. Also, the lead acid batteries are quite expensive and have a shorter life than the vehicle itself, typically needing replacement every 3 years. The simulation of the discharge characteristics plot for lithium-ion batteries obtained in MATLAB/Simulink also depicts its performance (see Fig. 4). For the nominal self-discharge current of 1.32 A, nominal voltage of 26 V, the performance of the battery for current of 6.5 A, 13 A and 32.5 A is plotted. From the discharge characteristics, we find lithium-ion batteries to have larger area of section (nominal area) indicating better SOC performance in comparison to other batteries. Also, these batteries being operated to discharge at 1.32 A, its nominal discharge capability is 2.8696 A which means that it can deliver comparatively higher values of current especially required for EV application. These factors have led to lithium-ion batteries leading as the

Table 1 Charger specifications of popular brands of EV

EV Brands	Charger	Charge time	Charging standard
AUDI E-E-TRON	95 kWh Li-ion pack	60 min or less	CCS
BMW I3	42.2 kWh Li-ion pack	50 min	CCS
Chevrolet Bolt	60 kWh Li-ion pack	30 min	CCS
Nissan Leaf Plus	64 kWh Li-ion pack	45 min or less	CHAdeMO
Volkswagen E-Golf	35.8 kWh Li-ion pack	60 min or less	CCS
Tesla Model X	75 kWh Li-ion pack	75 min or less	Tesla super charger

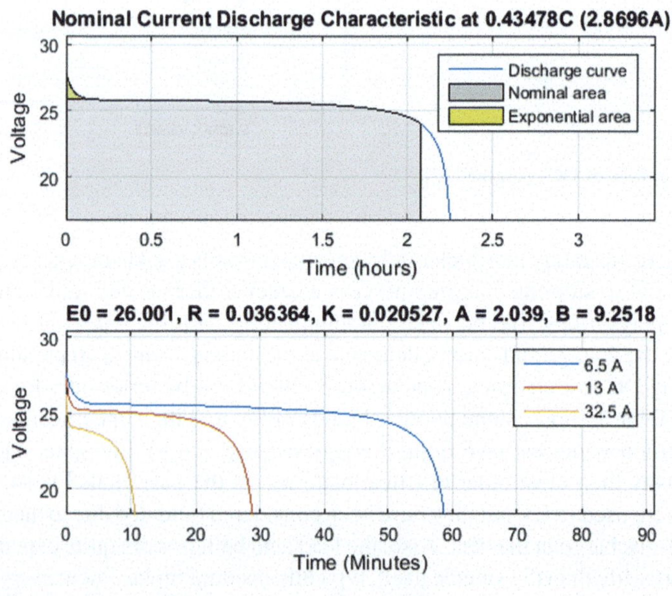

Fig. 4 Discharge characteristics plot of lithium-ion battery

predominant battery for EVs. Even more, the price of these batteries is constantly decreasing thus making EV more affordable and attractive in the market.

Despite having so many advantages there are few shortcomings of using li-ion batteries as the sole charge storage device. These batteries have quite low energy density and often subjected to high losses. Continuously being exposed to dc fast charging also degrades the lifespan of the batteries and they should be used within safe temperature and voltage ranges for efficient operations [14]. They take more time in charging and only 70–80% SOC can be charged with fast charging and they are more

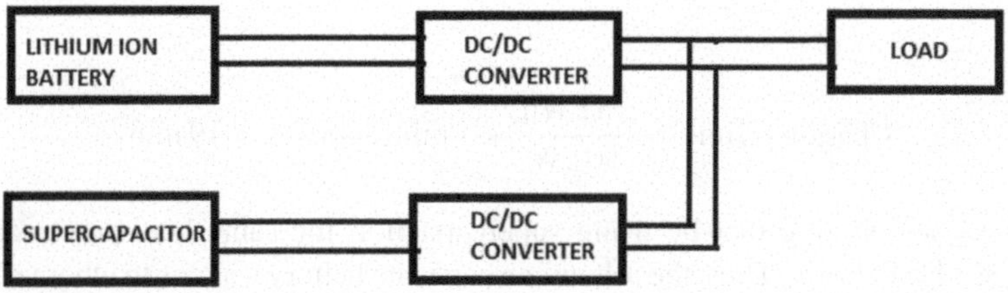

Fig. 5 Parallel active hybrid topology

subjected to overcharging which is not suitable for this model. Hence to overcome all these disadvantages, it is popularly believed that supercapacitors can be connected in parallel with the battery [15], thus acting as an auxiliary energy storage device, resulting in fast charge and discharge, improving the charge time and acceleration, protecting the battery, and reducing load on it so that it lasts longer because they expensively expire well within the life of the vehicle. If the battery tolerates deep discharge the supercapacitor achieves this resulting in 10% more energy being made available from the battery. This provides 10% more range to the EV in relatively small amount of charging time. The Parallel Active Hybrid Topology (PAH) as shown in Fig. 5 is the most optimal connection that results in a reduction of current peaks in the battery and improvement of battery life and efficiency [16]. These results are improved further when ultracapacitor is connected through a converter [17] and its voltage is allowed to a much wider range. If voltage is restricted by battery most stored energy becomes unavailable.

3 Supercapacitor-Based Fast Charging Method

In this Volkswagen model [9] being developed, the rated output capacity of the quick charger is 50 kW, while the power of the fully charged Li-ion battery pack in the battery wagon is 25 kWh. Hence according to the formula for estimating charging time, the total time needed to charge the EV is found as,

$$\text{Charging Time} = \frac{\text{power of battery}}{\text{power of load}} = \frac{25\,\text{kWh}}{50\,\text{kW}} = 30\,\text{min} \tag{1}$$

Here, instead of using a 25 kWh Li-ion battery pack, if we use lets say a 15 kWh Li-ion battery pack and a 10 kWh Supercapacitor pack for the same load of 50 kW, then it will take relatively less time to charge 15 kWh pack while charging the supercapacitor pack will take approximately 1 min or even less. According to the formula (1),

$$\text{Charging Time} = \text{charge time}_{\text{battery pack}} + \text{charge time}_{\text{SCpack}} \tag{2}$$

or,

$$\text{Charging Time} = \frac{15\,\text{kWh}}{50\,\text{kW}} + 1\,\text{min(approx)} \approx 19\,\text{min} \tag{3}$$

Hence, we observe that by using supercapacitor, the same load power can be achieved 11 min faster. Thus, the robots can carry the battery wagons to other vehicles that are in need of charge. As a result, within a short span of time the battery wagons can attend many vehicles. This less time in charging will also affect the costs to charge at public charging stations whose rates are based on time taken to completely charge the EV. If an EV gets charged in less time means it will significantly reduce the charging rates. So, this proposed method can be considered in this Volkswagen model to increase its charging efficiency. Based on this a control algorithm for effective charging of the EV is proposed in this study which determines the charging of both batteries and supercapacitor packs effectively. Figure 6 shows this proposed battery-supercapacitor hybrid charging system.

The AC power provided by the grid is converted to DC power in the charging station which in turn charges the battery wagon. The robots have indicators that detect when the battery wagon needs to be charged. At that point of time, it carries the battery wagon to the charging point and connects the input port to the charging circuit that regulates the voltage and current to charge the li-ion battery in the wagon. When it gets charged as per the capacity then the indicators indicate, and robot detects that the battery wagon is ready to use. Simultaneously, it connects the discharging circuit to the output port to charge the vehicle. This discharging circuit controls the current that is drawn from the battery. In case there is too large discharge current then the battery lifetime reduces. So, to control these excess current supercapacitors are connected which makes sure that the required amount of discharge current is drawn by the electric vehicle. In turn, it acts as a backup energy storage device supporting the li-ion battery in emergency/fast charging cases thus reducing the load on it and improving its lifetime.

4 Simulation Results

The simulation works have been carried out using the battery and Supercapacitor model available in MATLAB/Simulink. Realizing li-ion battery packs and supercapacitor packs as in the battery wagon discussed previously requires additional circuitry involving battery management system which have not been focused on this work. This analysis is mainly focused on investigating the effects of adding Supercapacitor to the battery configuration and comparing the results with only battery configuration. Figure 7 shows the only Battery configuration using a Boost Converter and PI control. Figure 8 presents the response for Battery only configuration wherein battery power obtained does not fulfill the condition of the required power and is unable to handle the peak power during transient operation. The error is high when

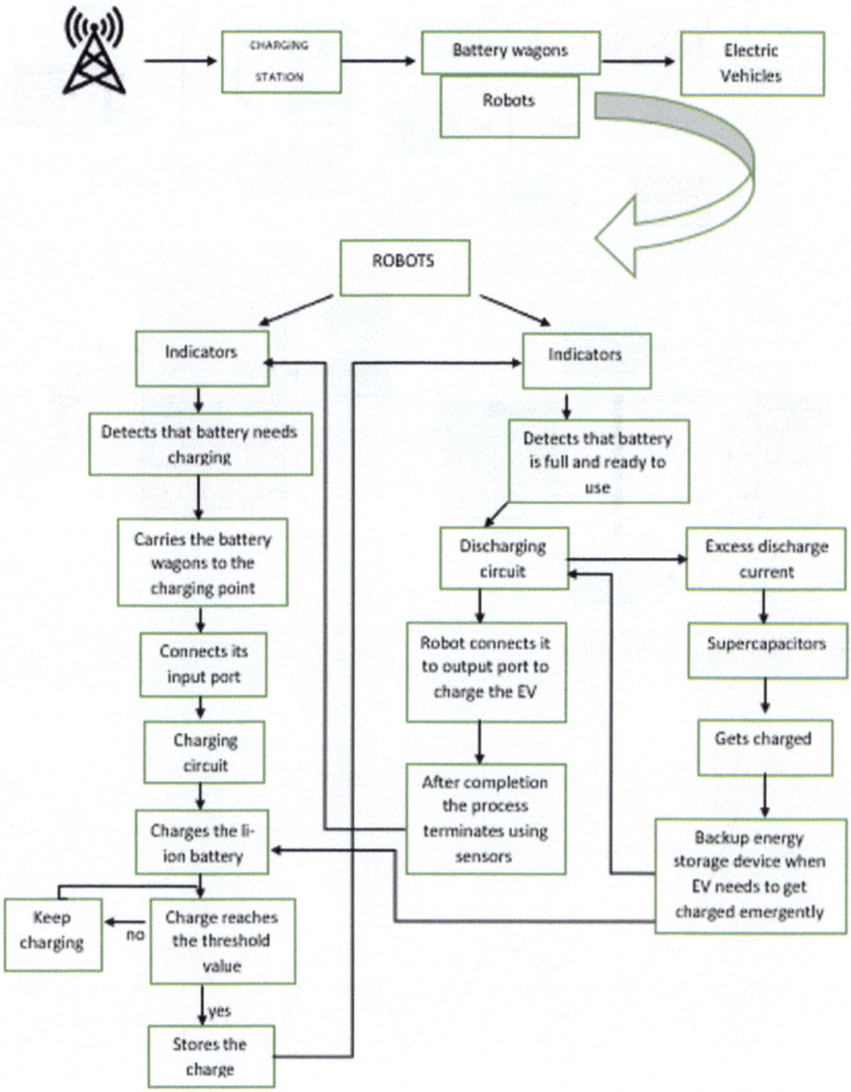

Fig. 6 Proposed method for battery and supercapacitor-based charging system

sudden power demand occurs. The important parameters of Battery such as state of charge (SOC), current, and voltage waveforms are shown in Figs. 9, 10 and 11 respectively.

Next, the Supercapacitor in parallel with battery has been considered and Simulink model is displayed in Fig. 12. The flow through Supercapacitor and battery are controlled using DC-DC converters. The PI control structure has been implemented for the Boost converter with PI control. The Supercapacitor power flow is regulated

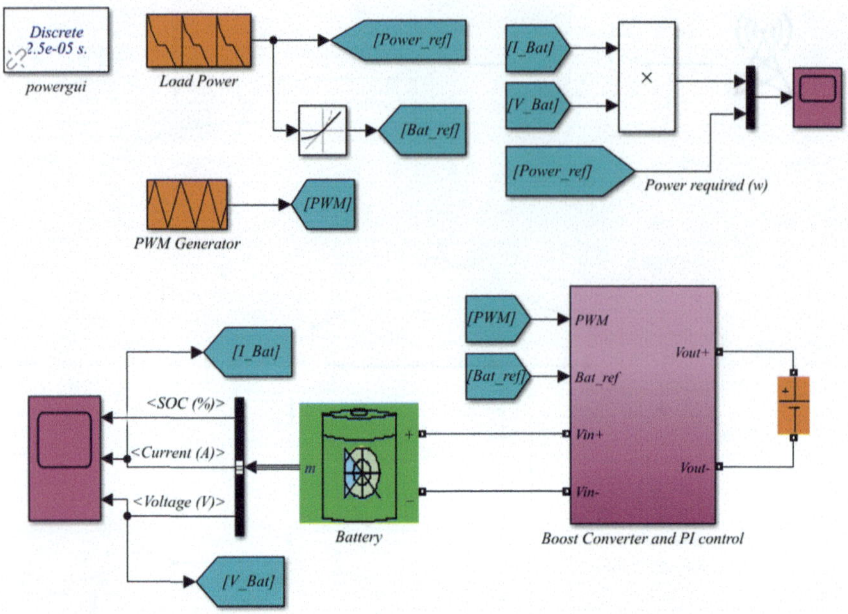

Fig. 7 Battery only configuration

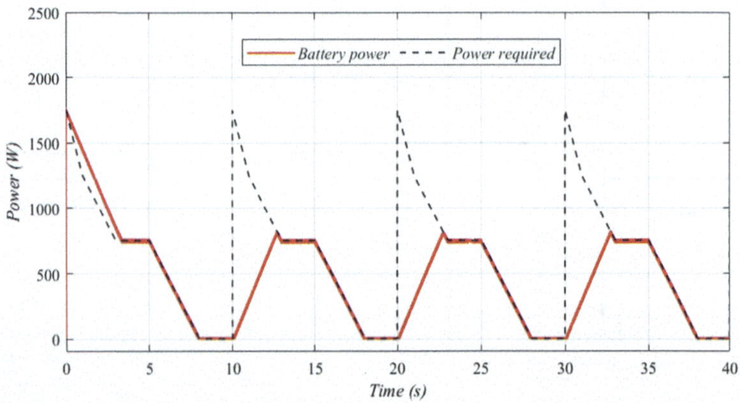

Fig. 8 Required power and battery power obtained

through the Buck/Boost converter that employs two PI controllers. The peak power requirements during the transient operation are handled well by the Supercapacitor while the battery continues with its normal operations. Hence the actual power meets the required power as shown in Fig. 13. Figure 14 shows the battery and Supercapacitor power waveforms separately. The important parameters of Supercapacitor such as SOC, current, and voltage waveforms are shown by Figs. 15, 16 and 17 respec-

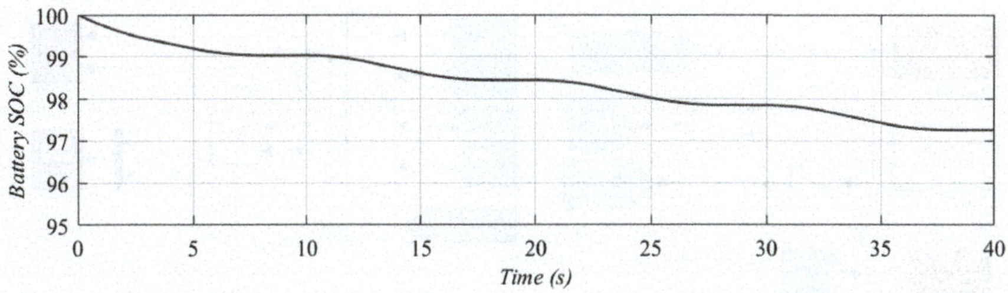

Fig. 9 Battery SOC (%)

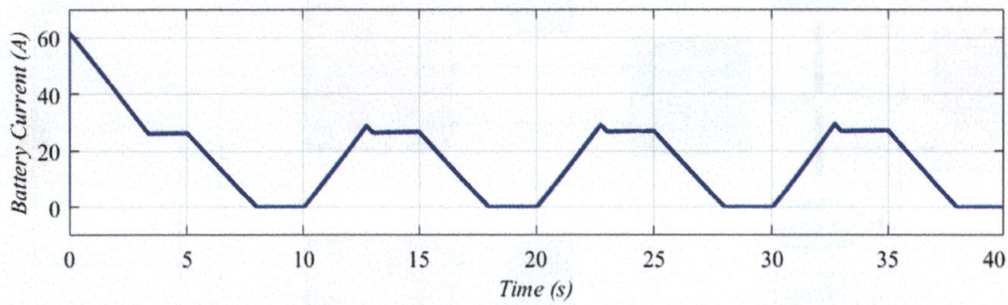

Fig. 10 Battery current

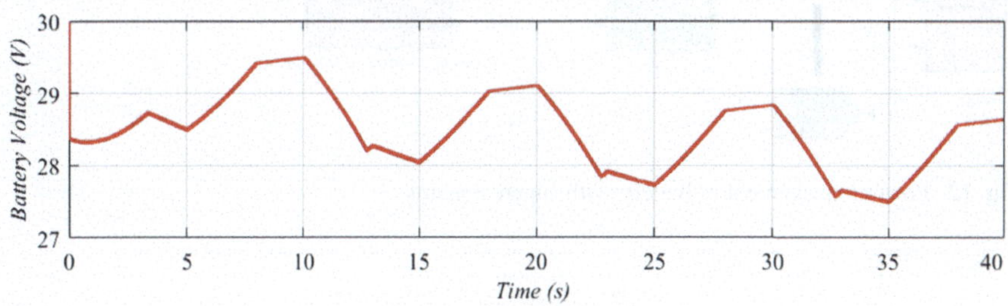

Fig. 11 Battery voltage

tively. Therefore, the battery and Supercapacitor parallel configuration improves the charging and discharging response as compared to battery-only configuration.

In previous two configurations, only discharging profiles of battery wagon have been discussed and compared. However, it is very important to see the charging behavior of the battery while maintaining the power demand. Figure 18 presents the MATLAB simulation diagram wherein a charging circuit has been integrated to maintain the battery SOC always greater than 98%. The charging circuit consists of two switches to directly connect the charging source with the battery whenever SOC goes below 98%. This function has been implemented in Fig. 18b and charging pulse becomes high. In order to avoid any sudden jump in current level a resistor of 1 Ω in series is connected with switch on positive terminal.

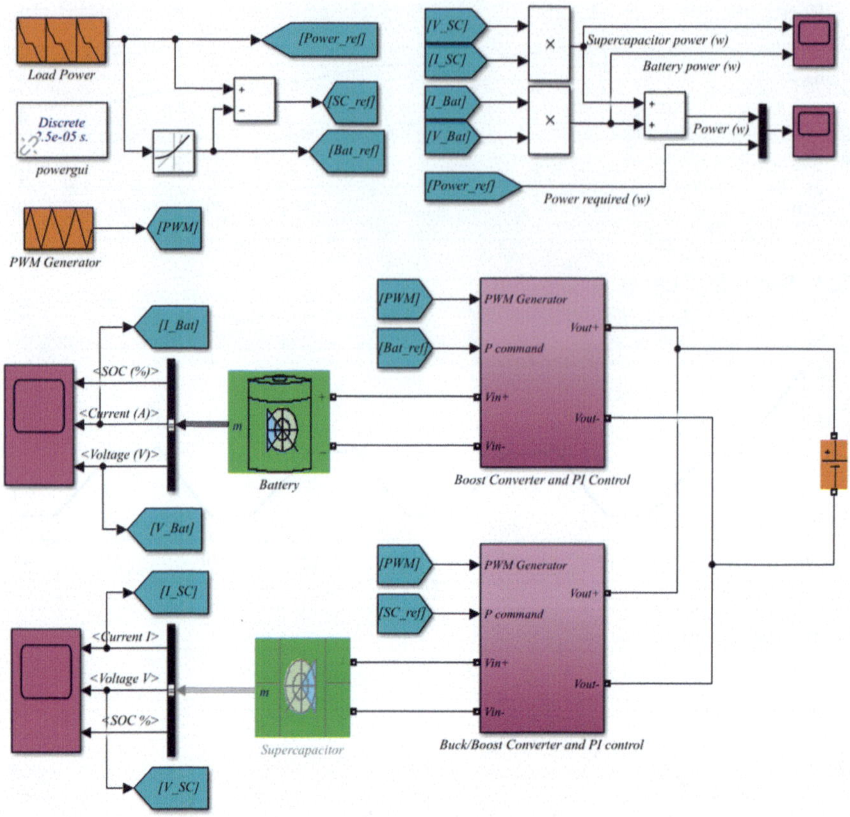

Fig. 12 Power management of battery with supercapacitor

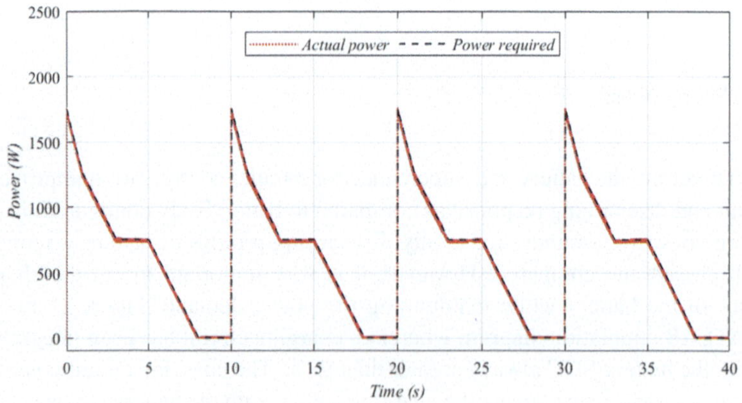

Fig. 13 Required power and actual power

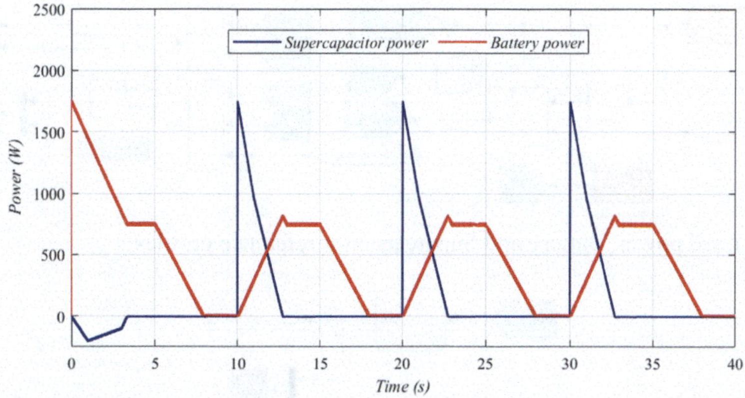

Fig. 14 Supercapacitor power and battery power

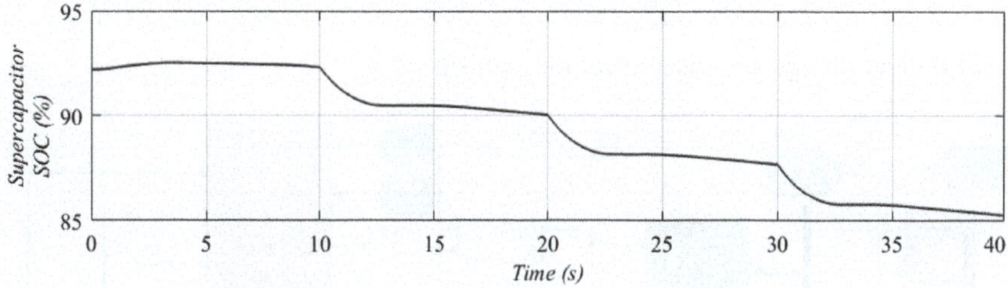

Fig. 15 Supercapacitor SOC (%)

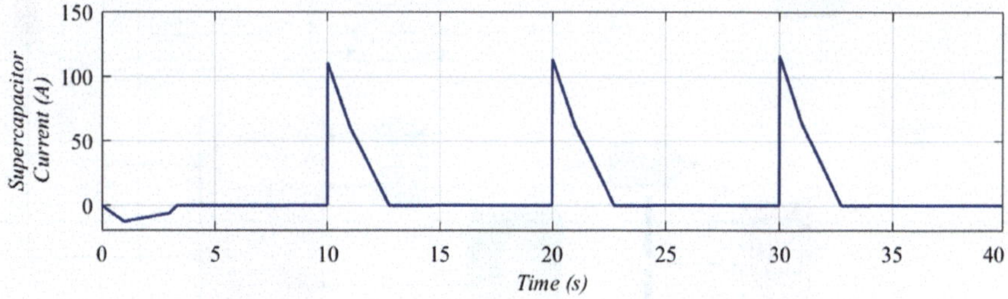

Fig. 16 Supercapacitor current

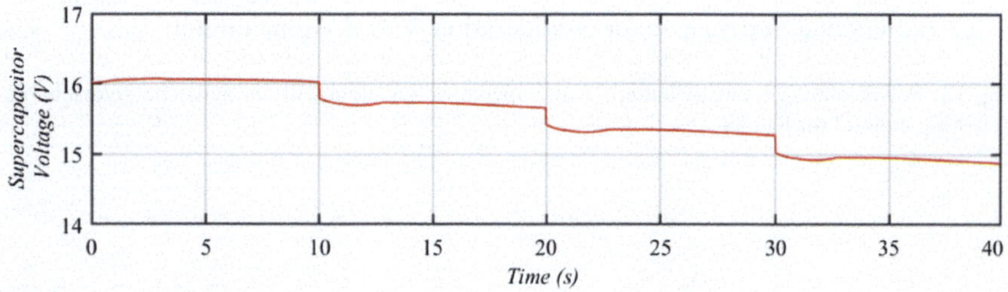

Fig. 17 Supercapacitor voltage

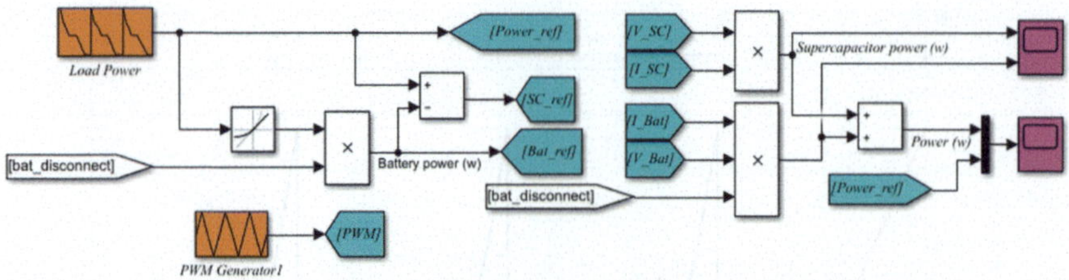

(a) Load power, Battery and Supercapacitor reference profiles

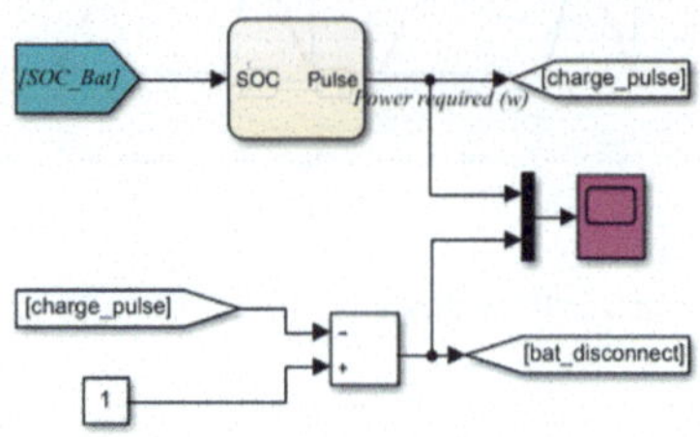

(b) Battery charge and disconnect mechanism

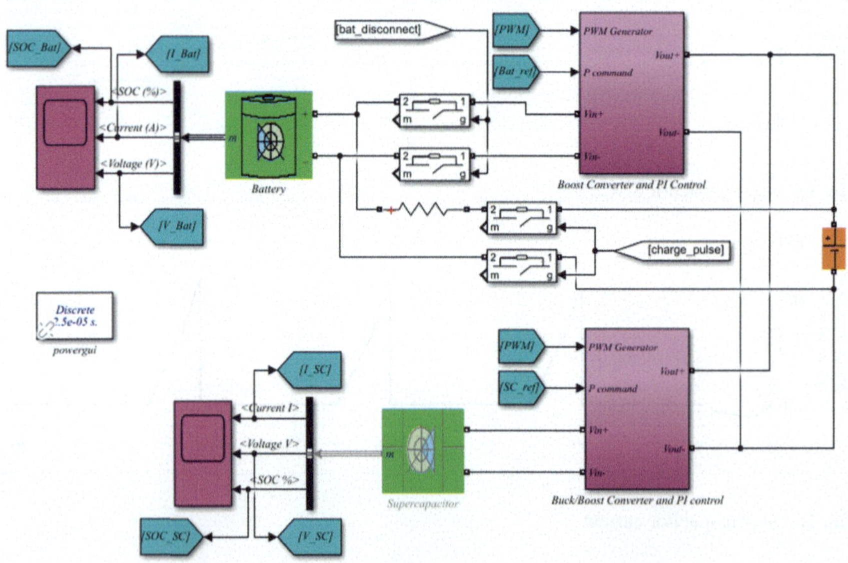

(c) Battery and Supercapacitor configuration with charging circuit

Fig. 18 Power management of battery with supercapacitor alongwith an auto charge/discharge switching method for battery

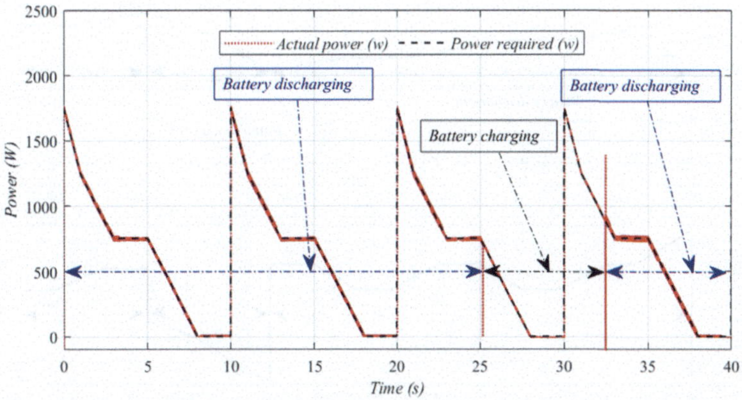

Fig. 19 Actual and required power waveforms with simultaneous discharging/charging

As shown in Fig. 18a, b, the battery disconnect signal is generated to remove the battery temporarily from the operation during charging mode. The battery disconnect signal is the inverted form of charging signal. When charging of the battery is completed, battery again resumes the operation. Figure 19 displays the actual and required power when charging and discharging of the battery happens simultaneously. It can be observed that there are spikes in the actual power waveform during switching between the two modes. During the battery charging period, the Supercapacitor compensates for power demand. When battery SOC reaches in the 100% state, it is again reconnected. During charging mode, as shown in Fig. 20, the battery starts drawing the current and battery terminal voltage also increases by more than 10 V.

5 Conclusion

With this concept of using mobile charging robots and flexible charging stations supplying to EVs can be provided in parking garages, underground parking garages, managed parking lots, rest areas, or even in any big events without charging infrastructure. Nevertheless, these technologies will continue to develop over time we will find substantial changes in EV, which can communicate well and charged in comparatively less time with more efficiency. Hence supercapacitor-based charging system can provide fast charging to the mobile robot-based autonomous charging system.

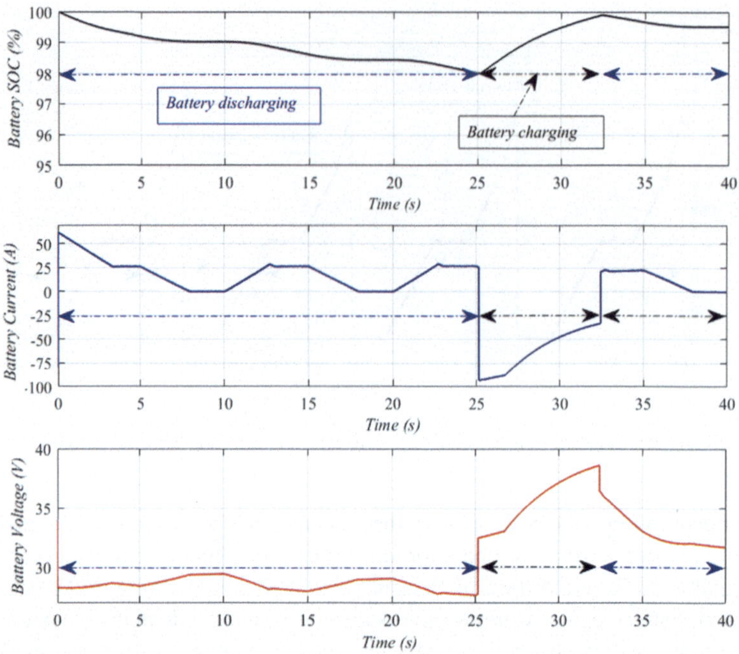

Fig. 20 Battery parameters with simultaneous discharging/charging

References

1. IEA: Global EV Outlook 2020, Technology Report—June 2020, IEA, Paris (2020). https://www.iea.org/reports/global-ev-outlook-2020
2. Walzel, B., Sturm, C., Fabian, J., Hirz, M.: Automated robot-based charging system for electric vehicles. In: 16 Internationales Stuttgarter Symposium, pp. 937–949. Springer, Wiesbaden, (2016)
3. Behl, M., DuBro, J., Flynt, T., Hameed, I., Lang G., Park, F.: Autonomous electric vehicle charging system. In: Systems and Information Engineering Design Symposium (SIEDS—2019), Charlottesville, VA, USA, 2019, pp. 1–6. https://doi.org/10.1109/SIEDS.2019.8735620
4. Zhang, Y., Liu, X., Wei, W., Peng, T., Hong, G., Meng, C.: Mobile charging: a novel charging system for electric vehicles in urban areas. Appl. Energy **278**, 115648 (2020)
5. Kong, P.: Autonomous robot-like mobile chargers for electric vehicles at public parking facilities. IEEE Trans. Smart Grid **10**(6), 5952–5963 (2019). https://doi.org/10.1109/TSG.2019.2893962
6. Liu, J., Wang, Y., Wu, Y., et al.: Research and implementation of new-type supercapacitor and battery hybrid energy storage system. J. Power Electron. (2020). https://doi.org/10.1007/s43236-020-00130-7
7. Miller, J.M.: Ultracapacitor Applications. The Institution of Engineering and Technology, (2011).
8. Kumar, M.R., Ghosh, S., Das, S.: Charge-discharge energy efficiency analysis of ultracapacitor with fractional-order dynamics using hybrid optimization and its experimental validation. AEU-Int. J. Electroni. Commun. **78**, 274–280 (2017)

9. Volkswagen lets its charging robots loose @2019 www.volkswagen-newsroom.com. [Online]. Available: https://www.volkswagen-newsroom.com/en/stories/volkswagen-lets-its-charging-robots-loose-5700
10. Wang, J., Shao, Y., Ge, Y., Yu, R.: A survey of vehicle to everything (V2X) testing. Sensors **19**(2), 334 (2019)
11. Car2X in the new Golf: a technological milestone @2020 www.volkswagen-newsroom.com. [Online]. Available: https://www.volkswagen-newsroom.com/en/stories/car2x-in-the-new-golf-a-technological-milestone-5919
12. Ronanki, D., Kelkar, A., Williamson, S.S.: Extreme fast charging technology—prospects to enhance sustainable electric transportation. Energies **12**(19), 1–17 (2019). https://doi.org/10.3390/en12193721
13. DC Fast Charging Explained—EV Safe Charge. https://evsafecharge.com/dc-fast-charging-explained/. Accessed 03 Sept 2020
14. is-dc-fast-charging-bad-for-your-electric-car @ www.myev.com. [Online]. Available: https://www.myev.com/research/interesting-finds/is-dc-fast-charging-bad-for-your-electric-car
15. Sher, H.A., Addoweesh, K.E.: Power storage options for hybrid electric vehicles—a survey. J. Renew. Sustain. Energy 4(5) (2012). https://doi.org/10.1063/1.4759457
16. Wang, G., Dai, Z., Guan, Y., Dong, P., Wu, L.: Power management of hybrid power systems with Li-Fe batteries and supercapacitors for mobile robots. Adv. Mech. Eng. **2014** (2014). https://doi.org/10.1155/2014/270537.
17. Artal, J.S., Bandrés, R., Fernández, G.: Ulises: Autonomous mobile robot using ultracapacitors-storage energy system. Renew. Energy Power Qual. J. **1**(9), 1105–1110 (2011). https://doi.org/10.24084/repqj09.558

Different Control Mechanisms of a PMSM Drive for Electrified Transportation—A Survey

Chiranjit Sain, Atanu Banerjee, Pabitra Kumar Biswas, Sudhakar Babu Thanikanti, and Karthik Balasubramanian

Abstract With the availability of permanent magnet materials, the concept of modern industrial electronics with some integrated signal processing applications Permanent Magnet Synchronous Motor Drive (PMSM) is an essential competitor in the area of electrified transportation. It offers numerous approaches to PMSM control techniques, inclusive of manipulating strategies primarily based on sensible management and digital control. The authors in this paper present a detailed review of different control schemes regarding sensorless control for low-cost, high volume applications are proposed. To meet the modern integrated energy-efficient environment, cost minimization and compactness are achievable in the electrified transportation because of the simple control circuit, low cost, lesser torque ripples, low acoustic noise and, improved performance, etc.

C. Sain (✉)
Department of Electrical Engineering, Siliguri Institute of Technology, Sukna, Siliguri, Darjeeling, India
e-mail: chiranjitsain@nitm.ac.in

A. Banerjee
Department of Electrical Engineering, National Institute of Technology Meghalaya, Bijni Complex, Laitumukhrah, Shillong, Meghalaya, India

P. K. Biswas
Department of Electrical and Electronics Engineering, National Institute of Technology Mizoram, Aizawl, India

S. B. Thanikanti
Department of Electrical and Electronics Engineering, Chaitanya Bharathi Institute of Technology (CBIT), Hyderabad 500075, India

K. Balasubramanian
Design and Engineering (Electrical), Offshore Technology Development Pte., Ltd., Keppel Offshore and Marine Ltd., Singapore 629353, Singapore

J. Gu et al. (eds.), *Communication and Control for Robotic Systems*, Smart Innovation, Systems and Technologies 229, https://doi.org/10.1007/978-981-16-1777-5_25

1 Introduction

In present days, electrified transportation system has enriched as an emerging area to enhance energy-efficiency and for environmental concern. The utility of these recent trends rely on the victorious consolidation of electric vehicles into the infrastructures and assist to increase its capacity. Moreover, the topology electrified transportation decreases the use of oil, assuring some policy interests and, the latest environmental need in the regions. Electrified transportation system involves electric bicycle, electric scooter, light motor vehicles, electric bus, hybrid electric vehicles, and plug-in electric vehicles. Most of the electrical appliances utilize one or several motors for servo application. Induction motor, DC motors (brushed and brushless), stepper motors, etc., are the most common motors that are frequently used for the same. Most of the residential and commercial applications have conventional drive technologies. Conventionally, machines that are implemented in these appliances are generally single-phase induction motors or dc machines (brushed). Permanent magnets have many advantages, such as no excitation losses and faster dynamic performances. The speed, which is an important parameter decided by the frequency of the power grid is currently used by the present systems. From the 1970s, the evolution from single speed drives to variable speed drives leads to the preservation of energy and environment protection [1–3]. Industrial drives use permanent magnet synchronous motors. While the researchers have a strong interest to work on the analysis, design, and control of Permanent Magnet Motor drives in different industrial applications. In a variable speed drives or servo drive systems, speed or position feedback is used for precise control of motors. With the advent of power semiconductor devices, control of PMSM has become simpler and cost-effective. The efficiency and efficacy of electric vehicles addressed in depth by the authors in the literature. Clean energy sources can be encouraging to use such a vehicle due to environmental concern raised from conventional vehicles. The best choice of controllers is highly critical as it plays an important role for better performance. Control strategies which are intellectual in nature have the capacity to interact with the system and sustain the operation consistently [4, 5].

2 Types of Permanent Magnet Rotor

Various methods of permanent magnet rotor are detailed below.

2.1 Surface-Mounted PMSM

The rotor laminations are provided on the outer surface of the region. This configuration gives the maximum air-gap flux density without the interruption of any other medium. This arrangement has many flaws, such as low mechanical robustness.

2.2 Surface-Inset PMSM

In this arrangement, the outer surface is cylindrical and robust due to the magnets placed on the outer surface of the rotor laminations. This arrangement is precisely robust contrasted with surface mount construction as the magnets don't project out of the rotor covers as on account of the synchronous changeless magnet (SPM) and in this manner, they are completely and precisely installed in the rotor invigorating it mechanically from flying out.

2.3 Interior PMSM

The rotor covers in outspread and circumferential directions are watched. The inside PM rotor development is mechanically vigorous and appropriate for fast applications. The assembling of this course of action is greater mind-boggling than the floor mount or inset magnet rotors. The word that the percentage between the quadrature and direct axis inductances may be higher than that of the inset magnet rotor and for the maximum part is the scope of notwithstanding the reality that there have been cases of plenty better share for special different inside type pm rotor setups [1–3].

2.4 Line-Start PMSM

The motors with high efficiency have squirrel cage winding for the better performance of the torque. Rotor oscillations are observed in the same cage winding. When the motor is in synchronism, electrical torque is not observed. To conclude, the reason for the compact and smaller rotor is due to damper windings to offset hunting. The efficiency of the PM rotor is improved in comparison to the induction motor due to a change in the construction of the PM motor [4, 5].

3 PMSM Machine Configuration

According to the direction of the field, PMSM is classified into two ways, i.e. radial field machine and axial field machine. For radial field structure, flux path is along the radius of the machine, and for axial field configuration, flux path is parallel to the shaft. Moreover the radial field machines quite available however, the axial field machines are coming into practice due to higher power density and good acceleration properties by which they can be used in high-performance applications. Axial flux PMSM machine has several advantages like simple winding, low cogging torque, and lesser noise. Necessary technical details of PMSM used in a hybrid vehicle are illustrated in Table 1.

3.1 Flux Switching PMSM

The permanent magnets are placed in the middle in the middle of stator poles, and the adjacent permanent magnets have the opposite polarity. The orientation of the permanent magnets is in the circumferential direction. This encourages the utilization of low vitality thickness and eases magnets to give similar power densities as that of traditional PMSMs with high vitality thickness magnets. The schematic winding configuration is shown in Fig. 1.

3.2 Concentrated Coil Wound PMSM

This arrangement reduces the cost of winding, simpler construction, and high air-gap flux density. The concentrated coil wound PMSM has high direct axis flux inductance by which it is possible to have lower currents to counter the magnetic flux which establishes a wide flux weakening control operation is highly desired in electric vehicle applications. The schematic Winding configuration is shown in Fig. 2.

Table 1 Technical details of PMSM in a hybrid car

Design aspect	Technical rating
Maximum power rating	100 KW
Maximum torque rating	250 Nm
Rotational speed	13,500 rpm
PMSM rotor structure	Interior PMSM with V-shaped with laminated pole
Winding configuration	Distributed parallel
Type of cooling	Oil circulation with water

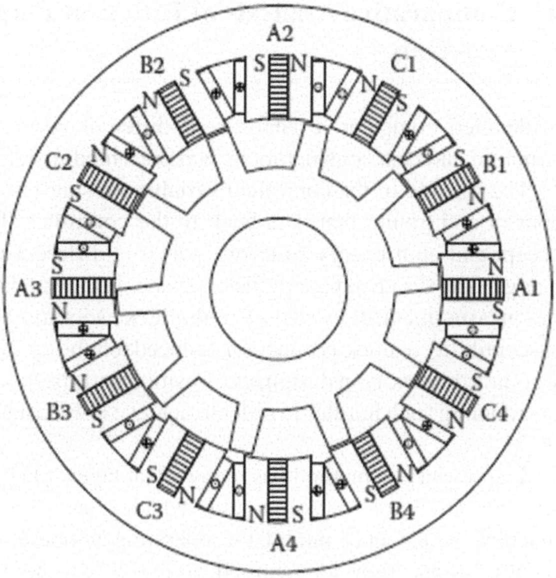

Fig. 1 Three phase flux switching PMSM

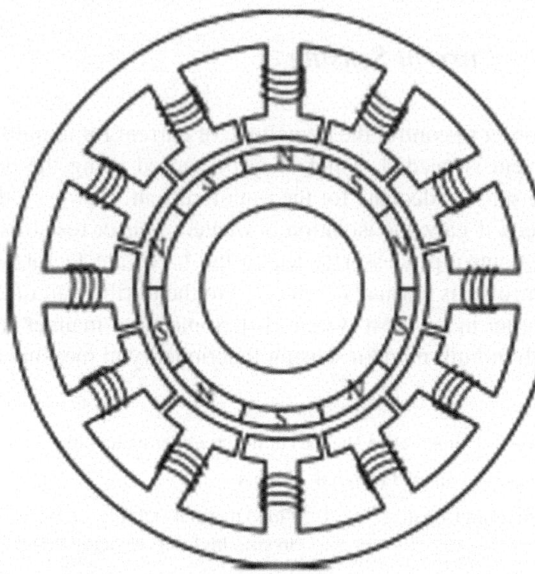

Fig. 2 Concentrated coil wound PMSM

4 Comparative Analysis of Different Permanent Magnet Materials

A detailed comparative analysis of different permanent materials regarding the structure and cost considerations is represented in Table 2.

The sensing of the controlled variable (or state) defines the concept of feedback control and giving that data back to the controller. The transducer or sensor will accept and emit energy from one part to a different part by changing the form of energy. The disadvantage of these sensors is they add noise to the system and affect the measurement. Presence of multiple sensors has several disadvantages such as susceptibility to noise production, reduced reliability, high cost, increased volume and weight, complexity in design, etc. Position and speed sensorless control is capable of overcoming such hurdles in order to achieve low-cost high-performance applications [6, 7].

The sensorless method has many advantages: (1) position detection can be done with the auxiliary control circuits, which does not seem to be placed next to the machine which may inhibit the operating temperature range; (2) only the main winding connections are required, so initial costs are minimized.

4.1 Current Sensing

In order to control the magnitude of current for a multi-phase machine, a two-phase current is needed, which may be sensed using the dc hyperlink, thereby making one sensor adequate for the run of current. However, these sensors may be steeply-priced if galvanic isolation is wanted. Sensor resistors need to be commonplace to the control processor to lessen the hardware factors. The commonplace from the controller is normally connected to the terrible rail of the dc bus, which delivers the inverter in low-cost systems [8]. Some other manner is to use the MOSFET gadgets with in-built modern sensing functionality to measure current.

Table 2 Comparative analysis of different permanent magnet materials

Permanent magnet material	Merits	Demerits
Samarium Cobalt	Good thermal stability, compact in size, high service temperature	High cost, Brittle, difficult to magnetize
Alnico	Low tooling cost, easy to magnetize	Very low coercivity, Robust, required thickness
Neodymium-Iron Boron	Compact, low energy cost, high coercivity, high remanence	Temperature-sensitive, difficult to magnetize, susceptible to corrosion
Ceramic	Low cost, easy to magnetize, high coercivity, high flexibility	Low remanence, limited shapes, low energy product

4.2 Position Estimation

Optical resolvers and encoders are very expensive but can give high resolution for the rotor position. Such methods can be used for estimation of commutation signals. Four proposed methods of sensorless rotor position control strategy are described as (1) sensing by external signal injection (2) current model adaptive scheme where the only difference in successive currents values is used [9].

4.2.1 Adaptive Scheme of the Current Model

This control strategy to be applied i.e. difference between the measured and true value of currents from the machine provides the discrepancy is considered and actual speed of the motor. Reduced current error ensures synchronous operation.

4.2.2 Signal Injection Sensor

The variation in the reluctance can be focused to extract rotor position information. The rate of change of current depends on the winding inductance. Inductance is a measure of the winding current and the rotor position estimation; therefore, we can extract the rotor position using this relationship. It can be applied even at zero speed due to lower value of the induced voltage.

4.2.3 Current Model-Based Injection Scheme

A model-based infusion conspire uses an outside voltage like the above plans. A voltage signal infusion into the d-axis twisting with an air conditioner square waveform of equivalent length makes incline like a sign with a low deviation of q-axis current. Since the d-axis current is upset, its impact on air hole torque isn't high, and the torque swell is low without undue apprehension for the acoustic commotion.

4.2.4 Position Estimation Using PWM Carrier Components

Function estimation the use of PWM provider uses the saliency of the gadget and not using a more signal injection. That is an awesome advantage. It makes use of the harmonic voltages within the PWM voltage supply and produces their harmonic modern-day reaction with the aid of setting apart essential and harmonics.

4.2.5 Sensing Based on Artificial Intelligence Control

An artificial neural network, genetic algorithm, etc. can use the variables like flux linkage and current to extract the rotor position. It's useful when the variables and commutation instances are latent. The ANN techniques do not require a mathematical model. They develop a better performance when tuned correctly and need less tuning effort. Genetic algorithms use a probabilistic method of solving optimization and search issues [10, 11].

5 Different Control Strategies for PMSM Drive

With the rapid development of power electronics and control theory, different control strategies are implemented in order to achieve more accurate and precise control in high-performance PMSM motor drive applications. Various control schemes such as vector control or decoupling control, constant torque angle control, unity power factor control (UPF), etc. are proposed in this section.

5.1 Vector Control

Vector control is one of the important topology in ac motor drives. For high-performance applications, control of flux and torque is crucial due to an input in the stator circuit [12, 13]. Moreover, the decoupling of the flux and torque channels establishes an advanced and faster dynamic behavior of the PMSM drives especially in the electrified transportation sector.

5.2 Constant Torque Angle Control

One of the most important parameters is the torque angle or power angle. The angle is fixed at 90°, and additionally, the magnitude of current along the d-axis is switched to zero. Below the base speed, this is the mode of operation, and this strategy is extremely effective in drive systems [14].

5.3 Power Factor Control

The input quality to the changeless magnet synchronous engine is very basic. The torque point is very basic as it is an element of the engine factors. The standardized torque per unit standardized current is under 1, showing that the control methodology isn't ideal as far as torque angle and, its productivity becomes lesser because of the expanded power dissipation for the angle of a similar torque [15–17].

5.4 Current Control Per Unit Optimum Torque

A suitable method to enlarge the magnitude of electromagnetic torque for a unit stator cutting-edge is important from the ideal gadget and, inverter use views. As indifferent control techniques, this control process is upheld with a torque part manage. Torque improves with higher saliency proportions are around 15–20% territory for appraised modern, and considerably greater for better flows. From a steady kingdom operational perspective, torque upgrade is ostensible and might not be a basic issue in picking this procedure for some packages [18].

5.5 Control of Power Loss

The most extreme torque versus speed envelope for the control procedures in the lower than base speed district is regularly found by restricting the stator current greatness to the appraised worth. In the higher than base speed area, the pole control is ordinarily restricted to appraised esteem. Current constraining limits copper losses; however, not really the centre losses. Correspondingly, restricting the pole control doesn't restrain control misfortunes legitimately. Restricting current and capacity to evaluated esteems, overlooks the warm strength of the machine that requires the all-out losses be obliged to an allowable worth. Evaluated current and power ensure satisfactory power losses just at appraised speed. Accordingly, these shortsighted confinements are substantial for movement control applications requiring activity at the evaluated speed [19–21].

6 Conclusion

This paper notably establishes a detailed overview of different control mechanisms employed in electric transportation sectors. To save the environment from the emission of greenhouse gasses and sustainable development, electric transportations are the optimistic solutions used in the developed countries. In this article a detailed discussion on the various PM materials, construction of PMSM machines, and the associated control mechanisms including sensorless control was discussed. Further, few necessary technical details of the PMSM used in a hybrid car was introduced. Additionally, for the development of PMSM drive system in electrified transportation manufactures/designers can select the fruitful control mechanism depending upon the need of the consumers/users in the relevant field. Finally, the outcome of this manuscript would be highly beneficial for readers as well as future researches to develop the suitable control mechanism of the PMSM drive to meet the better need of the environment.

References

1. T. Sebastian, G. Slemon, and M. Rahman.: Modelling of permanent magnet Synchronous motors. IEEE Transactions on Magnetics, 22, 1069-1071 (1986).
2. P. Pillay and R. Krishnan.: Modeling of permanent magnet motor drives. IEEE Transactions on Industrial Electronics, 35, 537-541 (1988).
3. S. Morimoto, Y. Tong, Y. Takeda, and T. Hirasa.: Loss minimization control of permanent magnet synchronous motor drives., IEEE Transactions on Industrial electronics, 41, 511–517 (1994).
4. A. H. Wijenayake and P. B. Schmidt.: Modeling and analysis of permanent magnet synchronous motor by taking saturation and core loss into account. 2nd International Conference on Power Electronics and Drive Systems, 2, 530–534 (1997).
5. R. Krishnan Electric Motor Drives- Modelling, Analysis and Control PHI, New Jersey, (2001)
6. F. Niu, B. Wang, A. S. Babel, K. Li, and E. G. Strangas.: Comparative evaluation of direct torque control strategies for permanent magnet synchronous machines IEEE Trans. Power Electron., vol. 31 (2), 1408–1424 (2016)
7. T. Turker, U. Buyukkeles, and A. Bakan.: A Robust Predictive Current Controller for PMSM Drives IEEE Trans. Ind. Electronics. 63 (6), 3906–3914 (2016)
8. S. Onoda and A. Emadi.: PSIM-based modelling of automotive power systems: Conventional, electric, and hybrid electric vehicles. Vehicular Technology, IEEE Transactions on, 53, 390–400, (2004).
9. Zare, J Askari .: Vector control of permanent magnet synchronous motor with surface magnet using genetic algorithm fuzzy logic controller. International Universities Power Engineering Conference (UPEC), 1–4, (2010) http://ieeexplore.ieee.org
10. Cheng-Kai Lin; Tian-Hua Liu; Jen-te Yu; Li-Chen Fu; Chieh-Fu Hsiao.: Model-Free Predictive Current Control for Interior Permanent-Magnet Synchronous Motor Drives Based on Current Difference Detection Technique IEEE transactions on Industrial Electronics, 61 (2), 667–681 (2014)
11. Sain, C., Biswas, P.K., Banerjee, A., Padmanaban, S.: An Efficient Flux Weakening Control Strategy of a Speed Controlled Permanent Magnet Synchronous Motor Drive for Light Electric Vehicle Applications. *IEEE-CALCON Conf.*, 1–5 (2017)
12. Zeraoulia, M., Benbouzid, M., Diallo, D. .: Electric motor drive selection issues for HEV propulsion systems: A comparative study. IEEE Trans. Veh. Technol., 1756–1764 (2006)
13. Sain, C., Banerjee, A., Biswas, P.K., Balas, V.E.: Performance optimization for closed loop control strategies towards simplified model of a PMSM drive by comparing with different classical and fuzzy intelligent 'controllers.' Int. J. Automation and Control **14**(4), 469–493 (2020)
14. Chiranjit Sain, A Banerjee, P K Biswas, P Sanjeevikumar.: A Sate of the Art Review on Solar Powered Energy Efficient PMSM Drive Smart Electric Vehicle for Sustainable Development Book Chapter: Advances in Greener Energy Technologies & Springer Book Series: *Green Energy and Technology (ISSN: 1865–3529)*, 231–258, *Springer* (2020).
15. C. Xia, J. Zhao, Y. Yan, and T. Shi .: A novel direct torque and flux control method of matrix converter-fed PMSM drives IEEE Trans. Power Electronics. 29 (10), 5417–5430, Oct. 2014.
16. Chiranjit Sain, A Banerjee, P K Biswas .: A Comprehensive Study on Various Topologies of Permanent Magnet Motor Drives for Electric Vehicles Application"- Book Title: Artificial Intelligent Techniques for Electric and Hybrid Electric Vehicles, *Scrivener Publishing, Wiley*, ISBN: 978-1-119-68190-8 (2020).

17. Chiranjit Sain, A Banerjee, P K Biswas: Comparative performance study for closed loop operation of an adjustable speed permanent magnet synchronous motor drive with different controllers. Int. J. Power Electroni. Drive Syst (IJPEDS), **7**(4) (2016).
18. Katsuhiko Ogata, Modern Control Engineering PHI, Third Edition, (1998)
19. Krause, P.C.: Analysis of Electric Machinery. McGraw Hill, New York (1987)
20. Bose, B.K.: Modern Power Electronics and AC Drives. Prentice Hall (2002)
21. Sain, C., Banerjee, A., Biswasm, P.K.: Modelling and comparative dynamic analysis due to demagnetization of a torque controlled permanent magnet synchronous motor drive for energy-efficient electric vehicle. ISA Trans. **97**, 384–400 (2020)

Direct Torque Control of BOOST-VSI Fed Induction Motor Using SVPWM-PAM Technique

Sudharani Potturi and Rajashekar P. Mandi

Abstract This paper proposes a Direct Torque Control (DTC) method for a Boost-VSI fed three-phase Induction Motor using Space Vector Pulse Width Modulation(SVPWM) and Pulse Amplitude Modulation(PAM) techniques, which reduces the torque ripple, speed, and DC Link Voltage distortions and improves the dynamic response of the Induction Motor. For VSI, SVPWM technique is used, where as for Boost Converter, both PWM (Pulse width Modulation) and PAM techniques are implemented. Simulation is done using MATLAB/SIMULINK and the results are discussed and analyzed for both the techniques. A considerable improvement in the performance of the Drive is observed when compared to the existing SVPWM-PWM based DTC schemes. A reduction in torque ripples is also observed with the proposed model.

1 Introduction

Direct Torque Control method of Induction motor [1, 2] gained its significance because of its ease of implementation, minimum torque response time, absence of coordinate transformation, robust for rotor parameter variation et cetera [1]. And also it abandons the traditional stator current control, instead it controls the stator flux. There are different DTC schemes available separately to provide good dynamic response, to enhance the efficiency [3–5] of the drive and to reduce the ripple in the developed torque [6]. But the proposed work aims to provide better performance with respect to all these features as a single entity. This paper proposes DTC control of a BOOST- VSI fed Induction Motor [7] using SVPWM-PAM method to get better dynamic response and to reduce the torque ripples and also reduce the input current drawn by the drive so that the copper losses can be reduced [5]. Pulse

S. Potturi (✉) · R. P. Mandi
REVA University, Bangalore, India
e-mail: chekuri.rani@gmail.com

R. P. Mandi
e-mail: dir.eee@reva.edu.in

J. Gu et al. (eds.), *Communication and Control for Robotic Systems*, Smart Innovation, Systems and Technologies 229, https://doi.org/10.1007/978-981-16-1777-5_26

amplitude Modulation is used for DC-DC converter which can address the DC link voltage distortions and also the capacitance needed for Inverter DC link is less when compared to conventional PWM method. So that the size of the system is reduced [8]. The amplitude of each pulse is controlled by the instantaneous amplitude of the modulation signal to generate PAM signals. Principle and Implementation of DTC [1, 9] control are discussed in Sect. 2: Methodology, Sect. 3: Simulation results and Sect. 4: Conclusion.

2 Methodology

The block diagram of DTC control of BOOST VSI fed Induction motor is shown in Fig. 1. In this method, the actual torque and the stator flux are calculated in DTC control block using the DC Link (capacitor) voltage, stator currents values. These calculated values will be compared with the respective reference values and the errors are estimated. The position of the voltage vector in state space can be determined from the value of the space angle which is also calculated in DTC Block. Depending upon the error values of the torque and stator flux and the space vector sector, the triggering pulses for VSI are generated and fed to the VSI to generate the required voltage. And also the PAM signals are generated to trigger the BOOST Converter using DC Link voltage as reference. As the Voltage is boosted to maintain the required Input to the Induction Motor, the current drawn by it will be reduced. Hence the amount of copper losses can also be reduced.

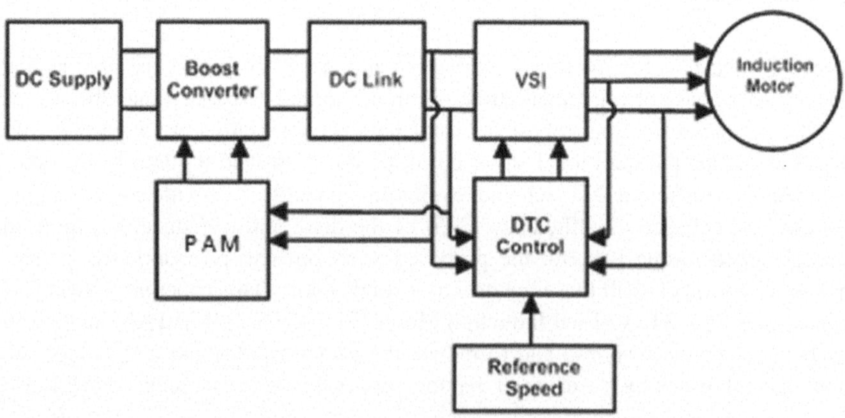

Fig. 1 Block diagram of proposed control method

2.1 Boost-VSI with PAM Technique

Depending upon the required speed, output voltage of the VSI needs to be adjusted and hence the Stator voltage of the Induction Motor. In the proposed model SVPWM Technique is applied for Voltage Source Inverter whereas for Boost Converter PAM technique is implemented [10]. When PWM technique is used for BOOST converter, the DC Link voltage is taken as constant. In case of PAM based converters, the DC link voltage is variable, and hence the inverter input [10]. The DC link variation results in change in Inverter output [11].

PAM signals are generated based on the polarity of modulated signals. PAM techniques are classified as single polarity PAM and Double polarity PAM [12]. If there are only positive signals, it is known as Single Polarity Pulse Amplitude Modulation and if it consists of both positive and negative signals, it is known as double polarity Pulse Amplitude Modulation. The PAM techniques are further classified as Flat Top PAM and Natural PAM. As the name indicates, the top of the pulse remains flat in case of Flat top PAM signals, where as the top of the pulse follows the amplitude of the modulating signal in case of Natural PAM. Single polarity Natural PAM is used for the proposed model. Here the PAM generator generates the gating signals for the Boost Converter by taking the DC Link voltage as reference. The flat top and natural PAM waveforms are shown in Figs. 2. and 3. respectively.

Fig. 2 Flat Top PAM [12]

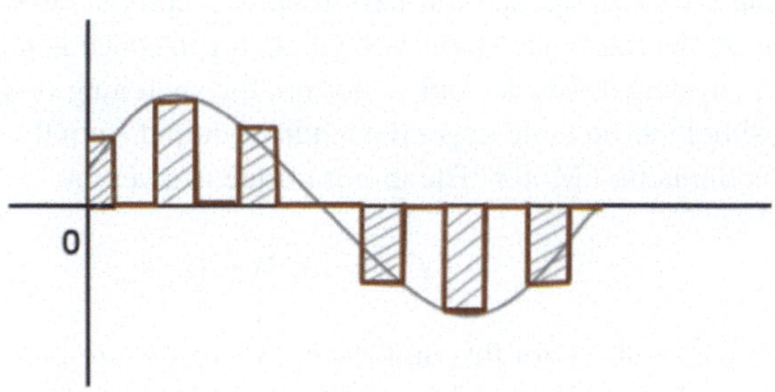

Fig. 3 Natural PAM [12]

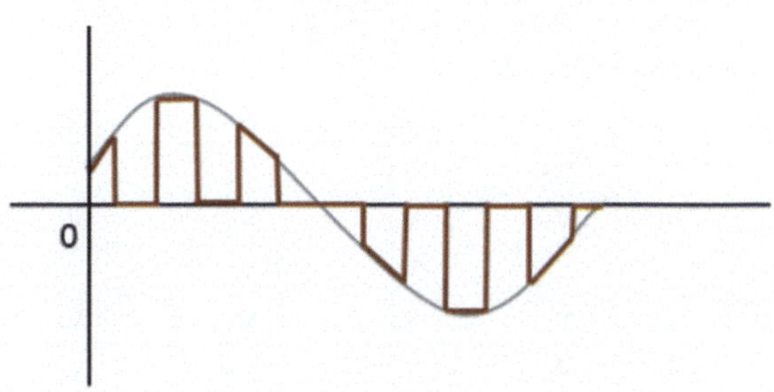

2.2 DTC Control

DTC control method consists of torque and stator flux loops. From the values of any two Stator phase Currents and DC Link voltage, actual values of flux and the motor torque are calculated. Then these values are compared with the reference values and the errors are sent to the respective hysteresis controllers [2].

The hysteresis controllers' outputs and the stator flux vector position are fed to the look-up table to release the required switching pattern. Depending upon the switching sequence, the inverter output voltage is varied. The basic switching table for DTC Control is given in Table 1.

If the value of flux need to be increased, the value of $\Delta_\phi = 1$ and if it needs to be decreased $\Delta_\phi = 0$ is considered. With respect to torque, there will be three cases, either torque needs to be increased or decreased or held constant, ie $\Delta T = 1$, $\Delta T = -1$ $\Delta T = 0$. The function of comparators is to maintain or limit the magnitudes of the torque and flux vectors within a narrow hysteresis band. This is the main reason for DTC to provide fast torque response without overshoot.

The currents from the two phases of the Induction Motor is sensed and from that the values of stator currents in α–β reference frame values is α and is β values are calculated. These values along with the Vsα and Vsβ are used to determine the stator flux and Torque estimations. Then these calculated values (ψs and Te respectively) are compared with the reference values (ψs* and Te* respectively) and errors are passed through hysteresis comparators. From the values of ψsα and ψsβ, the reference stator voltage sector in space is identified. Using the outputs of the hysteresis blocks and sector no, the switching pattern of the Inverter is decided using look-up table to get the required output from the inverter to correct the sped of the Induction Motor. The motor torque is given by

$$T_e = (3/2).p.\left(i_{s\beta}\Psi_{s\alpha} - i_{s\alpha}\Psi_{s\beta}\right) \tag{2.1}$$

where the stator flux and current vectors are represented in stationary α–β frame, and p denotes the number of poles.

Table 1 Switching table

Δ_ϕ	Δ_T	Sector I	Sector II	Sector III	Sector IV	Sector V	Sector VI
1	1	V_2	V_3	V_4	V_5	V_6	V_1
	0	V_7	V_0	V_7	V_0	V_7	V_0
	−1	V_6	V_1	V_2	V_3	V_4	V_5
0	1	V_3	V_4	V_5	V_6	V_1	V_2
	0	V_0	V_7	V_0	V_7	V_0	V_7
	−1	V_5	V_6	V_1	V_2	V_3	V_4

3 Simulation Results

Simulation of the proposed model i.e. Direct Torque Control of BOOST VSI fed 5.4HP Induction Motor is carried out and the SIMULINK model is shown in Fig. 4.

The simulation is carried by using PWM as well as PAM techniques for Boost Converter. While presenting results, the terminology, "without PAM "is used for PWM. Induction Motor parameters chosen for the simulation study are given in Table 2.

3.1 Torque Response

The simulation is carried out by taking different load torque values of 15, 25, 20, and 10 N m. The actual and reference torques with and without PAM technique are given in Figs. 5 and 7. respectively. It is observed that

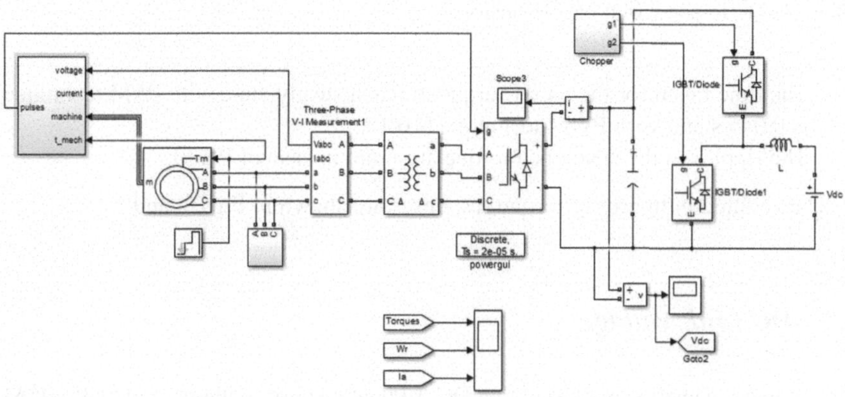

Fig. 4 DTC control of Boost VSI fed induction motor

Table 2 Motor parameters

S. No	Parameter	Value
1	HP	5.4
2	Voltage	400 V
3	Frequency	50 Hz
4	Stator resistance	1.405 Ω
5	Stator inductance	0.005839 H
6	Rotor resistance	1.395 Ω
7	Rotor inductance	0.005839 H
8	Mutual inductance	0.1772 H

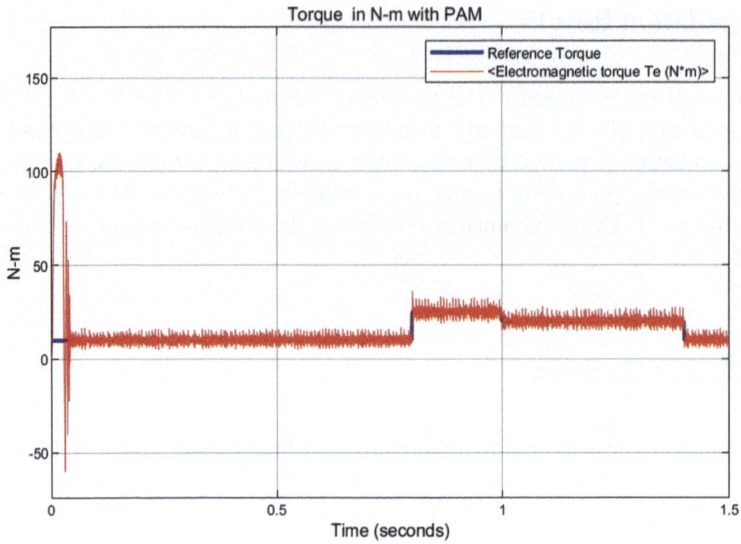

Fig. 5 Torque response with PAM technique

i. The time taken for the motor torque to reach steady-state with PAM technique is 0.045 s and with PWM technique is 0.1 s.
ii. The ripples in the developed torque are more in case of PWM.

The zoomed in figures of torque responses are shown in Figs. 6 and 8.

3.2 DC Link Voltage

The voltage waveforms across the DC Link capacitor without and with PAM technique are shown in Figs. 9 and 10. respectively.

For an Input voltage of 200 V DC, the output voltage of around 380 V is obtained. But with PWM technique the distortion in voltages are more when compared to the results with PAM technique. During starting of the Motor, the time taken for the DC Link voltage to reach steady state is 0.14 s in case of PWM technique and is 0.04 s with PAM technique. With PAM technique, the value of DC link voltage is 400 V approx. for a reference rotor speed of 150 rad/s and load torque of 15 N m and this voltage can be varied as per the required output of VSI.

3.3 Rotor Speed Response

The rotor speed response with and without PAM technique is represented in Figs. 11

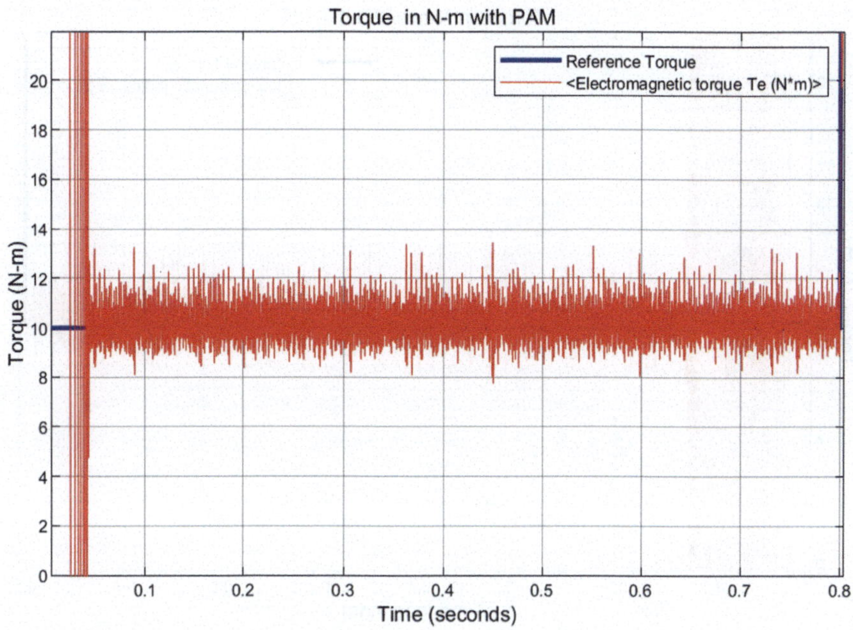

Fig. 6 Torque response with PAM technique

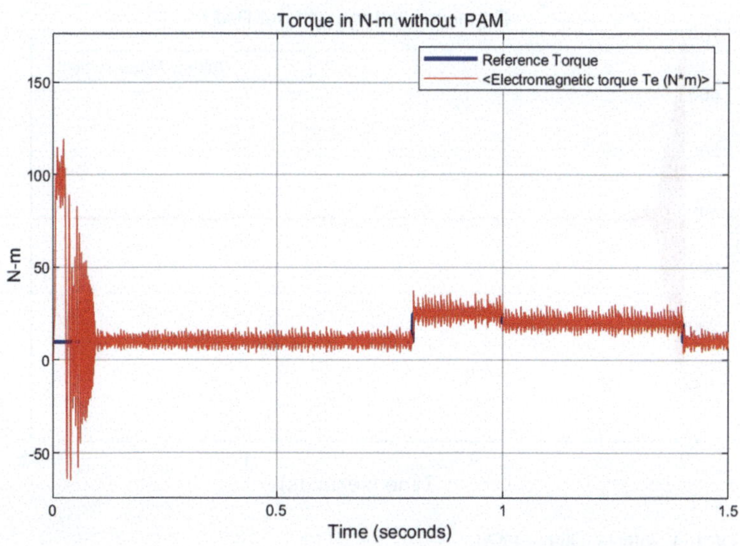

Fig. 7 Torque response without PAM technique

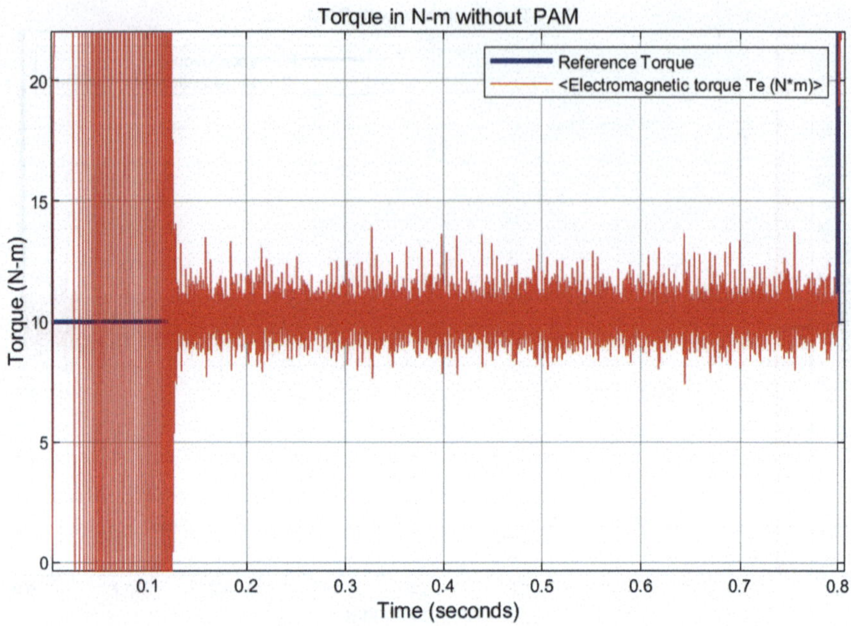

Fig. 8 Torque response without PAM technique

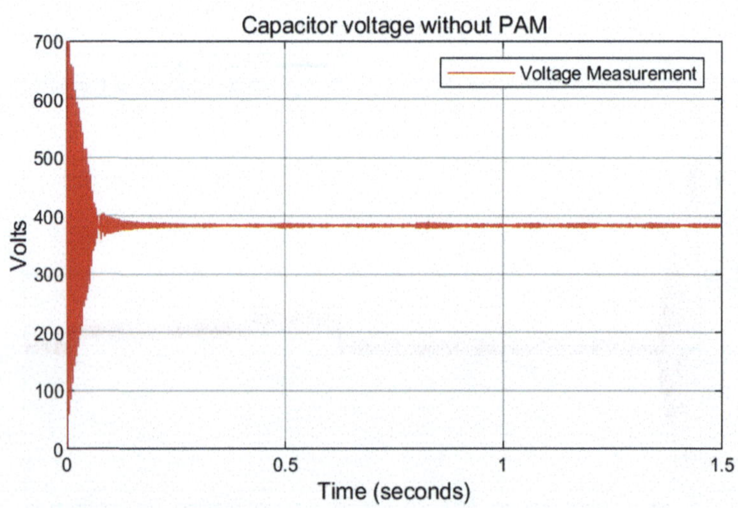

Fig. 9 DC link voltage without PAM

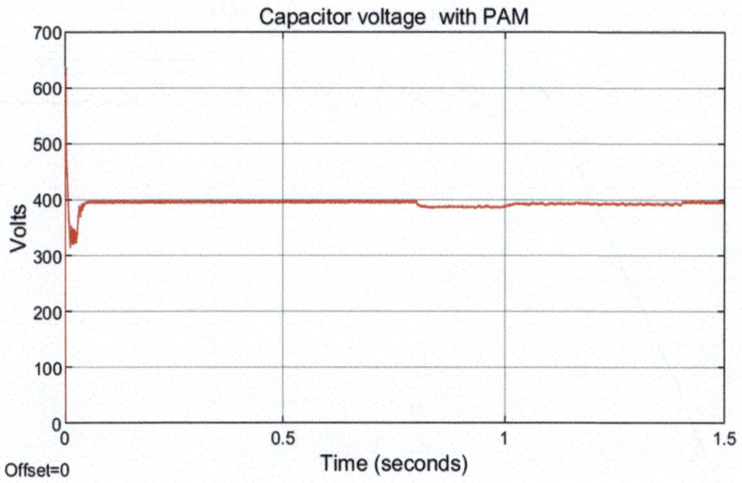

Fig. 10 DC link voltage with PAM

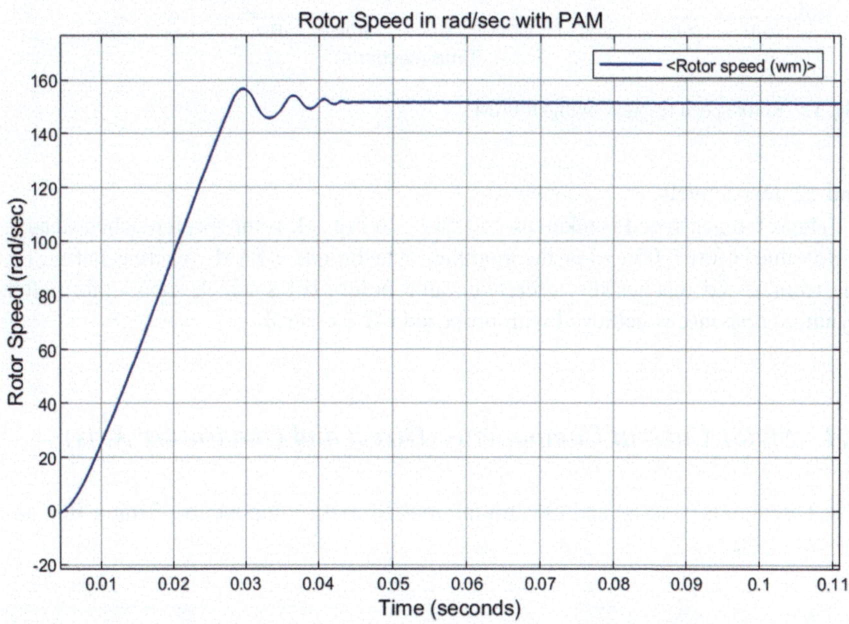

Fig. 11 Rotor speed response with PAM

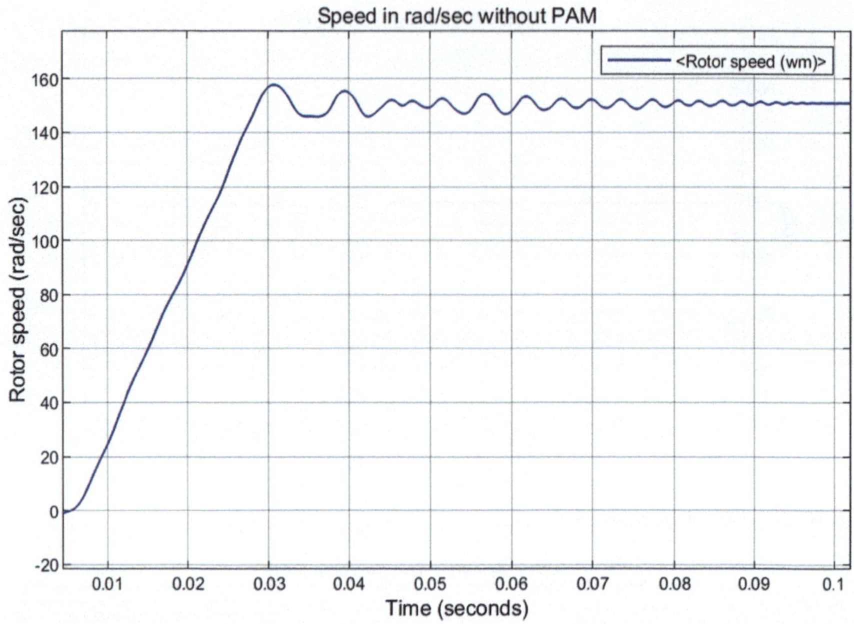

Fig. 12 Rotor speed response without PAM

and 12, respectively.

The reference speed is taken as 150 rad/s. In Fig. 11. rotor speed reaches steady-state value before 0.05 s when the modulation technique is PAM. Whereas in Fig. 12, the rotor speed reaches the reference value before 0.1 s which shows that better dynamic response is achieved with proposed DTC control.

3.4 Stator Current Components (Direct and Quadrature Axis)

The Direct axis (d-axis) and Quadrature axis (q-axis) components of stator flux are shown in Fig. 13.

3.5 Performance Comparison

The superiority of DTC control of Boost VSI fed Induction motor with PAM over PWM is given in Table 3.

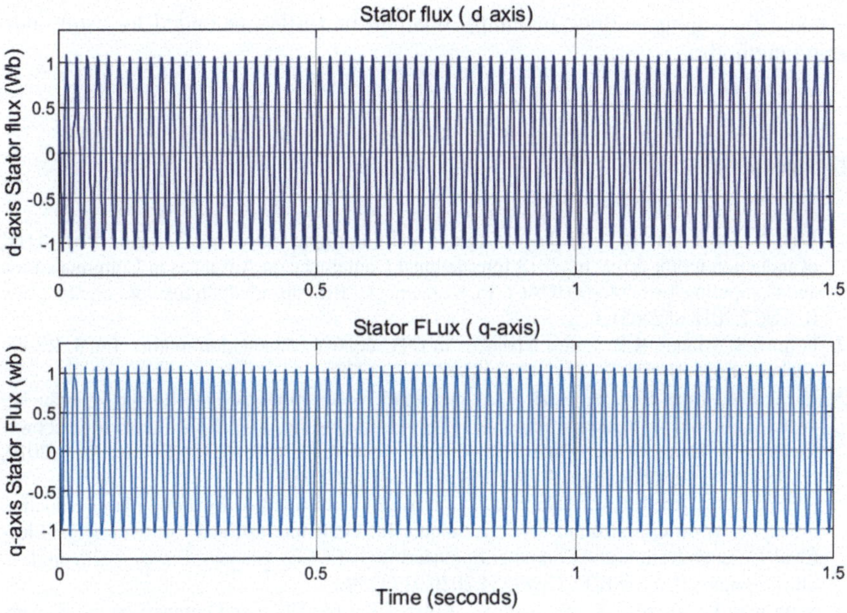

Fig. 13 D-axis and Q axis components of stator current

Table 3 Motor parameters

S. No	Parameter	PWM	PAM
1	Capacitor voltage	380 V	400 V
2	Time to reach steady-state speed	0.1 s	0.042 s
3	Time to develop required load torque during starting	0.15 s	< 0.1 s
4	Motor input current for load torque of 25 N m	7.4 A	6.9 A

4 Conclusion

Closed-loop DTC control of Boost-VSI Fed Induction Motor using SVPWM-PAM control is simulated. Feedback signals of DC Link Voltage, speed and Induction motor currents are utilized to get the switching signals for the inverter. DC link voltage reference is fed back to the PAM generator to generate switching pulses for Boost Converter. Also, the detailed comparison of the PWM and PAM-based DTC control method for Induction motor is presented. It has been proven that the proposed PAM control gives better speed response and reduced torque ripples compared to the PWM control. The performance at high constant speed is analyzed for different load torques and observed that the speed oscillations are minimized using PAM control. The proposed work can be extended by losses calculation and efficiency estimation.

PI controller tuning is done manually this can be further extended by using auto tuning methods.

References

1. Kale, S., Pawaskar, R.C.: Analysis of torque and flux ripple factor for DTC and SVM—DTC of induction motor drive. In: 2018 International Conference on Advances in Communication and Computing Technology (ICACCT), Sangamner, 2018, pp. 45–49. https://doi.org/10.1109/ICACCT.2018.8529651.J
2. Potturi, S., Mandi, R.P.: Latest advances in DTC control of induction motors. Int. J. Recent Technol. Eng. (IJRTE) **8**(6S) (2020). ISSN: 2277-3878
3. Vaezi, S.A., Iman-Eini, H., Razi, R.: A new space vector modulation technique for reducing switching losses in induction motor DTC-SVM scheme In: 2019 10th International Power Electronics, Drive Systems and Technologies Conference (PEDSTC), Shiraz, Iran, 2019, pp. 184–188. https://doi.org/10.1109/PEDSTC.2019.8697889
4. Fadhil, G.M., Abed, I.A., Jasim, R.S.: Improved control method for three phase induction motor based on optimization DTC technique. In: 2020 International Conference on Electrical, Communication, and Computer Engineering (ICECCE), Istanbul, Turkey, 2020, pp. 1–5. https://doi.org/10.1109/ICECCE49384.2020.9179239
5. Schwager, L., Tüysüz, A., Zwyssig, C., Kolar, J.W.: Modeling and comparison of machine and converter losses for PWM and PAM in high-speed drives. IEEE Trans. Ind. Appl. **50**(2), 995–1006. (2014). https://doi.org/10.1109/TIA.2013.2272711
6. Ohyama, R., Hoshi, N.: Efficiency comparison of PAM-PWM inverter with partially boost converter for motor drive system. In: 2019 22nd International Conference on Electrical Machines and Systems (ICEMS), Harbin, China, 2019, pp. 1–6. https://doi.org/10.1109/ICEMS.2019.8921682.
7. Lei, Q., Peng, F.Z.: Space vector pulsewidth amplitude modulation for a buck-boost voltage/current source inverter. IEEE Trans. Power Electron. **29**(1), 266–274 (2014). https://doi.org/10.1109/TPEL.2012.2225847
8. Ghoreishy, H., Varjani, A.Y., Mohamadian, M., Farhangi S., Zhang, Z.: A new selective harmonic elimination pulse-width and amplitude modulation (SHE-PWAM) for drive applications. In: IECON 2013—39th Annual Conference of the IEEE Industrial Electronics Society, Vienna, 2013, pp. 234–239. https://doi.org/10.1109/IECON.2013.6699141
9. Punith Kumar, M.R., Potturi, S.: Boost-VSI based on space vector pulse width amplitude modulation. IJSRD—Int. J. Sci. Res. Dev. 3(02) (2015). ISSN (online): 2321-0613
10. Yu, X., Lei, Q., Peng, F.Z.: Boost converter—Inverter system using PWAM for HEV/EV motor drive. In: 2012 twenty-seventh annual IEEE applied power electronics conference and exposition (APEC), Orlando, FL, 2012, pp. 946–950. https://doi.org/10.1109/APEC.2012.6165932
11. Lai, Y., Lee, K., Tseng, J., Chen, Y., Hsiao, T.: Efficiency comparison of PWM-controlled and PAM-controlled sensorless BLDCM drives for refrigerator applications. In: 2007 IEEE Industry y Applications Annual Meeting, New Orleans, LA, 2007, pp. 268–273. https://doi.org/10.1109/07IAS.2007.5
12. https://www.elprocus.com/pulse-amplitude-modulation/

Innovations in Robotic Systems

Launch Vehicle Autopilot Design Using H-Infinity Control Technique

Chintapalli Vaishnavi and M. V. Dhekane

Abstract In this paper, we try to understand the dynamics involved in the launch vehicle. Then, to meet the target trajectory, we have to design a controller using modern control technique which is H-infinity control technique. The major concept of this technique is the selection of weight functions, their augmentation with the plant and developing its state-space model. Various performance metrics used to analyse the performance of the system are gain margin, aero margin, rise time, settling time, overshoot and steady state error.

Nomenculture

T_c	Control thrust in pitch plane
T_s	Ungimballed thrust
I	Moment of inertia of vehicle about pitch axis
m	Total mass of the launch vehicle
m_0	Reduced mass of the launch vehicle
L_α	Aerodynamic load on launch vehicle per unit angle of attack
l_c	Distance from centre of mass of vehicle to engine swivel point
l_α	Distance from the centre of mass to centre of pressure
δ	Thrust deflection angle
α	Angle of attack
V	Forward velocity of vehicle
V_w	Wind velocity
α_w	Wind angle of attack
D	Drag
K_a	Servoamplifier gain
K_r	Rate gyro gain
l_{pi}	Distance from hinge point of ith pendulum to origin on body axis
L	Length of the vehicle
L_{pi}	Length of ith pendulum

C. Vaishnavi (✉) · M. V. Dhekane
IIST, Thiruvananthapuram, India

J. Gu et al. (eds.), *Communication and Control for Robotic Systems*, Smart Innovation, Systems and Technologies 229, https://doi.org/10.1007/978-981-16-1777-5_27

q^i	Generalised coordinate of ith bending mode
U_0	Forward velocity of vehicle
μ_c	Control moment coefficient: $= \frac{Tc*l_c}{I_{yy}}$
μ_α	Aerodynamic moment coefficient: $= \frac{L_\alpha*l_\alpha}{I_{yy}}$
μ_{pj}	$\frac{m_{pj}*l_{pj}*\dot{U}_0}{I_{yy}}$
ω_a	Actuator frequency
ω_{slosh}	Slosh frequency
ω_i	Frequency of ith bending mode. $i = 1,2$
ρ	Damping ratio
ρ_a	Damping ratio of actuator
ρ_i	Damping ratio of ith bending mode. $i = 1,2$

1 Introduction

Design of attitude controllers for launch vehicle using modern control techniques has been attempted for over four decades. All the important characteristics of launch vehicle are needed to be taken into consideration while designing a controller. Firstly, the vehicle dynamics is time varying due to change in mass, moment of inertia, slosh and flexibility characteristics during the thrusting phase. Typical launch vehicles are aerodynamically unstable. Moreover, proximity of flexible modes with rigid body can cause undesirable control structure interactions. Autopilot is a closed loop attitude control system designed to stabilise the vehicle, track the guidance command and also meet the performance specifications under parameter perturbations. Major assumption made in attitude control of launch vehicle is that there is minimal coupling between pitch, yaw and roll dynamics. Dynamics frozen at an instant on the trajectory is considered, and it is linearised. Controller is designed to meet performance specifications satisfying aeromargin (gain reduction margin), phase margin and gain margin. Designing an autopilot in high dynamic pressure region is a challenging task for which several techniques are being explored. In this paper, we discuss dynamics involved in the launch vehicle, formulate the problem for implementation of H-infinity technique and design a controller to achieve stable system with good performance specifications. The work presented in this paper is carried out as a part of final year project in IIST.

2 Dynamics Involved in the Launch Vehicle

The dynamics involved for attitude control of a launch vehicle are rigid body, actuator, bending modes and slosh dynamics. These are discussed in [3] the equations of motion for these dynamics are as follows:

2.1 Rigid Body Dynamics:

From Fig. 1a, we can get the moment equation about centre of mass as follows:

$$I_{yy}\ddot{\theta} = T_c\delta * l_c + L_\alpha\alpha * l_\alpha \tag{1}$$

$$\implies \ddot{\theta} = \mu_c\delta + \mu_\alpha\alpha \tag{2}$$

where $\mu_c = \frac{T_c l_c}{I_{yy}}$ and $\mu_\alpha = \frac{L_\alpha l_\alpha}{I_{yy}}$. Longitudinal acceleration = $\frac{T_t - D}{m}$ and lateral acceleration = $\frac{T_c\delta - L_\alpha\alpha}{m}$. Then, acceleration of the vehicle along Z_i is

$$\ddot{Z}_i = (\frac{T_c\delta - L_\alpha\alpha}{m}) * cos\theta - (\frac{T_t - D}{m}) * sin\theta \tag{3}$$

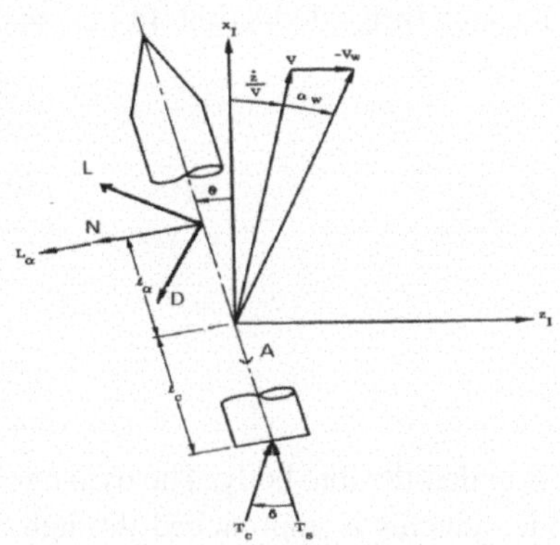

(a) Forces acting on vehicle in pitch plane

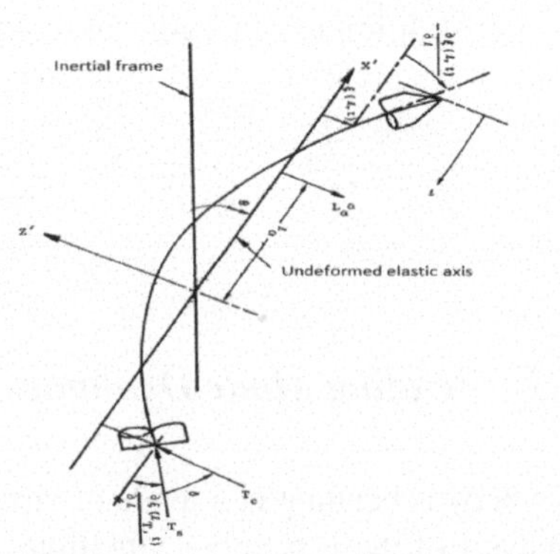

(b) Flexible Body dynamics of launch vehicle

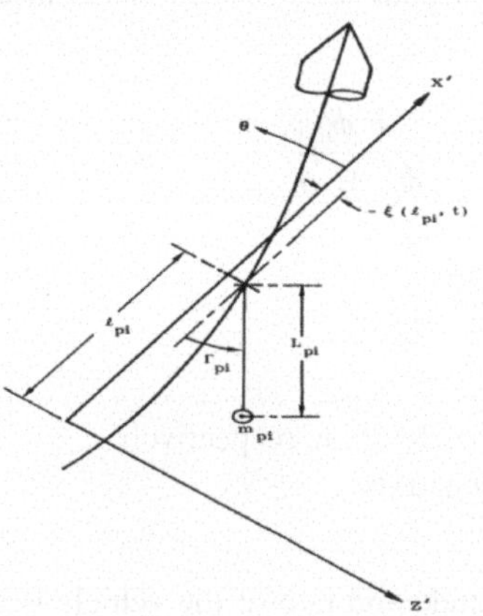

(c) Slosh dynamics of launch vehicle

Fig. 1 Dynamics involved in launch vehicle

since θ is very small, cosθ = 1 and sinθ = θ $\Longrightarrow$

$$\frac{\ddot{Z}_i}{V} = (\frac{T_c\delta - L_\alpha\alpha}{mV}) - (\frac{T_t - D}{mV}) * \theta \tag{4}$$

As seen from Fig. 1a, angle of attack

$$\alpha = \theta + \alpha_w + \frac{\dot{Z}}{V} \tag{5}$$

2.2 Actuator Dynamics

The function of actuator is an electromechanical/electrohydraulic system to gimbal the thrust. It implements control requirement through thrust deflection. In classical control, we prefer a second-order actuator as follows:

$$\frac{\delta}{\delta_c} = \frac{\omega_a^2}{s^2 + 2\rho_a\omega_a s + \omega_a^2} \tag{6}$$

2.3 Bending Mode Dynamics

If a body is bending in a mode of vibration, it is called flexible body. The dynamics analysis of these bending vibrations of flexible vehicles is approached through a wave-transmission concept as opposed to the traditional normal-mode method. In Fig. 1b for one bending mode, additional slope introduced is shown. Generalised force equation:

$$\ddot{q}_i + 2\rho_i\omega_i\dot{q}_i + \omega_i^2 q_i = \frac{GF_i}{GM_i} = \frac{\mathbf{F}.\boldsymbol{\phi}_\mathbf{i}}{\boldsymbol{\phi}_\mathbf{i}^\mathbf{T}.M.\boldsymbol{\phi}_\mathbf{i}} \tag{7}$$

where GF_i is generalised force of ith bending mode
GM_i is generalised mass of ith bending mode

q_i is generalised coordinate of ith bending mode
$\dot{q}_i$, $\ddot{q}_i$ are generalised velocity and generalised acceleration, respectively
F is resultant force vector along body coordinate z-axis
$\boldsymbol{\phi}$ is generalised mode shape

The effect of bending mode dynamics on the attitude and rate of the vehicle is as follows.

$$\theta_s = \theta + \sum_i \sigma_{iEB} * q_i \tag{8}$$

$$\dot{\theta}_s = \dot{\theta} + \sum_i \sigma_{iRGP} * q_i \tag{9}$$

In this paper, we include only two bending modes. Therefore, i= 1,2.

2.4 *Slosh Dynamics*

In launch vehicle, the slosh dynamics are approximated to linearised equations of motion of mechanical simple pendulum as shown in Fig. 1c. The effect of sloshing is significant in regions of flight where some amount of fuel is already burnt giving the fuel in the tank enough space for movement. The equations of motion that help in the analysis of the slosh dynamics are as follows:

$$\ddot{\theta} = \mu_c\delta + \mu_\alpha\alpha + \mu_p\Gamma_p \tag{10}$$

$$\ddot{\Gamma}_{pi} = \frac{1}{L_{pi}}[-\frac{\Sigma F_z}{m_0} + \frac{l_{pi} - L_{pi}}{I_{yy}}\Sigma M_y - \dot{U}_0\Gamma_{pi}] \tag{11}$$

where

$$\Sigma F_z = T_c\delta - L_\alpha\alpha + \sum_j m_{pj}\dot{U}_0\Gamma_{pj} \tag{12}$$

$$\Sigma M_y = T_c l_c\delta + L_\alpha l_\alpha - \sum_j m_{pj} l_{pj}\dot{U}_0\Gamma_{pj} \tag{13}$$

Here, we assume there is only one fuel tank for simplified analysis. Now put the values of ΣF_z , ΣM_y in (11)

$$\ddot{\Gamma}_p = -\frac{1}{L_p}[\frac{T_c}{m_0} - (l_p - L_p)\mu_c]\delta - \frac{1}{L_p}[(1 + \frac{m_p}{m_0})\dot{U}_0 + (l_p - L_p)\mu_p]\Gamma_p \tag{14}$$

3 H-∞ Control Technique

H-infinity control technique is used to synthesise controllers with robust stability and with guaranteed performance. The concept of H-infinity control is to minimise a suitably weighted norm of closed loop transfer function matrix which also means minimising the peak values (H-∞ norm) of the transfer function over frequency range. It is a mathematical optimisation problem where the controller designed will give optimised solution as long as the disturbance has bounded energy. The resulting controller is only optimal with respect to the cost function(i.e. H-∞ norm), but this need not be the best controller for the system in terms of performance measures such as settling time and overshoot. Important features of H-∞ control are: firstly, it is a robust design method for linear time invariant systems. Secondly, it is readily applicable for multivariable system with cross-coupling between the channels. Thirdly, it directly handles the uncertainity. The major drawback in modern control techniques is that the resulting controller has the same order as generalised plant which is larger than original (or nominal) plant due to inclusion of frequency-dependent weight functions to achieve desired performance.

3.1 Problem Statement

The standard closed loop representation of the problem in H-∞ control is as follows: where w is exogenous vector input signal which can be reference signal, disturbances and sensor noises, z is the exogenous vector output signal which can be the signals we wish to control like tracking error, y is the measurement signal from system to controller, and u is the control signal which is the output of controller K and given as a input to augmented plant P. The presence of w, z is the difference from the SISO feedback system where w is used to model disturbances and noises, while z is used to create design objectives.

The state-space equations for the above system shown in Fig. 2 are as follows:

$$\begin{bmatrix} z \\ y \end{bmatrix} = \begin{bmatrix} P_{11} & P_{12} \\ P_21 & P_{22} \end{bmatrix} \begin{bmatrix} w \\ u \end{bmatrix} = P \begin{bmatrix} w \\ u \end{bmatrix} \tag{15}$$

where P is the state-space matrix for open loop system. To analyse the performance of the system, the required transfer function is $\frac{z}{w}$ which is the closed loop transfer function. Solve the (15) to find this transfer function as follows:

$$z = P_{11}w + P_{12}u \tag{16}$$

$$y = P_{21}w + P_{22}u \tag{17}$$

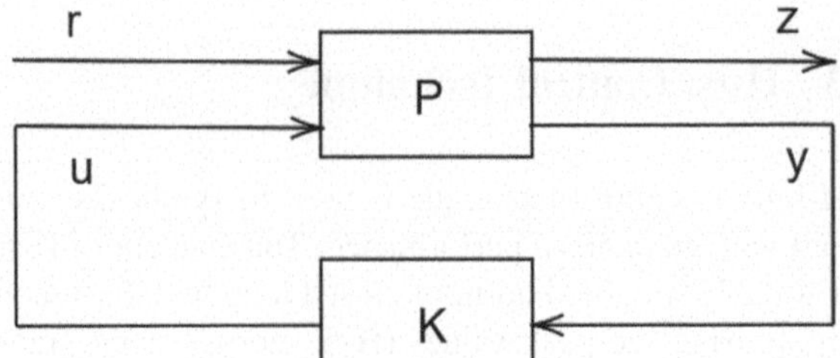

Fig. 2 Standard close loop configuration for system with H-∞ control

and from Fig. 2, we can say $u = Ky$

$$\implies y = P_{21}w + P_{22}Ky \tag{18}$$

$$\implies y = (I - P_{22}K)^{-1}P_{21}w \tag{19}$$

$$\implies z = P_{11}w + P_{12}K(I - P_{22}K)^{-1}P_{21}w \tag{20}$$

$$\frac{z}{w} = P_{11} + P_{21}K(I - P_{22}K)^{-1}P_{21} = \|F_l(P, K)\| = T_{zw} \tag{21}$$

For H-∞ optimisation problem, we need to find or synthesise a stabilising controller K such that $\|F_l(P, K)\|_\infty < \gamma$ and $F_l(P, K)$ is internally stable where

$$\|F_l(P, K)\|_\infty = \|T_{zw}\|_\infty = \sup_{w \in \mathbb{R}} \bar{\sigma}(T_{zw}(j\omega)) \tag{22}$$

is H-infinity norm and σ is maximum singular value.

3.2 Methodology

- **Step 1:** Know all the equations of motion of the dynamics involved in nominal plant and make a state-space model of this.
- **Step 2:** Selection of weight functions.
- **Step 3:** Compute the state-space model of the plant augmented with the weight functions
- **Step 4:** Synthesise a H-∞ controller K using $hinfsyn$ and $hinfsynOptions$ which are inbuilt functions in MATLAB.

3.3 Model Used for the Analysis

Here, r1 = θ_c which is reference signal and r2 = α_w is disturbance input. G is the nominal plant. We, Wu, Wd are weight functions, and K is the synthesised controller.

3.4 Selection of Weight Functions

The selected weight functions correspond to the stability and robustness of the system, and as they are frequency-dependent, they can be tuned to achieve certain frequency domain specifications like system bandwidth, and low frequency gain also adjust the overshoot of the step response obtained.

The control weight W_u is weight function for input control signal, and it needs to limit the high frequency activity and allow low frequency tracking. Therefore, W_u needs to be a high-pass filter to satisfy these requirements.

$$W_u = h\frac{ks + \omega}{s + k\omega} \tag{23}$$

ω is chosen so that the high frequency control is minimised above certain threshold frequency and k is chosen to allow low frequency control effort and minimise the high frequency control effort simultaneously. If k is kept low, the actuators operate in linear region.

At high frequency, the controller will not be able to counteract errors due to inertia of the vehicle. At low frequency where constant errors occur, more precise control is expected. Hence, W_e the weight function for tracking error is chosen to behave like a lag filter as follows:

$$W_e = h\frac{s + \omega_z}{s + \omega_p} \tag{24}$$

where $\omega_z \gg \omega_p$. The disturbances given as input to the plant are usually slower than launch vehicle dynamics, thereby having low frequency. Hence, we choose W_d the weight function for disturbance input to the plant as a low pass filter as follows:

$$W_d = \frac{k}{s + \omega_d} \tag{25}$$

3.5 *Plant Augmentation and State-Space Model*

The basic state-space equations for the standard closed loop configuration in Fig. 2 from which state-space model of plant P is derived is as follows:

$$\dot{x} = Ax + B_1 w + B_2 u \tag{26}$$

$$z = C_1 x + D_{11} w + D_{12} u \tag{27}$$

$$y = C_2 x + D_{21} w + D_{22} u \tag{28}$$

From these equations, we get $\mathrm{P} = \begin{bmatrix} A & B_1 & B_2 \\ C_1 & D_{11} & D_{12} \\ C_2 & D_{21} & D_{22} \end{bmatrix}$.

3.6 Derivation of State-Space Model of Augmented Plant

In the block diagram of the model in Fig. 3, G is the nominal plant. Let

$$G = \begin{bmatrix} A_G & B_G \\ C_G & D_G \end{bmatrix}, W_e = \begin{bmatrix} A_1 & B_1 \\ C_1 & D_1 \end{bmatrix}, W_u = \begin{bmatrix} A_2 & B_2 \\ C_2 & D_2 \end{bmatrix}, W_d = \begin{bmatrix} A_3 & B_3 \\ C_3 & D_3 \end{bmatrix} \quad (29)$$

The equations for a plant in Fig. 3 which has two reference signals r_1, r_2 and one control signal u are as follows:

$$\dot{x}_G = A_G x_G + Br_1.r_1 + Br_2.r_2 + Bu.u \quad (30)$$

$$y_G = C_G x_G + Dr_1.r_1 + Dr_2.r_2 + Du.u \quad (31)$$

And from the block diagram in Fig. 3

$$\dot{x}_1 = A_1.x_1 + B_1.(r_1 - y_G) \quad (32)$$

$$= A_1.x_1 + B_1.r_1 - B_1.y_G \quad (33)$$

$$= A_1.x_1 + B_1.r_1 - B_1.(C_G x_G + Dr_1.r_1 + Dr_2.r_2 + Du.u) \quad (34)$$

$$\implies \dot{x}_1 = A_1.x_1 + B_1.r_1 - B_1.C_G x_G - B_1.Dr_1 r_1 - B_1.Dr_2 r_2 - B_1.Du.u \quad (35)$$

$$z_1 = C_1 x_1 + D_1.(r_1 - y_G) \quad (36)$$

$$= C_1 x_1 + D_1.r_1 - D_1.y_G \quad (37)$$

$$= C_1 x_1 + D_1.r_1 - D_1.(C_G x_G + Dr_1.r_1 + Dr_2.r_2 + Du.u) \quad (38)$$

$$\implies z_1 = C_1 x_1 + D_1.r_1 - D_1.C_G x_G - D_1 Dr_1.r_1 - D_1 Dr_2.r_2 - D_1 Du.u \quad (39)$$

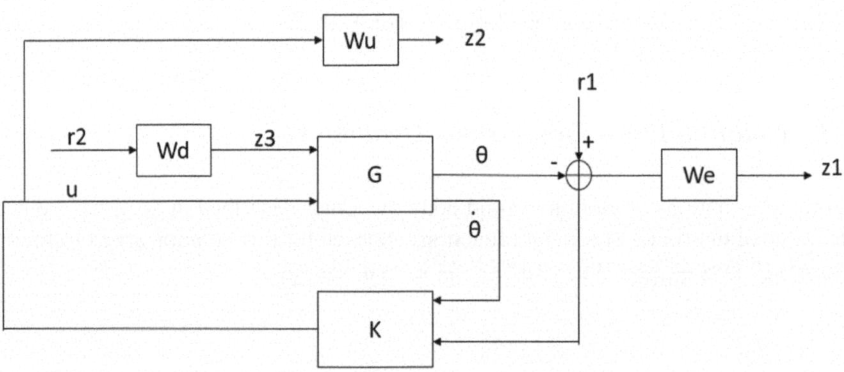

Fig. 3 Block diagram of model used to implement H-infinity control

$$\dot{x}_2 = A_2.x_2 + B_2.u \qquad \dot{x}_3 = A_3.x_3 + B_3.r_2 \tag{40}$$

$$z_2 = C_2.x_2 + D_2.u \qquad z_3 = C_3.x_3 + D_3.r_2 \tag{41}$$

Hence, the state-space model of plant augmented with weight funcions is as follows:

$$\begin{bmatrix} \dot{x}_G \\ \dot{x}_1 \\ \dot{x}_2 \\ \dot{x}_3 \\ z_1 \\ z_2 \\ y_{G1} \\ y_{G2} \end{bmatrix} = \begin{bmatrix} A_G & 0 & 0 & Br_2.C_3 & Br_1 & Br_2.D_3 & Bu \\ -B1.C_{G1} & A_1 & 0 & -B_1.Dr_2.C_3 & (B_1 - B_1.Dr_1) & -B_1.Dr_2.D_3 & -B_1.Du \\ 0 & 0 & A_2 & 0 & 0 & 0 & B_2 \\ 0 & 0 & 0 & A_3 & 0 & B_3 & 0 \\ -D_1.C_{G1} & C_1 & 0 & -D_1.Dr_2.C_3 & (D_1 - D_1.Dr_2) & -D_!.Dr_2.D_3 & -D_1.Du \\ 0 & 0 & C_2 & 0 & 0 & 0 & D_2 \\ C_{G1} & 0 & 0 & 0 & 0 & 0 & 0 \\ C_{G2} & 0 & 0 & 0 & 0 & 0 & 0 \end{bmatrix} \begin{bmatrix} x_G \\ x_1 \\ x_2 \\ x_3 \\ r_1 \\ r_2 \\ u \end{bmatrix} \tag{42}$$

4 Results

Using the above state-space model of an augmented plant, H-∞ control technique is implemented on launch vehicle by adding each dynamics one at a time. Initially, we start with a rigid body dynamics. While discussing the performance of the system using H-∞ controller, we note the time-domain characteristics like settling time, overshoot, rise time, steady state error, and in margins, we observe aeromargin and gain margin. Also, we observe the poles and zeros of the designed controller and ensure they are on left half of s-plane to achieve a stable system. In the following sections, we discuss the nyquist, bode and step response.

4.1 Nominal Plant : Rigid Body Dynamics

From the equations of motion of rigid body dynamics described in Sect. 2.1, we get the value of matrix G. The weight functions choosen for this iteration are as follows: We = $0.1235 * \frac{(s+0.1)}{(s+0.002)}$; Wu = $0.1 * \frac{(s+0.5333)}{(s+50)}$; Wd = $\frac{(0.1)}{(s+0.245)}$;

$$A_G = \begin{bmatrix} 0 & 1 & 0 \\ \mu_\alpha & 0 & \mu_\alpha \\ \frac{-(L_\alpha + Tt - D)}{mV} & 0 & \frac{-L_\alpha}{mv} \end{bmatrix}, Br_1 = \begin{bmatrix} 0 \\ 0 \\ 0 \end{bmatrix}, Br_2 = \begin{bmatrix} 0 \\ \mu_\alpha \\ \frac{-L_\alpha}{mV} \end{bmatrix}, Bu = \begin{bmatrix} 0 \\ \mu_c \\ \frac{Tc}{mV} \end{bmatrix}$$

$$C_G = \begin{bmatrix} 1 & 0 & 0 \\ 0 & 1 & 0 \end{bmatrix}, Dr_1 = 0, Dr_2 = 0, Du = \begin{bmatrix} 0 \\ 0 \end{bmatrix}$$

Few important parameters used in the analysis are $\mu_\alpha = 0.6361$ and $\mu_c = 0.6857$. The selected weight transfer functions W_e, W_u, W_d are converted to state-space model using $tf2ss$ an inbuilt MATLAB function. Then, put all the values of these matrices into the state-space model of augmented plant of equation (42). The weight functions are selected following the concept discussed in Sect. 3.4 and little trial and error to tune the weights to attain desired overshoot, stable controller with its poles and zeros in left half of s-plane. The nyquist plot is cutting fourth quadrant because of the phase lead provided by designed controller in frequency range 10^0 to 10^3(rad/s) as seen in bode plot of the controller transfer functions (Fig. 4; Table 1).

4.2 *Nominal Plant : Rigid Body + Actuator + 2 Bending Modes + 1 Slosh Dynamics*

In order to reduce the order of the system, we have chosen first-order equivalent as an actuator instead of a standard second-order equivalent. The actuator dynamics used is as follows: $\frac{\delta}{\delta_c} = \frac{50}{s+50}$. While reducing the system order, we try to match gain or phase characteristics of the system. Here, we do phase match equivalence because the gain of the system is 1 for large range of low frequency region but the phase changes.

As the launch vehicle is a continous body, the system will have elastic motion, thereby involving flexible body dynamics. We introduce two bending mode dynamics to the system with frequency of bending mode 1 is 13.188rad/s, bending mode 2 is 24.9944rad/s. The coefficient b_i of bending mode dynamics transfer function in Fig. 5 is $b_i = \frac{Tc*\phi_i}{GM_i}$ where i = 1,2.$\sigma_{1EB} = 0.0608$, $\sigma_{2EB} = 0.089028$, $\sigma_{1RGP} = 0.0009954$ and $\sigma_{2RGP} = -0.020813$. Since the liquid fuel undergoes sloshing, we include slosh dynamics with freqeuncy 6.4085rad/s to the system and analyse the response of the system (Fig. 6; Table 2).

The weight functions choosen for this iteration are as follows: Wu = $10*\frac{(s+1.5)}{(s+50)}$; We = $0.1235*\frac{(s+1)}{(s+0.075)}$; Wd = $\frac{(0.1)}{(s+0.245)}$;

The rise time is an indicator of the rapidity of the closed loop system response, and ideally, it shall assume values as small as possible. The smaller rise time means higher bandwidth which means better tracking of the reference signal, the settling time and the maximum overshoot are also desired to be minimal (Fig. 6; Table 2).

The aeromargin of the system is the gain reduction margin which is calculated by tedious process of iterating the gain value in the simulink block and noting the value of gain for which the time-domain response has sustainable oscillations. The desired

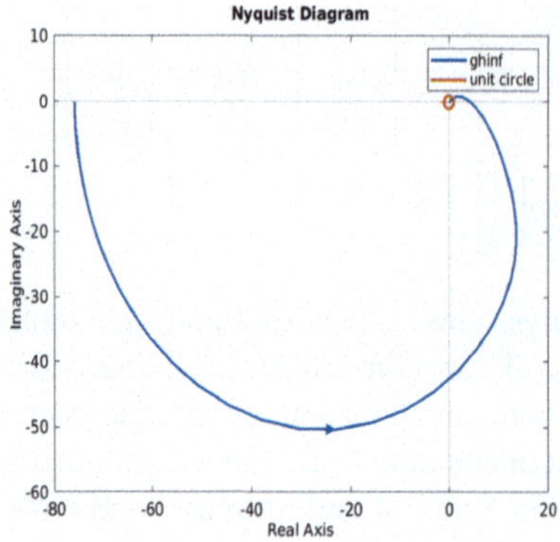

(a) Nyquist plot of the system with H-∞ control

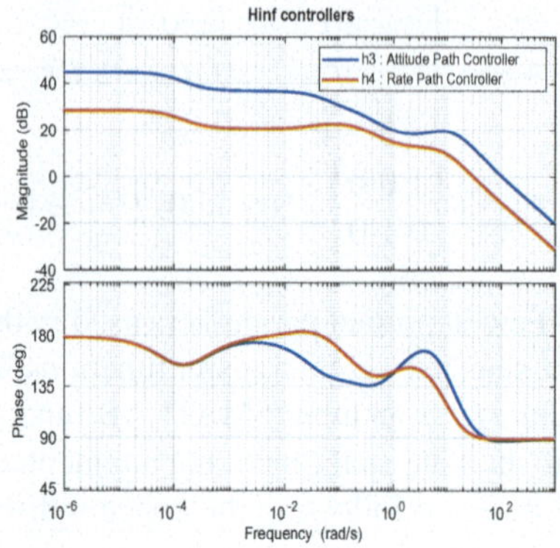

(b) Bode plot of the systhesized H-∞ controllers

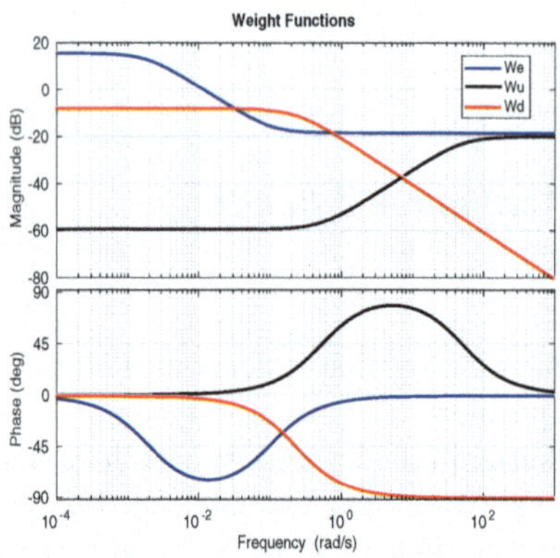

(c) Bode plot of the selected weight functions W_e, W_u, W_d

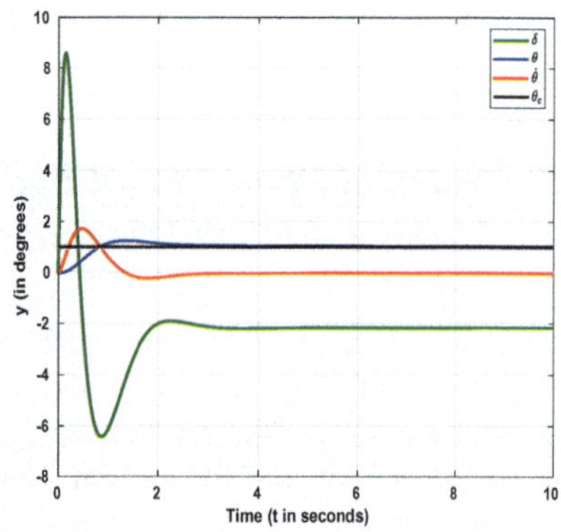

(d) Step response of the system with $\alpha_w = 1^\circ$

Fig. 4 Various plots analysing the H-∞ control technique with nominal plant having only rigid body dynamics

Table 1 Performance metrics of the augmented plant system involving only rigid body dynamics with H-∞ control

Performance metric	Value
Aeromargin (in dB)	12.39577
Gain margin (in dB)	24.6089
Settling time (in seconds)	13.464
Rise time (in seconds)	1.647
Peak value	1.23
Overshoot (%)	23%
Steady state value	1.034
Steady state error (e_{ss})	3.4%

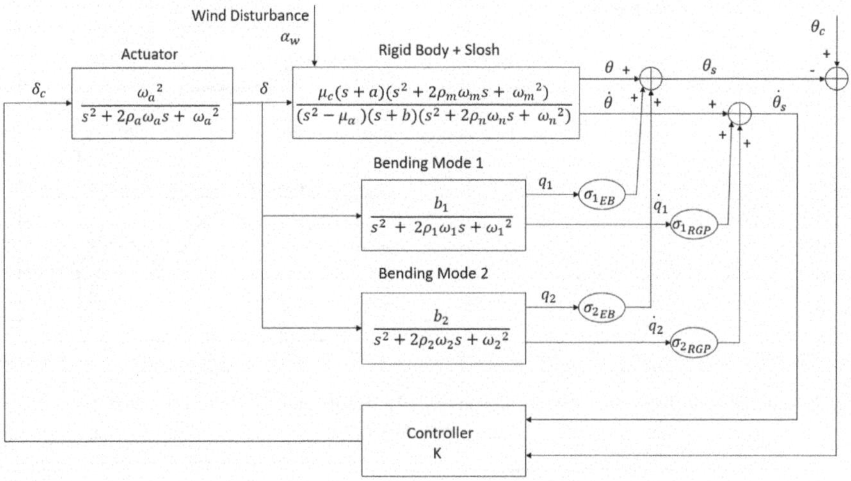

Fig. 5 Control loop block diagram without augmentation of weight functions

aeromargin for any system to be stable is a minimum of 6dB. Another important characteristic is steady state error which is desired to be less than 5% of the reference value. Also here, due to the presence of a pole very close to origin in the controller, this acts like an integrator, and hence, the model order reduction methods can be applied to reduce the order of the whole system. The presence of an integrator also helps in reduction of steady state error (Fig. 7).

5 Simulink Model

6 Conclusion and Future Work

In this paper, the H-∞ control technique is discussed in detail. The criteria for the selection of weights discussed in Sect. 3.4 are a way to get the initial assumption of the weights, but tuning of these weights through trial and error is done to achieve desirable overshoots, stability margins and a controller with all its poles and zeros in the left-half plane so that the system is stable. To implement this control technique, the state-space model of the plant augmented with the weight functions is derived in Sect. 3.6. The elements of the matrix are obtained from the dynamics involved in the launch vehicle and the weight functions. Gain margin, aeromargin are frequency domain performance metrics, whereas rise time, settling time, overshoot and steady state error are the time-domain performance metrics used to understand the stability and performance of the system. Frequency-dependent weight functions and aeromargin are used to analyse the robustness of the system using H-∞ control technique.

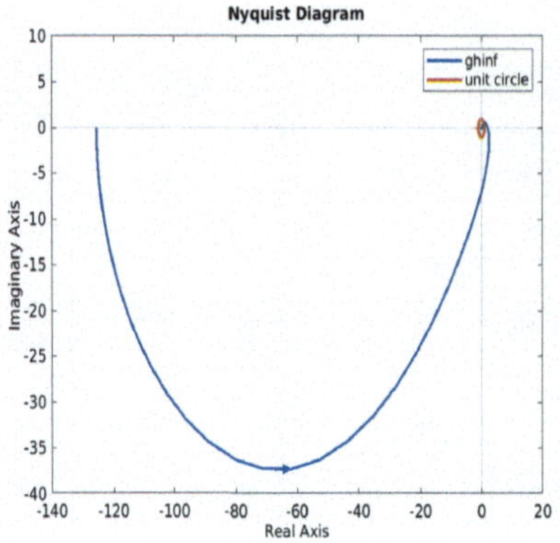

(a) Nyquist plot of the system with H-∞ control

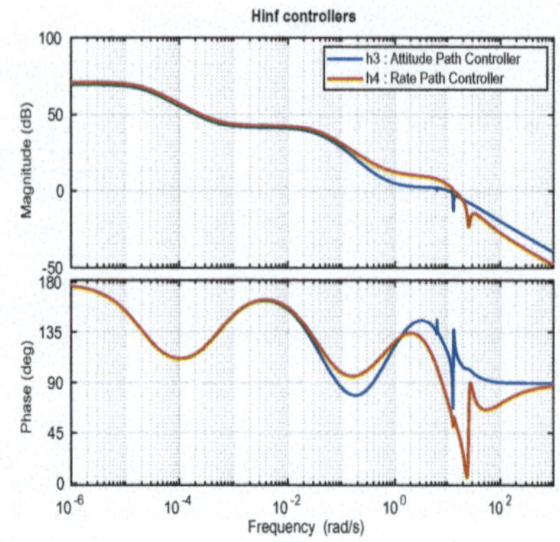

(b) Bode plot of the systhesized H-∞ controllers

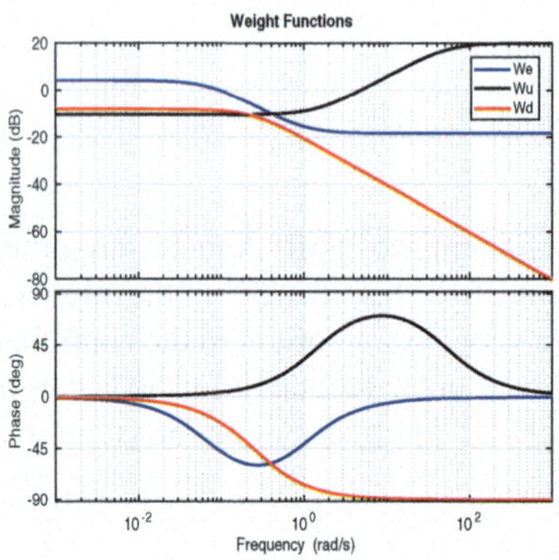

(c) Bode plot of the selected weight functions W_e, W_u, W_d

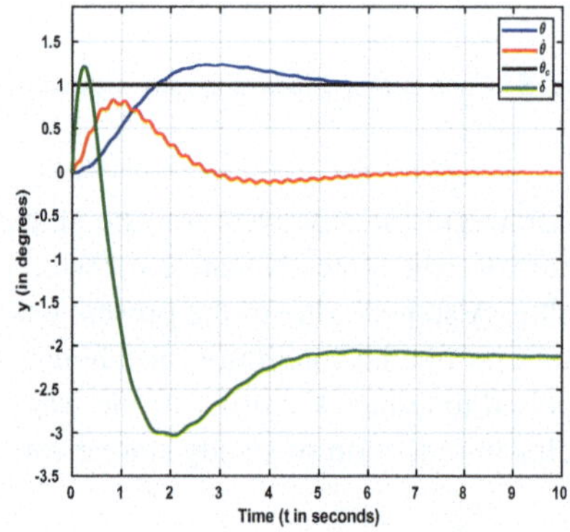

(d) Step response of the system with $\alpha_w = 1°$

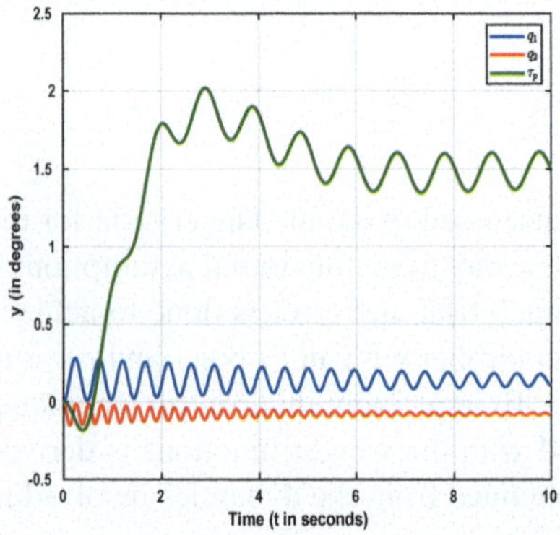

(e) Step response of the system with $\alpha_w = 1°$

Fig. 6 Various plots analysing the H-∞ control technique with nominal plant having rigid body + actuator + 2 bending mode + slosh dynamics

Table 2 Performance metrics of the augmented plant system involving rigid body + actuator + 2 bending mode + slosh dynamics with H-∞ control

Performance metric	Value
Aeromargin (in dB)	6.558
Gain margin (in dB)	14.32
Settling time (in seconds)	–
Rise time (in seconds)	1.098
Peak value	1.239
Overshoot (%)	23.9%
Steady state value	1.016
Steady state error (e_{ss})	1.6%

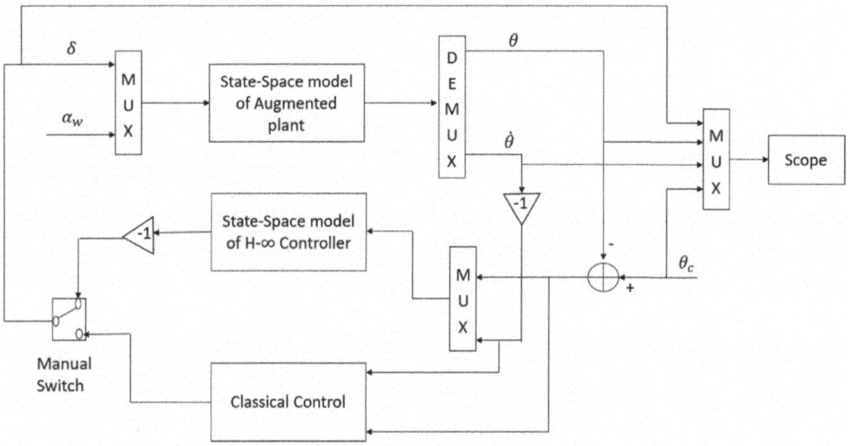

Fig. 7 Simulink model of system implementing both classical and H-∞ control

The present work can be extended by studying model order/ controller order reduction technique to conveniently implement H-∞ control technique for higher order systems.

References

1. Engelberg, S.: Tutorial 15: control theory, Part 1,2. IEEE (2008)
2. Bhat, M.S., Jaisimha, B.S., Kumar, S.R.V.: Optimal Digital Autopilot for Satellite Launch Vehicles During Atmospheric Phase (1992)
3. Arthur, L.: NASA CR-826 Analysis and Design of Space Vehicle Fight Control Systems. San Diego, California (1967)
4. Hespanha, J.P.: Linear Systems Theory. Princeton University

5. Khalate, A.A., Rao, K.K.: Robust Autopilot for Atmospheric flight of Launch Vehicle based on H H_∞ Approach, Centre for Artificial Intelligence and Robotics, Bangalore
6. Deodhare, G., Patel, V.V.: A "Modern" look at Gain and Phase margins: H_∞/μ Approach , Centre for Artificial Intelligence and Robotics, Bangalore
7. Wie, B., Mark, W.: Analysis and design of launch vehicle flight control systems (2008)

Design and Development of the Pineapple Harvesting Robotic Gripper

Francis Kurbah, Shemphang Marwein, Teiborlin Marngar, and Bikash Kumar Sarkar

Abstract Applications of robot in various fields like industrial automation, mining, space application, medical science, and agriculture application have been increased with recent need and technological development. Recently, various researches reported the development of agricultural robot with suitable gripper for particular applications, like tomato harvesting, strawberry harvesting, apple harvesting, cherry harvesting, eggplant harvesting, cabbage harvesting, cucumber harvesting, and crop cutting. A lot of scope still is there to improve agriculture root. Development of the robot component with 3D printing increased due to first production capability, complex shape can be manufactured very easily. Recently, fused deposition modeling method is widely used for 3D printing to manufacture complicated parts of robot to develop working model. It has been found polylactic acid printing material with controlled printing condition that provides good strength of the printed parts. The Arduino UNO-based controller for actuation of the robotic system has become popular due to low cost, less energy requirement, and its simplicity. The northeastern region of India cultivates a lot of pineapples. Pineapple cultivation increased rapidly last two decades. The regions are mainly covered by hills, so the access of pineapple field by human has difficulty. The pineapple leaves are also very rough, which may cause human injury. Development of the robot for pineapple cropping can improve productivity and human comfort. Special type of gripper design has to be done for pineapple cropping application. In the present study, three fingers gripper cum cutter has been developed for pineapple harvesting application. Model of the gripper has been developed in CREO software based on the dimensional survey of the pineapple. Mechanical analysis of the 3D printed gripper components has been done to obtain final design of the gripper component. The gripper component has been manufactured through FDM 3D printing. Finally, the 3D printed gripper components have been assembled to develop working model. The Arduino UNO and Bluetooth module are integrated with servo motors for controlling the finger components, cutter arm, and cutter rotor.

F. Kurbah · S. Marwein · T. Marngar · B. K. Sarkar (✉)
NIT Meghalaya, Shillong 793003, India
e-mail: bikash.sarkar@nitm.ac.in

J. Gu et al. (eds.), *Communication and Control for Robotic Systems*, Smart Innovation, Systems and Technologies 229, https://doi.org/10.1007/978-981-16-1777-5_28

1 Introduction

Robots are perceptive programmable machine, which can perform variety of tasks with dull, dirty, dangerous environment. Robots can be used in various fields like industrial automation, mining, space application, medical application, and agriculture application due to recent need and technological development. Robots can be used to perform various tasks in agricultural activities like cultivation, transplanting, spraying, and selective harvesting [1–3]. Some researcher reported the implementation of agricultural robot for reduction of labor cost [1, 3]. Labor shortage has been found as another aspect to develop agriculture robot to maintain productivity and to reduce production cost. Agricultural robots experienced an unstructured environment in the open space. It has a great impact on topography, soil structure, and landscape property [1]. Changes of environment like variation of temperature, humidity, wind are experienced in the outdoor environment. The agricultural robot design development and controller design have to be done judiciously to operate efficiently. The phases of the agriculture are soil preparation, planning, production, harvesting, and material handling. Each step may perform by different robot, or some of the task may perform by a special type of robot. Some of the agricultural robot design especially harvesting robot design needs to design with the consideration of special application. Mechanical method of fruit harvesting has disadvantages like of damage of fruits [2]. Fu et al. have developed [2] robot with image acquisition system to detect apple for harvesting application. Recently, various researches reported with agricultural robot and development of the suitable gripper with various applications, like tomato harvesting, strawberry harvesting [3], crop cutting. Agricultural robots, i.e., fruit harvesting robots, are found costly and less efficiency [4]. To improve efficiency, some researchers are proposed multi-arm mobile robot for harvesting application [5]. Nagochi et al. reported multi-cooperative robot along with master–slave robot for agricultural application [6]. Limited research reported as fruit harvesting robot development and implementation. Some researchers reported research on orange [7, 8] harvesting robot, apple harvesting robot [9, 10], date palm harvesting [11], tomato harvesting robot [4], strawberry harvesting root [3], cherry [12], eggplant [13], cabbage [14], cucumber [15].

Recently, 3D printing technology widely used with the replacement of conventional manufacturing [16–19] process. The 3D printing can be used manufacturing of complex shape parts applied to biomedical application, aerospace, automotive education, and [18] robotics. The 3D printing process has advantages like low cost, design flexibility, rapid manufacturing from [18] model, and environment friendly. The fused deposition modeling (FDM) is one of the popular methods of printing, which processed layer by layer deposition of material through heated nozzle [19]. The researchers are demonstrated that the mechanical property of printed material depends on the method of printing and material used for printing [20]. It has been observed that the FDM method with polylactic acid (PLA) material widely accepted. The printing parameter optimization was studied by various researchers to improve mechanical properties [19].

Implementations of the control through data acquisition and sensor integration have been found as another important issue for robotic system operation. Depending upon the force requirement, various actuators may be used to generate motion in the robot system. Hydraulic actuators are generally used for high force requirement. Electrical actuator used for low force requirement with precious control. Electrical servo motor with pulse width modulation (PWM) control has been widely used for position and direction control in robotic system. Recently, Arduino UNO-based controller for robotic system control becomes very popular due to simple, low cost, and its effectiveness.

Pineapple cropping has been increased significantly last two decades in India, especially northeast India. Pineapple cultivation required huge rainfall and 15 °C to 30 °C temperature, which is suitable for northeast India. The most of the parts of northeast India is mainly covered by hills, so the access of pineapple field by human has become very difficult. The pineapple leaves are also very rough, which may cause human injury. Development of the robot for pineapple cropping can improve productivity and human comfort. Special type of gripper design has to be done for pineapple cropping application.

In the present study, three fingers gripper cum cutter has been developed for pineapple harvesting application. Model of the gripper has been developed in CREO software based on the dimensional survey of the pineapple. The working model of the gripper has been developed by 3D printing of the gripper components. Finally, the 3D printed model integrated through Arduino UNO, Bluetooth module in Blynk platform for motion generation and control application. The performance of the gripper finger motion and cutter movement has been studied.

2 System Description and Modeling

Schematic diagram of a pineapple harvesting robot has been shown in Fig. 1. It has been depicted from Fig. 1 the considered pineapple harvesting manipulator combination of primary and secondary manipulator. The robot consists of tray to keep pineapple, which has been situated on a plate or base of main parallel manipulator. The tray along with the legged manipulator has been considered as primary manipulator. The secondary manipulator connected with a gripper. The gripper consists of three fingers and a cutter. The gripper has to hold the pineapple first, and then the cutter arm and cutter blade will be activated to cut the pineapple from its root. The gripper manipulator creates trajectory to keep the pineapple in the tray. After one successful plucking of pineapple, primary manipulator will move further to pluck another pineapple. Identification of the pose of the pineapple may be done through vision sensor, which did not consider in the present study. After several successful pineapples plucking, the primary manipulator will move to store the pineapple. The pineapple harvesting manipulator consists of (i) manipulator, (ii) gripper, (iii) sensors, (iv) controller unit, etc. The present study focused on the design of gripper only for pineapple harvesting robot.

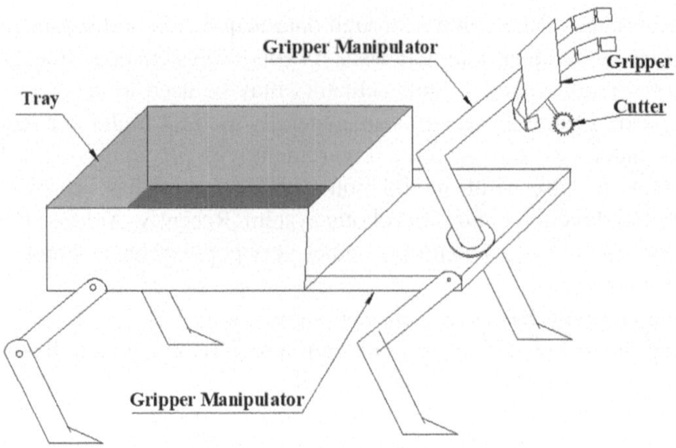

Fig. 1 Schematic of a pineapple harvesting robot

The three-dimensional (3D) model of the present pineapple harvesting gripper has been shown in Fig. 2. The 3D model gripper has been shown as keeping the palm horizontally for better visibility but in actual operating condition, the palm will be in vertical configuration as shown in Fig. 1. The model has been developed in the CREO software. The components of the gripper have been shown in Fig. 2. The gripper consists of three fingers. Each finger consists of three parts, i.e., lower part, middle part, and upper part of the finger. The parts are connected through pin joint to allow relative rotation motion between components to allow close and open of finger to hold and release the pineapple. The three components are connected through strings, which pass through ports having port. The end of string has been connected to the shaft of a servo motor. String from all three figures connected to the common shaft. The rotation of servo motor causes of close the fingers, which helps

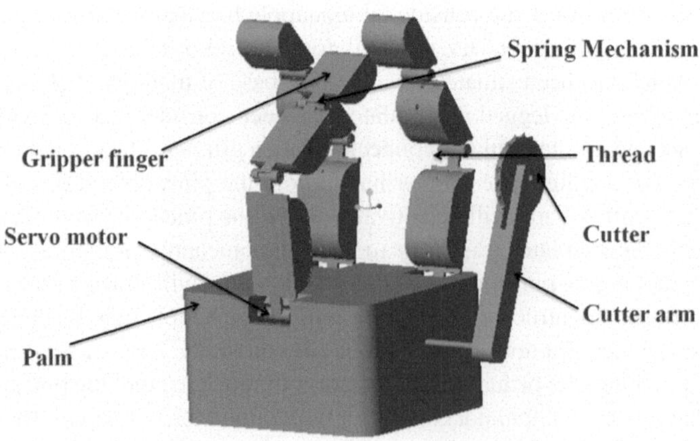

Fig. 2 Three-dimensional model of the pineapple harvesting gripper

Table 1 Dimensions of a standard pineapple cultivated in Meghalaya, India

Specimen / Parameter	Specimen 1	Specimen 2	Specimen 3	Average reading
Radius 1, $d_1/2$ (cm)	5.86	5.77	5.92	5.85
Radius 2, $d_2/2$ (cm)	6.30	6.20	6.28	6.26
Radius 3, $d_3/2$ (cm)	5.34	5.17	5.25	5.25
Height, h (cm)	21.34	22.86	22.34	22.18
Weigh, m (g)	1258.98	1268.35	1264.73	1263.33

for grasping. The back sides of each finger component are connected by a stainless steel compression spring, 41.9 mm × 4.32 mm, 0.08 N/mm. The spring force allows to the opening of the finger of the gripper, while motor released the string. The palm of gripper consists of hollow part to mount servo motors. Three servo motor mounted in the palm, i.e., finger servo, cutter arm servo, and cutter servo. These three servo motors are connected in the gripper for gripping and cutting action. The figure servo is used to generate motion of figure of the gripper for grasping action. The cutter arm servo provides orientation of the cutter blade, and cutter servo provides rotational motion of the cutter for cutting action. The cutter arm rotates by the second servo, and cutter blade rotates by the third servo motor.

To design the gripper for pineapple harvesting first has to identify the proper dimension. To understand the dimension and size of gripper component, comprehensive survey on pineapple has been carried out. Total 100 samples have been taken for dimension and weight measurement. The dimensions for three samples have been shown in Table 1. The dimension measured as three diameters and height of the pineapple as shown in Fig. 3 and Table 1. The average values of three radii $d_1/2$, $d_2/2$, and $d_3/2$ are found as 5.85 cm, 6.26 cm, and 5.25 cm, respectively. The average height, h, of the pineapple has been found as 22.18 cm. The average weight of the pineapple has been found as 1263.33 gm.

The dimension and weight of the pineapple were used to estimate the dimension of the gripper finger component, palm and cutter arm and cutter. Based on the geometric configuration and weight of the pineapple, a three-figure gripper has been chosen with mechanical grasping process. The outer part of the pineapple is hard with rough surface; hence, the mechanical grasping can be considered as suitable. Electrically actuated three-finger cable and spring-driven gripper have been considered for pineapple harvesting application. Most of the agriculture grippers are designed with two fingers with two degree of freedom (2DOF) mechanism, with point contact with the harvesting component. The present gripper has been designed with three degree of freedom finger with multiple point contact with harvesting item, i.e., pineapple. As the pineapple quite heavy compared to the other fruit, multiple point contact can provide better grasping. Further, the multiple point contact reduced change of mechanical damage of the fruit compare to the single point contact. The estimated component dimension and shape of the gripper and cutter component were finalized through structural analysis and geometrical optimization in CREO software [21]. The

Fig. 3 Sample pineapple for dimension estimation

details of the geometric configuration and mechanical analysis have been presented in Sects. 3 and 4, respectively.

The steps of operation for the manipulator and its gripper have been assumed described with respect to the flowchart as shown in Fig. 4a and b, respectively. The operation of the primary manipulator, secondary manipulator, and gripper has been shown in Fig. 4a, and the operation of the gripper only has been shown in Fig. 4b. The sequence as below,

(a) First the identification of pineapple to harvest through primary manipulator operation.
(b) Identification of the position and orientation of the pineapple.
(c) Activation of the secondary manipulator.
(d) Approach the gripper toward the pineapple.
(e) Finger of the gripper starts moving.
(f) Grasp the pineapple.
(g) Maximum force acting on the fingers for final grasping.
(h) Cutter arm start moving.
(i) Cutter blade starts rotating.
(j) Cutting process started.
(k) Reverse motion of the cutter arm.
(l) Movement of the secondary manipulator.
(m) Release of pineapple.

In the present study, the operation of gripper only has been studied without manipulator operation. Therefore, the step (e) to (k) and (m) has been taken care while studying the operation of the gripper.

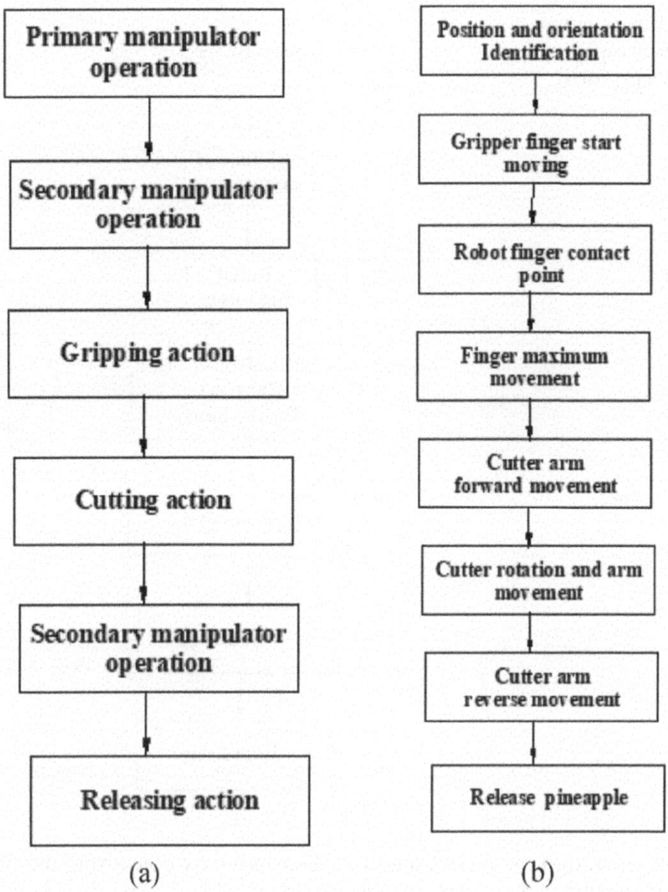

Fig. 4 Block diagram of the **a** manipulator operation **b** gripper operation for pineapple harvesting application

3 Design and Development of the Pineapple Harvesting Gripper

Block diagram of the pineapple harvesting gripper component design procedure has been shown in Fig. 5. Figure 5 depicts the survey data of the pineapple dimension, and its weight has been used for choose the type of the gripper. The proper choice of gripper is very much important for efficient harvesting process. The three-finger gripper has been chosen with cutter in the present case. Each finger has three parts, i.e., each finger with 3DOF configuration. Based on the choice of gripper finger configuration, initial dimension has been chosen. The initial dimension of the figure components used to make the model in CREO software. The developed model has been used for stress analysis of the component. If the obtained stress values in within

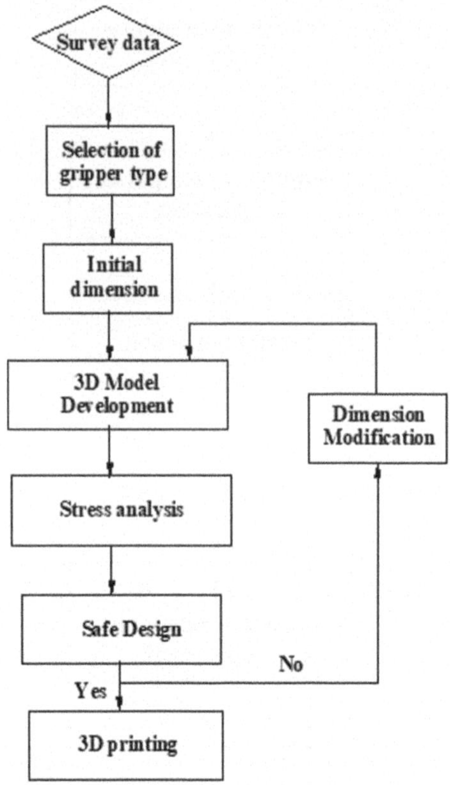

Fig. 5 Block diagram for design of the pineapple harvesting gripper parts

limit of safe operation the model send for 3D printing by converting the file format as '.stl' format. If the obtained stress is not in safe region or the dimension is over specified, then the dimension has to be further modified, and further, stress analysis needs to perform till to get suitable design.

The detailed component of gripper finger has been shown in Fig. 6. The gripper components are designed in CREO software. The bottom part, middle part, and top part of the fingers are shown in Fig. 6a–c, respectively. The middle portion of finger has been shown in Fig. 7. The cutter arm design has been shown in Fig. 8. The dimensions of gripper finger components and cutter arm have been shown in figures, which have been developed in the CREO software. All the indicated dimensions are in mm.

4 Result and Discussion

The designed components are analyzed with the consideration of the load, which has to be applied while operation of the gripper. The load is considered based on the

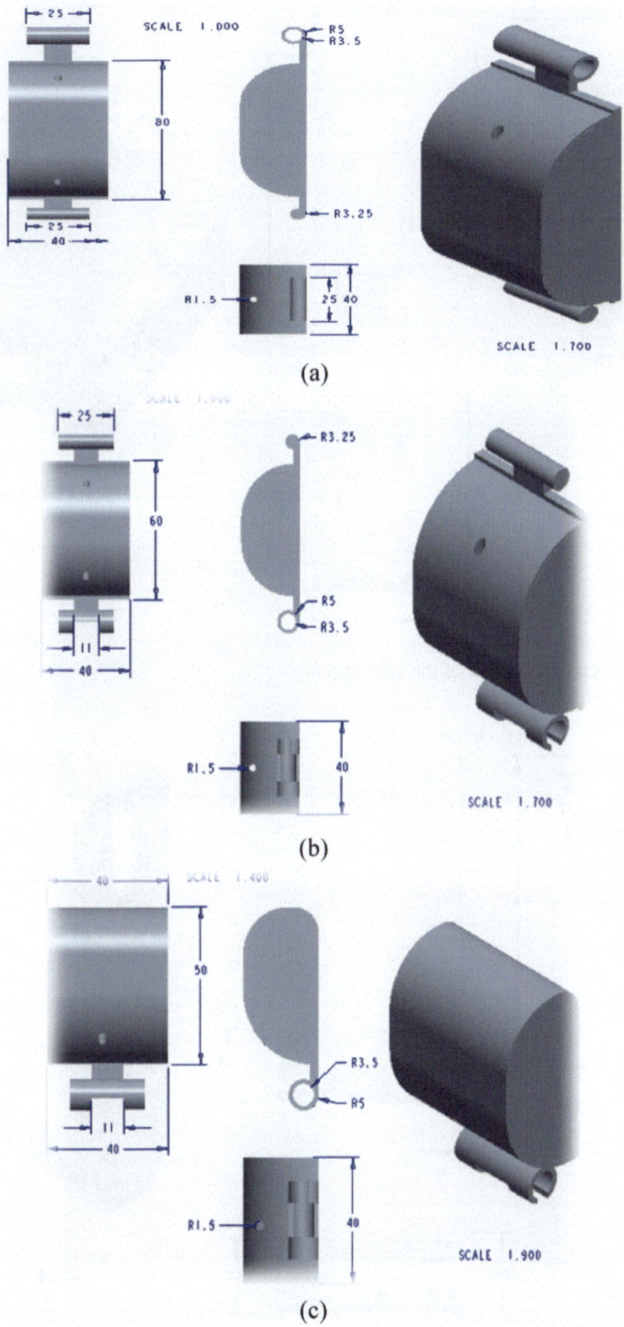

Fig. 6 Three-directional model of the gripper component **a** bottom part of the finger **b** middle part of the finger **c** top part of the finger

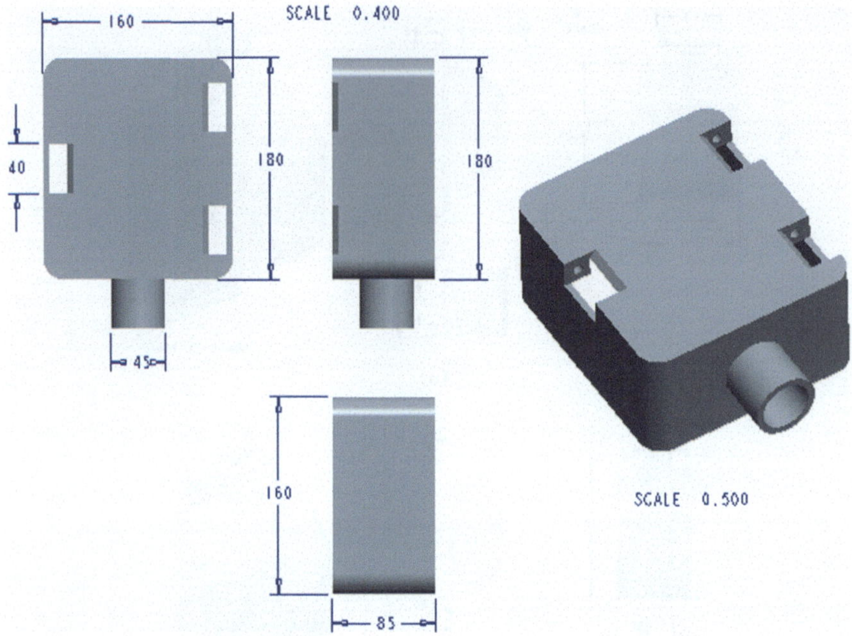

Fig. 7 Three-directional model of the gripper palm

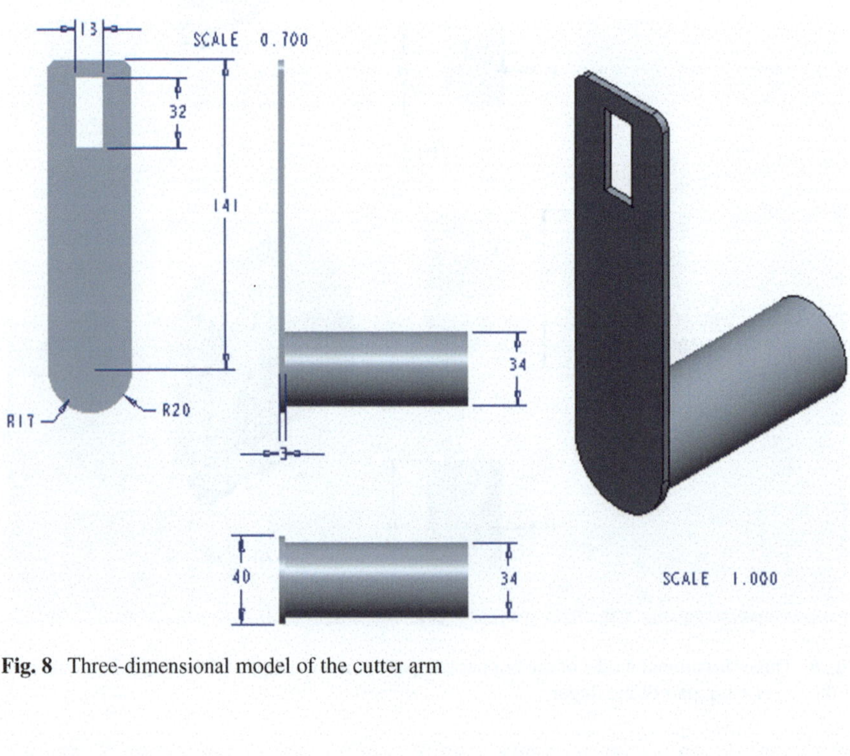

Fig. 8 Three-dimensional model of the cutter arm

approximated weight of the pineapple. The stress and strain of the gripper components are have been shown in Fig. 9 and Fig. 10, respectively. The stresses developed in the top part of the finger, middle part of the finger, lower part of the finger, and for cutter arm have been shown in Fig. 9a–d, respectively. It has been depicted from Fig. 9 the stress developed in the joint part move. Of course the highest stress developed in the joints of the finger parts as the joint components are having less cross sectional area. The stress analysis has been carried out in CREO software.

Strain developed in the gripper finger and cutter arm are shown in the Fig. 10(a)–(c). Figure 10 consists of three parts, the middle part of the finger represented as Fig. 10(a), the lower part of the gripper represented as Fig. 10(b) and the cutter arm of gripper is gripper represented as Fig. 10(c). It has been depicted from Fig. 10 the strain developed by the gripper component within the safe value. As the top part of the gripper finger does not experience by significant load compare to the other part so developed stress has been found significantly less as shown in Fig. 9a. Of course, the developed strain in the top part of the finger also less compares to the other part, i.e., the strain developed in the top part not provided in the paper.

The 3D printed gripper assembly has been shown in Fig. 11a, and metal cutter has been shown in Fig. 11b. The print has been done with PLA material. The 3D printing process was adopted as FVM method with PLA material. The 3D printing has been carried out in Robotics and Mechatronics Lab, NIT Meghalaya. The 3D printing has been done in climate condition with temperature of 23 ± 2 °C, humidity 65 ± 5 percent, printing speed 60 mm/s with infill regulation 0.02 mm and nozzle diameter 0.4 mm, nozzle temperature maintained 170 °C to 190 °C [16]. The maximum allowable stress of the 3D printed component with PLA material as per the parameter setting considered in the present study [16] has been found as 43 MPa. The present value of maximum stress has been found as 28Mpa, which is within the safe value of operation. Of course, the results obtained in Figs. 9 and 10 justify the design and dimension of the gripper components. The metal cutter has been developed for fast and easy cutting action.

The primary components and gripper control circuit are shown in Fig. 12a and b, respectively. The main components are used to build the circuit which are breadboard, Bluetooth module REES52 HC-05, Arduino UNO, Jumper wires, Servo motor MG 995, and SG 90 as shown in Fig. 12a. Apart from the primary component, one wire lead terminal lithium rechargeable battery, 1800mAh 1 53.5 × 35 × 10.4, and battery charger are also used. The components are assembled connected through the circuit as shown in Fig. 12b. The circuit has been shown to control three servo motors, i.e., two MG 995 and one SG 90 motor. The MG 995 servos have been used as gripper servo and cutter arm servo, because it can produce high torque with low rotational speed. The SG 90 servo has been used as cutter rotor, because it can produce high rotational speed. The high rotational speed helps to rotate cutter at high speed for cutting action. The circuit diagram connection between Bluetooth module, Arduino UNO and servo motors through breadboard has been shown in Fig. 12b. The red line represents as power supply, green lines as ground, and cyan lines as signal, which connected by jumper wires. The power supply comes to the circuit through lithium rechargeable

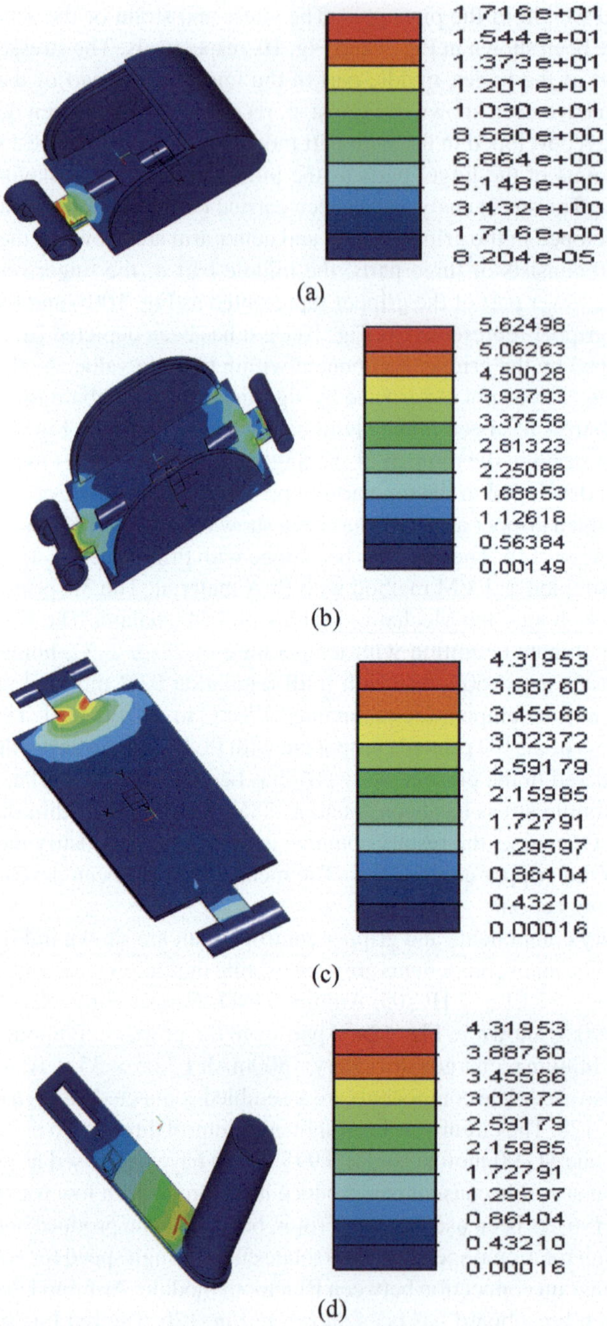

Fig. 9 Stress analysis of the gripper ginger components **a** upper part of finger **b** lower part of finger **c** lower part of finger **d** gripper arm

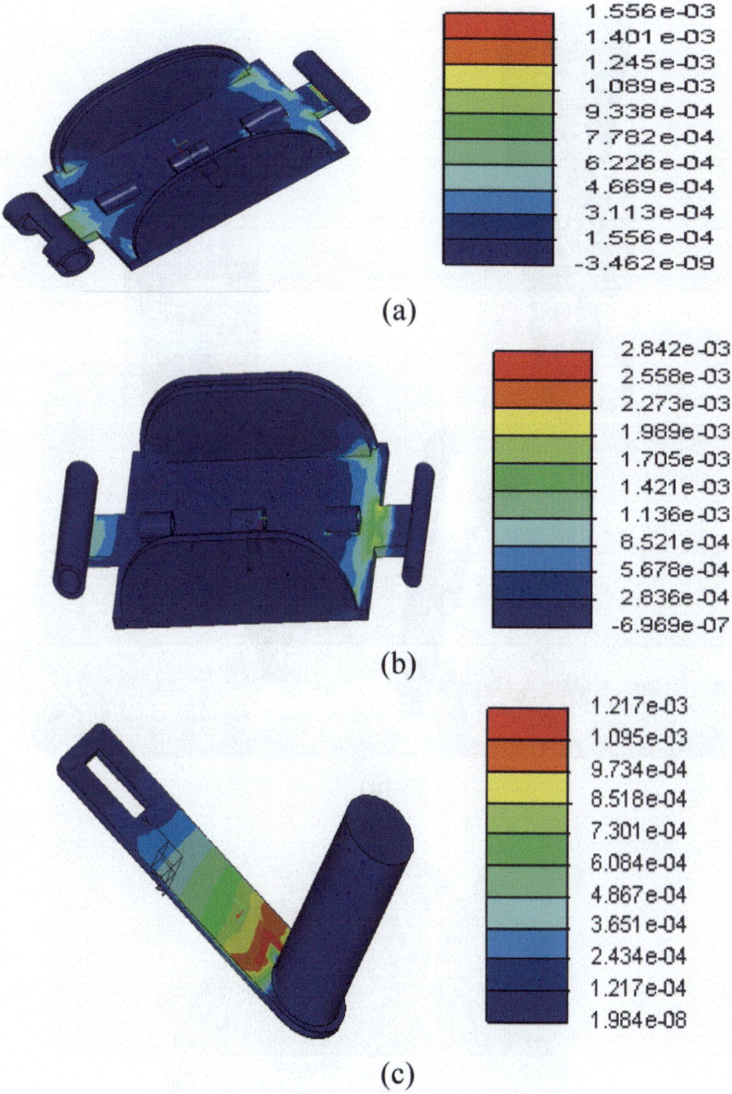

Fig. 10 **a** Strain of the gripper component **b** middle part of the finger

battery. This diagram shows connectivity of Arduino UNO microcontroller to a smart phone or PC via Bluetooth module.

The commands signal can be transmitted in both the directions, from smart phone or PC to Arduino UNO and vice versa. Bluetooth module REES52 HC-05, Android UNO and smart phone or PC synchronized together for control motion of gripper finger. The REES52 HC-05 has been considered as a transceiver that realizes the connection between the smart phone and PC to Arduino UNO. The REES52 HC-05

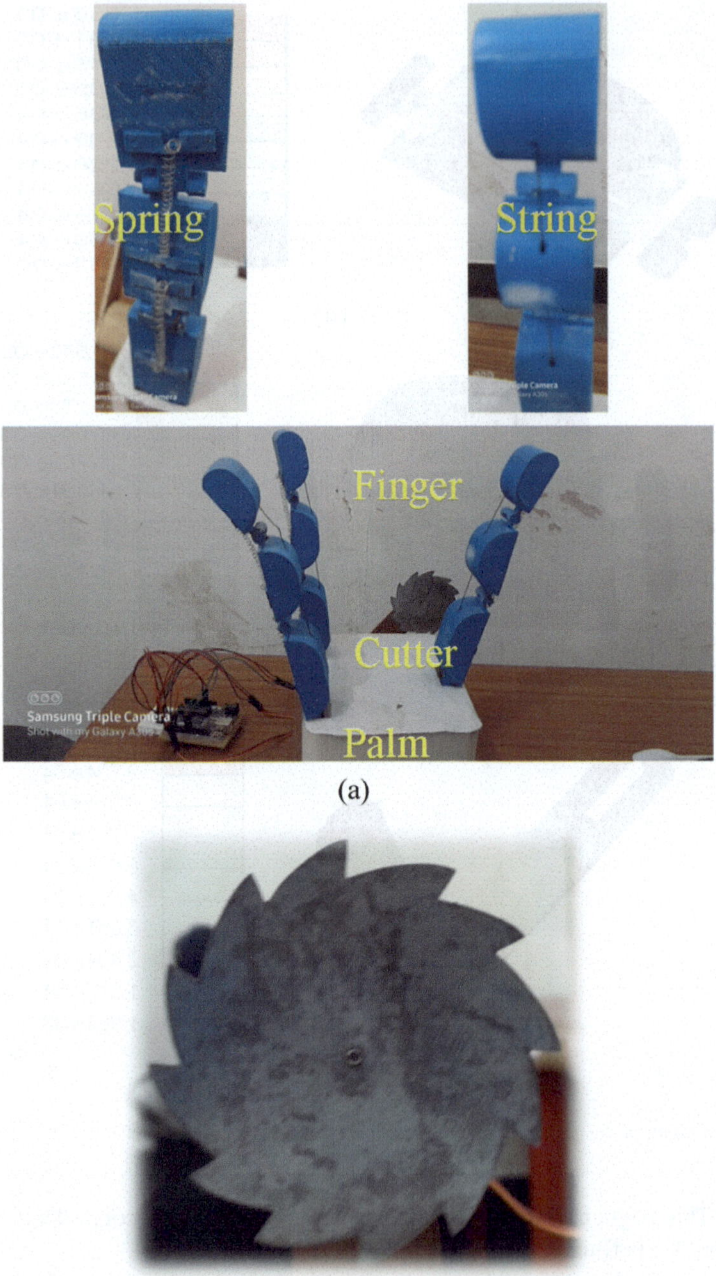

Fig. 11 **a** 3D printed gripper for pineapple harvesting **b** metal cutter

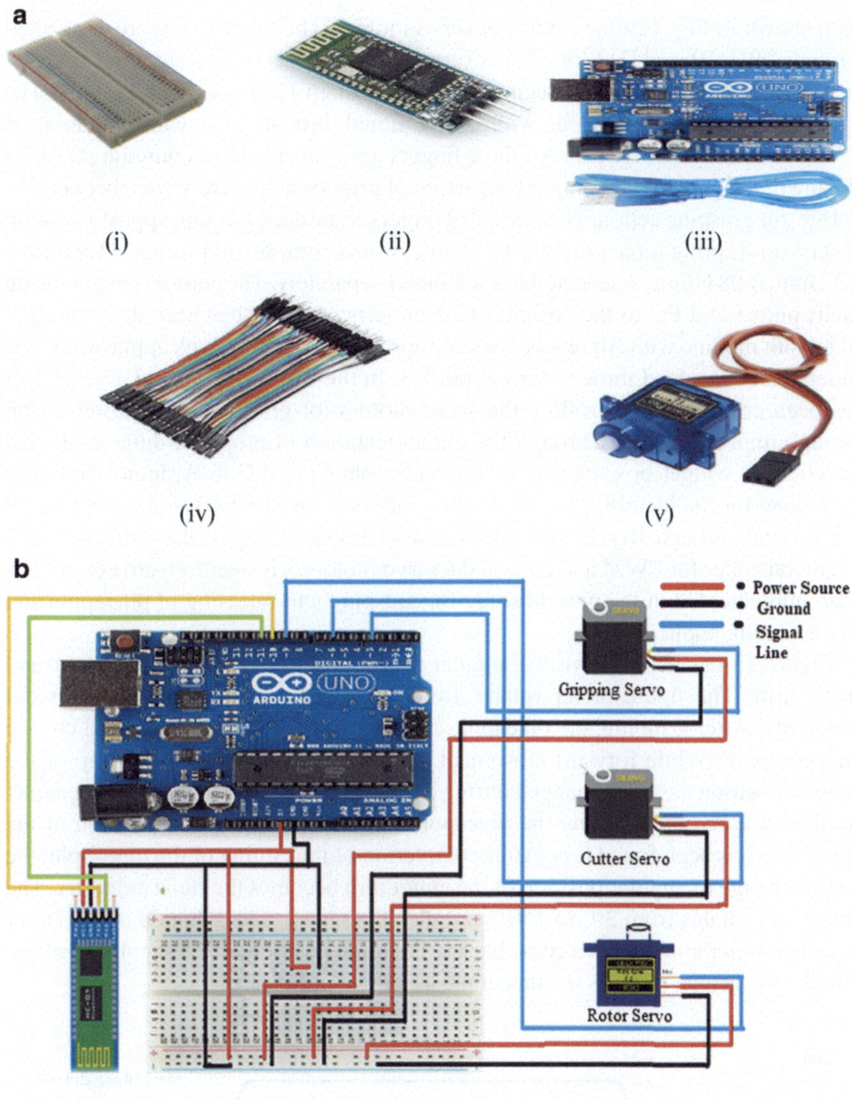

Fig. 12 **a** Primary component of the gripper control circuit (i) breadboard (ii) Bluetooth module (iii) Arduino UNO (iv) Jumper wires (v) Servo motor. **b** Circuit for controlling the pineapple harvesting gripper

works at 3.3 V, but it operates through 5 V signal received from common breadboard as it has internal regulation to convert 5 V to 3.3 V. However, the signals applied on its pins do not exceed 3.3 V as Arduino UNO sends signals on Tx pin with maximum 5 V magnitude. The signals on the out of REES52 HC-05, available on pin Tx, have maximum magnitude of 3.3 V so they can be applied directly to the Rx pin of Arduino UNO, which can support signals up to 5 V. The wiring has been shown in Fig. 12b for control of servo motors. The three servos are connected through D9, D10, and D11 for gripper servo, cutter arm servo, and cutter rotor servo, respectively. Each servo has been connected through GND, Rx and Tx pins. The control of servo has been done with + 5 V signal through pulse width modulation (PWM) as shown in Fig. 12b. All three fingers are connected to a common servo for closing the fingers for grasping. The rotation of gripper servo is restricted between 0° to 180° for grasping action. The release of finger can be done through opposite motion of servo and spring force provided by stainless steel compression spring, 41.9 mm × 4.32 mm, 0.08 N/mm, connected to each finger separately. The connectivity between smart phone and PC to the Arduino UNO microcontroller has been done through Bluetooth module with Blynk open-source application. In the Blynk application, one slider has been added through virtual pin V3. In the Blynk application, one project has been created for controlling the servo motors for grasping and cutting action through importing code. Through the implementation of proper Arduino code and selection of widget box, pairing of the smart phone or PC to Arduino UNO has been done for the identification of devices and its connectivity. The required signal can be sent through Blynk application and REES52 HC-05 to the Arduino UNO microcontroller for PWM and control the servo motors. The feedforward control has been implemented in the present study for grasping and releasing of pineapple and its harvesting application.

Figure 13 has been shown the angular position of the gripper finger driver and cutter arm. The figure driver rotates from 0° to 180° for grip a sample model pineapple. After gripping, the pineapple cutter arm starts movement toward cutting the pineapple. While forward movement of the arm along with cutter, slop of the angular position has been changed during cutting of the bottom part of the pineapple. As the force required cutting the pineapple rate of change angular position of the cutter arm has been found less. After completion of the cutting of the pineapple, the rate of change of angular position of the cutter arm becomes the same as earlier. The cutter arm rotates from 30° to 150°, and then possesses revere motion of the cutter arm. The cutter arm has been come back to its initial position as 30° angular position. Finally, the grippe releases the pineapple as shown in Fig. 13.

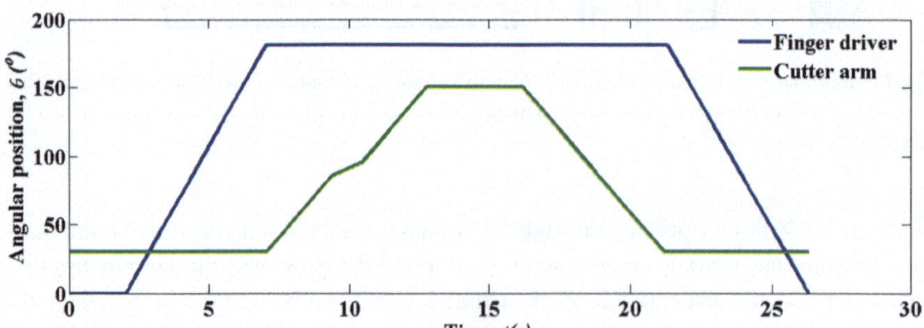

Fig. 13 Circuit for controlling the pineapple harvesting gripper

5 Conclusions

In the present study, the three-finger robotic gripper has been developed for pineapple harvesting application. The gripper has been designed based on the survey on dimension and weight of the pineapple. The gripper components are designed in CREO software. The designed components are performed mechanical analysis, i.e., stress and strain analysis, for design optimization of the gripper components. The maximum stress has been obtained from gripper component as 28 MPa as within the allowable range of 43 MPa for 3D printed PLA material. The gripper components are printed by a 3D printer through FDM process with PLA material. The 3D printing has been done with temperature of 23 ± 2 °C, humidity 65 ± 5 percent, printing speed 60 mm/s with infill regulation 0.02 mm and nozzle diameter 0.4 mm, nozzle temperature maintained 170 °C to 190 °C to ensure the mechanical strength of the printed part. The 3D printed components have been assembled in the form of gripper. The printed gripper is connected to the circuit for motion control of gripper finger, cutter arm, and cutter rotor. Finally, the gripper motion was generated through Blynk application with REES52 HC-05 Bluetooth device and Arduino UNO microcontroller. The feedforward control of the servo motor has been implemented through PWM. The gripper motion has been analyzed through model pineapple grasping, releasing, and cutting process.

References

1. Avital, B., Clement, V.: Agricultural robots for field operations. Part 2: Operations and systems. Biosyst. Eng. **153**, 110–128 (2017)
2. Longsheng, F., Yaqoob, M., Xin, Z., Manoj, K., Qin, Z.: Faster ReCNNebased apple detection in densefoliage fruiting-wall trees using RGB and depth features for robotic harvesting. Biosys. Eng. **197**, 245–256 (2020)
3. Ya, X., Cheng, P., Lars, G., Pål, J.F., Volkan, I.: Development and field evaluation of a strawberry harvesting robot with a cable-driven gripper. Comput. Electron. Agric. **157**, 392–402 (2019)
4. Zhao, Y., Gong, L., Liu, C., Huang, Y.: Dual-arm robot design and testing for harvesting tomato in greenhouse. IFAC-PapersOnLine **49–16**, 161–165 (2016)
5. Zion, B., Mann, M., Levin, D., Shilo, A., Rubinstein, D., Shmulevich, I.: Harvested-order planning for a multiarm robotic harvester. Comput. Electr. Agric. **103**, 75–81 (2014)
6. Noguchi, N., Will, J., Reid, J., Zhang, Q.: Development of a master-slave robot system for farm operations. Comput. Electr. Agr. **44**, 1–19 (2004)
7. Recce, M., Taylor, J., Plebe, A., Tropiano, G.: Vision and neural control for an orange harvesting robot, Proceedings of the 1996 International Workshop on Neural Networks for Identification, Control, Robotics, and Signal/Image Processing (NICROSP '96. p. 467 (1996)
8. Muscato, G., Prestifilippo, M., Abbate, N., Rizzuto, I.: A prototype of an orange picking robot: past history, the new robot and experimental results. Ind. Robot: Int. J. **32**(2), 128–138 (2005)
9. Zhaoxiang, L., Gang, L.: Apple maturity discrimination and positioning system in an apple harvesting robot. New Zealand J. Agric. Res. Abstracts 2007. **50**, 1103–1113 (2007)
10. Bulanon, D.M., Kataoka, T., Ota, Y., Hiroma, T.: A machine vision system for the apple harvesting robot, Agricultural Engineering International: the CIGR Journal of Scientific Research and Development. Manuscript PM 01 006, vol. III,2001.

11. Aljanobi A. A., Al-hamed S. A., and Al-Suhaibani S.A.: A setup of mobile robotic unit for fruit harvesting, 19th international workshop on robotics in Alpe-Adria-Danube Region – RAAD 2010, June 23–25, 2010, Budapest, Hungary
12. Tanigaki, K. Fujiura T., Akase, A. Imagawa J.: Cherry-harvesting robot. Comput. Electron. Agric. pp. 63:65–72 (2008)
13. Hayashi, S., Ganno, K., Ishii, Y.: Tanaka K Robotic harvesting system for eggplants. Jpn. Agric. Res.Q. **36**(3), 163–168 (2002)
14. Murakami, N., Otsuka, K., Inoue, K., Sugimoto, M.: Robotic cabbage harvester. In: Proceedings of international symposium on automation and robotics in bioproduction and processing, vol. 2. Kobe, Japan, pp. 25–32 (1995)
15. Ota, T., Bontsema, J., Hayashi, S., Kubota, K., Van Henten, E.J., Van Os, E.A., Ajiki, K.: Development of a cucumber leaf picking device for greenhouse production. Biosyst Eng **98**(4), 381–390 (2007)
16. Sang, L., Han, S., Li, Z., Yang, X., Hou, W.: Development of short basalt fiber reinforced polylactide composites and their feasible evaluation for 3D printing applications. Compos B Eng **164**, 629–639 (2019)
17. Parandoush, P., Lin, D.: A review on additive manufacturing of polymer-fiber composites. Compos. Struct. **182**, 36–53 (2017)
18. Guo, P., Zou, B., Huang, C., Gao, H.: Study on microstructure, mechanical properties and machinability of efficiently additive manufactured AISI 316L stainless steel by high-power direct laser deposition. J. Mater. Process. Technol. **240**, 12–22 (2017)
19. Peng, W., Bin, Z., Shouling, D., Chuanzhen, H., Zhenyu, S., Yongsheng, M., Peng, Y.: Preparation of short CF/GF reinforced PEEK composite filaments and their comprehensive properties evaluation for FDM-3D printing. Compos. B **198**, 108175 (2020)
20. Eutionnat-Diffo, P.A., Chen, Y., Guan, J., Cayla, A., Campagne, C., Zeng, X., Nerstrasz, V.: Stress, strain and deformation of poly-lactic acid filament deposited onto polyethylene terephthalate woven fabric through 3D printing process. Nat. Res. Sci. Rep. **9**, 14333 (2019)
21. CREO, tutorial, https://static.sdcpublications.com/pdfsample/978-1-63057-093-4-1.pdf.

Three-Axis Wireless Gesture-Controlled Car

Naiwrita Dey, Tridib Dey, Rishav Chakraborty, Sibam Jana, and Abul Hassan

Abstract A novel approach of three-axis gesture-controlled car has been illustrated in this paper. The propose scheme consists of three design modules: transmitter, receiver and the wireless network. The hand glove is integrated with a three-axis digital accelerometer ADXL 345 and NodeMCU-based transmitter unit which sends the reference command to the car unit associated with a Raspberry Pi-based receiver module. A Raspberry Pi 3B + module is used for the same. Human data glove is developed by incorporating six different human hand movements by using the accelerometer. Different features are obtained from the hand gesture data using statistical formulas. The car is driven using Raspberry Pi. Wireless application is based upon MQTT protocol. The wireless network is developed with same gateway address in which all the devices are connected. Hardware prototype of the proposed method is shown here.

1 Introduction

In recent years, a remarkable growth is noticed in the field of human–computer interaction techniques. In particular, hand gesture-based applications have become very popular in many sectors. Many things can be moved and controlled by means of different human gestures. Contactless control of any object driven by the movement of human hand is the basic idea behind this work. Several work has been reported in literature based upon gesture recognition [1]. Data glove is a assembled electronic device including motion sensors which gives the hand movement information. Hand gesture-based robotic surveillance system has been implemented. Raspberry Pi gives the command to move the robotic system in specified directions. ZigBee communication protocol is used to integrate the whole idea [2, 3]. Minimizing the cost,

N. Dey (✉)
Electronics and Communication Engineering, RCC Institute of Information Technology, Kolkata, West Bengal, India

T. Dey · R. Chakraborty · S. Jana · A. Hassan
Applied Electronics & Instrumentation Engineering, RCC Institute of Information Technology, Kolkata, West Bengal, India

J. Gu et al. (eds.), *Communication and Control for Robotic Systems*, Smart Innovation, Systems and Technologies 229, https://doi.org/10.1007/978-981-16-1777-5_29

smart phone sensor has been used to capture hand gesture movements for recognition [4]. This research is pioneered further with the application of nine-axis motion sensor for gesture data acquisition [5]. Feature extraction is an important aspect for this study. Different algorithms have been reported in literature [6, 7]. Apart from hand gesture data acquisition and recognition of it, control of the robotic system from remote location is also a vital thing to be considered [8]. Internet of thing is an essential part for connecting the car with the Internet by allowing the user to control it from anywhere [9, 10]. Using Wi-Fi, these wireless systems are regulated. MQTT protocol is implemented for wireless communication [11, 12]. Use of different controller like NodeMCU and ZigBee module has been reported. Raspberry Pi-based system integration is carried out [13, 14].

In this paper, the idea of hand gesture-controlled robotic car is implemented using Raspberry Pi module. This kind of application can be modified and used by the physically challenged people who are unable move on their own.

The paper is organized as follows. In Sect. 2, the overall methodology of the work is explained. In Sect. 3, the hardware prototype model is discussed including different subsections. In Sect. 4, the real-time data of hand gesture is shown followed by Sect. 5 with conclusive statement with a future extension idea of this work. References are mentioned at the end.

2 Methodology

Gesture-controlled car is controlled by human hand gestures. Basic idea for the same is that user needs to wear a smart device which includes a sensor known as data glove. The sensor will capture the specific hand movement, and depending upon hand gestures the car will be driven by the actuation signal accepted at the received end. In this present work, the data glove uses a three-axis accelerometer as motion sensor which allows the user to control the car movement in forward, backward, leftward and rightward direction. Transmitter which is data glove and receiver module will have two controllers, and they are connected with each other over Wi-Fi.

The proposed method can be divided into three sections as follows:

A. Data glove
B. Data transmission through WI-FI using MQTT protocol
C. Cart

In this paper, the methodology is explained into three parts. These are as follows:

A. Data glove
B. Data transmission through Wi-Fi using MQTT protocol.
C. Cart/ Car controller by hand gesture using Raspberry Pi.

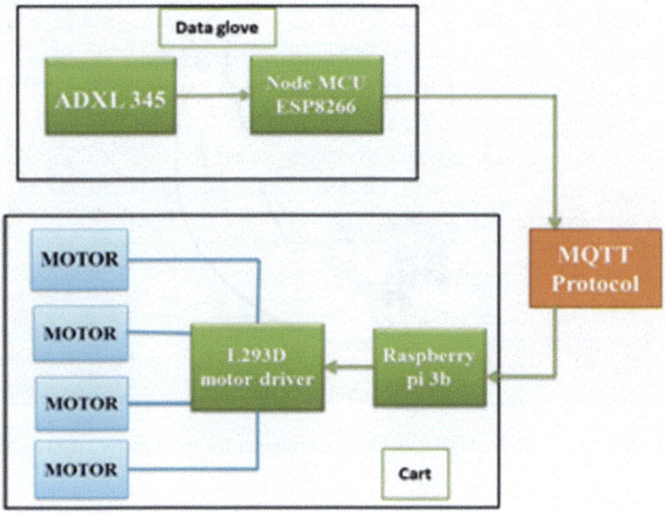

Fig. 1 Block diagram of total system

3 Hardware Prototype of the System

Prototyping of the system can be explained into three parts as follows.

3.1 Data Glove

A three-axis digital accelerometer ADXL345 is used here as motion sensor. The data glove has used a NodeMCU to connect the accelerometer, and the assembly has been placed on the back of the hand to acquire the motion data for six different hand gestures. Preprocessing is done on the captured data. Scatter plot has been made, and outliers have been removed. Several statistical features are calculated for six different hand gestures that have been considered for the present study, and proper gesture is recognized for specific car movement. Block diagram of the whole procedure for data acquisition and statistical feature extraction is shown in Fig. 2. In Fig. 3, the data acquisition part has been shown using accelerometer.

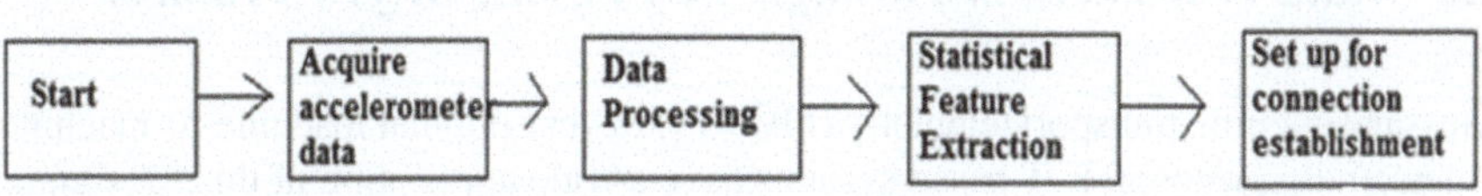

Fig. 2 Block diagram of gesture data acquisition using data glove

Fig. 3 Circuit connection of NodeMCU with accelerometer for hand gesture data acquisition

Mean, difference, standard deviation, variance and average of absolute deviation these five statistical parameters are calculated here for each of the hand gesture.

$$\text{Mean, } \overline{X} = \frac{\sum X_i}{n}, X_i \text{ denotes i}^{\text{th}} \text{no of data sample} \tag{1}$$

$$\text{Difference } X_d = X_1 - X_2 - \ldots\ldots X_n \tag{2}$$

$$\text{Standard deviation } \sigma = \sqrt{\frac{\sum (X - \overline{X})^2}{n}} \tag{3}$$

$$\text{Variance is } \sigma^2 = \frac{\sum (X - \overline{X})^2}{n} \tag{4}$$

$$\text{Average of absolute deviation } D = \frac{\sum |(X - \overline{X})|}{n} \tag{5}$$

Six different hand gestures used for this particular problem are given in Table 1.

3.2 *Data Transmission Through Wi-Fi Using MQTT Protocol*

Message queuing transport telemetry (MQTT) is a very popular machine-to-machine connectivity protocol. IOT-based systems have a wide application of this. Design of this protocol is very lightweight and used for publish/subscribe messaging transport with remote location at the cost of minimum network bandwidth. A small code foot-print is required to implement this. In this paper, MQTT protocol is used to establish

Table 1 Several used hand gestures

Gesture	Hand Shape	Action
G1	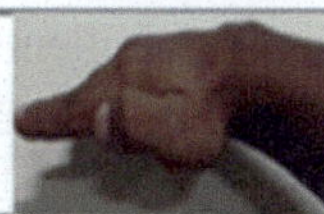	The motor moves forward
G2		The motor moves backward
G3	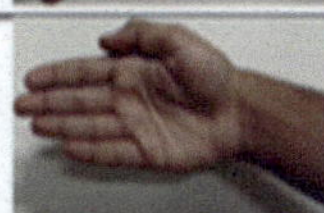	The motor movement stops
G4	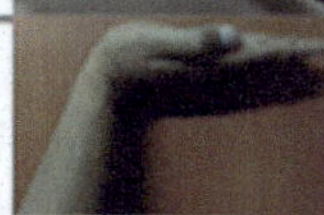	The motor moves right
G5		The motor moves left

the connection between data glove and the car. Figure 4 depicts the implementation of the same in this present work.

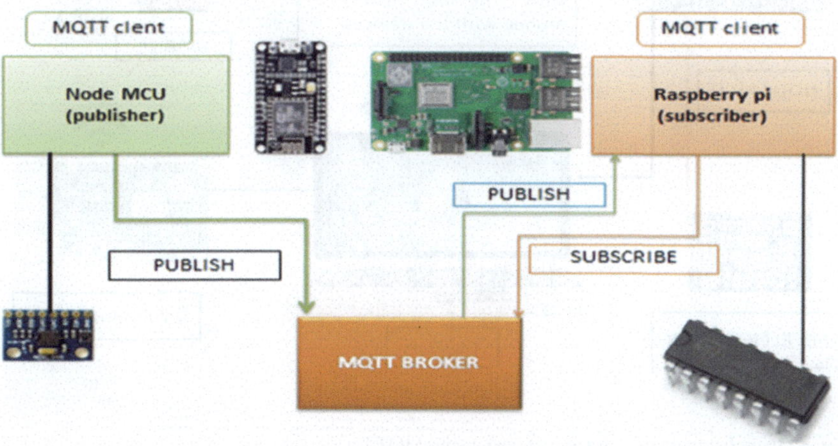

Fig. 4 Block diagram of MQTT protocol used for proposed scheme

3.3 MQTT Broker

The broker plays the vital role in any publish/subscribe protocol. Subjected to specific application, the broker is able to handle several concurrently connected MQTT clients. Authentication and authorization of clients is done by the broker. Figure 5 shows the establishment of wireless network using MQTT.

IP addresses used here for the establishment of Wi-Fi connection among the devices of the proposed system are as follows:

1. IP1 = IP address of Raspberry Pi
2. IP2 = IP address of NodeMCU
3. IP3 = IP address of laptop
4. IP4 = IP address of mobile set

- There are severel steps to follow:
- Step 1: Switch on the mobile hotspot.
- Step 2: By giving the command IP config to the command window, IP configuration has to be checked and must be connected to mobile hotspot.
- Step 3: NodeMCU is also to be configured with the same hotspot.
- Step 4: NodeMCU will ping the Raspberry Pi and only after getting acknowledgement from it actuation signal is to be transmitted to the car.

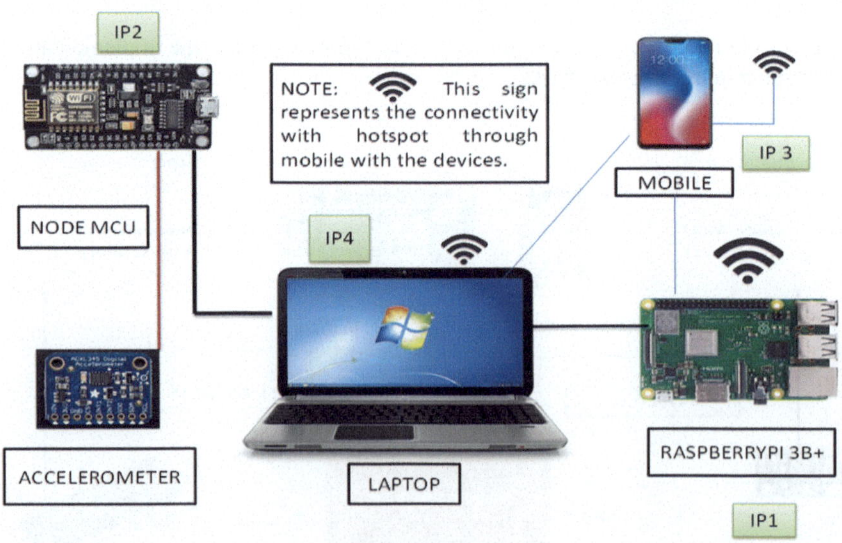

Fig. 5 Establishment of wireless network using MQTT

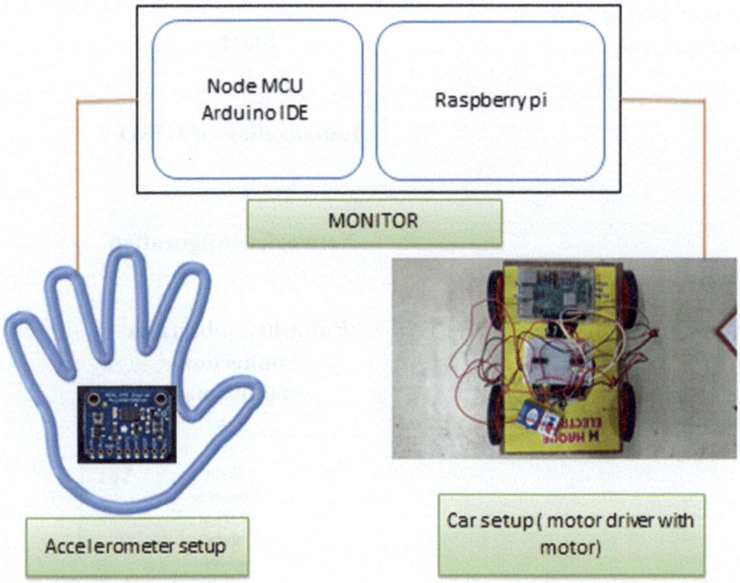

Fig. 6 Raspberry Pi-based car set-up

3.4 *Cart /Car Controller by Hand Gesture Using Raspberry Pi*

The mechanical model of cart has been created using one piece of wooden board, four DC motors are tied by the iron wire below the board, and four motors are connected with four wheels. The upper part of the frame contains Raspberry Pi 3 B, L293d motor driver; both devices are connected, and the motor is connected with motors. In Fig. 6, the car set-up using Raspberry Pi is shown which is controlled by the data glove.

The working algorithm of the receiving end is given in Fig. 7.

3.5 *Experimental Data*

In this present work, six different hand gesture movements have been taken into consideration to control car movement. Three-axis motion data captured for G1 gesture has been shown in Fig. 8a–c.

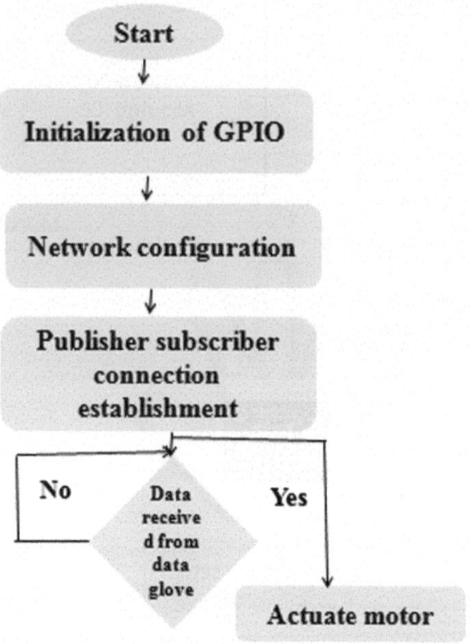

Fig. 7 Algorithm flow chart of receiver end (cart/ car)

4 Conclusion

This work is mainly focused to develop a human hand gesture-based control of car. A three-axis accelerometer-based data glove has been proposed and implemented to detect the significant hand gestures to control the car. Six different hand gestures are used to drive the car in four different directions. From the previous observations, it can be concluded that depending upon different statistical features, different gesture is recognized and control command has been sent to the cart module. MQTT protocol is implemented for wireless transmission. In future this work can be extended with the help of machine learning for more precise feature extraction and control of the cart.

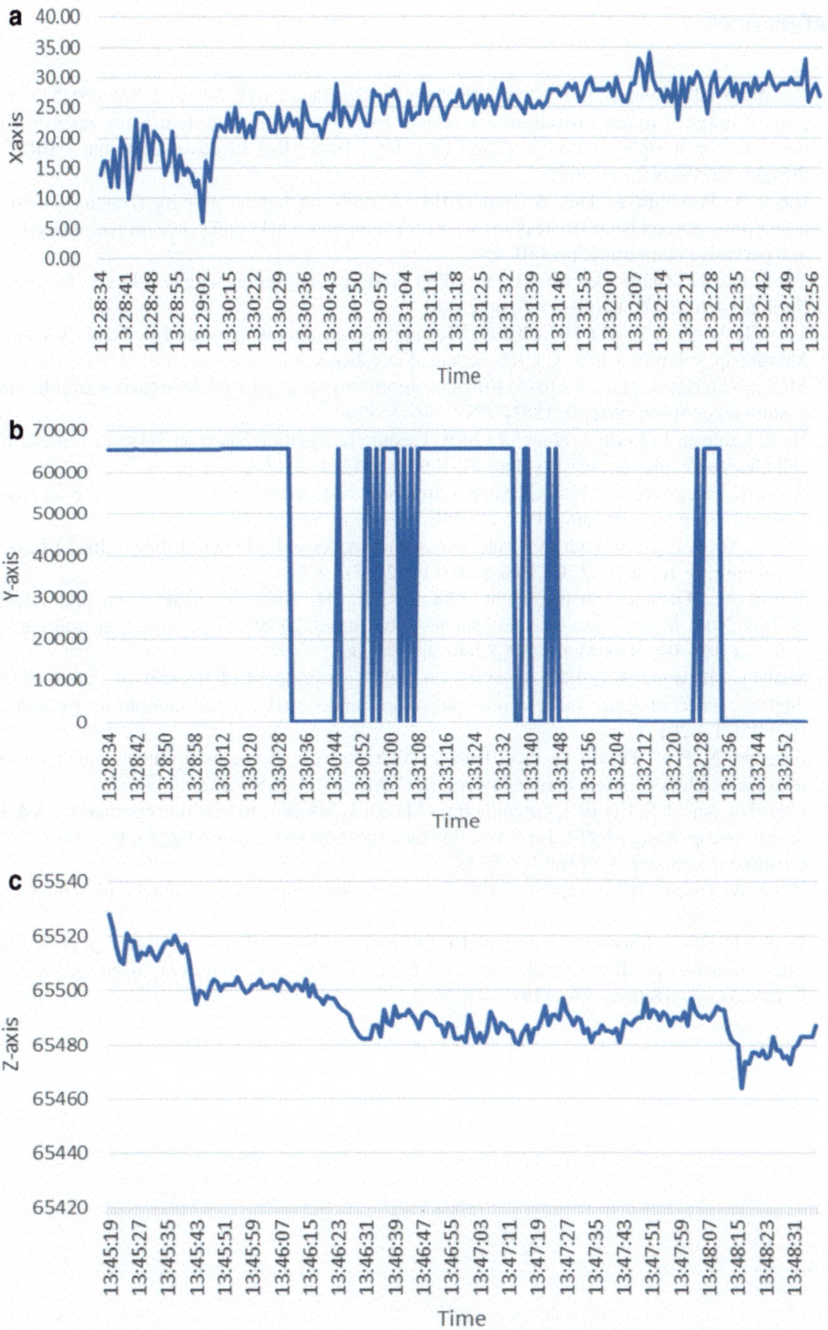

Fig. 8 **a**. Raw data of X-axis in gesture G1. **b** Raw data of Y-axis in gesture G1. **c** Raw data of Z-axis in gesture G1.

References

1. Kausthub NP, Rounak S, Bhakthavathsalam R, Gowranga KH, Saqquaf SM (2015) Hand gesture initiated motor operations for remote robot surveillance system using raspberry pi processor on a zigbee communication link. Int J Innov Res Electron Commun (IJIREC), 2(5):55–62, ISSN 2349–4042
2. Bayat A, Pomplun A, Duc A Tran (2014) A study on human activity recognition using accelerometer data from smart phones. The 11th international conference on mobile systems and pervasive computing, pp 450–457
3. Wu J, Pan G, Zhang D, Qi G, Li S (2009) Gesture recognition with a 3-D accelerometer. Springer, UIC 2009, LNCS 5585, pp 25–38
4. Liu FT, Wang YT, Ma HP (2017) "Gesture recognition with wearable 9-axis sensors". Manuscript submitted June 1, 2016; accepted October 3
5. Marqués G, Basterretxea K (2015) Efficient algorithms for accelerometer-based wearable hand gesture recognition systems. ISSN 978–1–4673–8299
6. He Z, Lianwen J., Lixin Z, Huang J (2008) Gesture recognition based on 3Daccelerometer for cell phones interaction, IEEE xplore ISSN 978–1–4244–2342–2
7. Arras K, Siegwart R (1997) Feature-extraction and scene interpretation for map-based navigation and map-building. Proc of SPIE, Mobile Robotics XII
8. Mehta, M.: A breakthrough in wireless sensor networks and internet of things. Int J Electron Commun Eng Technol (IJECET) **6**(8), 07–11 (2015)
9. Todica M.: Distance measurement with SR HC 04, NodeMcu ESP 8266 and Blynk. 24 July 2018. https://www.researchgate.net/publication/326580937_Distance_measurement_with_SR_HC_04_NodeMcu_ESP_8266_and_Blynk
10. Soni D, Makwana A (2017) A survey on MQTT: a protocol of internet of things (IOT), international conference on telecommunication, power analysis and computing techniques (ICTPACT—2017)
11. Kalambe N, Prof. Thakur D, Prof. Paul S (2015) Review of microstrip patch antenna using uwb for wireless communication devices. (IJCSMC), 4(1), ISSN 2320–088X
12. Ullas BS, Anush S, Roopa J, Govinda Raju M (2014) Machine to machine communication for smart systems using MQTT. Int J Adv Res Elec Electron Instrument Eng (An ISO 3297: 2007 Certified Organization) 3(3):8242–8248
13. Kyaw A, Truong HP, Joseph J (2018) Low-cost computing using raspberry pi 2 model B. 13(3):287–299
14. Patel NR (2017) Interactive interface for DC motor using GUI with raspberry pi controller. Interactive interface for DC motor using GUI with raspberry pi controller,(An ISO 3297: 2007 Certified Organization) 5(6):11917–11922

Design of Robust Controller for Enhanced Performance of 2-DOF Torsion System

Priyam Rai and Bhanu Pratap

Abstract This paper presents the design of a robust control law for stabilization of the 2-DOF torsion system, which is an underactuated system, characterized by a higher number of degrees of freedom than the number of actuators. The output of the given servo dynamical system with two flexible torsional couplings should be positioned properly with minimum vibrations and response time. Control design focuses on application of nonlinear control technique, hierarchical sliding mode control, for desired and robust performance. HSMC is a robust controller which is capable of rejecting matched uncertainties like its conventional counterpart, sliding mode control. It has the added benefit of having a simpler subsystem-wise sliding mode design, which makes it more convenient to apply on underactuated system. The HSMC technique is applied by choosing the angular position error as the sliding surface. The accuracy and robustness of the controller are presented and compared with conventional linear quadratic regulator (LQR) and proportional, integral, derivative (PID) controller. The study of performance of the proposed scheme is based on bounded matched disturbance, randomly varying viscous damping coefficient, parameter uncertainties, and varying frictional force. MATLAB simulations have been carried out to verify the above proposition. The results show the superiority of HSMC over conventional LQR and PID controller in robustness.

1 Introduction

The 2-DOF torsion system is a good representation of torsional dynamics in real-world control applications. It is an underactuated system that simulates the effect of a flexible coupling between an actuator and a load encountered in complex industrial processes. Various control techniques that have been applied to the torsion system

P. Rai · B. Pratap (✉)
Department of Electrical Engineering, National Institute of Technology Kurukshetra, Haryana 136119, India
e-mail: bhanu@nitkkr.ac.in

P. Rai
e-mail: priyam_31910113@nitkkr.ac.in

J. Gu et al. (eds.), *Communication and Control for Robotic Systems*, Smart Innovation, Systems and Technologies 229, https://doi.org/10.1007/978-981-16-1777-5_30

for the control of vibration are reported in the literature. A state feedback control with high-accuracy integration method, modern Taylor Series method, for higher quality results, has been presented in [1]. Feedback control using the linear quadratic regulator (LQR) weight selection method has been previously applied to the torsion system in [2]. A fractional order PID controller has been designed to control the angular position of the 1-DOF torsion module [3].

Variable structure control with sliding mode control is a powerful nonlinear feedback control method. The inherent stability and robustness of this method, due to the total rejection of disturbances and insensitivity of parameter variations, are the result of the discontinuous nature of its fast switching control action [4]. But in the case of underactuated systems, application of conventional sliding mode control can be problematic and cumbersome. Such systems have a higher relative degree; therefore, the parameters of sliding surface cannot be obtained directly by using the Hurwitz condition. Thus, to obtain the reference control, inconvenient calculations might have to be undertaken. To counter such problem, hierarchical sliding mode control is used. In this paper, hierarchical sliding mode control is developed for 2-DOF torsion system, which consists of three subsystems. Therefore, three individual sliding surfaces are defined for each subsystem [5]. The Lyapunov theorem is used to derive the hierarchical control law [6]. Further, Barbalat's theorem is used to prove that the L_2 and L_∞ gains for the obtained closed-loop system are bounded and therefore proving that sliding surfaces obtained are asymptotically stable [7]. The validity of the proposed controller for the 2-DOF torsion system has been proved by the simulation results as shown in Sect. 4.

LQR controller with weight matrix selection by ARE solution and the PID controller are used for comparitive study of the performance of proposed controller for achieving the desired control requirement. LQR is a well-known optimal control solution in which only the figure of merit needs to be decided by the designer. It generates static gain matrix K and therefore keeping the order of closed-loop system the same as the plant. LQR achieves infinite gain margin and phase margin ≥ 60 degrees. These days various intelligent optimization techniques are being applied for selection of weighting matrices Q and R [8–10]. The problem with these methods is that they take high computation power as well as a long time for optimization of Q and R; therefore, these cannot be applied on an online systems. So in this paper, the control law is calculated by solving the algebraic Riccati equation [11]. The stability of the closed-loop system obtained is proved by using Lyapunov theorem [12].

This widespread popularity of PID controller is owed to the fact that it only contains three parameters to be tuned and therefore making it easier to understand and implement [13]. The robustness of the PID controller can be taken care of by paying mind to relative stability and sensitivity functions while designing the controller parameters [14]. In this paper, the PID tuning for adequate performance is done in SIMULINK, MATLAB 2017a.

To the best of our knowledge, the robust control design for torsional system using sliding mode control has not received much attention as available in the literature. The proposed control design is focused on the development of hierarchical sliding mode controller for the underactuated 2-DoF torsion system. The Gaussian uncertainty and matched disturbances are considered while study of the robust performance of the proposed control scheme developed for the considered model of the torsion system. The mean square of error and control input is calculated for each controller to compare their performances. Also, the percentage change in these values for matched disturbances and parameter uncertainties is calculated and compared.

In the remainder of the paper, the other sections are given as follows. The problem statement is formulated in Sect. 2, whereas the proposed controllers are designed in Sect. 3. Section 4 presents the simulation studies, and the conclusion is given in Sect. 5 followed by the references.

2 Problem Formulation and Preliminaries

2.1 System Modelling

This system consists of electrical and mechanical subsystems. The dynamical behaviour of the system can be represented by equations of motion. Figure 1 shows the 2-DOF torsion system.

As shown, the assembly consists of a torsional spring elements with stiffness $k_s i$, damping coefficient B_i and moment of inertia J_i where $i = 1, 2, 3$. The input to the system is given by an external torsional torque τ , which is generated by a servo drive with input voltage V . This is described by the following equation.

$$J_1\ddot{\theta}_1 = rk_m \frac{V - rk_m\dot{\theta}_1}{R_m} - \tau \tag{1}$$

Applying conservation of angular momentum on the assembly of torsion elements, following equation is obtained:

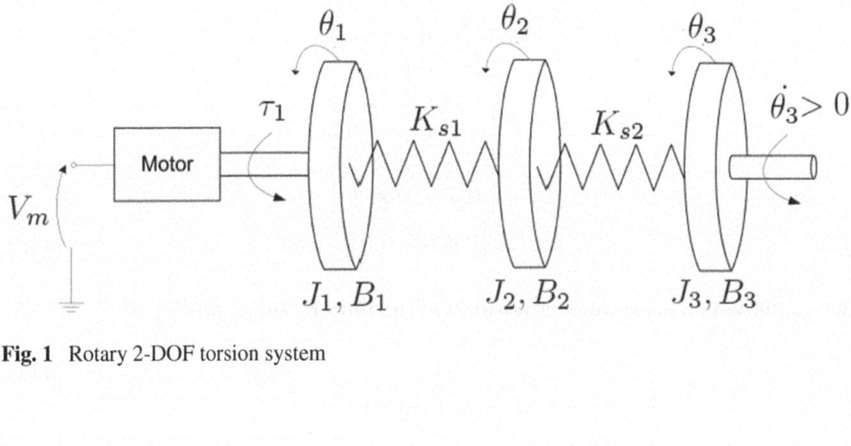

Fig. 1 Rotary 2-DOF torsion system

$$J_i\ddot{\theta}_i = k_{s(i-1)}[\theta_{(i-1)} - \theta_i] + B_{(i-1)}[\dot{\theta}_{(i-1)} - \dot{\theta}_i] \\ +k_{si}[\theta_{(i+1)} - \theta_i] + B_i[\dot{\theta}_{(i+1)} - \dot{\theta}_i] \quad (2)$$

Using equations (1) and (2), we obtain the following system equations:

$$\ddot{\theta}_1 = \frac{1}{J_1}[k_{s1}(\theta_2 - \theta_1) + B_1(\dot{\theta}_2 - \dot{\theta}_1) + V\frac{k_m r}{R_m} \\ -\dot{\theta}_1\frac{k_m^2 r^2}{R_m}] \quad (3)$$

$$\ddot{\theta}_2 = \frac{k_{s1}}{J_2}[\theta_1 - \theta_2] + \frac{k_{s2}}{J_2}[\theta_3 - \theta_2] + \frac{B_1}{J_2}[\dot{\theta}_1 - \dot{\theta}_2] \\ +\frac{B_2}{J_2}[\dot{\theta}_3 - \dot{\theta}_2] \quad (4)$$

$$\ddot{\theta}_3 = \frac{k_{s2}}{J_3}[\theta_2 - \theta_3] + \frac{B_2}{J_3}[\dot{\theta}_2 - \dot{\theta}_3] \quad (5)$$

2.2 State-Space Model

From the system equations (3), (4) and (5), the following state space model is obtained:

$$\dot{x} = Ax + bu \quad (6)$$

$$u = V \quad (7)$$

$$x = \left[\theta_1\ \theta_2\ \theta_3\ \dot{\theta}_1\ \dot{\theta}_2\ \dot{\theta}_3\right]^T \quad (8)$$

$$A = \begin{bmatrix} 0 & 0 & 0 & 1 & 0 & 0 \\ 0 & 0 & 0 & 0 & 1 & 0 \\ 0 & 0 & 0 & 0 & 0 & 1 \\ a & b & 0 & c & d & 0 \\ e & f & g & h & i & j \\ 0 & k & l & 0 & m & n \end{bmatrix} \quad (9)$$

$$b = \left[0\ 0\ 0\ o\ 0\ 0\right]^T \quad (10)$$

The output which is the angular position of the third torsional disc θ_3 is:

$$y = cx \quad (11)$$

$$c = \left[0\ 0\ 1\ 0\ 0\ 0\right] \quad (12)$$

where

$$a = -\frac{k_{s1}}{J_1}$$
$$b = \frac{k_{s1}}{J_1}$$
$$c = -\frac{{k_m}^2{k_{s1}}^2 + B_1 R_m}{J_1 R_m}$$
$$d = \frac{B_1}{J_1}$$
$$e = \frac{k_{s1}}{J_2}$$
$$f = -\frac{k_{s1} + k_{s2}}{J_2}$$
$$g = \frac{k_{s2}}{J_2}$$
$$h = \frac{B_1}{J_2}$$
$$i = -\frac{B_1 + B_2}{J_2}$$
$$j = \frac{B_2}{J_2}$$
$$k = \frac{k_{s2}}{J_3}$$
$$l = -\frac{k_{s2}}{J_3}$$
$$m = \frac{B_2}{J_3}$$
$$n = -\frac{B_2}{J_3}$$
$$o = \frac{k_m r}{J_1 R_m}$$

2.3 *Desired Control Requirement*

The 2-DOF torsion system consists of three components: the servo drive and two torsional elements. The goal here is to position the final output, that is, the position variable of third disc, θ_3 at desired position with minimum vibration and response time. The control design should reduce the error to zero asymptotically. Analytically, θ_3 and $\dot{\theta}_3$ should tend to zero.

Model parameters description

3 Various Controllers Applied

3.1 Hierarchical Sliding Mode Control

Controller Design The given system consists of various subsystems. Due to this physical structure, the hierarchical structure of sliding mode control is defined. Sliding surface of each subsystem is defined separately and then layered together till the entire subsystems are included. Thereafter, the control law is decided based on Lyapunov theorem.

Defining the sliding surfaces of each subsystem:

$$S_1 = \lambda_1\theta_1(t) + \dot{\theta_1}(t) \tag{13}$$

$$S_2 = \lambda_2\theta_2(t) + \dot{\theta_2}(t) \tag{14}$$

$$S_3 = \lambda_3\theta_3(t) + \dot{\theta_3}(t) \tag{15}$$

The constants λ_1 , λ_2 and λ_3 should satisfy the Hurwitz condition. The combined sliding surface of the complete system is obtained as below:

$$\begin{aligned} S = {} & \epsilon_1[\lambda_1\theta_1(t) + \dot{\theta_1}(t)] \\ & +\epsilon_2[\lambda_2\theta_2(t) + \dot{\theta_2}(t)] \\ & +\epsilon_3[\lambda_3\theta_3(t) + \dot{\theta_3}(t)] \end{aligned} \tag{16}$$

Let

$$\dot{S} = -kS - \eta|S| \tag{17}$$

Differentiating 16 with respect to time t:

$$\begin{aligned} \dot{S} = {} & (a+e)\theta_1 + (b+f+k)\theta_2 + (g+l)\theta_3 \\ & +(\lambda_1 + c + h)\dot{\theta_1} + (d + +i + m + \lambda_2)\dot{\theta_2} \\ & +(j + n + \lambda_3)\dot{\theta_3} + ou \end{aligned} \tag{18}$$

Assuming $\epsilon_1 = \epsilon_2 = \epsilon_3 = 1$, the control law can be derived using Lyapunov stability theorem. Lyapunov function for the given sliding surface can be chosen as:

$$V(t) = \frac{S^2}{2} \tag{19}$$

Differentiating (19) with respect to time t, we obtain following equation:

$$\dot{V}(t) = S\dot{S} \tag{20}$$

From equations (17), (18) and (20), the following control law is obtained:

$$u = -\frac{[(a+e)\theta_1 + (b+f+k)\theta_2 + (g+l)\theta_3]}{o} - \frac{[(\lambda_1 + c + h)\dot{\theta}_1 + (d+i+m+\lambda_2)\dot{\theta}_2 + (j+n+\lambda_3)\dot{\theta}_3]}{o} - \frac{[kS + \lambda_4 \tanh(S)]}{o} \tag{21}$$

Stability analysis The stability of the system is proved by proving that the asymptotic stability of each layer of sliding surface for every subsytem is individually stable. Thereafter, the asymptotic stability of the consolidated sliding surface of the entire system is proved.

Part 1:

Taking underactuated system given by equations (3), (4) and (5). If the control law applied to the consolidated system is given by (21) and individual sliding surfaces as (13), (14) and (15), then each sliding surface, say ith sliding surface, is asymptotically stable. Therefore, proving S is asymptotically stable.

Proof:

Substituting (17) into the equation (20), following is obtained:

$$\dot{V}_i = S_i\dot{S}_i \tag{22}$$

$$= S_i(-k_i S_i - \eta_i sgn S_i) \tag{23}$$

$$= -k_i S_i^2 - \eta_i |S_i| \tag{24}$$

Calculating the energy of the fuction by integrating equation (24) on both sides, following is obtained:

$$\int_0^t \dot{V}_i d\tau = \int_0^t (-k_i S_i^2 - \eta_i |S_i|) d\tau \tag{25}$$

$$V_i(t) - V_i(0) = -\int_0^t (k_i S_i^2 + \eta_i |S_i|) d\tau \tag{26}$$

$$V_i(0) = V_i(t) + \int_0^t (k_i S_i^2 + \eta_i |S_i|) d\tau \tag{27}$$

$$\geq \int_0^t (k_i S_i^2 + \eta_i |S_i|) d\tau \tag{28}$$

Applying limit to the integrated term in above equation:

$$\lim_{t\to\infty} \int_0^t (k_i S_i^2 + \eta_i |S_i|) d\tau \leq V_i(0) < \infty \tag{29}$$

Barbalat's lemma states that supposing $f(t)\epsilon C^1(a\infty)$ and $\lim_{x\to\infty} f(t) = \alpha$ where $\alpha < \infty$. If $\dot{f}(t)$ is uniformly continuous, then $\lim_{x\to\infty} \dot{f}(t) = 0$.

According to Barbalat's lemma:

$$\lim_{t\to\infty} (k_i S_i^2 + \eta_i |S_i|) = 0 \tag{30}$$

According to 30 $\lim_{t\to\infty} S_i = 0$, which implies that ith sliding layer is asymptotically stable. Therefore, S is asymptotically stable too.

Part 2:

Taking the system represented by system equations (3), (4) and (5); with control law given by (21). Assuming the sliding surfaces as (13), (14) and (15). If the following equations are satisfied:

$$sgn(\epsilon_1 S_1) = sgn(\epsilon_2 S_2) = sgn(\epsilon_3 S_3) \tag{31}$$

if S_1 , S_2 , $S_3 \neq 0$

then S_1 , S_2 , S_3, and therefore, the layered sliding surface S represented by (16) is still asymptotically stable.

Proof:

This is achieved firstly by using L_2 gain, which shows that the ratio of output and input transfer function has a finite energy. It is proved that S_1, S_2 and S_3 have finite L_2 gain. From the result obtained in (30), the following is proved:

$$\int_0^\infty S^2 d\tau = \int_0^\infty (\epsilon_1 S_1 + \epsilon_2 S_2 + \epsilon_3 S_3)^2 d\tau \tag{32}$$

$$= \int_0^\infty ({\epsilon^2}_1 S_1^2 + \epsilon_2^2 S_2^2 + \epsilon_3^2 S_3^2) d\tau \tag{33}$$

$$+ 2 \int_0^\infty (\epsilon_1 \epsilon_2 S_1 S_2 + \epsilon_1 \epsilon_3 S_1 S_3 + \epsilon_2 \epsilon_3 S_2 S_3) d\tau \tag{34}$$

$$< \infty \tag{35}$$

Secondly, it is proved S_1, S_2, $S_3 \in L_\infty$, to prove that these belong to a space of bounded functions.

By (30), the following is proved:

$$S_i \in L_\infty \tag{36}$$

Thirdly, using (24), it is proved that:

$$\dot{S}_1, \dot{S}_2, \dot{S}_3 \in L_\infty.$$

Using (30) and (36), it is known that sliding surfaces converge asymptotically to zero if (31) is satisfied. When (31) is not specified, the signs of ϵ_1, ϵ_2 and ϵ_3 are changed.

3.2 LQR Controller

Controller Design The objective of LQR controller is to bring the states to desired trajectory by minimizing the cost function given below:

$$J(u) = \int_0^\infty [x^T(t) Q x(t) + u^T(t) R u(t)] dt \tag{37}$$

where Q and R are state and input weighting matrices. These are diagonal matrices, with Q being positive semi-definite matrix and R being positive definite matrix, for a stable response and as the input weighting matrix cannot be a zero matrix.

A quadratic performance index is used here, which is a weighted integral of states and inputs; Q and R are the tuning parameters of LQR controller.

The solution of following algebraic Ricatti equation (ARE), the transformation matrix P which is a symmetric matrix, gives the state feedback gain for obtaining the desired response from the system.

$$A^T P + PA + Q - PBR^{-1}B^T P = 0 \tag{38}$$

Optimal control of input is given by:

$$u = -kx \tag{39}$$

where

$$k = R^{-1}B^T P \tag{40}$$

Solving the ARE, taking performance parameters, overshoot less than 10% and settling time with 2% tolerance band as 0.1s, while taking $R = 63$, the transformation matrix (P) is calculated, using which the following Q matrix is obtained:

$$Q = \begin{bmatrix} 50 & 0 & 0 & 0 & 0 & 0 \\ 0 & 10 & 0 & 0 & 0 & 0 \\ 0 & 0 & 3 & 0 & 0 & 0 \\ 0 & 0 & 0 & 0.33 & 0 & 0 \\ 0 & 0 & 0 & 0 & 6 \times 10^{-3} & 0 \\ 0 & 0 & 0 & 0 & 0 & 4 \times 10^{-5} \end{bmatrix} \tag{41}$$

Using the lqr command in MATLAB, the following state feedback gain is obtained:

$$k = \begin{bmatrix} 0.6786 & 0.1684 & 0.1530 & 0.1317 & 0.0210 & 0.0205 \end{bmatrix} \tag{42}$$

Stability analysis

Given system:

$$\frac{d}{dt}x = Ax + bu \tag{43}$$

Feedback control law:

$$u = -kx \tag{44}$$

Therefore, the required closed-loop system with the real part of the eigenvalues of $A - bk$ negative is obtained as:

$$\frac{d}{dt}x = Ax - bkx \tag{45}$$

Let the positive definite Lyapunov function be defined such that its derivative function is negative semidefinite along the system trajectories, as:

$$V(x) = x^T Px \tag{46}$$

Now calculating the first derivative of equation (46):

$$\begin{aligned}\frac{d}{dt}V(x) &= \dot{x}^T P x + x^T P \dot{x} \\ &= (Ax - bkx)^T P x + x^T P (A - bkx) \\ &= x^T (A - bk)^T P x + x^T P (A - bk) x\end{aligned}$$

For the system to be stable in the sense of Lyapunov:

$$(A - bk)^T P + P(A - bk) = -F \tag{47}$$

where F is positive definite.
P matrix obtained is:

$$P = \begin{bmatrix} 6.33 \times 10^{-3} & 0 & 0 & 0 & 0 & 0 \\ 0 & 1.57 \times 10^{-3} & 0 & 0 & 0 & 0 \\ 0 & 0 & 1.43 \times 10^{-3} & 0 & 0 & 0 \\ 0 & 0 & 0 & 1.23 \times 10^{-3} & 0 & 0 \\ 0 & 0 & 0 & 0 & 1.96 \times 10^{-4} & 0 \\ 0 & 0 & 0 & 0 & 0 & 1.91 \times 10^{-4} \end{bmatrix} \tag{48}$$

P matrix is found to be positive definite by Sylvester's test.
Solving (47), matrix F obtained is found to be positive definite as well. Therefore, the closed-loop system obtained is globally asymptotically stable.

4 Simulation Results

This section contains the simulation results of the three controllers discussed above. For testing the accuracy, efficiency and robustness of these controllers, the plots of response for following matched disturbances and parameter uncertainties are taken and compared:

Gaussian matched disturbance:

$$z = 0.0002 \times \text{rand}(1, 1) * \sin(0.5) \tag{49}$$

Viscous friction uncertainty:

$$B' = B \times (1 + \text{rand}(1, 1)) \tag{50}$$

Moment of inertia uncertainty:

$$J' = J(1 \pm 20\%) \tag{51}$$

The analysis is done based on following three cases (Figs. 2, 3, 4, 5, 6; Tables 1, 2):
Case 1: where only the condition (49) is taken into consideration.
Case 2: where the conditions (49) and (50) are taken into consideration.
Case 3: where the conditions (49), (50) and (51) are taken into consideration.
The legend of the plots presented here based on these cases are as follows:
Case 1:
Case 2:
Case 3:

The plots are shown in Figs. 2a, 3b and 5a which represent the state response of each controller. Since the state of the system is taken as an error, it is observed from the figures that the error is bounded and tending to zero in each case. This implies that with given bounded matched disturbance and parametric uncertainties, each of the three controllers was capable of stabilizing the unstable open-loop system. From Table 3, it is observed that the root mean square error and control input in increasing order are obtained from LQR, HSMC and PID. From Table 4, the robustness of the applied controllers can be deduced in the face of parametric uncertainties. It is observed that HSMC as well as LQR shows high robustness properties, whereas PID

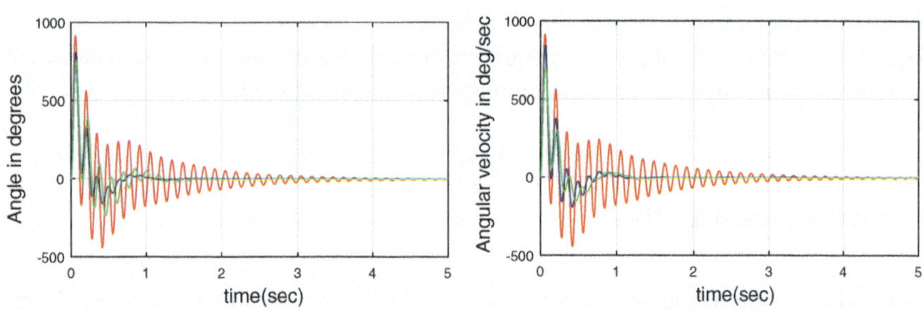

(a) Angular response (θ_3) of PID (b) Angular velocity ($\dot{\theta}_3$) PID

Fig. 2 (**a**) Angular response (θ_3) of PID; (**b**) Angular velocity (θ_3) of PID

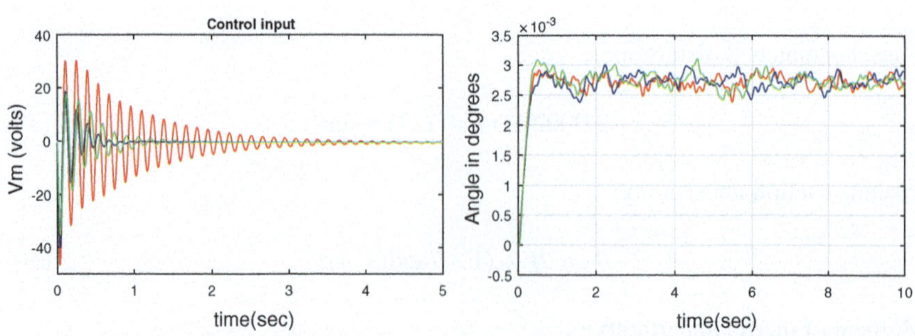

(a) Control input in PID (b) Angular response (θ_3) of LQR

Fig. 3 (**a**) Control input in PID; (**b**) Angular response (θ_3) of LQR

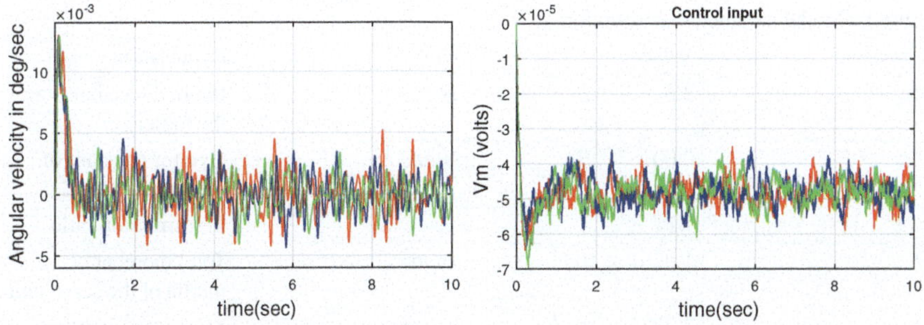

(a) Angular velocity ($\dot{\theta}_3$) of LQR

(b) Control input in LQR

Fig. 4 (**a**) Angular velocity (θ_3) of LQR; (**b**) Control input in LQR

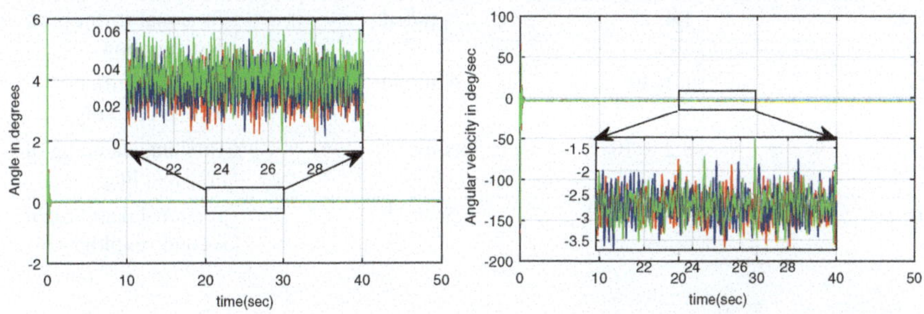

(a) Angular response (θ_3) of HSMC

(b) Angular velocity ($\dot{\theta}_3$) of HSMC

Fig. 5 (**a**) Angular response (θ_3) of HSMC; (**b**) Angular velocity (θ_3) of HSMC

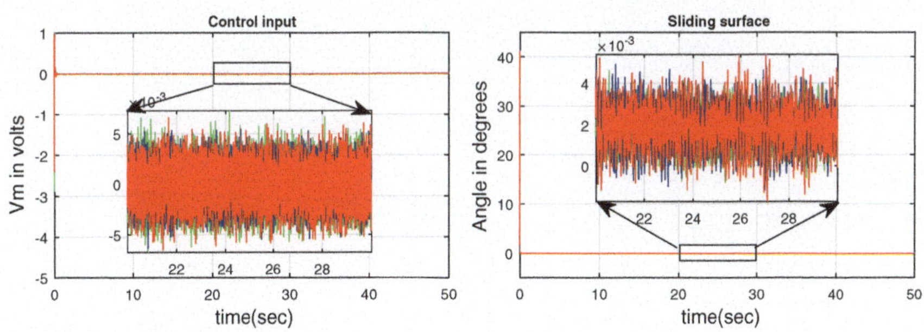

(a) Control input in HSMC

(b) Sliding surface in HSMC

Fig. 6 (**a**) Control input in HSMC; (**b**) Sliding surface in HSMC

Table 1 Model parameters description

Constant	Value	Unit	Explanation
R_m	2.6	Ω	Internal resistance of the motor
K_m	0.00767	NmA^{-1}	Gyrator constant of the motor
r	70	–	Transmission ratio
J_1	1.9261×10^{-3}	kg m^2	Total moment of inertia of the servo unit
J_2	3.2281×10^{-4}	kg m^2	Moment of inertia of first torsional disc
J_3	3.2281×10^{-4}	kg m^2	Moment of inertia of second torsional disc
k_{s1}	1.4927	N m/rad	Torsional stiffness of first coupling
k_{s2}	1.3639	N m/rad	Torsional stiffness of second coupling
B_1	8.2×10^{-3}	N ms^{-1}	Torsional damping of first coupling
B_2	7.8×10^{-3}	N ms^{-1}	Torsional damping of second coupling

Table 2 HSMC parameters found by trial and error

Parameter	Value
λ_1	50
λ_2	20
λ_3	10
ϵ_1	6
ϵ_2	4
ϵ_3	2
λ_4	250

shows relatively poor robustness. As per the results given in Table 3, the RMS value of error and control input without considering uncertainty is less in the LQR controller than the proposed HSMC controller designed for the torsion system. Further, robust performance analysis has been done with taking into account the 20% parametric uncertainties in the nominal plant model. However, the percent change in error and control effort has increased in the case of LQR controller, and these factors are lesser in the proposed HSMC as compared with PID as well as LQR as shown in Table 4.

Table 3 Error and control input tabulation

Controller	Analyzing factor	Case 1	Case 2	Case 3
PID	Error(RMS)	18.2907	12.119	12.2355
	Control input(RMS)	8.4223	3.9314	3.9070
LQR	Error(RMS)	0.0027	0.0028	0.0026
	Control input(RMS)	4.8128×10^{-5}	4.8931×10^{-5}	4.7322×10^{-5}
HSMC	Error(RMS)	0.0299	0.0369	0.0315
	Control input(RMS)	0.0020	0.0019	0.0022

Table 4 Absolute percentage variation in square mean error and control input in cases 2 and 3, taking case 1 as reference

Controller	$\|\Delta$ RMS error$\|$(%)		$\|\Delta$ RMS control input $\|$(%)	
	Case 2–Case 1	Case 3–Case1	Case 2–Case 1	Case 3–Case 1
PID	33.78	33.105	53.32	53.6
LQR	3.7	3.7	1.67	1.64
HSMC	0.57	1.64	0.325	1.9

5 Conclusions

A robust controller for the improved performance of a 2-DOF torsion system is presented in this paper. The gierarchical sliding mode control (HSMC) approach is proposed for the required control design. The stability of the system is analysed using the Lyapunov theory. The stabilization of the torsion system through HSMC techniques has been investigated, and the obtained results are compared with the conventional PID and the LQR controllers. The robustness issue of the proposed control scheme is addressed against the bounded matched disturbance, randomly varying viscous damping coefficient, parameter uncertainties and varying frictional forces. It can be concluded that the proposed controller is feasible for stabilization of the given system, while showing better robustness property than conventional PID and LQR.

References

1. Fuchs, G., Šátek, V., Vopěnka, V., Kunovský, J., Kozek, M.: Application of the modern taylor series method to a multi-torsion chain. Simul. Mod. Pract. Theory 33:89–101 (2013)
2. Elumalai, V.K., Subramanian, R.G.: A new algebraic LQR weight selection algorithm for tracking control of 2 DoF torsion system. Arch. Electr. Eng. 66(1), 55–75 (2017)

3. Rojas-Moreno, A., Salazar-Aquino, O., Pampamallco-Jara, J.: Control of the angular position of a rotary torsion plant using a 2dof fo pid controller. In: 2018 IEEE 38th Central America and Panama Convention (CONCAPAN XXXVIII), pp. 1–5 (2018, November)
4. Young, K.D., Utkin, V.I., Ozguner, U.: A control engineer's guide to sliding mode control. IEEE Trans. Control Syst. Technol. **7**(3), 328–342 (1999)
5. Wang, W.: Design of a stable sliding-mode controller for a class of second-order underactuated systems. IEE Proceedings—Control Theory and Applications, vol. 151, issue no. 7, pp. 683–690 (2004, November). https://digital-library.theiet.org/content/journals/10.1049/ip-cta-20040902
6. Qian, D., Yi, J., Zhao, D., Hao, Y.: Hierarchical sliding mode control for series double inverted pendulums system. In: 2006 IEEE/RSJ International Conference on Intelligent Robots and Systems. pp. 4977–4982 (2006, October)
7. Liu, J., Wang, X.: Advanced Sliding Mode Control for Mechanical Systems. Springer, Berlin, Heidelberg
8. Kukreti, S., Walker, A., Putman, P., Cohen, K.: Genetic Algorithm Based LQR forAttitude Control of a Magnetically Actuated CubeSat.https://arc.aiaa.org/doi/abs/10.2514/6.2015-0886
9. Howimanporn, S., Thanok, S., Chookaew, S., Sootkaneung, W.: Design and implementation of pso based lqr control for inverted pendulum through plc. In: 2016 IEEE/SICE International Symposium on System Integration (SII). pp. 664–669 (2016, December)
10. Vinodh Kumar, E., Raaja, G.S., Jerome, J.: Adaptive PSO for optimal LQR tracking control of 2 DoF laboratory helicopter. Appl. Soft Comput. 41, 77 – 90 (2016). http://www.sciencedirect.com/science/article/pii/S1568494615008029
11. Kumar, E.V., Jerome, J., Srikanth, K.: Algebraic approach for selecting the weighting matrices of linear quadratic regulator. In: 2014 International Conference on Green Computing Communication and Electrical Engineering (ICGCCEE). pp. 1–6 (March 2014)
12. Vadali, S.R., Kim, E.S.: Feedback control of tethered satellites using lyapunovstability theory. J. Guidance Control Dyn. 14(4),729–735 (1991). https://doi.org/10.2514/3.20706
13. Bennett, S.: Development of the PID controller. IEEE Control Syst. Mag. **13**(6), 58–62 (1993)
14. Silva, G.J., Datta, A., Bhattacharyya, S.P.: On the stability and controller robustness of some popular pid tuning rules. IEEE Transactions on Automatic Control **48**(9), 1638–1641 (2003)

The manufacturer's authorised representative in the EU is Springer Nature Customer Service Centre GmbH, Europaplatz 3, 69115 Heidelberg, Germany. If you have any concerns regarding our products, please contact ProductSafety@springernature.com

Printed and bound by CPI Group (UK) Ltd, Croydon, CR0 4YY
01/10/2026
02226185-0008